Life

The Science of Biology

Sixth Edition

Life

Sixth Edition

The Science of Biology

William K. Purves
Emeritus, Harvey Mudd College
Claremont, California

David Sadava
The Claremont Colleges
Claremont, California

Gordon H. Orians
Emeritus, The University of Washington
Seattle, Washington

H. Craig Heller
Stanford University
Stanford, California

 Sinauer Associates, Inc.

 W. H. Freeman and Company

The Cover

Giraffes (*Giraffa camelopardalis*) near Samburu, Kenya.
Photograph © BIOS/Peter Arnold, Inc.

The Opening Page

Soap yucca (*Yucca elata*), White Sands National Monument, New Mexico.
Photograph © David Woodfall/DRK PHOTO.

The Title Page

The endangered Florida panther (*Felis concolor coryi*).
Photograph © Thomas Kitchin/Tom Stack & Associates.

Life: The Science of Biology, Sixth Edition

Copyright © 2001 by Sinauer Associates, Inc. All rights reserved. This
book may not be reproduced in whole or in part without permission.

Address editorial correspondence to:
Sinauer Associates, Inc., 23 Plumtree Road, Sunderland, Massachusetts 01375 U.S.A.
www.sinauer.com

Email: publish@sinauer.com

Address orders to:
VHPS/W. H. Freeman & Co. Order Department, 16365 James Madison Highway,
U.S. Route 15, Gordonsville, VA 22942 U.S.A.
www.whfreeman.com

Examination copy information: 1-800-446-8923
Orders: 1-888-330-8477

Library of Congress Cataloging-in-Publication Data

Life, the science of biology / William K. Purves...[et al.].--6th ed.
        p.   cm.
    Includes index.
    ISBN 0-7167-3873-2 (hardcover) – ISBN 0-7167-4348-5 (Volume 1) –
    ISBN 0-7167-4349-3 (Volume 2) – ISBN 0-7167-4350-7 (Volume 3)
    1. Biology I. Purves, William K. (William Kirkwood), 1934–

QH308.2 .L565 2000
570--dc21   00-048235

Printed in U.S.A.

Third Printing 2002 Courier Companies Inc.

This book is dedicated to the memory of Angeline Douvas

About the Authors

Gordon Orians   Craig Heller   Bill Purves   David Sadava

William K. Purves is Professor Emeritus of Biology as well as founder and former chair of the Department of Biology at Harvey Mudd College in Claremont, California. He received his Ph.D. from Yale University in 1959 under Arthur Galston. A fellow of the American Association for the Advancement of Science, Professor Purves has served as head of the Life Sciences Group at the University of Connecticut and as chair of the Department of Biological Sciences, University of California, Santa Barbara, where he won the Harold J. Plous Award for teaching excellence. His research interests focused on the chemical and physical regulation of plant growth and flowering. Professor Purves elected early retirement in 1995, after teaching introductory biology for 34 consecutive years, in order to turn his skills to writing and producing multimedia for introductory biology students. That year, he was awarded the Henry T. Mudd Prize as an outstanding member of the Harvey Mudd faculty or administration.

David Sadava is now responsible for *Life*'s chapters on the cell (2–8), in addition to the chapters on genetics and heredity that he assumed in the previous edition. He is the Pritzker Family Foundation Professor of Biology at Claremont McKenna, Pitzer, and Scripps, three of the Claremont Colleges. Professor Sadava received his Ph.D. from the University of California, San Diego in 1972, and has been at Claremont ever since. The author of textbooks on cell biology and on plants, genes, and agriculture, Professor Sadava has done research in many areas of cell biology and biochemistry, ranging from developmental biology, to human diseases, to pharmacology. His current research concerns human lung cancer and its resistance to chemotherapy. Virtually all of the research articles he has published have undergraduates as coauthors. Professor Sadava has taught a variety of courses to both majors and nonmajors, including introductory biology, cell biology, genetics, molecular biology, and biochemistry, and he recently developed a new course on the biology of cancer. For the last 15 years, Professor Sadava has been a visiting professor in the Department of Molecular, Cellular, and Developmental Biology at the University of Colorado, Boulder, and is currently a visiting scientist at the City of Hope Medical Center.

Gordon H. Orians is Professor Emeritus of Zoology at the University of Washington. He received his Ph.D. from the University of California, Berkeley in 1960 under Frank Pitelka. Professor Orians has been elected to the National Academy of Sciences and the American Academy of Arts and Sciences, and is a Foreign Fellow of the Royal Netherlands Academy of Arts and Sciences. He was President of the Organization for Tropical Studies, 1988–1994, and President of the Ecological Society of America, 1995–1996. He is chair of The Board on Environmental Studies and Toxicology of the National Research Council and a member of the board of directors of World Wildlife Fund–US. He is a recipient of the Distinguished Service Award of the American Institute of Biological Sciences. Professor Orians is a leading authority in ecology, conservation biology, and evolution, with research experience in behavioral ecology, plant–herbivore interactions, community structure, the biology of rare species, and environmental policy. He elected early retirement to be able to devote more time to writing and environmental policy activities.

H. Craig Heller is the Lorry Lokey/Business Wire Professor of Biological Sciences and Human Biology at Stanford University. He has served as Director of the popular interdisciplinary undergraduate program in Human Biology and is now Chairman of Biological Sciences. Professor Heller received his Ph.D. from Yale University in 1970 and did postdoctoral work at Scripps Institute of Oceanography on how the brain regulates body temperature of mammals. His current research focuses on the neurobiology of sleep and circadian rhythms. Professor Heller has done research on a great variety of animals ranging from hibernating squirrels to exercising athletes. He teaches courses on animal and human physiology and neurobiology.

Preface

Biologists' understanding of the living world is growing explosively. This isn't the world that the four authors of this book were born into. We never dreamed, as we began our research careers as freshly minted Ph.D.'s, that our science could move so rapidly. Biology has now entered the post-genomic era, allowing biologists and biomedical scientists to tackle once-unapproachable challenges. We are also at the threshold of some experiments that raise ethical concerns so great that we must stand back and participate with others in determining what is right to do and what is not.

The enormous growth and changes in biology create a special challenge for textbook authors. How can a biology textbook provide the basics, keep up with the exciting new discoveries, and not become overwhelming. The increasing bulk of textbooks is of great concern to authors as well as to instructors and their students, who blanch at the prospect of too many pages, too many term papers, and too little sleep. Some reconsideration of what is essential and how that is best presented needs to be made if the proliferation of facts is not to obscure the fundamental principles.

Our major goals were brevity, emphasis on experiments, and better ways to help students learn

In writing the Sixth Edition of *Life*, we committed ourselves to reversing the pattern of ever increasing page lengths in new editions. We wanted a shorter book that brings the subject into sharper focus. We tried to achieve this by judicious reduction of detail, by more concise writing, and by more use of figures as primary teaching sources. It worked! Our efforts were successful. This edition is 200 pages shorter than its predecessor, yet it covers much exciting new material.

While working to tighten and shorten the text, we were also determined to retain and even increase our emphasis on *how* we know things, rather than just *what* we know. To that end, the Sixth Edition inaugurates 72 specially formatted figures that show how experiments, field observations, and comparative methods help biologists formulate and test hypotheses (the figure at right is an example). Another 26 figures highlight some of the many field and laboratory methods created to do this research. These Experiment and Research Methods illustrations are listed on the endpapers at the back of the book.

In the Fifth Edition, we introduced "balloon captions" that guide the reader through the illustrations (rather than having to wade through lengthy captions). This feature was widely applauded and we have worked to refine the balloons' effectiveness. In response to suggestions from users

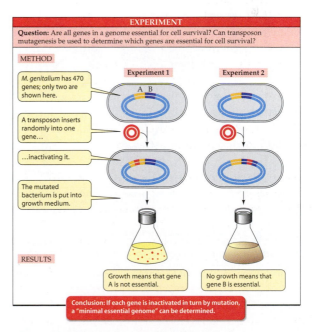

13.22 Using Transposon Mutagenesis to Determine the Minimal Genome
By inactivating genes one by one, scientists can determine which ones are essential for the cell's survival.

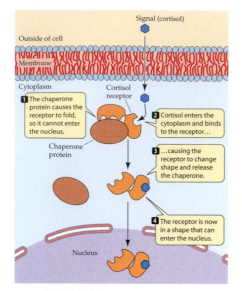

15.9 A Cytoplasmic Receptor
The receptor for cortisol is bound to a chaperone protein. Binding of the signal (which diffuses directly through the membrane) releases the chaperone and allows the receptor protein to enter the cell's nucleus, where it functions as a transcription factor.

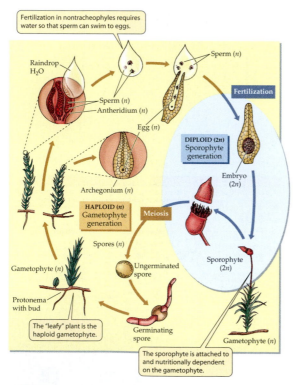

28.3 A Nontracheophyte Life Cycle
The life cycle of nontracheophytes, illustrated here by a moss, is dependent on an external source of liquid water. The visible green structure of nontracheophytes is the gametophyte; in nontracheophyte plants, the "leafy" structures are sporophytes.

of the Fifth Edition, in the Sixth Edition we now number many of the balloons, emphasizing the flow of the figure and making the sequence easier to follow (Figure 15.9 at left is an example).

This edition is accompanied by a comprehensive website, www.thelifewire.com (and an optional CD-ROM that contains the same material) that reinforces the content of every chapter. A key component of the website is a combination of animated tutorials and activities for each chapter, all of which include self-quizzes. Within each book chapter, this ⊕ icon refers students to a tutorial or an activity. An index of the icons begins in the front endpapers of the book. Figure 28.3 (left, below) shows a typical web icon placement.

As part of the ongoing challenge of keeping the writing and illustrations as clear as possible, we frequently employ bulleted lists. We think these lists will help students sort through what is, even after pruning, a daunting amount of material. And we have continued to provide plenty of interim summaries and bridges that link passages of text.

In all the introductory textbooks, the chapters end with summaries. In ours, we have organized the material within the chapter's main headings. In most cases, we tie key concepts to the figure (or figures) that illustrate it. For visual learners, this provides an efficient mode of reviewing the chapter.

From our many decades in the classroom, we know how important it is to motivate students. Each chapter begins with a brief description of some event, phenomenon, or idea that we hope will engage the reader while conveying a sense of the significance and purpose of the chapter's subject.

Evolution Continues to be the Dominant Theme

Evolution continues to be the most important of the themes that link our chapters and provide continuity. As we have written the various editions of the book, however, the emergence of *genomics* as a new paradigm in the late twentieth century has developed, revolutionizing most areas of biology. In this new century, understanding the workings of the genome is of paramount importance in almost any biological discussion.

In this edition, we have moved further toward updating the evolutionary theme to encompass the postgenomic era. Just two examples are the addition of a section on genomic evolution to our coverage of molecular evolution, and a section on "evo/devo" in the chapter on molecular biology of development. In addition, the chapters on the diversity of life reflect the vast changes in our understanding of systematics and phylogenetic relationships thanks to the genomic perspective.

In fact, each chapter of the book has undergone important changes.

The Seven Parts: Content, Changes, and Themes

In Part One, The Cell, the emphasis in the discussions of biological molecules and thermodynamics has shifted more decisively toward biological aspects and away from pure chemistry. We have made our discussions of enzymes, cell respiration, and photosynthesis less detailed and more focused on the biological applications.

A major addition to Part Two, Information and Heredity, is a new chapter (Chapter 15) on cell signaling and communication, introduced at a place where the students have the necessary grounding in cell biology and molecular genetics. That chapter leads logically into an updated chapter (Chapter 16) on the molecular biology of development, which includes a new section on the intersection of evolutionary and developmental biology—"evo-devo" in the modern jargon. Several chapters incorporate the exciting new work in genomics of prokaryotes, humans, and other eukaryotes.

We have updated all the chapters in Part Three, Evolutionary Processes. In particular, Chapter 24 ("Molecular and Genomic Evolution") reflects the rapid advances in this exciting field. The section on genomic evolution (on pages 446–447) is brand new and includes Figure 24.9 (shown at right).

Part Four, The Evolution of Diversity, now reflects some exciting changes. The chapter on the protists—which can no longer be treated as a single "kingdom"—reflects the continuing uncertainty over the origin and early diversification of eukaryotes. The equally great uncertainty over prokaryote phylogeny, as we deal with the implications of extensive lateral transfer of genes, is evident in the chapter on prokaryote phyla.

We have extended the coverage of the evolution and diversity of plants to two chapters, and that of the animals to three. Recent findings stemming largely from molecular research have led to modifications of the phylogenies of angiosperms and of the animal kingdom. These changes are reflected in the many simplified "trees" that give a broad overview of systematic relationships. Key evolutionary events that separate and unite the different groups are highlighted with red "hot spots" (see Figure 33.1 at right).

We have rearranged Part Five, "The Biology of Flowering Plants," to allow Chapter 39 ("Plant Responses to Environmental Challenges") to serve as a capstone to the whole part, drawing together some of the major threads. We have added sections on hormones and photoreceptors discovered in recent years, and on their signal transduction pathways. The opening chapter (Chapter 34) on "The Flowering Plant Body" has an increased emphasis on meristems.

Part Six, The Biology of Animals, continues to be a broad, comparative treatment of animal physiology with an emphasis on mechanisms of control and regulation. Much new material has been added, including a major revision of Animal

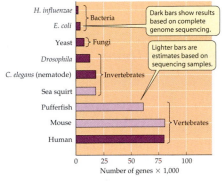

24.9 Complex Organisms Have More Genes than Simpler Organisms
Genome sizes have been measured or estimated in a variety of organisms, ranging from single-celled prokaryotes to vertebrates.

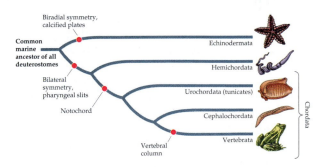

33.1 A Probable Deuterostomate Phylogeny
There are fewer major lineages and many fewer species of deuterostomes than of protostomes.

Development (Chapter 43) to complement and extend the earlier Chapter 16 (Development: Differential Gene Expression). Some other new topics are the role of melatonin in photoperiodism, the role of leptin in the control of food intake, and the discovery in fruit flies of a gene that controls male mating behavior. The extensive coverage of the fast moving field of neurobiology has been substantially updated.

Throughout Part Seven, Ecology and Biogeography, we have added examples of experimental approaches to understanding the dynamics of ecological systems. Some of the examples illustrate the use of experimental and comparative methods. As before, we conclude the book with a chapter on conservation biology (Chapter 58), emphasizing the use of scientific principles to help preserve Earth's vast biological diversity.

There Are Many People to Thank

The reviewing process for *Life*, once a single pass at the stage of draft manuscript, has become an ongoing phenomenon. When the Fifth Edition was still young, we received critiques that influenced our work on this Sixth Edition. The two most penetrating ones came from Zach Gertz, then an undergraduate at Harvard, and Joseph Vanable, a veteran introductory biology professor at Purdue.

Next, still during the Fifth Edition run, 18 instructors recorded their suggestions for improvements in *Life* while teaching from the book. We call these reviews Diary Reviews. The third stage was the Manuscript Reviews. Seventy-three dedicated teachers and researchers read the first-draft chapters and gave us significant and cogent advice. Still another stage has been added to the process and it turned out to be invaluable. We are indebted to 16 Accuracy Reviewers, colleagues who carefully reviewed the almost final page proofs of each chapter to spot lingering errors or imprecisions in the text and art that inevitably escape our weary eyes. Finally, we appreciate the advice given by several experts who reviewed the animations and activities that our publishers developed for the student Web Site/ CD-ROM that accompanies this edition of *Life*. We thank all these reviewers and hope this new edition measures up to their expectations. They are listed after this Preface.

J/B Woolsey Associates has again worked closely with each of us to improve an already excellent art program. They helped to refine the very successful "balloon captions" that were introduced in the Fifth Edition. With their creative input we introduced the Experiment and Research Method illustrations found throughout the text.

James Funston joined us again as the developmental editor for the Sixth Edition. As always, James enforced a rigorous standard for clear writing and illustrating. And he contributed significantly to the process of shortening the book. Norma Roche also suggested cuts, and provided incisive copy editing from beginning to end. Her many astute queries often led to rewrites that enhanced the clarity of the presentation. From first draft to final pages, Susan McGlew was tireless in arranging for expert academic reviews of all of the chapters. Since the First Edition, we have profited immeasurably from the work of Carol Wigg, who again coordinated the pre-production process, including illustration editing and copy editing. She wrote many figure captions, suggested several of the chapter-opening stories, orchestrated the flow of the text and art, kept us mostly on schedule, enforced—sometimes with her red pen—the mandate to be concise, and what's more, did it all with good humor, even under pressure. David McIntyre, photo researcher, found many wonderful new photographs to enhance the learning experience and enliven the appearance of the book as a whole.

We again wish to thank the dedicated professionals in W. H. Freeman's marketing and sales group. Their enthusiasm has helped bring *Life* to a wider audience with each edition. We appreciate their continuing support and valuable input on ways to improve the book. A large share of *Life's* success is due to their efforts in this publishing partnership.

We have always respected Sinauer Associates for their outstanding list of biology books at all levels and we have enjoyed having them lead and assist us through yet another edition. Andy Sinauer has been the guiding spirit behind the development of *Life* since two of us first began to write the First Edition. Andy never ceases helping his authors to achieve our goals, while remaining gentle but firm about his agendas. It has been a very satisfying experience for us to work with him yet again, and we look forward to a continuing association.

Bill Purves   David Sadava   Gordon Orians   Craig Heller

November, 2000

Reviewers for the Sixth Edition

Diary Reviewers

Carla Barnwell, University of Illinois

Greg Beaulieu, University of Victoria

Gordon Fain, University of California, Los Angeles

Ruth Finkelstein, University of California, Santa Barbara

Steve Fisher, University of California, Santa Barbara

Alice Jacklet, SUNY, Albany

Clare Hasenkampf, University of Toronto, Scarborough

Werner Heim, Colorado College

David Hershey, Hyattsville, MD

Hans-Willi Honegger, Vanderbilt University

Durrell Kapan, University of Texas, Austin

Cheryl Kerfeld, University of California, Los Angeles

Michael Martin, University of Michigan, Ann Arbor

Murray Nabors, Colorado State University

Ronald Poole, McGill University

Nancy Sanders, Truman State University

Susan Smith, Massasoit Community College

Raymond White, City College of San Francisco

Manuscript Reviewers

John Alcock, Arizona State University

Allen V. Barker, University of Massachusetts, Amherst

Andrew R. Blaustein, Oregon State University

Richard Brusca, University of Arizona

Matthew Buechner, University of Kansas

Warren Burggren, University of North Texas

Jung Choi, Georgia Institute of Technology

Andrew Clark, Pennsylvania State University

Carla D'Antonio, University of California, Berkeley

Alan de Queiroz, University of Colorado

Michael Denbow, Virginia Tech

Susan Dunford, University of Cincinnati

William Eickmeier, Vanderbilt University

John Endler, University of California, Santa Barbara

Gordon L. Fain, University of California, Los Angeles

Stu Feinstein, University of California, Santa Barbara

Danilo Fernando, SUNY, Syracuse

Steve Fisher, University of California, Santa Barbara

Doug Futuyma, SUNY, Stony Brook

Scott Gilbert, Swarthmore College

Janice Glime, Michigan Technological University

Elizabeth Godrick, Boston University

Robert Goodman, University of Wisconsin, Madison

Nancy Guild, University of Colorado

Jessica Gurevitch, SUNY, Stony Brook

Jeff Hardin, University of Wisconsin, Madison

Joseph Heilig, University of Colorado

David Hershey, Hyattsville, MD

Mark Johnston, Dalhousie University

Walter Judd, University of Florida

Thomas Kane, University of Cincinnati

Laura Katz, Smith College

Elizabeth Kellogg, University of Missouri, St. Louis

Peter Krell, University of Guelph

Thomas Kursar, University of Utah

Wayne Maddison, University of Arizona

William Manning, University of Massachusetts, Amherst

Michael Marcotrigiano, Smith College

Lloyd Matsumoto, Rhode Island College

Stu Matz, The Evergreen State College

D. Jeffrey Meldrum, Idaho State University

Mike Millay, Ohio University (Southern Campus)

David Mindell, University of Michigan, Ann Arbor

Deborah Mowshowitz, Columbia University

Laura Olsen, University of Michigan, Ann Arbor

Guillermo Orti, University of Nebraska

Constance Parks, University of Massachusetts, Amherst

Jane Phillips, University of Minnesota

Ronald Poole, McGill University

Warren Porter, University of Wisconsin, Madison

Thomas Poulson, University of Illinois, Chicago

Loren Rieseberg, Indiana University

Ian Ross, University of California, Santa Barbara

Nancy Sanders, Truman State University

Paul Schroeder, Washington State University

Jim Shinkle, Trinity University

Mitchell Sogin, Marine Biological Laboratory, Woods Hole

Wayne Sousa, University of California, Berkeley

Charles Staben, University of Kentucky

James Staley, University of Washington

Steve Stanley, The Johns Hopkins University

Barbara Stebbins-Boaz, Willamette University

Antony Stretton, University of Wisconsin, Madison

Steven Swoap, Williams College

Gerald Thrush, California State University, San Bernardino

Richard Tolman, Brigham Young University

Mary Tyler, University of Maine

Michael Wade, Indiana University

Bruce Walsh, University of Arizona

Steven Wasserman, University of California, San Diego

Alex Weir, SUNY, Syracuse

Mary Williams, Harvey Mudd College

Jonathan Wright, Pomona College

Accuracy Reviewers

Andrew Clark, Pennsylvania State University

Joanne Ellzey, University of Texas, El Paso

Tejendra Gill, University of Houston, University Park

Paul Goldstein, University of Texas, El Paso

Laura Katz, Smith College

Hans Landel, North Seattle Community College

Sandy Ligon, University of New Mexico

Peter Lortz, North Seattle Community College

Roger Lumb, Western Carolina University

Coleman McCleneghan, Appalachian State University

Janie Milner, Santa Fe Community College

Zack Murrell, Appalachian State University

Ben Normark, University of Massachusetts, Amherst

Mike Silva, El Paso Community College

Phillip Snider, University of Houston, University Park

Steven Wasserman, University of California, San Diego

Media Reviewers

Karen Bernd, Davidson College

Mark Browning, Purdue University

William Eldred, Boston University

Joanne Ellzey, University of Texas, El Paso

Randall Johnson, University of California, San Diego

Coleman McCleneghan, Appalachian State University

Melissa Michael, University of Illinois

Tom Pitzer, Florida International University

Kenneth Robinson, Purdue University

To the Student

Welcome to the study of life! In our student days—and ever since—we have enjoyed studying the fascinating and fast-changing field of biology, and we hope that you will, too.

Getting the Most Out of the Book

There are a few things you can do to help you get the most from this book and from your course. For openers, read the book actively—don't just read passively, but do things that force you to think as you read. If we pose questions, stop and think about them. Ask questions of the text as you go. Do you understand what is being said? Does it relate to something you already know? Is it supported by experimental or other evidence? Does that evidence convince you? How does this passage fit into the chapter as a whole? Annotate the book—write down comments in the margins about things you don't understand, or about how one part relates to another, or even when you find an idea particularly interesting. People remember things they think about much better than they remember things they have read passively. Highlighting is passive; copying is drudge work; questioning and commenting are active and well worthwhile.

"Read" the illustrations actively too. You will find the balloon captions in the illustrations especially useful—they are there to guide you through the complexities of some topics and to highlight the major points.

The chapter summaries will help you quickly review the high points of what you have read. A summary identifies particular illustrations that you should study to help organize the material in your mind. Add concepts and details to the framework by reviewing the text. A way to review the material in slightly more detail after reading the chapter is to go back and look at the boldfaced terms. You can use the boldfaced terms to pose questions—and see if you can answer those questions. The boldfacing will probably be more useful on a second reading than on the first.

Use the "For Discussion" questions at the end of each chapter. These questions are usually open-ended and are intended to cause you to reflect on the material.

The glossary and the index can help you a great deal. When you are uncertain of the meaning of a term, check the glossary first—there are more than 1,500 definitions in it. If you don't find a term in the glossary, or if you want a more thorough discussion of the term, use the index to find where it's discussed.

The Web Site

Use the student Web Site/CD-ROM to help you understand some of the more detailed material and to help you sort out the information we have laid before you. An illustrated guide to the learning resources found on the Web Site/CD-ROM is in the front of this book. Pay particular attention to the activities and animated tutorials on key concepts, and to the self-quizzes. The self-quizzes provide extensive feedback for each correct and incorrect answer, and include hot-linked references to text pages. If you'd like to pursue some topics in greater detail, you'll find a chapter-by-chapter annotated list of suggested readings. We have tried to choose readings from books and magazines, especially *Scientific American*, that should be available in your college library.

What If the Going Gets Tough?

Most students occasionally have difficulty in courses, including biology courses. If you find that you are slipping behind in the course, or if a particular topic is giving you an unreasonable amount of trouble, here are some useful steps you might take. First, the basics: attend class, take careful lecture notes, and read the textbook assignments. Second, note that one of the most important roles of studying is to discover what you *don't* know, so that you can do something about it. Use the index, the glossary, the chapter summaries, and the text itself to try to answer any questions you have and to help you organize the material. Make a habit of looking over your lecture notes within 24 hours of when you take them—find out right away what points are unclear, and get them straightened out in your mind. The web site can help by providing a different perspective.

If none of these self-help remedies does the trick, get help! Other students are often a good source of help, because they are dealing with the material at the same level as you are. Study groups can be very useful, as long as the participants are all committed to learning the material. Tutors are almost always helpful, as are faculty members. The main thing is to *get help when you need it*. It is not a good idea to be strong and silent and drift into a low grade.

But don't make the grade the point of this or any other course. You are in college to learn, to pursue interesting subjects, and to enjoy the subjects you are pursuing. We hope you'll enjoy the pursuit of biology.

Bill Purves   David Sadava   Gordon Orians   Craig Heller

Life's Supplements

For the Student

Web Site/CD-ROM

Student Web Site at www.thelifewire.com
Life 6.0 CD-ROM (optionally bundled with the text)

The Web Site and CD-ROM each support the entire text, offering:

► Over 65 **Animated Tutorials** clarifying key topics from the text

► **Activities**, including flashcards for key terms and concepts, and drag-and-drop exercises

► **Self-quizzes** with extensive feedback, references to the Study Guide, and hot-linked references to *Life: The Science of Biology*, Sixth Edition

► **Glossary** of key terms and concepts

► **End-of-chapter Online Quizzes** (see "Online Quizzing" under "For the Instructor")

► **Lifelines**

Study Skills (Jerry Waldvogel, *Clemson University*) provides class-tested practical advice on time management, test-taking, note-taking, and how to read the textbook

Math for Life (Dany Adams, *Smith College*) helps students learn or reacquire basic quantitative skills

► **Suggested Readings** for further study
Order ISBN 0-7167-3874-0, *Life 6.0* CD-ROM, or
ISBN 0-7167-3875-9, Text/CD-ROM bundle

Study Guide

Christine Minor, *Clemson University*, Edward M. Dzialowski and Warren W. Burggren, *University of North Texas*, Lindsay Goodloe, *Cornell University*, and Nancy Guild, *University of Colorado at Boulder*.

For each chapter of the text, the study guide offers clearly defined learning objectives, summaries of key concepts, references to *Life* and to the student *Web/CD-ROM*, and review and exam-style self-test questions with answers and explanations.
Order ISBN 0-7167-3951-8

Lecture Notebook

This new tool presents black and white reproductions of all the Sixth Edition's line art and tables (more than 1000 images, with labels). The *Notebook* provides ample ruled spaces for note-taking.
Order ISBN 0-7167-4449-X

For the Instructor

Instructor's Teaching Kit

This **new** comprehensive teaching tool (in a three-ring binder) combines:

1. Instructor's Manual

Erica Bergquist, *Holyoke Community College*

The Manual includes:
► Chapter overviews
► Chapter outlines
► A "What's New" guide to the Sixth Edition
► All the bold-faced key terms from the text
► Key concepts and facts for each chapter
► Overviews of the animated tutorials from the Student Web Site/CD-ROM
► Custom lab ordering information (see "Custom Labs")

2. Enriched Lecture Notes, with diagrams
Charles Herr, *Eastern Washington University*

3. A PowerPoint® Thumbnail Guide to the PowerPoint® presentations on the Instructor's CD-ROM

Test Bank

Charles Herr, *Eastern Washington University*

The test bank, available in both computerized and printed formats, offers more than 4000 multiple-choice and sentence-completion questions.

The easy-to-use computerized test bank on CD-ROM includes Windows and Macintosh versions in a format that lets instructors add, edit, and resequence questions to suit their needs. From this same CD-ROM, instructors can access *Diploma* Online Testing from the Brownstone Research Group. *Diploma* allows instructors to easily create and administer secure exams over a network and over the Internet, with questions that incorporate multimedia and interactive exercises. More information about *Diploma* is available at http://www.brownstone.net

Online Quizzing

The online quizzing function is accessed via the Student Web Site at www.thelifewire.com. Using Question Mark's *Perception*, instructors can easily and securely quiz students online using multiple-choice questions for each text chapter and its media resources.

Instructor's Resource CD-ROM

The Instructor's Resource CD-ROM employs **Presentation Manager** and includes:

▶ All four-color line art and tables from the text (more than 1000 images), resized and reformatted to maximize large-hall projection

▶ More than 1500 photographic images, including electron micrographs, from the Biological Photo Service collection—all keyed to *Life* chapters

▶ More than 60 animations from the Student Web Site/CD-ROM

▶ Exceptional video microscopy from Jeremy Pickett-Heaps and others

▶ Chapter outlines and lecture notes from the Instructor's Teaching Kit in editable Microsoft® Word documents

PowerPoint® Presentations

The PowerPoint® slide set for *Life* follows the chapter summaries provided in the Instructor's Teaching Kit and can be used directly or customized. Each slide incorporates a figure from *Life*.

PowerPoint® Tutorials

QuickTime™ movies demonstrate how to use PowerPoint®.

Classroom Management

As a service for adopters using WebCT, we will provide a fully-loaded WebCourselet, including the instructor and student resources for this text. The files can then be customized to fit your specific course needs, or can be used as is. Course outlines, pre-built quizzes, activities, and a whole array of materials are included, eliminating hours of work for instructors interested in creating WebCT courses. For more information and a demo of the WebCourselet for this text, please visit our Web Site (http://bfwpub.com/mediaroom/Index.html) and click "WebCT".

Overhead Transparencies

The transparency set includes all four-color line art and tables from the text (more than 1000 images) in a convenient three-ring binder. Balloon captions (and some labels) are deleted to enhance projection and allow for classroom quizzing. Labels and images have been resized for maximum readability.

Slide Set

The slide set includes selected four-color figures from the text. Labels and images have been resized for maximum readability.

Laboratory Manuals

Biology in the Laboratory, Third Edition

Doris Helms, Robert Kosinski, and John Cummings, *all of Clemson University*

The revised edition of this popular lab manual, which includes a CD-ROM, is available to accompany the Sixth Edition of *Life*.
Order ISBN 0-7167-3146-0

Laboratory Outlines in Biology VI

Peter Abramoff and Robert G. Thomson, *Marquette University*
Order ISBN 0-7167-2633-5

The following manuals are available in a bound volume or as separates:
Anatomy and Dissection of the Rat, Third Edition
Warren F. Walker, Jr., *Oberlin College*, and Dominique Homberger, *Louisiana State University*
Order ISBN 0-7167-2635-1
Anatomy and Dissection of the Fetal Pig, Fifth Edition
Warren F. Walker, Jr., *Oberlin College*, and Dominique Homberger, *Louisiana State University*
Order ISBN 0-7167-2637-8
Anatomy and Dissection of the Frog, Second Edition
Warren F. Walker, Jr., *Oberlin College*
Order ISBN 0-7167-2636-X

Custom Labs

Custom Publishing for Laboratory Manuals at www.custompub.whfreeman.com

With this custom publishing option, instructors can build and order customized lab manuals in just minutes, choosing material from Freeman's acclaimed biology laboratory manuals—lab-tested experiments that have been used successfully by hundreds of thousands of students. Instructors determine the manual's content (with the option to incorporate their own material or blank pages), table of contents or index styles, and cover design, and submit the order. A streamlined production process provides a quick turnaround to meet crucial deadlines.

Contents in Brief

Contents

Part One
THE CELL

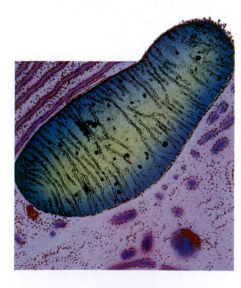

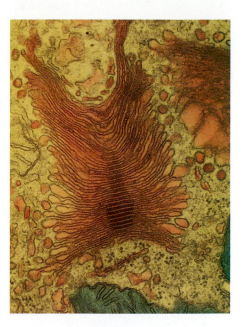

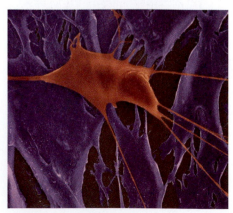

Part Two
INFORMATION AND HEREDITY

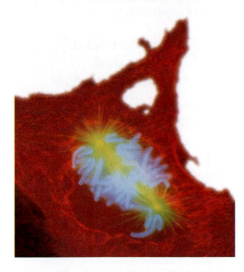

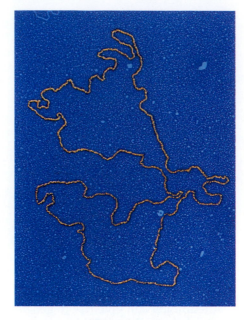

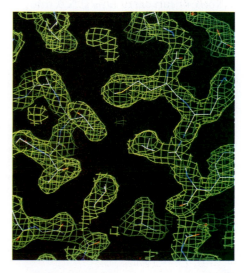

Part Three
EVOLUTIONARY PROCESSES

Part Four
THE
EVOLUTION
OF DIVERSITY

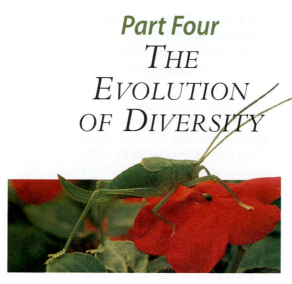

Part Five

THE BIOLOGY OF FLOWERING PLANTS

Part Six
THE BIOLOGY OF ANIMALS

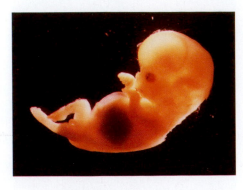

Part Seven
ECOLOGY AND BIOGEOGRAPHY

1 An Evolutionary Framework for Biology

AT MIDNIGHT ON DECEMBER 31, 1999, MAS-sive displays of fireworks exploded in many places on Earth as people celebrated a new millennium—the passage from one thousand-year time frame into the next—and the advent of the year 2000. One such millennial display took place above the Egyptian pyramids.

We are impressed with the size of the pyramids, how difficult it must have been to build them, and how ancient they are. The oldest of these awe-inspiring monuments to human achievement was built more than 4,000 years ago; in the human experience, this makes the Egyptian pyramids very, very old. Yet from the perspective of the age of Earth and the time over which life has been evolving, the pyramids are extremely young. Indeed, if the history of Earth is visualized as a 30-day month, recorded human history—the dawn of which coincides roughly with the construction of the earliest pyramids—is confined to the last *30 seconds* of the final day of the month (Figure 1.1).

The development of modern biology depended on the recognition that an immense length of time was available for life to arise and evolve its current richness. But for most of human history, people had no reason to suspect that Earth was so old. Until the discovery of radioactive decay at the beginning of the twentieth century, no methods existed to date prehistoric events. By the middle of the nineteenth century, however, studies of rocks and the fossils they contained had convinced geologists that Earth was much older than had generally been believed. Darwin could not have conceived his theory of evolution by natural selection had he not understood that Earth was very ancient.

In this chapter we review the events leading to the acceptance of the fact that life on Earth has evolved over several billion years. We then summarize how evolutionary mechanisms adapt organisms to their environments, and we review the major milestones in the evolution of life on Earth. Finally, we briefly describe how scientists generate new knowledge, how they develop and test hypotheses, and how that knowledge can be used to inform public policy.

A Celebration of Time
One millennial fireworks display celebrating the year 2000 took place over the ancient pyramids of Egypt, structures that represent more than 4,000 years of human history but an infinitesimal portion of Earth's geologic history.

Organisms Have Changed over Billions of Years

Long before the mechanisms of biological evolution were understood, some people realized that organisms had changed over time and that living organisms had evolved from organisms no longer alive on Earth. In the 1760s, the French naturalist Count George-Louis Leclerc de Buffon (1707–1788) wrote his *Natural History of Animals*, which contained a clear statement of the possibility of evolution. Buffon originally believed that each species had been divinely created for a particular way of life, but as he studied animal anatomy, doubts arose. He observed that the limb bones of all mammals, no matter what their way of life, were re-

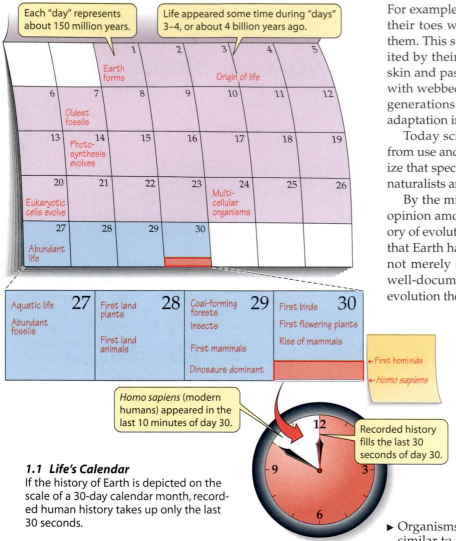

1.1 Life's Calendar
If the history of Earth is depicted on the scale of a 30-day calendar month, recorded human history takes up only the last 30 seconds.

For example, Lamarck suggested that aquatic birds extend their toes while swimming, stretching the skin between them. This stretched condition, he thought, could be inherited by their offspring, which would in turn stretch their skin and pass this condition along to their offspring; birds with webbed feet would thereby evolve over a number of generations. Lamarck explained many other examples of adaptation in a similar way.

Today scientists do not believe that changes resulting from use and disuse can be inherited. But Lamarck did realize that species change with time. And after Lamarck, other naturalists and scientists speculated along similar lines.

By the middle of the nineteenth century, the climate of opinion among many scholars was receptive to a new theory of evolutionary processes. By then geologists had shown that Earth had existed and changed over millions of years, not merely a few thousand years. The presentation of a well-documented and thoroughly scientific argument for evolution then triggered a transformation of biology.

The theory of evolution by natural selection was proposed independently by Charles Darwin and Alfred Russel Wallace in 1858. We will discuss evolutionary theory in detail in Chapter 21, but its essential features are easy to understand. The theory rests on two facts and one inference drawn from them. The two facts are:

▶ The reproductive rates of all organisms, even slowly reproducing ones, are sufficiently high that populations would quickly become enormous if mortality rates did not balance reproductive rates.

▶ Organisms of all types are variable, and offspring are similar to their parents because they inherit their features from them.

The inference is:

▶ The differences among individuals influence how well those individuals survive and reproduce. Traits that increase the probability that their bearers will survive and reproduce are more likely to be passed on to their offspring and to their offspring's offspring.

Darwin called the differential survival and reproductive success of individuals **natural selection**. The remarkable features of all organisms have evolved under the influence of natural selection. Indeed, *the ability to evolve by means of natural selection clearly separates life from nonlife.*

Biology began a major conceptual shift a little more than a century ago with the general acceptance of long-term evolutionary change and the recognition that differential survival and reproductive success is the primary process that adapts organisms to their environments. The shift has taken a long time because it required abandoning many components of an earlier worldview. The pre-Darwinian view held that the world was young, and that organisms had been created in their current forms. In the Darwinian view,

markably similar in many details (Figure 1.2). Buffon also noticed that the legs of certain mammals, such as pigs, have toes that never touch the ground and appear to be of no use. He found it difficult to explain the presence of these seemingly useless small toes by special creation.

Both of these troubling facts could be explained if mammals had not been specially created in their present forms, but had been modified over time from an ancestor that was common to all mammals. Buffon suggested that the limb bones of mammals might all have been inherited, and that pigs might have functionless toes because they inherited them from ancestors with fully formed and functional toes. Buffon's idea was an early statement of evolution (descent with modification), although he did not attempt to explain how such changes took place.

Buffon's student Jean Baptiste de Lamarck (1744–1829) was the first person to propose a mechanism of evolutionary change. Lamarck suggested that lineages of organisms may change gradually over many generations as offspring inherit structures that have become larger and more highly developed as a result of continued use or, conversely, have become smaller and less developed as a result of disuse.

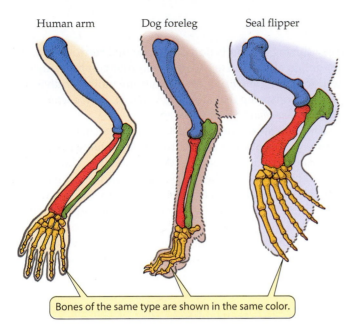

Human arm   Dog foreleg   Seal flipper

Bones of the same type are shown in the same color.

1.2 Mammals Have Similar Limbs
Mammalian forelimbs have different purposes, but the number
and types of their bones are similar, indicating that they have
been modified over time from a common ancestor.

the world is ancient, and both Earth and its inhabitants
have been continually changing. In the Darwinian view of
the world, organisms evolved their particular features because individuals with those features survived and reproduced better than individuals with different features.

Adopting this new view of the world means accepting
not only the processes of evolution, but also the view that
the living world is constantly evolving, and that evolutionary change occurs without any "goals." The idea that evolution is not directed toward a final goal or state has been
more difficult for many people to accept than the process of
evolution itself. But even though evolution has no goals,
evolutionary processes have resulted in a series of profound changes—milestones—over the nearly 4 billion years
life has existed on Earth.

Evolutionary Milestones

The following overview of the major milestones in the evolution of life provides both a framework for presenting the
characteristics of life that will be described in this book and
an overview of how those characteristics evolved during
the history of life on Earth.

Life arises from nonlife

All matter, living and nonliving, is made up of chemicals.
The smallest chemical units are atoms, which bond together
into molecules; the properties of those molecules are the
subject of Chapter 2. The processes leading to life began
nearly 4 billion years ago with interactions among small
molecules that stored useful information.

The information stored in these simple molecules eventually resulted in the synthesis of larger molecules with

complex but relatively stable shapes. Because they were
both complex and stable, these units could participate in
increasing numbers and kinds of chemical reactions. Some
of these large molecules—carbohydrates, lipids, proteins,
and nucleic acids—are found in all living systems and perform similar functions. The properties of these complex
molecules are the subject of Chapter 3.

Cells form from molecules

About 3.8 billion years ago, interacting systems of molecules came to be enclosed in compartments surrounded by
membranes. Within these membrane-enclosed units, or
cells, control was exerted over the entrance, retention, and
exit of molecules, as well as the chemical reactions taking
place within the cell. Cells and membranes are the subjects
of Chapters 4 and 5.

Cells are so effective at capturing energy and replicating
themselves—two fundamental characteristics of life—that
since the time they evolved, they have been the unit on
which all life has been built. Experiments by the French
chemist and microbiologist Louis Pasteur and others during the nineteenth century convinced most scientists that,
under present conditions on Earth, cells do not arise from
noncellular material, but must come from other cells.

For 2 billion years, cells were tiny packages of molecules
each enclosed in a single membrane. These **prokaryotic
cells** lived autonomous lives, each separate from the other.
They were confined to the oceans, where they were
shielded from lethal ultraviolet sunlight. Some prokaryotes
living today may be similar to these early cells (Figure 1.3).

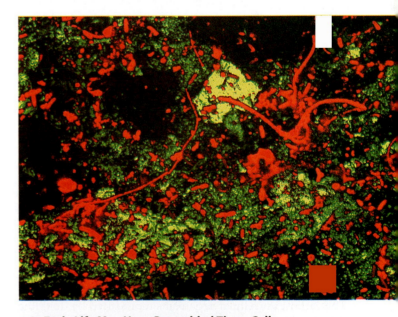

1.3 Early Life May Have Resembled These Cells
"Rock-eating" bacteria, appearing red in this artificially colored
micrograph, were discovered in pools of water trapped between
layers of rock more than 1,000 meters below Earth's surface.
Deriving chemical nutrients from the rocks and living in an environment devoid of oxygen, they may resemble some of the earliest prokaryotic cells.

To maintain themselves, to grow, and to reproduce, these early prokaryotes, like all cells that have subsequently evolved, obtained raw materials and energy from their environment, using these as building blocks to synthesize larger, carbon-containing molecules. The energy contained in these large molecules powered the chemical reactions necessary for the life of the cell. These conversions of matter and energy are called **metabolism**.

All organisms can be viewed as devices to capture, process, and convert matter and energy from one form to another; these conversions are the subjects of Chapters 6 and 7. *A major theme in the evolution of life is the development of increasingly diverse ways of capturing external energy and using it to drive biologically useful reactions.*

Photosynthesis changes Earth's environment

About 2.5 billion years ago, some organisms evolved the ability to use the energy of sunlight to power their metabolism. Although they still took raw materials from the environment, the energy they used to metabolize these materials came directly from the sun. Early photosynthetic cells were probably similar to present-day prokaryotes called cyanobacteria (Figure 1.4). The energy-capturing process they used—**photosynthesis**—is the basis of nearly all life on Earth today; it is explained in detail in Chapter 8. It used new metabolic reactions that exploited an abundant source of energy (sunlight), and generated a new waste product (oxygen) that radically changed Earth's atmosphere.

The ability to perform photosynthetic reactions probably accumulated gradually during the first billion years or so of evolution, but once this ability had evolved, its effects were dramatic. Photosynthetic prokaryotes became so abundant that they released vast quantities of oxygen gas (O_2) into the atmosphere. The presence of oxygen opened up new avenues of evolution. Metabolic reactions that use O_2, called **aerobic metabolism**, came to be used by most organisms on Earth. The oxygen in the air we breathe today would not exist without photosynthesis.

Over a much longer time, the vast quantities of oxygen liberated by photosynthesis had another effect. Formed from O_2, ozone (O_3) began to accumulate in the upper atmosphere. The ozone slowly formed a dense layer that acted as a shield, intercepting much of the sun's deadly ultraviolet radiation. Eventually (although only within the last 800 million years of evolution), the presence of this shield allowed organisms to leave the protection of the oceans and establish new lifestyles on Earth's land surfaces.

Sex enhances adaptation

The earliest unicellular organisms reproduced by doubling their hereditary (genetic) material and then dividing it into two new cells, a process known as mitosis. The resulting progeny cells were identical to each other and to the parent. That is, they were clones. But **sexual reproduction**—the combining of genes from two cells in one cell—appeared early during the evolution of life. Sexual reproduction is advantageous because an organism that combines its genetic information with information from another individual produces offspring that are more variable. *Reproduction with variation is a major characteristic of life.*

Variation allows organisms to adapt to a changing environment. **Adaptation** to environmental change is one of life's most distinctive features. An organism is adapted to a given environment when it possesses inherited features that enhance its survival and ability to reproduce in that environment. Because environments are constantly changing, organisms that produce variable offspring have an advantage over those that produce genetically identical "clones," because they are more likely to produce some offspring better adapted to the environment in which they find themselves.

Eukaryotes are "cells within cells"

As the ages passed, some prokaryotic cells became large enough to attack, engulf, and digest smaller cells, becoming the first predators. Usually the smaller cells were destroyed within the predators' cells. But some of these smaller cells survived and became permanently integrated into the operation of their hosts' cells. In this manner, cells with complex internal compartments arose. We call these cells **eukaryotic cells**. Their appearance slightly more than 1.5 billion years ago opened more new evolutionary opportunities.

Prokaryotic cells—the Bacteria and Archaea—have no membrane-enclosed compartments. Eukaryotic cells, on the

1.4 Oxygen Produced by Prokaryotes Changed Earth's Atmosphere
These modern cyanobacteria are probably very similar to early photosynthetic prokaryotes.

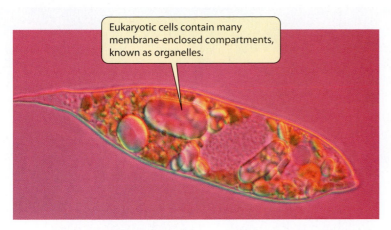

Eukaryotic cells contain many membrane-enclosed compartments, known as organelles.

1.5 Multiple Compartments Characterize Eukaryotic Cells
The nucleus and other specialized organelles probably evolved from small prokaryotes that were ingested by a larger prokaryotic cell. This is a photograph of a single-celled eukaryotic organism known as a protist.

other hand, are filled with membrane-enclosed compartments. In eukaryotic cells, genetic material—genes and chromosomes—became contained within a discrete nucleus and became increasingly complex. Other compartments became specialized for other purposes, such as photosynthesis. We refer to these specialized compartments as **organelles** (Figure 1.5).

Multicellularity permits specialization of cells

Until slightly more than 1 billion years ago, only single-celled organisms existed. Two key developments made the evolution of multicellular organisms—organisms consisting of more than one cell—possible. One was the ability of a cell to change its structure and functioning to meet the challenges of a changing environment. This was accomplished when prokaryotes evolved the ability to change from rapidly growing cells into resting cells called **spores** that could survive harsh environmental conditions. The second development allowed cells to stick together in a "clump" after they divided, forming a multicellular organism.

Once organisms could be composed of many cells, it became possible for the cells to specialize. Certain cells, for example, could be specialized to perform photosynthesis. Other cells might become specialized to transport chemical materials such as oxygen from one part of an organism to another. Very early in the evolution of multicellular life, certain cells began to be specialized for sex—the passage of new genetic information from one generation to the next.

With the presence of specialized sex cells, genetic transmission became more complicated. Simple nuclear division—mitosis—was and is sufficient for the needs of most cells. But among the sex cells, or gametes, a whole new method of nuclear division—meiosis—evolved. Meiosis allows gametes to combine and rearrange the genetic infor-

mation from two distinct parent organisms into a genetic package that contains elements of both parent cells but is different from either. The recombinational possibilities generated by meiosis had great impact on variability and adaptation and on the speed at which evolution could occur.

Mitosis and meiosis are covered in detail in Chapter 9.

Controlling internal environments becomes more complicated

The pace of evolution, quickened by the emergence of sex and multicellular life, was also heightened by changes in Earth's atmosphere that allowed life to move out of the oceans and exploit environments on land. Photosynthetic green plants colonized the land, providing a rich source of energy for a vast array of organisms that consumed them. But whether it is made up of one cell or many, an organism must respond appropriately to its external environment. Life on land presented a new set of environmental challenges.

In any environment, external conditions can change rapidly and unpredictably in ways that are beyond an organism's control. An organism can remain healthy only if its internal environment remains within a given range of physical and chemical conditions. Organisms maintain relatively constant internal environments by making metabolic adjustments to changes in external and internal conditions such as temperature, the presence or absence of sunlight, the presence or absence of specific chemicals, the need for nutrients (food) and water, or the presence of foreign agents inside their bodies. Maintenance of a relatively stable internal condition—such as a constant human body temperature despite variation in the temperature of the surrounding environment—is called **homeostasis**. *A major theme in the evolution of life is the development of increasingly complicated systems for maintaining homeostasis.*

Multicellular organisms undergo regulated growth

Multicellular organisms cannot achieve their adult shapes or function effectively unless their growth is carefully regulated. Uncontrolled growth—one example of which is cancer—ultimately destroys life. *A vital characteristic of living organisms is regulated growth.* Achieving a functional multicellular organism requires a sequence of events leading from a single cell to a multicellular adult. This process is called **development**.

The adjustments that organisms make to maintain constant internal conditions are usually minor; they are not obvious, because nothing appears to change. However, at some time during their lives, many organisms respond to changing conditions not by maintaining their status, but by undergoing major cellular and molecular reorganization. An early form of such developmental reorganization was the prokaryotic spores that were generated in response to environmental stresses. A striking example that evolved much later is **metamorphosis**, seen in many modern in-

1.6 Organisms May Change Dramatically During Their Lives
The caterpillar, pupa, and adult are all stages in the life cycle of a monarch butterfly. The transition from one stage to another is triggered by internal signals.

sects, such as butterflies. In response to internal chemical signals, a caterpillar changes into a pupa and then into an adult butterfly (Figure 1.6).

The activation of gene-based information within cells and the exchange of signal information among cells produce the well-timed events that are required for the transition to the adult form. Genes control the metabolic processes necessary for life. The nature of the genetic material that controls these lifelong events has been understood only within the twentieth century; it is the story to which much of Part Two of this book is devoted.

Altering the timing of development can produce striking changes. Just a few genes can control processes that result in dramatically different adult organisms. Chimpanzees and humans share more than 98 percent of their genes, but the differences between the two in form and in behavioral abilities—most notably speech—are dramatic (Figure 1.7). When we realize how little information it sometimes takes to create major transformations, the still mysterious process of **speciation** becomes a little less of a mystery.

Speciation produces the diversity of life

All organisms on Earth today are the descendants of a kind of unicellular organism that lived almost 4 billion years ago. The preceding pages described the major evolutionary events that have led to more complex living organisms. The course of this evolution has been accompanied by the storage of larger and larger quantities of information and increasingly complex mechanisms for using it. But if that were the entire story, only one kind of organism might exist

on Earth today. Instead, Earth is populated by many millions of kinds of organisms that do not interbreed with one another. We call these genetically independent groups of organisms **species**.

As long as individuals within a population mate at random and reproduce, structural and functional changes may occur, but only one species will exist. However, if a population becomes divided into two or more groups, and individuals can mate only with individuals in their own group, differences may accumulate with time, and the groups may evolve into different species.

The splitting of groups of organisms into separate species has resulted in the great variety of life found on Earth today, as described in Chapter 20. How species form is explained in Chapter 22. From a single ancestor, many species may arise as a result of the repeated splitting of populations. How biologists determine which species have descended from a particular ancestor is discussed in Chapter 23.

1.7 Genetically Similar Yet Very Different
By looking at the two, you might be surprised to learn that chimpanzees and humans share more than 98 percent of their genes.

1.8 Adaptations to the Environment
(*a*) The long, pointed wings of the peregrine falcon allow it to accelerate rapidly as it dives on its prey. (*b*) The action of a hummingbird's wings allows it to hover in front of a flower while it extracts nectar. (*c*) In a water-limited environment, this saguaro cactus stores water in its fleshy trunk. Its roots spread broadly to extract water immediately after it rains. (*d*) The aboveground root system of mangroves is an adaptation that allows these plants to thrive while inundated by salt water—an environment that would kill most terrestrial plants.

Sometimes humans refer to species as "primitive" or "advanced." These and similar terms, such as "lower" and "higher," are best avoided because they imply that some organisms function better than others. In this book, we use the terms "ancestral" and "derived" to distinguish characteristics that appeared earlier from those that appeared later in the evolution of life.

It is important to recognize that *all* living organisms are successfully adapted to their environments. The wings that allow a bird to fly and the structures that allow green plants to survive in environments where water is either scarce or overabundant are examples of the rich array of adaptations found among organisms (Figure 1.8).

The Hierarchy of Life

Biologists study life in two complementary ways:

▶ They study structures and processes ranging from the simple to the complex and from the small to the large.

▶ They study the patterns of life's evolution over billions of years to determine how evolutionary processes have resulted in lineages of organisms that can be traced back to recent and distant ancestors.

These two themes of biological investigation help us synthesize the hierarchical relationships among organisms and the role of these relationships in space and time. We first describe the hierarchy of interactions among the units of biology from the smallest to the largest—from cells to the biosphere. Then we turn to the hierarchy of evolutionary relationships among organisms.

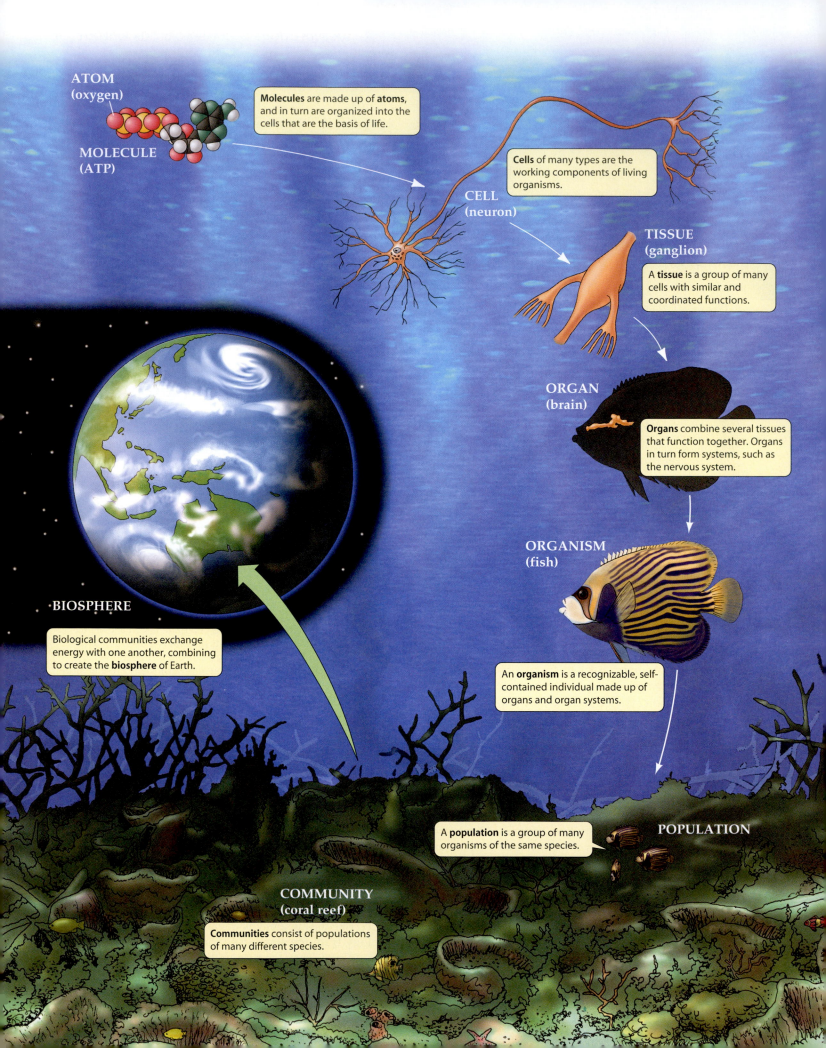

ATOM
(oxygen)

MOLECULE
(ATP)

Molecules are made up of **atoms**, and in turn are organized into the cells that are the basis of life.

CELL
(neuron)

Cells of many types are the working components of living organisms.

TISSUE
(ganglion)

A **tissue** is a group of many cells with similar and coordinated functions.

ORGAN
(brain)

Organs combine several tissues that function together. Organs in turn form systems, such as the nervous system.

ORGANISM
(fish)

An **organism** is a recognizable, self-contained individual made up of organs and organ systems.

BIOSPHERE

Biological communities exchange energy with one another, combining to create the **biosphere** of Earth.

POPULATION

A **population** is a group of many organisms of the same species.

COMMUNITY
(coral reef)

Communities consist of populations of many different species.

1.9 The Hierarchy of Life

The individual organism is the central unit of study in biology, but understanding it requires a knowledge of many levels of biological organization both above and below it. At each higher level, additional and more complex properties and functions emerge.

Biologists study life at different levels

Biology can be visualized as a hierarchy in which the units, from the smallest to the largest, include atoms, molecules, cells, tissues, organs, organisms, populations, and communities (Figure 1.9).

The organism is the central unit of study in biology. Parts Five and Six of this book discuss organismal biology in detail. But to understand organisms, biologists must study life at all its levels of organization. Biologists study molecules, chemical reactions, and cells to understand the operations of tissues and organs. They study organs and organ systems to determine how organisms function and maintain internal homeostasis. At higher levels in the hierarchy, biologists study how organisms interact with one another to form social systems, populations, ecological communities, and biomes, which are the subjects of Part Seven of this book.

Each level of biological organization has properties, called **emergent properties**, that are not found at lower levels. For example, cells and multicellular organisms have characteristics and carry out processes that are not found in the molecules of which they are composed.

Emergent properties arise in two ways. First, many *emergent properties of systems result from interactions among their parts.* For example, at the organismal level, developmental interactions of cells result in a multicellular organism whose adult features are vastly richer than those of the single cell from which it grew. Other examples of properties that emerge through complex interactions are memory and emotions. In the human brain, these properties result from interactions among the brain's 10^{12} (trillion) cells with their 10^{15} (quadrillion) connections. No single cell, or even small group of cells, possesses them.

Second, *emergent properties arise because aggregations have collective properties* that their individual units lack. For example, individuals are born and they die; they have a life span. An individual does not have a birth rate or a death rate, but a population (composed of many individuals) does. Birth and death rates are emergent properties of a population. Evolution is an emergent property of populations that depends on variation in birth and death rates, which emerges from the different life spans and reproductive success of individuals in the various populations.

Emergent properties do not violate the principles that operate at lower levels of organization. However, emergent properties usually cannot be detected, predicted, or even suspected by studying lower levels. Biologists could never discover the existence of human emotions by studying single nerve cells, even though they may eventually be able to explain it in terms of interactions among many nerve cells.

Biological diversity is organized hierarchically

As many as 30 million species of organisms inhabit Earth today. Many times that number lived in the past but are now extinct. If we go back four billion years, to the origin of life, all organisms are believed to be descended from a single *common ancestor*. The concept of a common ancestor is crucial to modern methods of classifying organisms. *Organisms are grouped in ways that attempt to define their evolutionary relationships, or how recently the different members of the group shared a common ancestor.*

To determine evolutionary relationships, biologists assemble facts from a variety of sources. Fossils tell us where and when ancestral organisms lived and what they looked like. The physical structures different organisms share—toes among mammals, for example—can be an indication of how closely related they are. But a modern "revolution" in classification has emerged because technologies developed in the past 30 years now allow us to compare the genomes of organisms: We can actually determine how many genes different species share. The more genes species have in common, the more recently they probably shared a common ancestor.

Because no fossil evidence for the earliest forms of life remains, the decision to divide all living organisms into three major **domains**—the deepest divisions in the evolutionary history of life—is based primarily on molecular evidence (Figure 1.10). Although new evidence is constantly being brought to light, it seems clear that organisms belonging to a particular domain have been evolving separately from organisms in the other two domains for more than a billion years.

Organisms in the domains **Archaea** and **Bacteria** are prokaryotes—single cells that lack a nucleus and the other internal compartments found in the Eukarya. Archaea and Bacteria differ so fundamentally from each other in the chemical reactions by which they function and in the products they produce that they are believed to have separated into distinct evolutionary lineages very early during the evolution of life. These domains are covered in Chapter 26.

Members of the third domain have eukaryotic cells containing nuclei and complex cellular compartments called organelles. The **Eukarya** are divided into four groups—the protists and the classical kingdoms Plantae, Fungi, and Animalia (see Figure 1.10). Protists, the subject of Chapter 27, are mostly single-celled organisms. The remaining three kingdoms, whose members are all multicellular, are believed to have arisen from ancestral protists.

Some bacteria, some protists, and most members of the kingdom Plantae (plants) convert light energy to chemical energy by photosynthesis. The biological molecules that they produce are the primary food for nearly all other living organisms. The Plantae are covered in Chapters 28 and 29.

The Fungi, the subject of Chapter 30, include molds, mushrooms, yeasts, and other similar organisms, all of

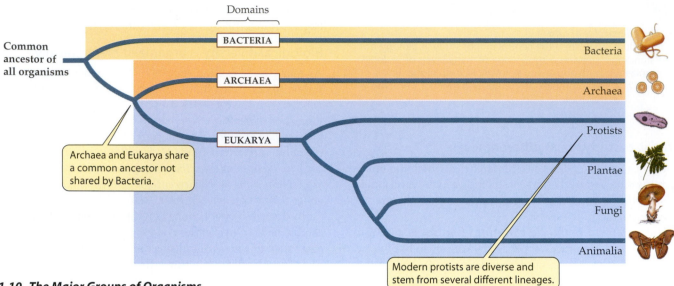

Domains

BACTERIA

Bacteria

ARCHAEA

Archaea

EUKARYA

Protists

Common ancestor of all organisms

Archaea and Eukarya share a common ancestor not shared by Bacteria.

Plantae

Fungi

Animalia

Modern protists are diverse and stem from several different lineages.

1.10 The Major Groups of Organisms
The classification system used in this book divides Earth's organisms into three domains. The domain Eukarya contains numerous groups of unicellular and multicellular organisms. This "tree" diagram gives information on evolutionary relationships among the groups, as described in Chapter 23.

which are **heterotrophs**: They require a food source of energy-rich molecules synthesized by other organisms. Fungi absorb food substances from their surroundings and break them down (digest them) within their cells. They are important as decomposers of the dead bodies of other organisms.

Members of the kingdom Animalia (animals) are also heterotrophs. These organisms ingest their food source, digest the food outside their cells, and then absorb the products. Animals get their raw materials and energy by eating other forms of life. Perhaps because we are animals ourselves, we are often drawn to study members of this kingdom, which is covered in Chapters 31, 32, and 33.

The biological classification system used today has many hierarchical levels in addition to the ones shown in Figure 1.10. We will discuss the principal levels in Chapter 23. But to understand some of the terms we will use in the intervening chapters, you need to know that each species of organism is identified by two names. The first identifies the **genus**—a group of species that share a recent common ancestor—of which the species is a member. The second name is the species name. To avoid confusion, a particular combination of two names is assigned to only a single species. For example, the scientific name of the modern human species is *Homo sapiens*.

Asking and Answering "How?" and "Why?"

Because biology is an evolutionary science, biological processes and products can be viewed from two different but complementary perspectives. Biologists ask, and try to answer, functional questions: How does it work? They also

ask, and try to answer, adaptive questions: Why has it evolved to work that way?

Suppose, for example, that some marine biologists walking on mudflats in the Bay of Fundy, Nova Scotia, Canada, observe many amphipods (tiny relatives of shrimps and lobsters) crawling on the surface of the mud (Figure 1.11). Two obvious questions they might ask are

▶ *How* do these animals crawl?
▶ *Why* do they crawl?

To answer the "how" question, the scientists would investigate the molecular mechanisms underlying muscular contraction, nerve and muscle interactions, and the receipt of stimuli by the amphipods' brains. To answer the "why" question, they would attempt to determine why crawling on the mud is adaptive—that is, why it improves the survival and reproductive success of amphipods.

Is either of these two types of questions more basic or important than the other? Is any one of the answers more fundamental or more important than the other? Not really. The richness of possible answers to apparently simple questions makes biology a complex field, but also an exciting one. Whether we're talking about molecules bonding, cells dividing, blood flowing, amphipods crawling, or forests growing, we are constantly posing both how and why questions. To answer these questions, scientists generate hypotheses that can be tested.

Hypothesis testing guides scientific research

The most important motivator of most biologists is curiosity. People are fascinated by the richness and diversity of life, and they want to learn more about organisms and how they function and interact with one another. Curiosity is probably an adaptive trait. Humans who were motivated to learn about their surroundings are likely to have survived and reproduced better, on average, than their less curious relatives. We hope this book will help you share in the ex-

1.11 An Amphipod from the Mud Flats
Scientists studied this tiny crustacean (whose actual size of approximately 1 centimeter is shown by the scale bar) in an attempt to see whether its behavior changes when it is infected by a parasitic worm. The female of this amphipod species is at the top; the lower specimen is a male.

citement biologists feel as they develop and test hypotheses. There are vast numbers of how and why questions for which we do not have answers, and new discoveries usually engender questions no one thought to ask before. Perhaps *your* curiosity will lead to an important new idea.

Underlying all scientific research is the **hypothetico-deductive (H-D) approach** by which scientists ask questions and test answers. The H-D approach allows scientists to modify and correct their beliefs as new observations and information become available. The method has five stages:

▶ Making observations.
▶ Asking questions.
▶ Forming **hypotheses,** or tentative answers to the questions.
▶ Making predictions based on the hypotheses.
▶ Testing the predictions by making additional observations or conducting experiments.

The data gained may support or contradict the predictions being tested. If the data support the hypothesis, it is subjected to still more predictions and tests. If they continue to support it, confidence in its correctness increases, and the hypothesis comes to be considered a **theory**. If the data do not support the hypothesis, it is abandoned or modified in accordance with the new information. Then new predictions are made, and more tests are conducted.

Applying the hypothetico-deductive method

The way in which marine biologists answered the question "Why do amphipods crawl on the surface of the mud rather than staying hidden within?" illustrates the H-D approach. As we saw above, the biologists observed something occurring in nature and formulated a question about it. To begin answering the question, they assembled available information on amphipods and the species that eat them.

They learned that during July and August of each year, thousands of sandpipers assemble for four to six weeks on the mudflats of the Bay of Fundy, during their southward migration from their Arctic breeding grounds to their wintering areas in South America (Figure 1.12). On these mud-

1.12 Sandpipers Feed on Amphipods
Migrating sandpipers crowd the exposed tidal flats in search of food. By consuming infected amphipods, the sandpipers also become infected, serving as hosts and allowing the parasitic worm to complete its life cycle.

flats, which are exposed twice daily by the tides, they feed vigorously, putting on fat to fuel their next long flight. Amphipods living in the mud form about 85 percent of the diet of the sandpipers. Each bird may consume as many as 20,000 amphipods per day!

Previous observations had shown that a nematode (roundworm) parasitizes both the amphipods and the sandpipers. To complete its life cycle, the nematode must develop within both a sandpiper and an amphipod. The nematodes mature within the sandpipers' digestive tracts, mate, and release their eggs into the environment in the birds' feces. Small larvae hatch from the eggs and search for, find, and enter amphipods, where they grow through several larval stages. Sandpipers are reinfected when they eat parasitized amphipods.

GENERATING A HYPOTHESIS AND PREDICTIONS. Based on the available information, biologists generated the following hypothesis: *Nematodes alter the behavior of their amphipod hosts in a way that increases the chance that the worms will be*

1.13 Collecting Field Data
Amphipods are collected from the mud to be tested for infection by parasites. Some of these crustaceans will be used in laboratory experiments.

passed on to sandpiper hosts. From this general hypothesis they generated two specific predictions.

▶ First, they predicted that amphipods infected by nematodes would increase their activity on the surface of the mud during daylight hours, when the sandpipers hunted by sight, but not at night, when the sandpipers fed less and captured prey by probing into the mud.

▶ Second, they predicted that only amphipods with late-stage nematode larvae—the only stage that can infect sandpipers—would have their behavior manipulated by the nematodes.

For each hypothesis proposing an effect, there is a corresponding **null hypothesis**, which asserts that the proposed effect is absent. For the hypothesis we have just stated, the null hypothesis is that nematodes have no influence on the behavior of their amphipod hosts. The alternative predictions that would support the null hypothesis are (1) that infected amphipods show no increase their activity either during the day or at night and (2) that all larval stages affect their hosts in the same manner. It is important in hypothesis testing to generate and test as many alternate hypotheses and predictions as possible.

TESTING PREDICTIONS.   Investigators collected amphipods in the field, taking them from the surface and from within the mud, during the day and at night (Figure 1.13). They found that during the day, amphipods crawling on the surface were much more likely to be infected with nematodes than were amphipods collected from within the mud. At night, however, there was no difference between the proportion of infected amphipods on the surface and those burrowing within the mud. This evidence supported the first prediction.

The field collections also showed that a higher proportion of the amphipods collected on the surface than of those collected from within the mud were parasitized by late-stage nematode larvae. However, amphipods crawling on the surface were no more likely to be infected by early-stage nematode larvae than were amphipods collected from the mud. These findings supported the *second* prediction.

To test the prediction that nematode larvae are more likely to affect amphipod behavior once they become infective, biologists performed laboratory experiments. They artificially infected amphipods with nematode eggs they obtained from sandpipers collected in the field. The infected amphipods established themselves in mud in laboratory containers.

By examining infected amphipods, investigators determined that it took about 13 days for the nematode larvae to reach the late, infective stage. By monitoring the behavior of the amphipods in the test tubes, the researchers determined that the amphipods were more likely to expose themselves on the surface of the mud once the parasites had reached the infective stage (Figure 1.14). This finding supported the second prediction.

Thus a combination of field and laboratory experiments, observation, and prior knowledge all supported the hypothesis that nematodes manipulate the behavior of their amphipod hosts in a way that decreases the survival of the amphipods, but increases the survival of the nematodes.

As is common practice in all the sciences, the researchers gathered all their data and collected them in a report, which they submitted to a scientific journal. Once such a report is published,* other scientists can evaluate the data, make their own observations, and formulate new ideas and experiments.

Experiments are powerful tools

The key feature of **experimentation** is the control of most factors so that the influence of a single factor can be seen clearly. In the laboratory experiments with amphipods, all individuals were raised under the same conditions. As a result, the nematodes reached the infective stage at about the same time in all of the infected amphipods.

Both laboratory and field experiments have their strengths and weaknesses. The advantage of working in a laboratory is that control of environmental factors is more

*In the case illustrated here, the data on amphipod behavior were published in the journal *Behavioral Ecology*, Volume 10, Number 4 (1998). D. McCurdy et al., "Evidence that the parasitic nematode *Skrjabinoclava* manipulates host *Corophium* behavior to increase transmission to the sandpiper, *Calidris pusilla*."

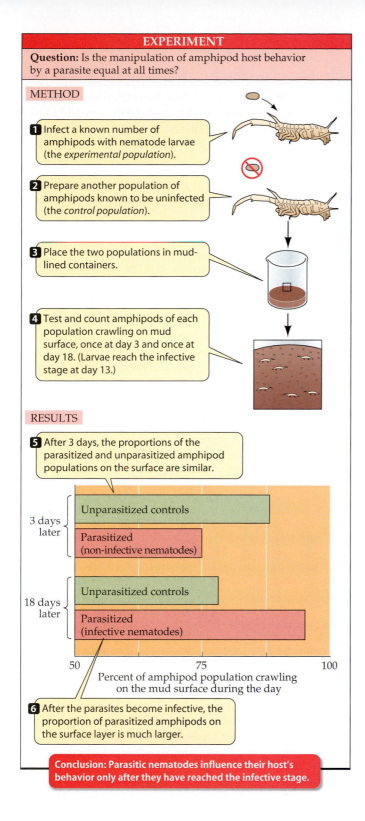

EXPERIMENT

Question: Is the manipulation of amphipod host behavior by a parasite equal at all times?

METHOD

1 Infect a known number of amphipods with nematode larvae (the *experimental population*).

2 Prepare another population of amphipods known to be uninfected (the *control population*).

3 Place the two populations in mud-lined containers.

4 Test and count amphipods of each population crawling on mud surface, once at day 3 and once at day 18. (Larvae reach the infective stage at day 13.)

RESULTS

5 After 3 days, the proportions of the parasitized and unparasitized amphipod populations on the surface are similar.

3 days later
- Unparasitized controls
- Parasitized (non-infective nematodes)

18 days later
- Unparasitized controls
- Parasitized (infective nematodes)

50   75   100
Percent of amphipod population crawling on the mud surface during the day

6 After the parasites become infective, the proportion of parasitized amphipods on the surface layer is much larger.

Conclusion: Parasitic nematodes influence their host's behavior only after they have reached the infective stage.

1.14 An Experiment Demonstrates that Parasites Influence Amphipod Behavior
Amphipods are more likely to crawl on the surface of the mud, exposing themselves to being captured by sandpipers, when their parasitic nematodes have reached the stage at which they can infect a sandpiper.

and field experiments are needed to test most hypotheses about what organisms do.

A single piece of supporting evidence rarely leads to widespread acceptance of a hypothesis. Similarly, a single contrary result rarely leads to abandonment of a hypothesis. Results that do not support the hypothesis being tested can be obtained for many reasons, only one of which is that the hypothesis is wrong. Incorrect predictions may have been made from a correct hypothesis. A negative finding can also result from poor experimental design, or because an inappropriate organism was chosen for the test. For example, a species of sandpiper that fed only by probing in the mud for its prey would have been an unsuitable subject for testing the hypothesis that nematodes alter their hosts in a way to make them more visible to predators.

Accepted scientific theories are based on many kinds of evidence

A general textbook like this one presents hypotheses and theories that have been extensively tested, using a variety of methods, and are generally accepted. When possible, we illustrate hypotheses and theories with observations and experiments that support them, but we cannot, because of space constraints, detail all the evidence. Remember as you read that statements of biological "fact" are mixtures of observations, predictions, and interpretations.

No amount of observation could possibly substitute for experimentation. However, this does not mean that scientists are insensitive to the welfare of the organisms with which they work. Most scientists who work with animals are continually alert to finding ways of getting answers that use the smallest number of experimental subjects and that cause the subjects the least pain and suffering.

Not all forms of inquiry are scientific

If you understand the methods of science, you can distinguish science from non-science. Recently some people have claimed that "creation science," sometimes called "scientific creationism," is a legitimate science that deserves to be taught in schools together with the evolutionary view of the world presented in this book. In spite of these claims, creation science is not science.

Science begins with observations and the formulation of hypotheses that can be tested and that will be rejected if significant contrary evidence is found. Creation science begins with the assertions, derived from religious texts, that Earth is only a few thousand years old and that all species of organisms were created in approximately their present forms. These assertions are not presented as a hypothesis

complete. Field experiments are more difficult because it is usually impossible to control more than a small number of environmental factors. But field experiments have one important advantage: Their results are more readily applicable to what happens where the organisms actually live and evolve. Just because an organism does something in the laboratory does not mean that it behaves the same way in nature. Because biologists usually wish to explain nature, not processes in the laboratory, combinations of laboratory from which testable predictions can be derived. Advocates of creation science assume their assertions to be true and that no tests are needed, nor are they willing to accept any evidence that refutes them.

In this chapter we have outlined the hypotheses that Earth is about 4 billion years old, that today's living organisms evolved from single-celled ancestors, and that many organisms dramatically different from those we see today lived on Earth in the remote past. The rest of this book will provide evidence supporting this scenario. To reject this view of Earth's history, a person must reject not only evolutionary biology, but also modern geology, astronomy, chemistry, and physics. All of this extensive scientific evidence is rejected or misinterpreted by proponents of "creation science" in favor of their particular religious beliefs.

Evidence gathered by scientific procedures does not diminish the value of religious accounts of creation. Religious beliefs are based on faith—not on falsifiable hypotheses, as science is. They serve different purposes, giving meaning and spiritual guidance to human lives. They form the basis for establishing values—something science cannot do. The legitimacy and value of both religion and science is undermined when a religious belief is presented as scientific evidence.

Biology and Public Policy

During the Second World War and immediately thereafter, the physical sciences were highly influential in shaping public policy in the industrialized world. Since then, the biological sciences have assumed increasing importance. One reason is the discovery of the genetic code and the ability to manipulate the genetic constitution of organisms. These developments have opened vast new possibilities for improvements in the control of human diseases and agricultural productivity. At the same time, these capabilities have raised important ethical and policy issues. How much, and in what ways, should we tinker with the genetics of people and other species? Does it matter whether organisms are changed by traditional breeding experiments or by gene transfers? How safe are genetically modified organisms in the environment and in human foods?

Another reason for the importance of the biological sciences is the vastly increased human population. Our use of renewable and nonrenewable natural resources is stressing the ability of the environment to produce the goods and services upon which society depends. Human activities are causing the extinction of a large number of species and are resulting in the spread of new human diseases and the resurgence of old ones. Biological knowledge is vital for determining the causes of these changes and for devising wise policies to deal with them.

Therefore, biologists are increasingly called upon to advise governmental agencies concerning the laws, rules, and regulations by which society deals with the increasing number of problems and challenges that have at least a par-

tial biological basis. We will discuss these issues in many chapters of this book. You will see how the use of biological information can contribute to the establishment and implementation of wise public policies.

Chapter Summary

▶ If the history of Earth were a month with 30 days, recorded human history would occupy only the last 30 seconds. **Review Figure 1.1**

Organisms Have Changed over Billions of Years

▶ Evolution is the theme that unites all of biology. The idea of, and evidence for, evolution existed before Darwin. **Review Figure 1.2**

▶ The theory of evolution by natural selection rests on two simple observations and one inference from them.

Evolutionary Milestones

▶ Life arose from nonlife about 3.8 billion years ago when interacting systems of molecules became enclosed in membranes to form cells.

▶ All living organisms contain the same types of large molecules—carbohydrates, lipids, proteins, and nucleic acids.

▶ All organisms consist of cells, and all cells come from preexisting cells. Life no longer arises from nonlife.

▶ A major theme in the evolution of life is the development of increasingly diverse ways of capturing external energy and using it to drive biologically useful reactions.

▶ Photosynthetic single-celled organisms released large amounts of oxygen into Earth's atmosphere, making possible the oxygen-based metabolism of large cells and, eventually, multicellular organisms.

▶ Reproduction with variation is a major characteristic of life. The evolution of sexual reproduction enhanced the ability of organisms to adapt to changing environments.

▶ Complex eukaryotic cells evolved when some large prokaryotes engulfed smaller ones. Eukaryotic cells evolved the ability to "stick together" after they divided, forming multicellular organisms. The individual cells of multicellular organisms became modified for specific functions within the organism.

▶ A major theme in the evolution of life is the development of increasingly complicated systems for responding to changes in the internal and external environments and for maintaining homeostasis.

▶ Regulated growth is a vital characteristic of life.

▶ Speciation resulted in the millions of species living on Earth today.

▶ Adaptation to environmental change is one of life's most distinctive features and is the result of evolution by natural selection.

The Hierarchy of Life

▶ Biology is organized into a hierarchy of levels from molecules to the biosphere. Each level has emergent properties that are not found at lower levels. **Review Figure 1.9**

▶ Species are classified into three domains: Archaea, Bacteria, and Eukarya. The domains Archaea and Bacteria consist of prokaryotic cells. The domain Eukarya contains the protists and the kingdoms Plantae, Fungi, and Animalia, all of which have eukaryotic cells. **Review Figure 1.10**

Asking and Answering "How?" and "Why?"

▶ Biologists ask two kinds of questions. "How" questions ask how organisms work. "Why" questions ask why they evolved to work that way.

▶ Both how and why questions are usually answered using a hypothetico-deductive (H-D) approach. Hypotheses are tentative answers to questions. Predictions are made on the basis of a hypothesis. The predictions are tested by observations and experiments, the results of which may support or refute the hypothesis. **Review Figure 1.13**

▶ Science is based on the formulation of testable hypotheses that can be rejected in light of contrary evidence. The acceptance on faith of already refuted, untested, or untestable assumptions is not science.

Biology and Public Policy

▶ Biologists are often called upon to advise governmental agencies on the solution of important problems that have a biological component.

For Discussion

1. According to the theory of evolution by natural selection, a species evolves certain features because they improve the chances that its members will survive and reproduce. There is no evidence, however, that evolutionary mechanisms have foresight or that organisms can anticipate future conditions. What, then, do biologists mean when they say, for example, that wings are "for flying"?

2. Why is it so important in science that we design and perform tests capable of rejecting a hypothesis?

3. One hypothesis about the manipulation of a host's behavior by a parasite was discussed in this chapter, and some tests of that hypothesis were described. Suggest some other hypotheses about the ways in which parasites might change the behavior and physiology of their hosts. Develop some critical tests for one of these alternatives. What are the appropriate associated null hypotheses?

4. Some philosophers and scientists believe that it is impossible to prove any scientific hypothesis—that we can only fail to find a reason to reject it. Evaluate this view. Can you think of reasons why we can be more certain about rejecting a hypothesis than about accepting it?

5. Discuss one current environmental problem whose solution requires the use of biological knowledge. How well is biology being used? What factors prevent scientific data from playing a more important role in finding a solution to the problem?

Part Three

EVOLUTIONARY PROCESSES

20 *The History of Life on Earth*

WHEN YOU WANT TO KNOW WHAT TIME IT IS, you probably look at your watch, or at the clock on the wall or on your computer. You could also listen to the radio or watch television to hear some announcement of time. But suppose the electric power system failed and you lost your watch. How could you tell time then? You would use the cues that people used during most of human history—the cycle of day and night. We are so accustomed to having time-measuring devices all around us that we forget these devices are recent inventions. When Galileo studied the motion of a ball rolling down an inclined plane 350 years ago, he used his pulse to mark off equal intervals of time.

The science of biology is intimately linked to concepts of time. Biology as we know it could not and did not develop very far until an appreciation of the age of Earth was provided by geologists more than 150 years ago. Until that time, most people believed that Earth was only a few thousand years old. Darwin could not have developed his theory of evolution by natural selection if he had not read the works of Charles Lyell, England's leading geologist, who believed that Earth was ancient. As we pointed out in Chapter 1, Darwin's theory was based on the assumption that Earth was very old and that life had existed for a very long time, during which it had steadily evolved.

The goals of Part Three are to document the history of life on Earth, to describe the processes of evolutionary change, and to discuss the agents that cause them. We begin in this chapter by asking the following questions: How do we know that Earth is ancient? What is the evidence that life evolved early during Earth's history and has continued to evolve since then? In the following chapter we discuss the processes by which life evolved. In subsequent chapters, we will see how biologists determine the evolutionary histories of organisms, and how the millions of species that live today (as well as those that became extinct) formed from a single common ancestor. Finally, in Chapter 25, we will examine how life probably arose from nonliving matter several billion years ago.

Understanding biological evolution is important because evolutionary changes are taking place all around us. These changes have powerful implications for human welfare. Our own attempts to control populations of undesirable species and increase populations of desirable ones make human beings powerful agents of evolutionary change. In addition to producing the results we desire, these efforts often cause undesirable outcomes, such as the evolution of resistance to medicines by pathogens and to pesticides by pests. Medicine and agriculture can respond creatively to the evolutionary changes they are causing only if their practitioners understand how and why those changes happen. But what exactly is biological evolution?

Biological evolution is a change over time in the genetic composition of a population. Changes that happen during the lifetimes of species constitute **microevolution**. Plant and animal breeding and changes occurring in response to environmental shifts over decades provide good examples of microevolution. Changes that involve the appearance of new species and evolutionary lineages are called **macro-**

The Hands of Time
London's Big Ben, perhaps the most recognizable timepiece in the world, epitomizes our worldview of the importance of hours and minutes. Geological time, however, is much more difficult to grasp but is essential to understanding biological evolution.

evolution. The fossil record provides the best evidence of macroevolutionary changes among organisms. Many of these changes are dramatic.

To understand the long-term patterns of evolutionary change that we will document in this chapter, we must think in time frames spanning many millions of years and imagine events and conditions very different from those we now observe. The Earth of the distant past is, to us, a foreign planet inhabited by strange organisms. The continents were not where they are today, and climates were different. One of the remarkable achievements of modern science has been the development of sophisticated techniques for inferring past conditions and dating them accurately.

In this chapter, we first examine how events in the distant past can be dated. Then we review the major changes in physical conditions on Earth during the past 4 billion years, look at how those changes affected life, and discuss the major patterns in the evolution of life.

How Do We Know Earth Is Ancient?

It is difficult to age rocks because a rock of a particular type could have been formed at any time during Earth's history. It is easier to determine the ages of rocks relative to one an-

other. The first person to recognize that this could be done was the seventeenth-century Danish physician Nicolaus Steno. Steno realized that in an undisturbed sequence of sedimentary rocks, the oldest strata lie at the bottom and successively higher strata are progressively younger (Figure 20.1).

Geologists subsequently combined Steno's insights with their observations of the **fossils**—remains of ancient organisms—contained within rocks. They discovered that fossils of similar organisms were found in widely separated places on Earth, that certain organisms were always found in younger rocks than others, and that organisms in the most recent strata were more similar to modern organisms than those found in lower, more ancient strata. With this information, they were able to determine much about the relative ages of sedimentary rocks and about patterns in the evolution of life. But they still could not tell how old the rocks were. A method of dating rocks did not become available until the discovery of radioactivity at the turn of the twentieth century.

Radioactivity provides a way to date rocks

Radioactive isotopes decay in a regular pattern during successive, equal periods of time. During each successive time

20.1 Earth's Geological History

RELATIVE TIME SPAN	ERA	PERIOD	ONSET	MAJOR PHYSICAL CHANGES ON EARTH
Precambrian	Cenozoic	Quaternary	1.8 mya	Cold/dry climate; repeated glaciations
		Tertiary	65 mya	Continents near current positions; climate cools
	Mesozoic	Cretaceous	144 mya	Northern continents attached; Gondwana begins to drift apart; meteorite strikes Yucatán Peninsula
		Jurassic	206 mya	Two large continents form: Laurasia (north) and Gondwana (south); climate warm
		Triassic	245 mya	Pangaea slowly begins to drift apart; hot/humid climate
	Paleozoic	Permian	290 mya	Continents aggregate into Pangaea; large glaciers form; dry climates form in interior of Pangaea
		Carboniferous	354 mya	Climate cools; marked latitudinal climate gradients
		Devonian	409 mya	Continents collide at end of period; asteroid probably collides with Earth
		Silurian	440 mya	Sea levels rise; two large continents form; hot/humid climate
		Ordovician	500 mya	Gondwana moves over South Pole; massive glaciation, sea level drops 50 m
		Cambrian	543 mya	O_2 levels approach current levels
	Precambrian		600 mya	O_2 level at >5% of current level
			2.5 bya	O_2 level at >1% of current level
			3.8 bya	O_2 first appears in atmosphere
			4.5 bya	

20.1 Young Rocks Lie on Top of Old Rocks
The oldest rocks at the bottom of this photo of the North Rim of the Grand Canyon formed about 540 million years ago. The youngest rocks on top are about 500 million years old.

MAJOR EVENTS IN THE HISTORY OF LIFE

Humans evolve; large mammals become extinct

Radiation of birds, mammals, flowering plants, and insects

Dinosaurs continue to radiate; flowering plants and mammals diversify. **Mass Extinction** at end of period (≈76% of species disappear)

Diverse dinosaurs; first birds; two minor extinctions

Early dinosaurs; first mammals; marine invertebrates diversify. **Mass Extinction** at end of period (≈65% of species disappear)

Reptiles radiate; amphibians decline; **Mass Extinction** at end of period (≈96% of species disappear)

Extensive "fern" forests; first reptiles; insects radiate; earliest flowering plants

Fishes diversify; first insects and amphibians. **Mass Extinction** at end of period (≈75% of species disappear)

Jawless fishes diversify; first bony fishes; plants and animals colonize land

Mass Extinction at end of period (≈75% of species disappear)

Most animal phyla present; diverse algae

Ediacaran fauna
Eukaryotes evolve; several animal phyla appear
Origin of life; prokaryotes flourish

interval, an equal fraction of the remaining radioactive material of any radioisotope decays, either changing to another element or becoming the stable isotope of the same element. For example, in 14.3 days, one-half of any sample of phosphorus-32 (^{32}P) decays to its stable isotope, phosphorus-31 (^{31}P). During the next 14.3 days, one-half of the remaining half decays, leaving one-fourth of the original ^{32}P. The time it takes for half of an isotope to decay is that isotope's **half-life**. After 42.9 days, three half-lives have passed, so one-eighth (that is, ½ × ½ × ½) of the original ^{32}P remains.

Each radioisotope has a characteristic half-life. Which isotope is used to estimate the age of an ancient material depends on how old the material is thought to be. Tritium (^{3}H) has a half-life of 12.3 years, and carbon-14 (^{14}C) has a half-life of about 5,700 years. The half-life of potassium-40 (^{40}K) is 1.3 billion years; the decay of potassium-40 to argon-40 has been used to date most of the ancient events in the evolution of life.

To use a radioisotope to date a past event, we must know or estimate the concentration of the isotope at the time of that event. In the case of carbon, we know that the production of new ^{14}C in the upper atmosphere (by the reaction of neutrons with ^{14}N) just balances the natural radioactive decay of ^{14}C. Therefore, the ratio of ^{14}C to its stable isotope, ^{12}C, exists in a more or less steady state in living organisms and their environment.

However, as soon as an organism dies, it ceases to exchange carbon compounds with the rest of the world. Its decaying ^{14}C is no longer replenished, and the ratio of ^{14}C to ^{12}C decreases. The ratio of ^{14}C to ^{12}C in fossil organisms can be used to date fossils (and thus the sedimentary rocks that contain those fossils) that are less than 50,000 years old with a fair degree of certainty.

Dating rocks more ancient than 50,000 years requires estimating isotope concentrations that exist in volcanic (but not in sedimentary) rocks. To date ancient sedimentary rocks, geologists search for places where volcanic ash or lava flows have intruded into beds of sedimentary rock. Radiometric dating, combined with observations of fossils, is the most powerful method of determining the ages of rocks.

But there are many places where sedimentary rocks do not contain suitable volcanic intrusions and few fossils are present. In these areas, dating methods other than radiometry must be used. One such method is based on the fact that Earth's magnetic poles move and occasionally reverse themselves. Because both sedimentary and igneous rocks preserve a record of Earth's magnetic field at the time they were formed, *paleomagnetism* helps determine the ages of those rocks. Other "time machines," which we will describe later, include continental drift, sea level changes, and molecular clocks.

Using these methods, geologists have divided Earth's history into eras, which in turn are subdivided into periods (Table 20.1). The boundaries between these divisions, which are based on major differences in the fossils contained in successive layers of rocks, were established before the actual ages of the eras and periods were known.

Earth has undergone many physical changes that have influenced the evolution of life. The physical events we describe in this chapter, along with the most important milestones in the history of life, are listed in Table 20.1. Most of these biological milestones have taken place since the sudden explosion of new life forms that characterized the early Cambrian period, about 543 million years ago. The scale at the left of Table 20.1 gives a relative sense of geological time and the vast span of the **Precambrian**, during which early life evolved amid stupendous physical changes on Earth.

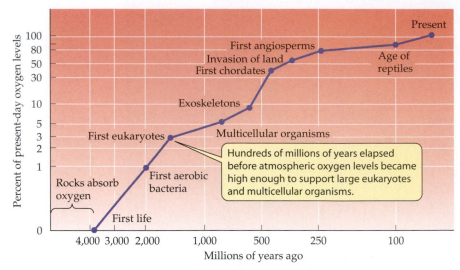

20.2 Large Cells Need More Oxygen
Although aerobic prokaryotes can flourish with less, large eukaryotic cells with lower surface area-to-volume ratios require at least 2 to 3 percent of current atmospheric O_2 concentrations. (Both axes of the graph are on logarithmic scales.)

How Has Earth Changed over Time?

Two important physical changes on Earth have been unidirectional. First, Earth is cooling because it is continuously losing the heat that was generated when it formed. Second, the radioactive furnace in Earth's core is steadily weakening, generating less and less heat to replace the heat that is lost to space. As a result, the processes that cause the continents to move about on Earth's surface have also weakened during Earth's history.

Earth's atmosphere has also changed unidirectionally. The atmosphere of early Earth probably had little or no free oxygen (O_2). Oxygen concentrations in the atmosphere began to increase markedly about 2.5 billion years ago, when certain sulfur bacteria evolved the ability to use water as a source of hydrogen ions for photosynthesis. The cyanobacteria that evolved from these sulfur bacteria liberated enough O_2 to open the way for the evolution of oxidation reactions as the energy source for the synthesis of ATP.

An atmosphere rich in oxygen also made possible larger cells and more complicated organisms. Small, unicellular aquatic organisms can obtain enough O_2 by simple diffusion even when O_2 concentrations are very low. Larger unicellular organisms have lower surface area-to-volume ratios (see Figure 4.2). In order to obtain enough O_2 by simple diffusion, they must live in an environment with a relatively high concentration of O_2. Bacteria can thrive on 1 percent of the current atmospheric O_2 level, but eukaryotic cells require oxygen levels that are at least 2 to 3 percent of current atmospheric concentrations.

About 1,500 million years ago (mya), O_2 concentrations became high enough for large eukaryotic cells to flourish and diversify (Figure 20.2). Further increases in atmospheric O_2 levels 700 to 570 mya enabled multicellular organisms to evolve. The fact that it took many millions of years for Earth to develop an oxygenated atmosphere probably explains why only unicellular prokaryotes lived on Earth for more than a billion years.

Unlike the largely unidirectional changes in Earth's temperature and atmospheric O_2 concentrations, most physical changes on Earth have involved irregular oscillations in the planet's internal processes, such as volcanic activity and the shifting and colliding of continents. External events such as collisions with meteorites have also left their mark. In some cases, these events caused **mass extinctions**, wiping out a large proportion of the species living at the time.

The continents have changed position

The maps and globes that adorn our walls, shelves, and books give an impression of a static Earth. It would be easy for us to assume that the continents have always been where they are. But we would be wrong.

Earth's crust consists of solid plates approximately 40 km thick that float on a fluid mantle. The mantle fluid circulates because heat produced by radioactive decay in Earth's core sets up convection patterns. The plates move because material from the mantle rises and pushes them aside, resulting in seafloor spreading along ocean ridges. Where plates are pushed together, either they move sideways past each other, or one plate moves under the other, creating mountain ranges. The movement of the plates and the continents they contain—a process known as **continental drift**—has had enormous effects on climate, sea levels, and the distributions of organisms.

At times, the drifting of the plates has brought the continents together; at other times, they have drifted apart. The positions and sizes of the continents influence oceanic circulation patterns, sea levels, and global climate patterns. Mass extinctions of species, particularly marine organisms, have usually accompanied major drops in sea level (Figure 20.3).

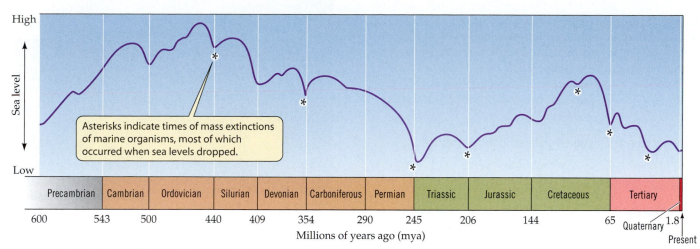

Asterisks indicate times of mass extinctions of marine organisms, most of which occurred when sea levels dropped.

20.3 Sea Levels Have Changed Repeatedly
Most mass extinctions of marine organisms have coincided with periods of low sea levels.

Earth's climate has shifted between hot/humid and cold/dry conditions

Through much of its history, Earth's climate was considerably warmer than it is today, and temperatures decreased more slowly toward the poles. At other times, however, Earth was colder than it is today. Large areas were covered with glaciers toward the end of the Precambrian and during the Carboniferous, Permian, and Quaternary periods, but these cold periods were separated by long periods of milder climates (Figure 20.4). Because we are living in one of the colder periods in the history of Earth, it is difficult for us to imagine the mild climates that were found at high latitudes during much of the history of life.

Usually climates change slowly, but major climatic shifts have taken place over periods as short as 5,000 to 10,000 years, primarily as a result of changes in Earth's orbit around the sun. A few climatic shifts appear to have been even more rapid. For example, during one Quaternary interglacial period, the Antarctic Ocean changed from being ice-covered to being nearly ice-free in less than 100 years.

Such rapid changes are usually caused by sudden shifts in ocean currents. Climates have sometimes changed so rapidly that extinctions caused by them appear "instantaneous" in the fossil record.

Volcanoes have disrupted evolution

On the morning of August 27, 1883, Krakatau, an island the size of Manhattan located in the Sunda Strait between Sumatra and Java, was devastated by a series of volcanic eruptions. Tidal waves caused by the eruption hit the shores of Java and Sumatra, demolishing towns and villages and killing 40,000 people. As impressive as this eruption was, however, its effects were local and short-lived. It did not cause major changes in patterns of the evolution of life. But much larger volcanic eruptions have occurred several times during Earth's history and have had major consequences for life.

During the late Permian period (about 275 mya), the continents came together to form a single gigantic land mass, Pangaea. This collision of continents caused massive

20.4 Hot/Humid and Cold/Dry Conditions Have Alternated over Earth's History
Throughout Earth's history, periods of cold climates and glaciations have been separated by long periods of milder climates.

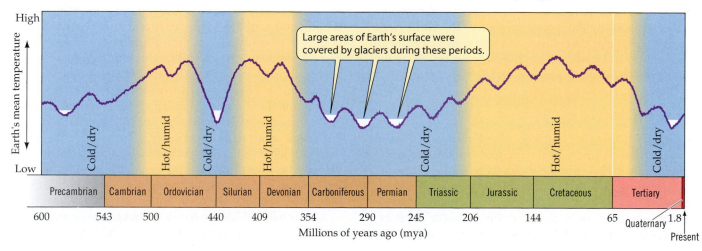

Large areas of Earth's surface were covered by glaciers during these periods.

volcanic eruptions. The ash the volcanoes ejected into Earth's atmosphere reduced the penetration of sunlight to Earth's surface, lowering temperatures, reducing photosynthesis, and triggering massive glaciation. Massive volcanic eruptions also occurred as the continents drifted apart during the late Triassic period and again at the end of the Cretaceous.

External events have triggered other changes on Earth

At least 30 meteorites between the sizes of baseballs and soccer balls hit Earth each year, but collisions with large meteorites are rare. In 1980, Luis Alvarez and several of his colleagues at the University of California, Berkeley, proposed that the mass extinction at the end of the Cretaceous period, about 65 mya, might have been caused by the collision of Earth with a large meteorite. These scientists based their hypothesis on the finding of abnormally high concentrations of the element iridium in a thin layer separating the rocks deposited during the Cretaceous period from those of the Tertiary (Figure 20.5). Iridium is abundant in some meteorites, but is exceedingly rare on Earth's surface.

To account for the estimated amount of iridium in this layer, Alvarez postulated that a meteorite 10 km in diameter collided with Earth at a speed of 72,000 km per hour. The force of such an impact would have ignited massive fires, created great tidal waves, and sent up an immense dust cloud that blocked the sun, thus cooling the planet. As it settled, the dust would have formed the iridium-rich layer.

This hypothesis generated a great deal of controversy and stimulated much research. Some scientists searched for the site of impact of the supposed meteorite. Others worked to improve the precision with which events of that age could be dated. Still others tried to determine more exactly the speed with which extinctions occurred at the Cretaceous–Tertiary boundary. Progress on all three fronts has favored the meteorite theory.

The theory was supported by the discovery of a circular crater 180 km in diameter buried beneath the northern coast of the Yucatán Peninsula of Mexico, thought to have been formed 65 mya. Recent fossil evidence also suggests that there may have been a sudden extinction of organisms 65 mya, as required by the meteorite theory. Therefore, most scientists accept that the collision of Earth with a large meteorite contributed importantly to the mass extinctions at the boundary between the Cretaceous and Tertiary periods.

The Fossil Record

Geological evidence is a major source of information about changes on Earth during the remote past. But the fossils preserved in the rocks—not the rocks themselves—are what have enabled geologists to order those events in time. What are fossils, and what do they tell us about the influence of physical events on the evolution of life on Earth? After examining the conditions that preserve the remains of organ-

A thin band rich in iridium marks the boundary between rocks deposited in the Cretaceous and Tertiary periods.

20.5 Evidence of a Meteorite Collision with Earth
Iridium is a metal common in some meteorites, but rare on Earth. Its high concentrations in sediments deposited about 65 million years ago suggest the impact of a large meteorite.

isms, we will consider the completeness of the fossil record, and how that record reveals patterns in life's history.

An organism is most likely to become a fossil if its dead body is deposited in an environment that lacks oxygen. However, most organisms live in oxygenated environments and decompose when they die. Thus many fossil assemblages are collections of organisms that were transported by wind or water to sites that lacked oxygen. Occasionally, however, organisms, or imprints of them, are preserved where they lived. In such cases—especially if the environment in question was a cool, anaerobic swamp, where conditions for preservation were excellent—we can obtain a picture of communities of organisms that lived together.

How complete is the fossil record?

About 300,000 species of fossil organisms have been described, and the number is growing steadily. However, this number is only a tiny fraction of the species that have ever lived. We do not know how many species lived in the past, but we have ways of making reasonable estimates. Of the present-day biota—that is, all living species of all kinds—approximately 1.6 million species have been named. The actual number of living species is probably at least 10 million. It is possibly higher than 50 million, because most species of insects (the animal group with the largest number of species; see Chapter 32) have not yet been described. So the number of known fossil species is less than 2 percent of the probable minimum number of living species.

Because life has existed on Earth for about 3.8 billion years, and because most species exist, on average, for fewer than 10 million years, the species living on Earth must have turned over many times during geological history, and the total number of species that have lived over evolutionary time must vastly exceed the number living today.

The number of known fossils, although small in relation to the total number of extinct species, is higher for some

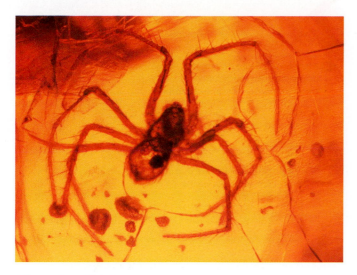

20.6 A Fossil Spider
Trapped in sap of a tree in what is now Arkansas about 50 mya, this spider is exquisitely preserved in the amber formed from the sap. The details of its external anatomy are clearly visible.

groups of organisms than for others. The record is especially good for marine animals that had hard skeletons. Among the nine major animal groups with hard-shelled members, approximately 200,000 species have been described from fossils, roughly twice the number of living marine species in these same groups. Paleontologists lean heavily on these groups in their interpretations of the evolution of life in the past. Insects and spiders are also relatively well represented in the fossil record (Figure 20.6).

The fossil record demonstrates several patterns

Despite its incompleteness, the fossil record reveals several patterns that are unlikely to be altered by future discoveries. First, great regularity exists. For example, organisms of particular types are found in rocks of specific ages, and new organisms appear sequentially in younger rocks. Second, as we move from ancient periods of geological time toward the present, fossil species increasingly resemble species living today. The fossil record also tells us that extinction is the eventual fate of all species.

The fossil record contains many series of fossils that demonstrate gradual change in lineages of organisms over time. A good example is the series of fossils showing the pathway by which whales evolved from hoofed terrestrial mammals, beginning about 50 mya. Fossils that are intermediate between whales and their terrestrial ancestors illustrate the major changes by which whales became adapted for aquatic existence and lost their hind limbs (Figure 20.7).

Interestingly, whales retain the genetic potential for developing legs; occasionally, living whales have been found with small hind legs that extend outside their bodies. The claim (made repeatedly by scientific creationists) that the fossil record does not contain examples of such intermediates is false. Intermediates abound, and more and more of them are being discovered.

But the incompleteness of the fossil record can mislead us when we try to interpret it. Organisms may have evolved in places where their fossils have not been discovered. Moreover, when a species that evolved in one place appears among the fossils at another site, it gives the false impression that it evolved very rapidly from one of the species that already lived there.

Horses, for example, evolved at varying rates over millions of years in North America. Many different lineages arose and died out (Figure 20.8). Ancestors of horses crossed the Bering land bridge into Asia at several different times, the most recent one only several million years ago. Evidence of each crossing appears suddenly in the Asian fossil record as a major new type of horse. If we lacked fossil evidence of horse evolution in North America, we might conclude that horses evolved very rapidly somewhere in Asia. On the other hand, an incomplete fossil record can also hide rapid changes.

By combining data about physical events during Earth's history with evidence from the fossil record, scientists can compose pictures of what Earth and its inhabitants looked like at different times. We know in general where the continents were and how life changed over time, but many of the details are poorly known, especially for events in the more remote past. In the next section we provide an overview of the major patterns in the history of life on Earth.

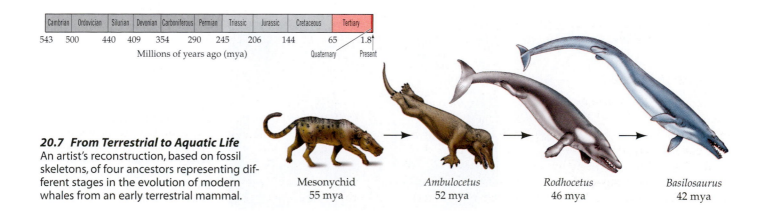

Cambrian	Ordovician	Silurian	Devonian	Carboniferous	Permian	Triassic	Jurassic	Cretaceous	Tertiary
543	500	440	409	354	290	245	206	144	65	1.8

Millions of years ago (mya)   Quaternary   Present

20.7 From Terrestrial to Aquatic Life
An artist's reconstruction, based on fossil skeletons, of four ancestors representing different stages in the evolution of modern whales from an early terrestrial mammal.

Mesonychid
55 mya

Ambulocetus
52 mya

Rodhocetus
46 mya

Basilosaurus
42 mya

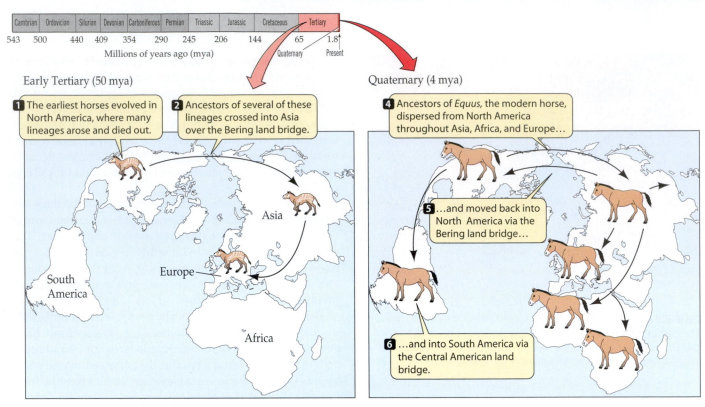

Early Tertiary (50 mya)

1 The earliest horses evolved in North America, where many lineages arose and died out.

2 Ancestors of several of these lineages crossed into Asia over the Bering land bridge.

Asia

Europe

South America

Africa

Quaternary (4 mya)

4 Ancestors of *Equus*, the modern horse, dispersed from North America throughout Asia, Africa, and Europe…

5 …and moved back into North America via the Bering land bridge…

6 …and into South America via the Central American land bridge.

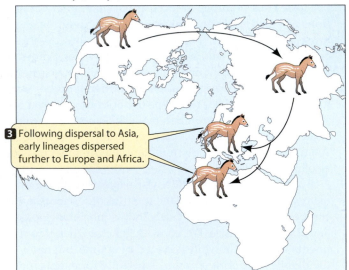

Mid-Tertiary (20 mya)

3 Following dispersal to Asia, early lineages dispersed further to Europe and Africa.

20.8 Horses Have a Complex Evolutionary History
Ancestors of horses crossed the Bering land bridge into Asia several times, the last one only a few million years ago. If we lacked the earlier fossil evidence of horse evolution in North America, we might reach the false conclusion that horses evolved rapidly somewhere in Asia.

Life in the Remote Past

Life first evolved about 3.8 billion years ago (bya). The major groups of eukaryotic organisms evolved during the Precambrian, about 2.5 bya. The fossil record of organisms that lived prior to the Cambrian period is fragmentary, but shows that the volume of organisms increased dramatically in late Precambrian times, about 650 mya (see Table 20.1). The shallow Precambrian seas teemed with life. Protists and small multicellular animals fed on floating algae. Living plankton and plankton remains were devoured by animals that filtered food from the water or ingested sediments and digested the organic material in them.

20.9 Ediacaran Animals
These fossils of soft-bodied invertebrates, excavated at Ediacara in southern Australia, formed 600 million years ago. They illustrate the diversity of life that evolved in Precambrian times.

Cambrian | Ordovician | Silurian | Devonian | Carboniferous | Permian | Triassic | Jurassic | Cretaceous | Tertiary

543 500 440 409 354 290 245 206 144 65 1.8

Millions of years ago (mya)

Quaternary   Present

The best fossil assemblage of Precambrian animals, all soft-bodied invertebrates, was discovered at Ediacara, in southern Australia (Figure 20.9). The Ediacaran fauna is very different from any assemblage of animals living today. Some of its members may represent animal lineages that have no living descendants.

Diversity exploded during the Cambrian

By the early Cambrian period (543–510 mya), oxygen levels in Earth's atmosphere approached their current concentrations, and the continental plates came together in several masses, the largest of which was Gondwana (Figure 20.10*a*). The three great evolutionary lineages of animals separated and began to radiate during this period. **Evolutionary radiation**—the proliferation of species within a single lineage—during this time resulted in the dramatic increase in diversity known as the Cambrian explosion. All of the major groups of animals that have species living today appeared during the Cambrian, as did animals belonging to many lineages that have left no surviving descendants.

The most extensive fossil evidence from the Cambrian period comes from an unusually well preserved fauna recently discovered in China (Figure 20.10*b*). Arthropods are the most diverse group in the Chinese fauna; some of them were large carnivores. A mass extinction occurred at the end of the Cambrian.

Major changes continued during the Paleozoic era

Because they have excellent fossil evidence and can date events relatively precisely, geologists have divided the remainder of the Paleozoic era into five periods: the Ordovician, Silurian, Devonian, Carboniferous, and Permian periods (see Table 20.1).

THE ORDOVICIAN (510–440 MYA). During the Ordovician period, the continents were located primarily in the Southern Hemisphere. Evolutionary radiation of marine organisms was spectacular during the early Ordovician, especially among animals (such as brachiopods and mollusks) that filter small prey from the water. All animals lived on the seafloor or burrowed in its sediments. Ancestors of club mosses and horsetails colonized wet terrestrial environments, but they were still relatively small. At the end of the Ordovician, sea levels dropped about 50 meters as massive glaciers formed over Gondwana, and ocean temperatures dropped. About 75 percent of the marine animal species became extinct, probably because of these major environmental changes.

THE SILURIAN (440–409 MYA). During the Silurian period, the northern continents coalesced, but the general positions of the continents did not change much. Marine life rebounded from the mass extinction at the end of the Ordovician. Animals able to swim and feed above the ocean bottom appeared for the first time, but no major new groups of marine organisms evolved. The tropical sea was uninterrupted by land barriers, and most marine genera were widely distributed. On land, the first terrestrial arthropods—scorpions and millipedes—appeared.

THE DEVONIAN (409–354 MYA). Rates of evolutionary change accelerated in many groups of organisms during the Devonian period. Northern and southern land masses slowly moved northward (Figure 20.11*a*). There was a great evolutionary radiation of corals and shelled squidlike cephalopods (Figure 20.11*b*). Fishes diversified as jawed forms replaced jawless ones, and heavy armor gave way to the less rigid outer coverings of modern fishes. All current major groups of fishes were present by the end of the period.

Terrestrial communities also changed dramatically during the Devonian. Club mosses, horsetails, and tree ferns became common, and some reached the size of trees. Their deep roots accelerated the weathering of rocks, resulting in the development of the first forest soils. Distinct floras

(a)

Cambrian	Ordovician	Silurian	Devonian	Carboniferous	Permian	Triassic	Jurassic	Cretaceous	Tertiary
543	500	440	409	354	290	245	206	144	65	1.8

Millions of years ago (mya)   Quaternary   Present

North Pole

The view of Earth has been distorted here so that you can see both poles.

South Pole

This group of land masses is gradually moving together to form Gondwana.

(b)

20.10 Cambrian Continents and Animals
(*a*) Positions of the continents during mid-Cambrian times (543–510 mya). (*b*) Fossil beds in China have yielded excellent remains of Cambrian animals including *Jianfangia*, a predatory arthropod.

Cambrian	Ordovician	Silurian	Devonian	Carboniferous	Permian	Triassic	Jurassic	Cretaceous	Tertiary
543	500	440	409	354	290	245	206	144	65	1.8

Millions of years ago (mya)

Quaternary Present

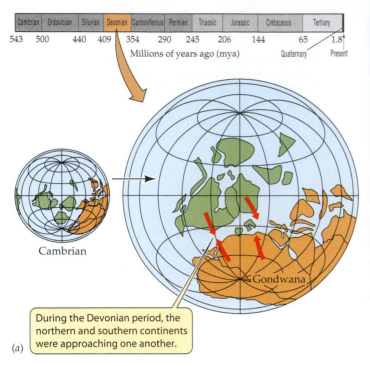

Cambrian

Gondwana

During the Devonian period, the northern and southern continents were approaching one another.

(a)

(b)

20.11 Devonian Continents and Marine Communities

(a) Positions of the continents during the Devonian period (409–354 mya). (b) This artist's reconstruction shows how a Devonian reef may have appeared.

evolved on the two land masses toward the end of the period, and the first gymnosperms appeared. The first known fossils of centipedes, spiders, pseudoscorpions, mites, and insects date to this period, and fishlike amphibians began to occupy the land.

An extinction of about 75 percent of all marine species marked the end of the Devonian. Paleontologists disagree on the cause of this mass extinction. Some believe that it was triggered by the collision of the two continents, which destroyed much of the existing shallow, warm-water marine environment. This hypothesis is supported by the fact that extinction rates were much higher among tropical than among cold-water species.

THE CARBONIFEROUS (354–290 MYA). Large glaciers formed over high-latitude Gondwana during the Carboniferous period, but extensive swamp forests grew on the tropical continents. These forests were not made up of the kinds of trees we know today, but were dominated by giant tree ferns and horsetails (see Figure 28.9). Fossilized remains of those "trees" formed the coal that we now mine for energy.

The diversity of terrestrial animals increased greatly. Snails, scorpions, centipedes,

20.12 A Carboniferous "Crinoid Meadow"

Crinoids, which were dominant marine animals during the Carboniferous (354–290 mya), may have formed communities that looked like this. Sharks and bony fishes were important members of these communities.

and insects were abundant and diverse. Insects evolved wings, which gave them access to tall plants; plant fossils from this period show evidence of insect damage. Amphibians became larger and better adapted to terrestrial existence. From one amphibian stock, the first reptiles evolved late in the period. In the seas, crinoids reached their greatest diversity, forming meadows on the seafloor (Figure 20.12).

THE PERMIAN (290–245 MYA). During the Permian period, the continents coalesced into a supercontinent—Pangaea. Massive volcanic eruptions resulted in outpourings of lava that covered large areas of Earth (Figure 20.13). The ash they produced blocked the sunlight, cooling the climate and resulting in the largest glaciers in Earth's history.

Permian deposits contain representatives of most modern groups of insects. By the end of the period, reptiles greatly outnumbered amphibians. Late in the period, the lineage leading to mammals diverged from one reptilian lineage. In fresh waters, the Permian period was a time of extensive radiation of bony fishes.

Cambrian	Ordovician	Silurian	Devonian	Carboniferous	Permian	Triassic	Jurassic	Cretaceous	Tertiary
543	500	440	409	354	290	245	206	144	65	1.8

Millions of years ago (mya)

Quaternary Present

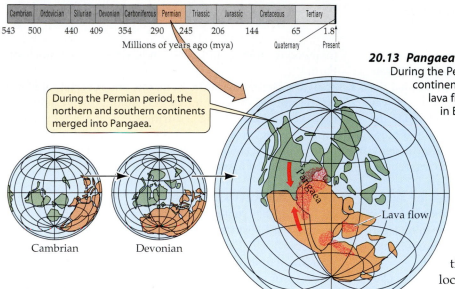

Cambrian	Ordovician	Silurian	Devonian	Carboniferous	Permian	Triassic	Jurassic	Cretaceous	Tertiary
543	500	440	409	354	290	245	206	144	65	1.8

Millions of years ago (mya)   Quaternary   Present

During the Permian period, the northern and southern continents merged into Pangaea.

Cambrian   Devonian

Pangaea   Lava flow

20.13 Pangaea Formed in the Permian Period
During the Permian (290–245 mya), the interior of the "super-continent" Pangaea experienced harsh climates. Massive lava flows spread over Earth, and the largest glaciers in Earth's history formed during this period.

During the Mesozoic, Earth's biota, which until that time had been relatively homogeneous, became increasingly provincialized. Distinct terrestrial floras and faunas evolved on each continent. The biotas of the shallow waters bordering the continents also diverged from one another. The localization that began during the Mesozoic continues to influence the geography of life today.

Toward the end of the Permian period, two events may have caused separate mass extinctions. The first event was the massive outpouring of volcanic lava, which drastically reduced the oxygen content of deep ocean waters. The second was a rapid turnover of the oceans that brought oxygen-depleted deep waters to the surface. These waters released toxic concentrations of carbon dioxide and hydrogen sulfide into surface waters and the atmosphere, poisoning most species.

Geographic differentiation increased during the Mesozoic era

At the start of the Mesozoic era (250 mya), the few surviving organisms found themselves in a relatively empty world. As Pangaea slowly separated into individual continents, the glaciers melted, and the oceans rose and reflooded the continental shelves, forming huge, shallow inland seas. Life again proliferated and diversified, but different lineages came to dominate Earth. The large plants that dominated the great coal-forming forests, for example, were replaced by new plant lineages in which seeds had evolved.

THE TRIASSIC (245–206 MYA).   During the Triassic period, many invertebrate lineages became more diverse, and many burrowing forms evolved from groups living on the surfaces of bottom sediments. On land, conifers and seed ferns became the dominant trees. The first frogs and turtles appeared. A great radiation of reptiles began, which eventually gave rise to dinosaurs, crocodilians, and birds. The end of the Triassic was marked by a mass extinction that eliminated about 65 percent of species on Earth. Why they went extinct is not known, but a meteor impact is suspected.

THE JURASSIC (206–144 MYA).   The mass extinction at the close of the Triassic was followed by another period of evolutionary diversification during the Jurassic period. Bony fishes began the great radiation that culminated in their dominance of the oceans. Salamanders and lizards first appeared. Flying reptiles evolved, and dinosaur lineages evolved into bipedal predators and large quadrupedal herbivores (Figure 20.14). Several groups of mammals first appeared during this time.

Cambrian	Ordovician	Silurian	Devonian	Carboniferous	Permian	Triassic	Jurassic	Cretaceous	Tertiary
543	500	440	409	354	290	245	206	144	65	1.8

Millions of years ago (mya)   Quaternary   Present

20.14 Mesozoic Dinosaurs
The dinosaurs of the Mesozoic era continue to capture our imagination. This painting illustrates some of the large species from the Jurassic period (206–144 mya).

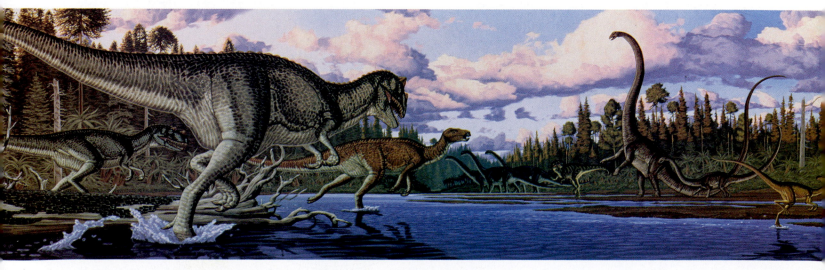

THE CRETACEOUS (144–65 MYA). By the early Cretaceous period, the northern continents were completely separate from the southern ones, and a continuous sea encircled the Tropics (Figure 20.15). Sea levels were high, and Earth was warm and humid. Life proliferated both on land and in the oceans. Marine invertebrates increased in variety and number of species. On land, dinosaurs continued to diversify. The first snakes appeared during the Cretaceous, though their lineage did not radiate until much later.

Early in the Cretaceous, flowering plants—the angiosperms—evolved from gymnosperm ancestors and began the radiation that led to their current dominance on land. By the end of the period, many groups of mammals had evolved, but these mammals were generally small.

Another mass extinction took place at the end of the Cretaceous period. On land, all vertebrates larger than about 25 kg in body weight apparently became extinct. In the seas, many planktonic organisms and bottom-dwelling invertebrates became extinct. This mass extinction was probably caused by the large meteorite that collided with Earth off the Yucatán Peninsula, as described on page 384.

The modern biota evolved during the Cenozoic era

By the early Cenozoic era (65 mya), the positions of the continents resembled those of today, but Australia was still attached to Antarctica, the Atlantic Ocean was much narrower, and the northern continents were connected. The Cenozoic era was characterized by an extensive radiation of mammals, but other groups were also undergoing important changes. Flowering plants diversified extensively and dominated world forests, except in cool regions.

THE TERTIARY (65–1.8 MYA). During the Tertiary period, Australia began its northward drift. By 20 mya it had nearly reached its current position. The map of the world during this period looks familiar to us. In the middle of the Tertiary, the climate became considerably drier and cooler. Many lineages of flowering plants evolved herbaceous (nonwoody) forms, and grasslands spread over much of Earth.

By the beginning of the Cenozoic era, invertebrate faunas resembled those of today. It is among the vertebrates that evolutionary changes during the Tertiary were most rapid. Living groups of reptiles, including snakes and lizards, underwent extensive radiations during this period, as did birds and mammals.

THE QUATERNARY (1.8 MYA–PRESENT). The current geological period, the Quaternary period, is subdivided into two epochs, the Pleistocene and the Holocene (also known as the Recent). The Pleistocene epoch, which began about 1.8 mya, was a time of drastic cooling and climatic fluctuations. During four major and about twenty minor episodes, massive glaciers spread across the continents. Earth became much cooler, and animal and plant populations shifted toward the equator. The last of these glaciers retreated from temperate latitudes less than 15,000 years ago. Organisms of the current Holocene epoch are still adjusting to these changes; many high-latitude ecological communities have occupied their current locations for no more than a few thousand years.

Interestingly, these climate fluctuations resulted in few extinctions. However, the Pleistocene was the scene of hominid evolution and radiation, resulting in the species *Homo sapiens*—modern humans (see Chapter 33). Many large birds and mammals became extinct in North and South America and in Australia when *H. sapiens* arrived on those continents. Human hunting may have caused these extinctions, although the existing evidence does not convince all paleontologists.

Rates of Evolutionary Change

Following each mass extinction, the diversity of life rebounded. How fast did evolution proceed during those times? Why did some lineages evolve rapidly while others

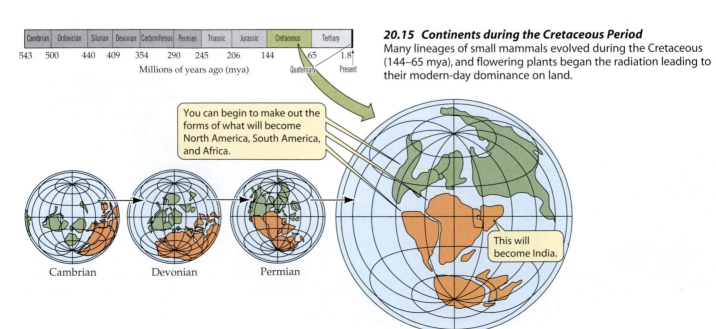

Cambrian	Ordovician	Silurian	Devonian	Carboniferous	Permian	Triassic	Jurassic	Cretaceous	Tertiary
543	500	440	409	354	290	245	206	144	65	1.8

Millions of years ago (mya)   Quaternary / Present

20.15 Continents during the Cretaceous Period
Many lineages of small mammals evolved during the Cretaceous (144–65 mya), and flowering plants began the radiation leading to their modern-day dominance on land.

You can begin to make out the forms of what will become North America, South America, and Africa.

This will become India.

Cambrian   Devonian   Permian

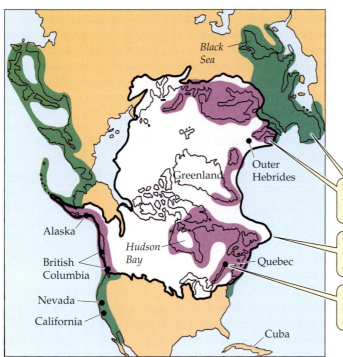

20.16 Natural Selection Acts on Stickleback Spines Three-spined stickleback populations with reduced spines are found principally in young lakes that were covered by ice during the most recent glacial period. These lakes lack large predatory fishes, but contain predatory insects that capture the fish by grasping their spines.

The current range of sticklebacks includes formerly glaciated areas (lavender) and unglaciated areas (green).

The region of the Northern Hemisphere that was once covered by Pleistocene glaciers is outlined in black.

Places where sticklebacks are known to have reduced spines are indicated by circles.

did not? Scientists have made enough progress in studying evolution to be able to give at least tentative answers to these questions.

Evolutionary rates vary

The fossil record shows that rates of evolution have been uneven. Many species have experienced times of **stasis**, long periods during which they changed very little. For example, many marine lineages have evolved slowly. Horseshoe crabs that lived 300 mya are almost identical in appearance to those living today, and the chambered nautiluses of the late Cretaceous are indistinguishable from living species. Such "living fossils" are found today in harsh environments that have changed relatively little for millennia. The sandy coastlines where horseshoe crabs spawn have extremes in temperature and salt concentration that are lethal to many other organisms. Chambered nautiluses spend their days in deep, dark ocean waters, ascending to feed in food-rich surface waters only under the protective cover of darkness. Their intricate shells provide little protection against today's visually hunting fishes.

Periods of stasis may be broken by times during which changes, either in the physical or the biological environment, create conditions that favor new traits. How new conditions favor rapid evolutionary change is illustrated by the spines of the three-spined stickleback (*Gasterosteus aculeatus*). This widespread marine fish has repeatedly invaded fresh water throughout its evolutionary history (Figure 20.16).

Sticklebacks are tiny fish, usually less than 10 cm long. All marine and most freshwater populations have well-developed pelvic girdles with prominent spines that make it difficult for other fishes to swallow them. However, large

predatory insects can readily grasp the stickleback's spines, and prey selectively on stickleback individuals with the largest spines. When stickleback populations invade freshwater habitats where predatory fish are absent but predatory insects are present, they rapidly evolve smaller spines. Populations with reduced spines are found primarily in young lakes that were covered by ice during the most recent glaciation, and hence do not have large predatory fishes.

The extensive fossil record of sticklebacks shows that spine reduction evolved many times in different populations that invaded fresh water. In addition, molecular data reveal that each freshwater population is most closely related to an adjacent marine population, not to other freshwater populations. Therefore, spine reduction has evolved rapidly many times in different places in response to the same ecological situation: the absence of predatory fish.

Extinction rates vary over time

More than 99 percent of the species that have ever lived are extinct. Species have become extinct throughout the history of life, but extinction rates have fluctuated dramatically over time; some groups had high extinction rates while others were proliferating.

Each mass extinction changed the flora and fauna of the next period by selectively eliminating some types of organisms, thereby increasing the relative abundance of others. For example, among the seashells of the Atlantic coastal plain of North America, species with broad geographic ranges were less likely to become extinct during normal periods (when no mass extinctions were taking place) than were species with small geographic ranges.

On the other hand, during the mass extinction of the late Cretaceous, groups of closely related species with large geographic ranges survived better than groups with small ranges, even if the individual species within the group had small ranges. Similar patterns are found in other molluscan groups elsewhere, suggesting that traits favoring long-term survival during normal times are often different from those that favor survival during times of mass extinctions.

At the end of the Cretaceous period, extinction rates on land were much higher among large vertebrates than among small ones. The same was true during the Pleistocene mass extinction, when extinction rates were high only among large mammals and large birds. During some mass extinctions, marine organisms were heavily hit while terrestrial organisms survived well. Other extinctions affected organisms living in both environments. These differences are not surprising, given that major changes on land and in the oceans did not always coincide.

Patterns of Evolutionary Change

Major new features, such as the feathers of birds or the legs of terrestrial vertebrates, that adapt organisms to a special way of life are called **evolutionary innovations**. How such novelties arise has been the subject of much debate from Darwin's time to the present. The variety of sizes and shapes among living organisms seems almost limitless, but the number of truly novel structures is remarkably small. As fiction writers often do, we can imagine unusual vertebrates with wings sprouting from their backs, but in reality the wings of vertebrates are always modified front legs. Modern mammals are highly varied in their shapes, but all

of their structures are modifications of structures found in ancestral mammals. As we saw earlier, even transforming a terrestrial mammal into a whale did not require a drastic reorganization of the mammalian body plan. Only a few evolutionary innovations, such as the notochord of chordates, do not appear to be modifications of a preexisting structure.

Three major faunas have dominated animal life on Earth

Only three events during the evolution of life have resulted in the evolution of major new faunas (Figure 20.17). The first one, the Cambrian explosion, took place about 540 mya. The second, about 60 million years later, resulted in the Paleozoic fauna. The great Permian extinctions 300 million years later were followed by the third event, the Triassic explosion, which led to our modern fauna.

During the Cambrian explosion, organisms representative of all major present-day lineages appeared, along with a number of lineages that subsequently became extinct. The Paleozoic and Triassic explosions greatly increased the number of families, genera, and species, but no new or dramatically different organismal body plans evolved. The later explosions resulted in many new organisms, but all of them were modifications of body plans that were already present when these great biological diversifications began.

Biologists have long puzzled over the striking differences between the Cambrian explosion and the two later explosions. A commonly accepted theory is that because the Cambrian explosion took place in a world that con-

20.17 Evolutionary Faunas
Representatives of the three major evolutionary faunas are shown together with a graphic illustration of the number of families in each fauna over time.

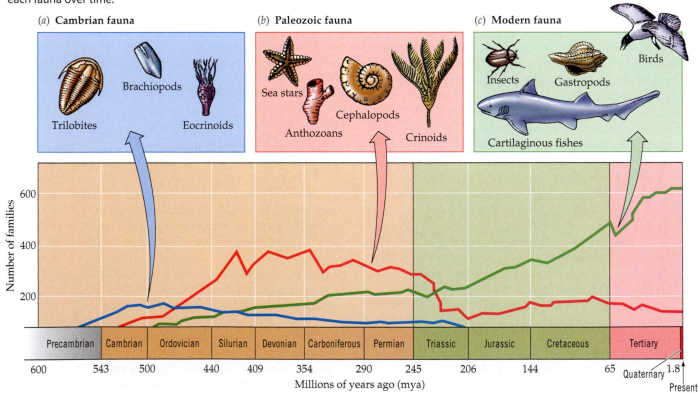

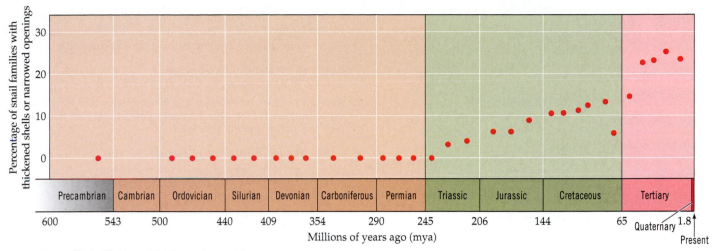

20.18 Snail Shells Have Thickened over Time
The percentage of families of snails that have internally thickened
or narrowed openings to their shells has increased with evolutionary time—evidence that predation on shelled animals intensified.

tained only a few species of organisms, all of which were
small, the ecological setting was favorable for the evolution
of many new body plans and different ways of life. Many
types of organisms were able to survive initially in this
world, but as competition intensified and new types of
predators evolved, many forms were unable to persist.

Although Earth was relatively poor in species at the time
of the two later explosions as well, the species that were already present included a wide array of body plans and
ways of life. As body plans became more specialized, major
transformations of form became increasingly less likely.
Therefore, major new innovations were less likely to evolve
at these times than in the Cambrian.

The size and complexity of organisms have increased

The earliest organisms were small prokaryotes. A modest
increase in size and a dramatic increase in structural complexity accompanied the evolution of the first eukaryotes
2.5 billion years ago. Since then, the maximum sizes of organisms in many lineages have increased, irregularly to be
sure. The most striking exception to this trend is insects,
which have remained relatively small throughout their evolutionary history.

The overall increase in body size is the result of two opposing forces. Within a species, selection often favors larger
size because larger individuals can dominate smaller ones.
But larger species on average survive for less time than
small species do, which is one reason why Earth is not populated primarily by large organisms.

Predators have become more efficient

Over time, predators have evolved increasingly efficient
methods of capturing prey, and prey, in turn, have evolved
better defenses. During the Cretaceous, for example, many
species of crabs with powerful claws evolved, and carnivorous marine snails able to drill holes in shells began to fill the
seas. Skates, rays, and bony fishes with powerful teeth capable of crushing mollusk shells also evolved, and large, powerful marine reptiles—the placodonts—fed heavily on clams.
The increasing thickness and narrowing openings of snail
shells during the Cretaceous is evidence that predation rates
intensified (Figure 20.18). Other evidence of heavy predation
pressure is the increase in the percentage of fossil shells that
show signs of having been repaired following an attack that
did not kill the owner.

Although shell thickness provided some protection from
predators, predators were so effective that clams disappeared from the surfaces of most marine sediments. The
survivors were species that burrowed into the substratum,
where they were more difficult to capture.

The Future of Evolution

The agents of evolution are operating today just as they
have been since life first appeared on Earth. However,
major changes are under way as a result of the dramatic increase of Earth's human population. Until recently, human-
caused extinctions affected mostly large vertebrates, but
these losses are now being compounded by increasing extinctions of small species, driven primarily by changes in
Earth's vegetation. Deliberately or inadvertently, people are
moving thousands of species around the globe, reversing
the provincialization of Earth's biota that evolved during
the Mesozoic era.

Humans have also taken charge of the evolution of certain valuable species by means of artificial selection and
biotechnology. Our ability to modify species has been enhanced by modern molecular methods that enable us to
move genes among species—even distantly related ones. In
short, humans have become the dominant evolutionary
agent on Earth today. How we handle our massive influence will powerfully affect the future of life on Earth.

Chapter Summary

▶ Changes that take effect during the lifetimes of species constitute microevolution. Changes that involve the appearance of new species and evolutionary lineages are called macroevolution.

How Do We Know that Earth is Ancient?

▶ The relative ages of rock layers in Earth's crust can be determined from their positions relative to one another and from their embedded fossils.

▶ Radioisotopes supplied the key for assigning absolute ages to rocks.

▶ Earth's geological history is divided into eras and periods. The boundaries between these units are based on differences between their fossil biotas. **Review Table 20.1**

How Has Earth Changed over Time?

▶ Unidirectional physical changes on Earth include gradual cooling and weakening of the forces that cause continental drift.

▶ Earth's early atmosphere lacked free oxygen. Oxygen accumulated after prokaryotes evolved the ability to use water as their source of hydrogen ions in photosynthesis. Increasing concentrations of atmospheric oxygen made possible the evolution of eukaryotes and multicellular organisms. **Review Figure 20.2**

▶ Throughout Earth's history the continents have moved about, sometimes separating from one another, at other times colliding. **Review Figures 20.10, 20.11, 20.13, 20.15**

▶ Earth has experienced periods of rapid climate change, massive volcanism, and major shifts in sea levels and ocean currents, all of which have had dramatic effects on the evolution of life. **Review Figures 20.3, 20.4**

▶ External events, such as collisions with meteorites, also have changed conditions on Earth. A meteorite may have caused the abrupt mass extinction at the end of the Cretaceous period.

The Fossil Record

▶ Much of what we know about the history of life on Earth comes from the study of fossils.

▶ The fossil record, although incomplete, reveals broad patterns in the evolution of life. About 300,000 fossil species have been described. The best record is that of hard-shelled marine animals.

▶ Fossils show that many evolutionary changes are gradual, but an incomplete record can falsely suggest or conceal times of rapid change. **Review Figures 20.7, 20.8**

Life in the Remote Past

▶ The fossil record for Precambrian times is fragmentary, but fossils from Australia show that many lineages that evolved then may not have left living descendants.

▶ Diversity exploded during the Cambrian period. **Review Figure 20.10**

▶ Geographic differentiation of biotas increased during the Mesozoic era.

▶ The modern biota evolved during the Cenozoic era.

Rates of Evolutionary Change

▶ Rates of evolutionary change have been very uneven.

▶ Rapid rates of evolution occur when changes to the physical or biological environment create conditions that favor new traits. **Review Figure 20.16**

Patterns of Evolutionary Change

▶ Truly novel features of organisms have evolved infrequently. Most evolutionary changes are the result of modifications of already existing structures.

▶ Three major faunas have dominated animal life on Earth. **Review Figure 20.17**

▶ Over evolutionary time, organisms have increased in size and complexity. Predation rates have also increased, resulting in the evolution of better defenses among prey species. **Review Figure 20.18**

The Future of Evolution

▶ The agents of evolution continue to operate today, but human intervention, both deliberate and inadvertent, now plays an unprecedented role in the history of life.

For Discussion

1. Some lineages of organisms have evolved to contain large numbers of species, whereas others have produced only a few species. Is it meaningful to consider the former more successful than the latter? What does the word "success" mean in evolution? How does your answer influence your thinking about *Homo sapiens*, the only surviving representative of the Hominidae—a family that never had many species in it?

2. Scientists date ancient events using a variety of methods, but nobody was present to witness or record those events. Accepting those dates requires us to believe in the accuracy and appropriateness of indirect measurement techniques. What other basic scientific concepts are based on the results of indirect measurement techniques?

3. Why is it useful to be able to date past events absolutely as well as relatively?

4. What factors favor increases in body size? Why might average body size among particular species in a lineage decrease even if natural selection favors larger body size in most species of that lineage?

5. The continents are still drifting today, but biologists ignore these movements when thinking about factors affecting current evolutionary changes. On what basis do they make that decision?

21 *The Mechanisms of Evolution*

MOST SPECIES OF CUCKOOS AND COWBIRDS LAY their eggs in the nests of other species of birds. This behavior is known as brood parasitism. The host birds often incubate the eggs and raise the parasite nestlings. Host birds that accept parasite eggs and raise parasite chicks are likely to produce fewer offspring than hosts that recognize parasite eggs and push them out of the nest.

To investigate the evolution of such defensive behaviors, biologists studied cuckoos and their hosts in areas where brood parasitism had been occurring for different periods of time. In one valley in southern Spain, great spotted cuckoos and common magpies have lived together for many centuries. Here 78 percent of the magpies removed artificial cuckoo eggs experimenters placed in their nests. However, in another Spanish valley, where cuckoos did not arrive until the early 1960s, only 14 percent of magpies ejected the eggs.

Ejection of parasites' eggs evolved rapidly in Japan, where the ranges of the common cuckoo and the azure-winged magpie have only recently overlapped. In a region where cuckoos have parasitized magpies for 10 years, none of the magpies ejected cuckoo eggs, but in areas where they have been parasitized for 20 years, 42 percent of the magpies ejected cuckoo eggs.

What explains these differences in magpie behavior? Charles Darwin's main contribution to biology was to propose a plausible and testable hypothesis of a mechanism of evolutionary change that could result in the adaptation of organisms to their environments. Keep in mind that the environment of an organism includes the physical environment, individuals of other species, and individuals of the same species. All of these components influence the survival and reproductive success of individuals.

In this chapter we will review how Darwin developed his ideas, and then turn to the advances in understanding of evolutionary processes since Darwin's time. We will discuss the genetic basis of evolution and show how genetic variation within populations is measured. We will describe the agents of evolution and show how biologists design studies to investigate them. Finally, we will discuss constraints on the pathways evolution can take. When you understand these processes, you will understand the mechanisms of evolution.

Charles Darwin and Adaptation

The term **adaptation** has two meanings in evolutionary biology. The first refers to *traits* that enhance the survival and reproductive success of their bearers. For example, we believe that wings are adaptations for flight, a spider's web is an adaptation for capturing flying insects, and so forth. The second refers to the *process* by which these traits are acquired—that is, the evolutionary mechanisms that produced them.

Biologists regard an organism as being adapted to a particular environment when they can imagine—or better still, measure the performance of—a slightly different organism that reproduces and survives less well in that environment. That is, adaptation is a relative concept; to understand adaptation, biologists compare the performance of individuals that differ in traits within and among species. For example, to investigate the adaptive nature of spiders' webs, we would try to determine the effectiveness of slightly different webs spun by a given species in capturing insects. We would also measure changes in the webs of the species in different situations. With these data, we could understand how variations in web structure influence the survival and reproductive success of their makers.

A Magpie and a Cuckoo Chick
In parts of Japan, the azure-winged magpie has only recently experienced brood parasitism by the common cuckoo. This adult magpie will care for the cuckoo chick at the expense of its own offspring.

Darwin proposed a mechanism to explain adaptation

Charles Darwin was a keen naturalist who observed many examples of structures and behaviors that seemed to be designed to assist the survival and reproductive success of their bearers. He was given an unprecedented opportunity to study the adaptations of organisms in various parts of the world when in 1831, his Cambridge University botany professor, John Henslow, recommended him as a naturalist to Captain Robert Fitzroy, who was preparing to sail around the world on the survey ship *H.M.S. Beagle* (Figure 21.1). Whenever possible during the voyage, Darwin (who was often seasick) went ashore to observe and collect specimens of plants and animals.

Darwin spent most of his time ashore in South America, where the species he saw differed strikingly from those of Europe. He also noted that the species of the temperate regions of South America (Argentina and Chile) were more similar to those of tropical South America (Brazil) than they were to European species. When he explored the Galápagos archipelago, west of Ecuador, he noted that most of its animal species were found nowhere else, but were similar to those of the mainland of South America, 1,000 kilometers to the east. Darwin also observed that the animals of the archipelago differed from island to island. He postulated that some animals had dispersed from mainland South America and then evolved differently on different islands.

When he returned to England in 1836, Darwin continued to ponder his observations. Within a decade he had developed the main features of his theory of evolution, which had two major components:

▶ Species are not immutable, but change, or *adapt*, over time. (In other words, Darwin asserted that evolution is a historical fact.)
▶ The agent that produces the changes is *natural selection*.

Darwin wrote a long essay on natural selection and the origin of species in 1844, but, despite urging from his wife and colleagues, he was reluctant to publish it, preferring to assemble more evidence first.

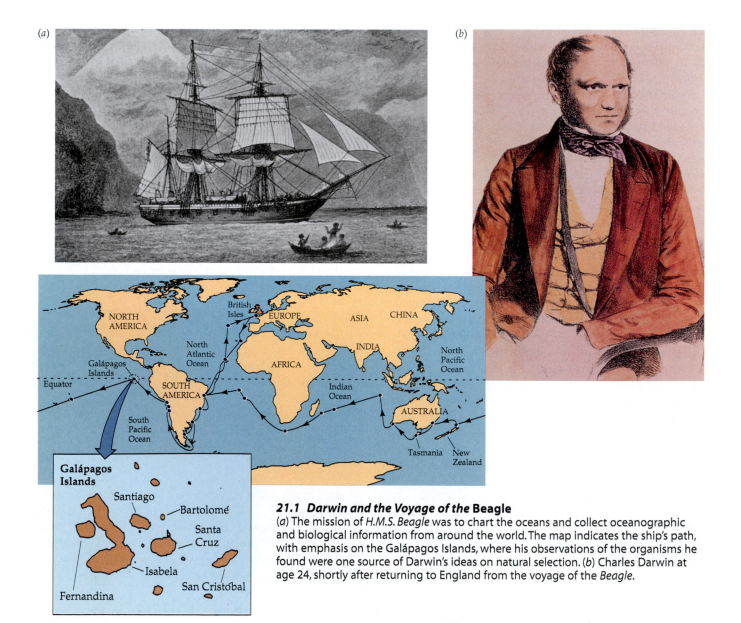

(a)   (b)

21.1 Darwin and the Voyage of the Beagle
(a) The mission of *H.M.S. Beagle* was to chart the oceans and collect oceanographic and biological information from around the world. The map indicates the ship's path, with emphasis on the Galápagos Islands, where his observations of the organisms he found were one source of Darwin's ideas on natural selection. (b) Charles Darwin at age 24, shortly after returning to England from the voyage of the *Beagle*.

21.2 Many Types of Pigeons Have Been Produced by Artificial Selection
Charles Darwin raised pigeons as a hobby, and he saw similar forces at work in artificial and natural selection. These are just some of over 300 varieties of pigeons that have been artificially selected by breeders to display different forms of traits such as color, size, and feather distribution.

Darwin's hand was forced in 1858 when he received a letter from Alfred Russel Wallace, who was studying plants and animals in the Malay Archipelago. Wallace asked Darwin to evaluate an enclosed manuscript, in which Wallace proposed a theory of natural selection almost identical to Darwin's. At first, Darwin was dismayed, believing that he had been preempted by Wallace. But extracts from Darwin's 1844 essay, together with Wallace's manuscript, were presented to the Linnaean Society of London on July 1, 1858, thereby giving credit for the idea to both men. Darwin then worked quickly to finish *The Origin of Species*, which was published the next year. Although both men conceived of natural selection independently, Darwin developed his ideas first, and his book provided a much more thorough justification of the concept—which is why natural selection is more closely associated with his name.

The facts that Darwin used to develop his theory of evolution by natural selection were familiar to most contemporary biologists. His insight was to perceive the significance of relationships among them. Darwin understood that populations of all species have the potential for exponential increases in numbers. To illustrate this point, he used the following example:

> Suppose … there are eight pairs of birds, and that only four pairs of them annually … rear only four young, and that these go on rearing their young at the same rate, then at the end of seven years (a short life, excluding violent deaths for any bird) there will be 2,048 birds instead of the original sixteen.

Yet such rates of increase are rarely seen in nature. Therefore, Darwin knew that death rates in nature must be high. Without high death rates, even the most slowly reproducing species would quickly reach enormous population sizes.

Darwin also observed that, although offspring tend to resemble their parents, the offspring of most organisms are not identical to one another or to their parents. He suggested that slight variations among individuals significantly affect the chance that a given individual will survive and the number of offspring it will produce. He called this differential reproductive success of individuals **natural selection**. Natural selection results from both differential survival and differential reproduction of individuals.

Darwin may have used the words "natural selection" because he was familiar with the artificial selection practices of animal and plant breeders. Many of Darwin's observations on the nature of variation came from domesticated plants and animals. Darwin was a pigeon breeder, and he knew firsthand the astonishing diversity in color, size, form, and behavior that could be achieved by humans selecting which pigeons to mate (Figure 21.2). He recognized close parallels between selection by breeders and selection in nature.

Darwin argued his case for natural selection in *The Origin of Species*:

> How can it be doubted, from the struggle each individual has to obtain subsistence, that any minute variation in structure, habits or instincts, adapting that individual better to the new conditions, would tell upon its vigour and health? In the struggle it would have a better chance of surviving; and those of its offspring which inherited the variation, be it ever so slight, would have a better chance.

That statement, written more than a hundred years ago, still stands as a good expression of the idea of evolution by natural selection.

Since Darwin wrote these words, biologists have developed a much deeper understanding of the genetic basis of evolutionary change and have assembled a rich array of examples of natural selection in action.

What have we learned about evolution since Darwin?

When Darwin proposed his theory of natural selection, he had no examples of selection operating in nature. He based his arguments on the results of selection on domesticated species. Since Darwin's time, many studies of the action of natural selection have been conducted; we will discuss some of them in this chapter.

We now know that biological evolution is a change over time in the genetic composition of a population. Darwin understood the importance of heredity for his theory, but he knew nothing of the mode of inheritance. He devoted considerable time to an attempt to develop a theory of heredity, but he failed to discover the laws of heredity, and he failed to understand the significance of Gregor Mendel's paper (see Chapter 10), which he apparently read.

Fortunately, the rediscovery of Mendel's publications in the 1900s paved the way for the development of **population genetics**, the field that provides a major underpinning for Darwin's theory. Population geneticists apply Mendel's laws to entire populations of organisms. They also study variation within and among species in order to understand the processes that result in evolutionary changes in species through time.

Genetic Variation within Populations

For a population to evolve, its members must possess variation, which is the raw material on which agents of evolution act. In everyday life, we do not directly observe the genetic composition of organisms or populations. What we do see in nature are *phenotypes*, the physical expressions of organisms' genes. The agents of evolution actually act on phenotypes, but for the moment we will concentrate on genetic variation within populations. We do so because genes are what is passed on to offspring via reproductive cells—eggs and sperm.

A *heritable trait* is a characteristic of an organism that is at least partly influenced by the organism's genes. The genetic constitution that governs a trait is called its *genotype*. *A population evolves when individuals with different genotypes survive or reproduce at different rates.*

Recall that different forms of a gene, called *alleles*, may exist at a particular locus. A single individual has only some of the alleles found in the population to which it belongs (Figure 21.3). The sum of all the alleles found in a population constitutes its **gene pool**. The gene pool contains the variation that produces the differing phenotypes on which agents of evolution act.

Fitness is the relative reproductive contribution of genotypes

The reproductive contribution of a genotype or phenotype to subsequent generations relative to the contribution of other genotypes or phenotypes in the same population is called **fitness**. The word "relative" is critical: The absolute number of offspring produced by an individual does not influence allele frequencies in the gene pool. Changes in absolute numbers of offspring are responsible for increases and decreases in the *size* of a population, but the relative success among genotypes within a population is what leads to changes in allele frequencies—that is, to evolution. When we discuss evolution, we talk about survival and reproductive success, because these rates determine how many genes different individuals contribute to subsequent generations.

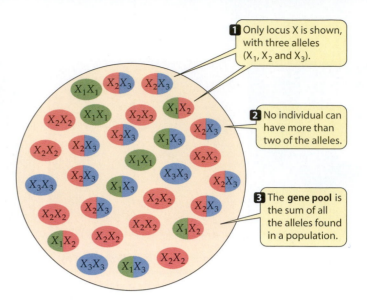

1 Only locus X is shown, with three alleles (X_1, X_2 and X_3).

2 No individual can have more than two of the alleles.

3 The **gene pool** is the sum of all the alleles found in a population.

21.3 A Gene Pool
The allele proportions in this gene pool are 0.20 for X_1, 0.50 for X_2, and 0.30 for X_3.

To contribute genes to subsequent generations, individuals must survive to reproductive age and produce offspring. The relative contribution of individuals of a particular genotype is determined by the probability that those individuals will survive times the average number of offspring they produce over their lifetimes. In other words, *the fitness of a genotype is determined by the average rates of survival and reproduction of individuals with that genotype.*

Most populations are genetically variable

Some level of genetic variation characterizes nearly all natural populations. Such variation has been demonstrated repeatedly for thousands of years by people attempting to develop desirable traits in plants and animals. For example, selection for different traits in a European wild mustard produced many important crop plants (Figure 21.4). Plant and animal breeders can achieve such results only if the original population has genetic variation for the traits of interest. Their success indicates that genetic variation is common, but it does not tell us how much variation there is.

Laboratory experiments also demonstrate that considerable genetic variation is present in most populations. In one such experiment, investigators chose as parents for subsequent generations of fruit flies (*Drosophila*) individuals with either high numbers or low numbers of bristles on their bodies. After 35 generations, flies in both lineages had bristle numbers that fell well outside the range found in the original population (Figure 21.5). These results show that there must have been considerable variation in the original fruit fly population for selection to act on.

To understand evolution, we need to know more precisely how much genetic variation populations contain, the sources of that genetic variation, and how genetic variation is maintained and expressed in populations in space and over time.

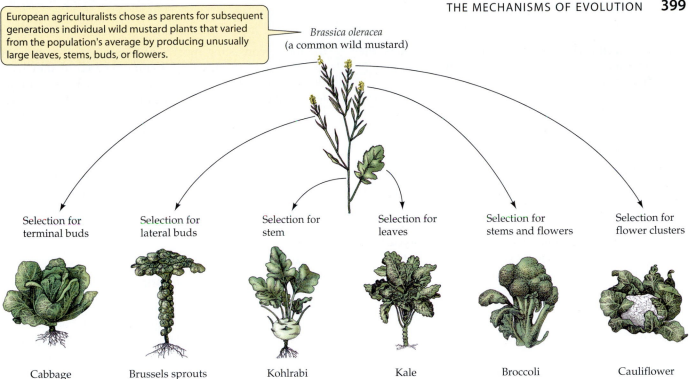

European agriculturalists chose as parents for subsequent generations individual wild mustard plants that varied from the population's average by producing unusually large leaves, stems, buds, or flowers.

Brassica oleracea
(a common wild mustard)

Selection for terminal buds	Selection for lateral buds	Selection for stem	Selection for leaves	Selection for stems and flowers	Selection for flower clusters
Cabbage	Brussels sprouts	Kohlrabi	Kale	Broccoli	Cauliflower

21.4 Many Vegetables from One Species
All of these crop plants have been derived from a single wild mustard species. They illustrate the vast amount of variation that can be present in a gene pool.

How do we measure genetic variation?

A locally interbreeding group within a geographic population is called a **Mendelian population**. Mendelian populations are often the subjects of evolutionary studies. To measure precisely the gene pool of a Mendelian population, we would need to count every allele at every locus in every organism in it. By measuring all the individuals, we could determine the relative proportions, or **frequencies**, of all alleles in the population.

Biologists can reliably estimate allele frequencies for a given locus by measuring numbers of alleles in a sample of individuals from a population. Measures of allele frequency range from 0 to 1; the sum of all allele frequencies at a locus is equal to 1. The frequencies of the different alleles at each locus and the frequencies of different genotypes in a Mendelian population describe its genetic structure.

An allele's frequency is calculated using the following formula:

$$p = \frac{\text{number of copies of the allele in the population}}{\text{sum of alleles in the population}}$$

If only two alleles (for example, *A* and *a*) for a given locus are found among the members of a diploid population, they may combine to form three different genotypes: *AA*,

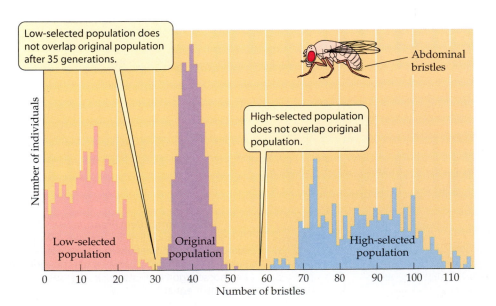

Low-selected population does not overlap original population after 35 generations.

Abdominal bristles

High-selected population does not overlap original population.

Number of individuals

Low-selected population

Original population

High-selected population

0   10   20   30   40   50   60   70   80   90   100   110
Number of bristles

21.5 Artificial Selection Reveals Genetic Variation
In laboratory experiments with *Drosophila*, changes in bristle number evolved rapidly when selected for artificially.

Aa, and *aa*. Using the formula above, we can calculate the relative frequencies of alleles *A* and *a* in a population of *N* individuals as follows:

▶ Let N_{AA} be the number of individuals that are homozygous for the *A* allele (*AA*)
▶ Let N_{Aa} be the number that are heterozygous (*Aa*)
▶ Let N_{aa} be the number that are homozygous for the *a* allele (*aa*)

Note that $N_{AA} + N_{Aa} + N_{aa} = N$, the total number of individuals in the population, and that the total number of alleles present in the population is $2N$ because each individual is diploid. Each *AA* individual has two *A* alleles, and each *Aa* individual has one *A* allele. Therefore, the total number of *A* alleles in the population is $2N_{AA} + N_{Aa}$, and the total number of *a* alleles in the population is $2N_{aa} + N_{Aa}$.

If *p* represents the frequency of *A*, and *q* represents the frequency of *a*, then

$$p = \frac{2N_{AA} + N_{Aa}}{2N}$$

and

$$q = \frac{2N_{aa} + N_{Aa}}{2N}$$

To show how this works, Figure 21.6 calculates the allele frequencies in two different populations, each containing 200 diploid individuals. Population 1 has mostly homozygotes (90 *AA*, 40 *Aa*, and 70 *aa*); population 2 has mostly heterozygotes (45 *AA*, 130 *Aa*, and 25 *aa*).

The calculations in Figure 21.6 demonstrate two important points. First, notice that for each population, $p + q = 1$. If there is only one allele in a population, its frequency is 1. If an allele is missing from a population, its frequency is 0, and the locus in that population is represented by one or more other alleles. Because $p + q = 1$, $q = 1 - p$, which means that when there are two alleles at a locus in a population, we can calculate the frequency of one allele and then easily obtain the second frequency by subtraction.

The second thing to notice in these calculations is that population 1 (consisting mostly of homozygotes) and population 2 (consisting mostly of heterozygotes) have the same allele frequencies for *A* and *a*. Therefore, they have the same gene pool for this locus. However, because the alleles in the gene pool are distributed differently, the *genotype frequencies* of the two populations differ.

Although we began our calculations with numbers of genotypes, for many purposes, genotypes, like alleles, are best thought of as frequencies. Genotype frequencies are calculated as the number of individuals that have the genotype divided by the total number of individuals in the population. In population 1 of our example, the genotype frequencies are 0.45 *AA*, 0.20 *Aa*, and 0.35 *aa*.

The Hardy–Weinberg Equilibrium

A population that is not changing genetically—that has the same allele and genotype frequencies from generation to generation—is said to be at **Hardy–Weinberg equilibrium**. The conditions that result in such an equilibrium population were discovered independently by the British mathematician Godfrey H. Hardy and the German physician Wilhelm Weinberg in 1908. Hardy wrote his equations in response to a question posed to him by the Mendelian geneticist Reginald C. Punnett (the inventor of the Punnett square) at the Cambridge University faculty club. Punnett was puzzled by the fact that although the allele for short fingers in humans was dominant and the allele for normal-length fingers was recessive, most people in Britain had normal-length fingers.

Hardy's equations explain why dominant alleles do not replace recessive alleles in populations. They also explain other features of the genetic structure of populations. The equations apply to sexually reproducing organisms. The particular example we will illustrate assumes that the organism in question is diploid, its generations are nonoverlapping, the gene under consideration has two alleles, and allele frequencies are identical in males and females. The equilibrium also applies if there are more than two alleles and generations overlap, but in those cases the mathematics are more complicated.

In any population:

$$\text{Frequency of allele } A = p = \frac{2N_{AA} + N_{Aa}}{2N} \qquad \text{Frequency of allele } a = q = \frac{2N_{aa} + N_{Aa}}{2N}$$

where *N* is the total number of individuals in the population.

For population 1 (mostly homozygotes):

$N_{AA} = 90$, $N_{Aa} = 40$, and $N_{aa} = 70$

so

$$p = \frac{180 + 40}{400} = 0.55$$

$$q = \frac{140 + 40}{400} = 0.45$$

For population 2 (mostly heterozygotes):

$N_{AA} = 45$, $N_{Aa} = 130$, and $N_{aa} = 25$

so

$$p = \frac{90 + 130}{400} = 0.55$$

$$q = \frac{50 + 130}{400} = 0.45$$

21.6 Calculating Allele Frequencies
The gene pool and allele frequencies are the same for both populations, but the alleles are distributed differently between heterozygous and homozygous genotypes. In all cases, $p + q$ must equal 1.

Generation I

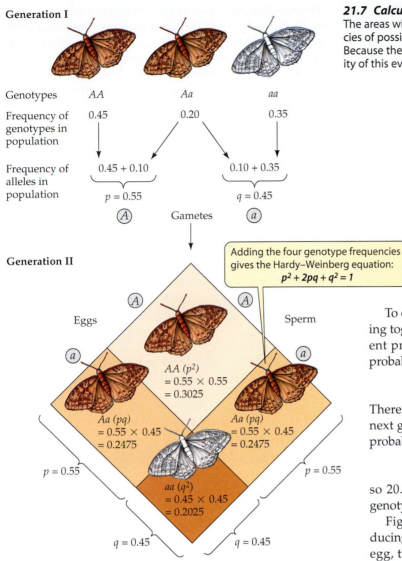

Genotypes	AA	Aa	aa

Frequency of genotypes in population: 0.45   0.20   0.35

Frequency of alleles in population:

0.45 + 0.10   0.10 + 0.35

$p = 0.55$ $q = 0.45$

(A)   Gametes   (a)

Generation II

Adding the four genotype frequencies gives the Hardy–Weinberg equation:
$p^2 + 2pq + q^2 = 1$

Eggs   Sperm

(A)   (A)

(a)   (a)

$AA\ (p^2)$
$= 0.55 \times 0.55$
$= 0.3025$

$Aa\ (pq)$
$= 0.55 \times 0.45$
$= 0.2475$

$Aa\ (pq)$
$= 0.55 \times 0.45$
$= 0.2475$

$aa\ (q^2)$
$= 0.45 \times 0.45$
$= 0.2025$

$p = 0.55$ $p = 0.55$

$q = 0.45$ $q = 0.45$

21.7 Calculating Hardy–Weinberg Genotype Frequencies

The areas within the squares are proportional to the expected frequencies of possible matings if mating is random with respect to genotype. Because there are two ways of producing a heterozygote, the probability of this event occurring is the sum of the two Aa squares.

To see why these results are true, consider population 1, used as an example in the previous section, in which the frequency of the A allele (p) is 0.55. Because we assume that individuals select mates at random, without regard to their genotype, gametes carrying A or a combine at random—that is, as predicted by the frequencies p and q. The probability that a particular sperm or egg in this example will bear an A allele rather than an a allele is 0.55. In other words, 55 out of 100 random selections of a sperm or an egg will bear an A allele. Because $q = 1 - p$, the probability of an a allele is $1 - 0.55 = 0.45$.

To obtain the probability of two A-bearing gametes coming together at fertilization, we multiply the two independent probabilities of drawing them (see the discussion of probability in Chapter 10):

$$p \times p = p^2 = (0.55)^2 = 0.3025$$

Therefore, 0.3025, or 30.25 percent, of the offspring in the next generation will have the AA genotype. Similarly, the probability of bringing together two a-bearing gametes is

$$q \times q = q^2 = (0.45)^2 = 0.2025$$

so 20.25 percent of the next generation will have the aa genotype (Figure 21.7).

Figure 21.7 also shows that there are two ways of producing a heterozygote: an A sperm may combine with an a egg, the probability of which is $p \times q$; or an a sperm may combine with an A egg, the probability of which is $q \times p$. Consequently, the overall probability of obtaining a heterozygote is $2pq$.

It is now easy to show that allele frequencies p and q remain constant for each generation. Notice that the total of $p^2 + pq$ represents the total of the A alleles. The fraction that this frequency constitutes of all alleles is

$$\frac{p^2 + pq}{p^2 + 2pq + q^2} = \frac{p(p+q)}{(p+q)(p+q)} = \frac{p}{p+q} = \frac{p}{p+(1-p)} = p$$

Similarly, the frequency of a in the next generation will be

$$\frac{q^2 + pq}{p^2 + 2pq + q^2} = \frac{q(p+q)}{(p+q)(p+q)} = \frac{q}{p+q} = \frac{q}{(1-q)+q} = q$$

Thus the original allele frequencies are unchanged, and the population is at Hardy–Weinberg equilibrium.

If some agent, such as nonrandom mating, were to alter the allele frequencies, the genotype frequencies would automatically settle into a predictable new set in the next generation. For instance, if only AA and Aa individuals bred, p and q would change, but there would still be aa individuals in the population.

The essential assumptions that must be met for Hardy–Weinberg equilibrium are:

▶ Mating is random.
▶ Population size is very large.
▶ There is no migration between populations.
▶ Mutation can be ignored.
▶ Natural selection does not affect the alleles under consideration.

If these conditions hold, two results follow. First, the frequencies of alleles at a locus will remain constant from generation to generation. And second, after one generation of random mating, the genotype frequencies will remain in the proportions

Genotype:	AA	Aa	aa
Frequency:	p^2	$2pq$	q^2

Stated another way, this is the equation for Hardy-Weinberg equilibrium:

$$p^2 + 2pq + q^2 = 1$$

Why is the Hardy–Weinberg equilibrium important?

The most important message of the Hardy–Weinberg equilibrium is that allele frequencies remain the same from generation to generation unless some agent acts to change them. Hence, simply because normal-length fingers in humans are recessive, we don't expect the frequency of normal-length fingers in the population to change unless a specific evolutionary force is acting on the underlying genes. The equilibrium also shows us what distribution of genotypes to expect for a population at genetic equilibrium at any value of p and q.

You may already have realized that populations in nature rarely meet the stringent conditions necessary to maintain them in Hardy–Weinberg equilibrium. Why, then, is it considered so important for the study of evolution? The answer is that without it, we cannot tell whether evolutionary agents are operating. More importantly, the pattern of deviations from the equilibrium tells us which assumptions are violated; thus we can identify the agents of evolutionary change on which we should concentrate our attention.

Microevolution: Changes in the Genetic Structure of Populations

Evolutionary agents are forces that change the allele and genotype frequencies in a population. In other words, they cause deviations from the Hardy–Weinberg equilibrium. Because such changes in the gene pool of a population constitute small-scale evolutionary changes, they are referred to as *microevolution*. The known evolutionary agents are mutation, gene flow, random genetic drift, nonrandom mating, and natural selection. Although only natural selection results in adaptation, to understand microevolutionary processes we need to discuss all five evolutionary agents before considering in detail how natural selection is studied.

Mutations are changes in genetic material

The origin of genetic variation is germ-line mutations (see Chapter 12). These mutations appear to be random with respect to the adaptive needs of organisms. Most mutations are harmful or neutral to (do not affect) their bearers. If the environment changes, however, previously harmful or neutral alleles may become advantageous.

Mutation rates are very low for most loci that have been studied. Rates as high as one mutation per locus in a thousand zygotes per generation are rare; one in a million is more typical. Nonetheless, these rates are sufficient to create considerable genetic variation because each of a large number of genes may mutate, and populations often contain a large number of individuals. For example, if the probability of mutation is 10^{-9} per nucleotide base pair per generation, then in each human gamete, the DNA of which contains 3×10^9 base pairs, there would be an average of one new mutation in each generation. Each newly fertilized egg would carry, on average, two new mutations. Therefore, the current human population of about 6 billion people would be expected to carry about 12 billion new muta-

tions that were not present one generation earlier. In addition, mutations can restore to populations alleles that other evolutionary agents remove. Thus mutations both create and help maintain variation within populations.

One condition for Hardy–Weinberg equilibrium is that there be no mutation. Although this condition is never strictly met, the rate at which mutations arise at single loci is usually so low that mutations result in only very small deviations from Hardy–Weinberg expectations. If large deviations are found, it is appropriate to dismiss mutation as the cause and to look for evidence of other evolutionary agents.

Migration of individuals followed by breeding produces gene flow

Because few populations are completely isolated from other populations of the same species, usually some migration between populations takes place. **Gene flow** happens when migrating individuals breed in their new location. Immigrants may add new alleles to the gene pool of a population, or may change the frequencies of alleles already present if they come from a population with different allele frequencies. For a population to be in Hardy–Weinberg equilibrium, there must be no immigration from other populations with different allele frequencies.

Random genetic drift may cause large changes in small populations

Chance events that alter allele frequencies result in **random genetic drift**. This process occurs at all loci in all populations, but has its greatest effect in small populations. If only a few individuals contribute genes to the next generation, the alleles they carry are not likely to be in the same proportions as alleles in the gene pool from which they were drawn.

In very small populations, random genetic drift may be strong enough to influence the direction of change of allele frequencies even when other evolutionary agents are pushing the frequencies in a different direction. Harmful alleles, for example, may increase because of random genetic drift, and rare advantageous alleles may be lost. As we will see later, even in large populations, random genetic drift can influence the frequencies of traits that do not influence the survival and reproductive rates of their bearers.

Even organisms that normally have large populations may pass through occasional periods when only a small number of individuals survive. During these **population bottlenecks**, genetic variation can be reduced by genetic drift. How this works is illustrated in Figure 21.8, in which allele frequencies are represented by red and yellow beans. In the small sample taken from the bean population, most of the beans that "survive" to germinate the next generation are, just by chance, red, so the new population has a much higher frequency of red beans than the previous generation had.

Suppose we have performed a cross of $Aa \times Aa$ individuals of a species of *Drosophila* to produce an offspring pop-

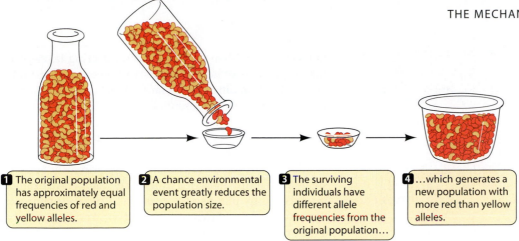

1 The original population has approximately equal frequencies of red and yellow alleles.

2 A chance environmental event greatly reduces the population size.

3 The surviving individuals have different allele frequencies from the original population…

4 …which generates a new population with more red than yellow alleles.

21.8 The Bottleneck Effect
Population bottlenecks occur when only a few individuals survive a random event, resulting in a shift in allele frequencies within the population.

ulation in which $p = q = 0.5$ and in which the genotype frequencies are 0.25 *AA*, 0.50 *Aa*, and 0.25 *aa*. If we randomly select four individuals from among the offspring to form the next generation, the allele frequencies in this small sample may differ markedly from $p = q = 0.5$. If, for example, we happen by chance to draw two *AA* homozygotes and two heterozygotes (*Aa*), the genotype frequencies in this "surviving population" are $p = 0.75$ and $q = 0.25$. If we replicate this sampling experiment 1,000 times, one of the two alleles will be missing entirely from about 8 of the 1,000 "surviving populations."

Populations in nature pass through bottlenecks for many different reasons. Predators may reduce populations of their prey to very small sizes. During the 1890s, hunting reduced the number of northern elephant seals to about 20 animals in a single population on the coast of Mexico. The actual breeding population may have been even smaller because in this species, only a few males mate with all the females and father all the offspring in any generation (Figure 21.9).

Using electrophoresis (see Chapter 17), investigators examined 24 proteins from tissues collected from the current California population of northern elephant seals. They found no evidence of variation in any of the 24 proteins. By contrast, the southern elephant seal, whose numbers were not severely reduced by hunting, has much more genetic variation. Currently, northern elephant seal populations are expanding rapidly, so their reduced genetic variation is not preventing high survival and reproductive rates. However, biologists worry that it may make them vulnerable to a disease outbreak or other sudden environmental change.

When a few pioneering individuals colonize a new re-gion, the resulting population will not have all the alleles found among members of its source population. The resulting pattern of genetic variation, called a **founder effect**, is equivalent to that in a large population reduced by a bottleneck. Because individuals of many plant species can reproduce sexually by self-fertilization, a new plant population may be started by a single seed—an extreme example of a founder effect.

Scientists were given an opportunity to study the genetic composition of a founding population when *Drosophila subobscura*, a well-studied European species of fruit fly, was discovered near Puerto Montt, Chile, in 1978 and at Port Townsend, Washington, in 1982. In both South and North America, populations of the flies grew rapidly and expanded their ranges. Today in North America, *D. subobscura* ranges from British Columbia, Canada, to central California. In Chile it has spread across 23° of latitude, nearly as wide a range as the species has in Europe (Figure 21.10).

The *D. subobscura* founders probably reached Chile and the United States from Europe aboard the same ship, because both populations are genetically very similar. For example, the North and South American populations have only 20 chromosomal inversions, 19 of which are the same on the two continents, whereas 80 inversions are known from European populations. New World populations also have lower enzyme diversity than Old World populations. Only alleles that have a frequency higher than 0.1 in Euro-

21.9 A Species with Low Genetic Variation
Because a few males sire most of the offspring in this northern elephant seal breeding colony, the size of the breeding population is smaller than the population as a whole. This pattern of non-random mating, together with a bottleneck that occurred when the seals were overhunted in the late nineteenth century, resulted in a population with very little genetic variation.

Mirounga angustirostris

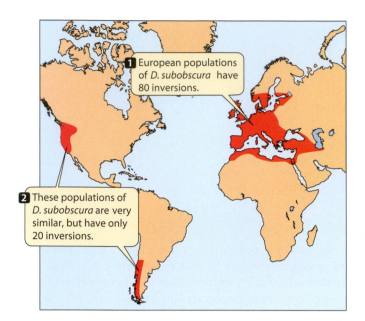

21.10 A Founder Effect
Populations of the fruit fly *Drosophila subobscura* in North and South America contain less genetic variation than the European populations from which they came, as shown by their numbers of chromosome inversions. Within two decates of arriving in the New World, the flies have increased dramatically and spread widely in spite of their reduced genetic variation.

pean populations are present in the Americas. Thus, as expected from a small founding population, only a small part of the total genetic variation found in Europe reached the Americas. Geneticists estimate that at least ten, but no more than a hundred, flies initially arrived in the New World.

Nonrandom mating changes the frequency of homozygotes

Another Hardy–Weinberg assumption is that mating is random. In many cases, however, individuals with certain genotypes mate more often with individuals of either the same or different genotypes than would be expected on a

Primula sinensus

random basis. When such **assortative mating** takes place, the proportions of homozygotes and heterozygotes in the next generation differ from Hardy–Weinberg expectations. If individuals mate preferentially with other individuals of the same genotype, homozygous genotypes are overrepresented and heterozygous genotypes underrepresented in the next generation.

Alternatively, individuals may mate primarily or exclusively with individuals of a different genotype. An example is provided by plant species such as *Primula* (primroses) that have flowers of two types. One type, known as *pin*, has a tall style (female reproductive organ) and short stamens (male reproductive organs). The other type, known as *thrum*, has a short style and tall stamens (Figure 21.11). Pollen grains from *pin* and *thrum* flowers are deposited on different parts of the bodies of insects that visit the flowers. When the insects visit other flowers, pollen grains from *pin* flowers are most likely to come into contact with stigmas of *thrum* flowers, and vice versa. In most species with this reciprocal arrangement, pollen from one flower type can fertilize only flowers of the other type.

Self-fertilization (selfing), another form of nonrandom mating, is common in many groups of organisms, especially plants. Selfing reduces the frequencies of heterozygous individuals below Hardy–Weinberg expectations. Under assortative mating and self-fertilization, genotype frequencies change, but allele frequencies remain the same.

 ## Natural selection produces variable results

As we have seen, individuals vary in heritable traits that determine the success of their reproductive efforts. Not all

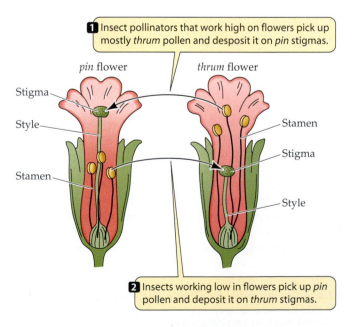

21.11 Floral Structure Fosters Assortative Mating
The structure of flowers in species such as the primroses (*Primula*) assures that fertilization usually occurs between individuals of different types.

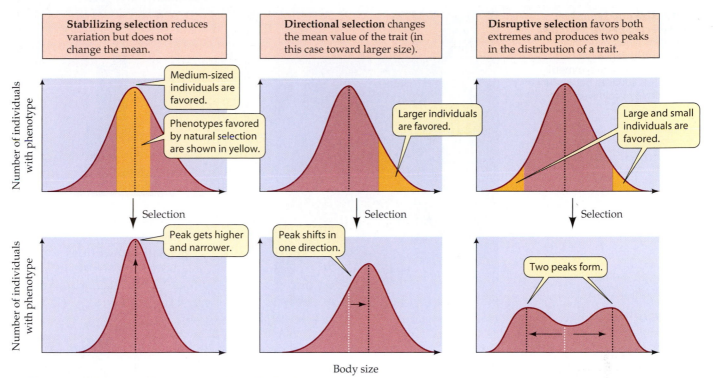

21.12 Natural Selection Operates on a Variable Trait
The curves plot the distributions of body size in a population before selection (top) and after selection (bottom). Natural selection may change the shape and position of the original curves.

individuals survive and reproduce equally well in a particular environment. Therefore, some individuals contribute more offspring to the next generation than do other individuals. This process is known as natural selection, and it causes allele frequencies in the population to change.

Depending on which traits are favored in a population, natural selection can produce any one of several quite different results.

▶ Selection may preserve the characteristics of a population by favoring average individuals.
▶ Selection may change the characteristics of a population by favoring individuals that vary in one direction from the mean of the population.
▶ Selection may change the characteristics of a population by favoring individuals that vary in both directions from the mean of the population.

Until now, we have been considering traits influenced by alleles at only a single locus. However, most traits are influenced by alleles at more than one locus. The size of an organism, for example, is likely to be controlled by many different loci. If many loci influence size—and there is no selection—then the distribution of sizes in a population should approximate the bell-shaped curve shown in the top row of Figure 21.12.

STABILIZING SELECTION. If both the smallest and the largest individuals contribute relatively fewer offspring to the

next generation than those closer to the average size do, **stabilizing selection** is operating (Figure 21.12*a*). Stabilizing selection reduces variation, but does not change the mean. Natural selection frequently acts in this way, countering increases in variation brought about by mutation or migration. We know from the fossil record that most populations evolve slowly most of the time. Rates of evolution are typically very slow because natural selection is usually stabilizing.

Biologists measured the results of the action of natural selection on cliff swallows in Nebraska. These birds nest in dense colonies (Figure 21.13*a*). They eat flying insects, so can feed only when weather conditions permit flying insects to be active. In 1996, a severe cold spell, which began on May 24 and lasted for 6 days, killed thousands of birds and reduced the local population by about 53 percent. During the first 3 days after the cold spell, the investigators collected nearly 2,000 dead cliff swallows underneath their nesting colonies, and also captured about 1,000 birds that had survived the cold spell.

By carefully measuring the sizes and shapes of these birds, the investigators were able to show that larger birds survived better during the cold spell than smaller birds (Figure 21.13*b*). They also found that birds whose wings and tails were more symmetrical survived better than individuals with greater wing and tail asymmetry (Figure 21.13*c*). Larger swallows probably survived better because they had more favorable surface-to-volume ratios and were able to store more fat. Birds with symmetrical wings and tails were probably more maneuverable, and hence more efficient at capturing flying insects. So stabilizing selection maintains a high level of wing and tail symmetry, and (because cliff

21.13 Size and Symmetry Are Selected in Cliff Swallows
(a) Cliff swallows build their mud nests in dense colonies. Larger (b) and more symmetrical (c) birds survived better during a period of cold weather.

(a)

Petrochelidon pyrrhonota

(b)

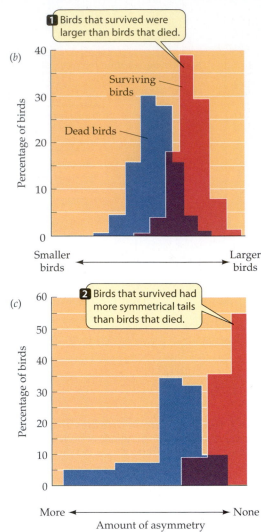

1 Birds that survived were larger than birds that died.

Surviving birds

Dead birds

Smaller birds ⟷ Larger birds

(c)

2 Birds that survived had more symmetrical tails than birds that died.

More ⟷ None
Amount of asymmetry

swallows are not evolving to become larger) some other selective force keeps the birds from becoming larger. That fact, together with cold weather (which keeps the the birds from becoming smaller), results in stabilizing selection.

DIRECTIONAL SELECTION.   If individuals at one extreme of the size distribution—the larger ones, for example (Figure 21.12b)—contribute more offspring to the next generation than other individuals do, then the mean size of individuals in the population will increase. In this case, **directional selection** is operating.

If directional selection operates over many generations, an evolutionary trend within the population results. Such directional evolutionary trends often continue for many generations, but they may be reversed if the environment changes and different phenotypes are favored, or they may be halted if an optimum is reached and the character then falls under stabilizing selection. The rapid evolution of rejection of parasite eggs by magpies, which we discussed at the beginning of the chapter, is an example of directional selection.

DISRUPTIVE SELECTION.   **Disruptive selection** is selection that simultaneously favors individuals at both extremes of the distribution (Figure 21.12c). This type of selection apparently is rare. When disruptive selection operates, individuals at the extremes contribute more offspring than those in the center, producing two peaks in the distribution of a trait.

The strikingly *bimodal* (two-peaked) distribution of bill sizes in the black-bellied seedcracker, *Pyrenestes ostrinus*, a West African finch (Figure 21.14), illustrates how disruptive selection can adapt populations in nature. Seeds of two types of sedges (a marsh plant) are the most abundant food source for the finches during part of the year. Birds with large bills can readily crack the hard seeds of the sedge *Scleria verrucosa*. Birds with small bills can crack *S. verrucosa*

seeds only with difficulty, but they can feed more efficiently on the soft seeds of the other sedge, *S. goossensii*, than birds with larger bills.

Young finches whose bills deviate markedly from the two predominant bill sizes do not survive as well as finches whose bills are close to one of the two sizes represented by the distribution peaks. Because there are few abundant food sources in the environment and because the seeds of the two sedges do not overlap in hardness, birds with intermediate-sized bills are inefficient in utilizing either one of the principal food sources. Disruptive selection therefore maintains a bill size distribution with two peaks.

Studying Microevolution

Biologists use several different methods to study microevolution, as illustrated by the examples we have just discussed. One method is to measure survival in the field under varying environmental conditions, as was done with cliff swallows. Another is to alter genotypes or phenotypes artificially and compare the performance of altered and normal individuals. A third is to use computer models to simulate natural selection. Here we discuss examples of the latter two methods.

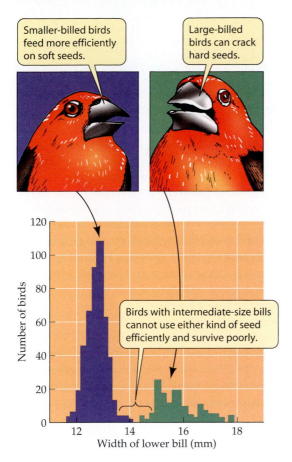

Smaller-billed birds feed more efficiently on soft seeds.

Large-billed birds can crack hard seeds.

Birds with intermediate-size bills cannot use either kind of seed efficiently and survive poorly.

21.14 Natural Selection Alters Bill Sizes

The bimodal distribution of bill sizes in the black-bellied seed-cracker of West Africa is an example of disruptive selection, which favors individuals with larger and smaller bill sizes over individuals with intermediate-sized bills.

ALTERING GENOTYPES AND COMPARING PERFORMANCE. Modern farmers attempt to control weeds by applying herbicides to crops. The success of this method is often poor because natural selection favors plants that produce chemicals that confer resistance to herbicides. But producing these chemicals is costly for the plant. Estimating the cost of defense against herbicides is difficult because individuals that differ in the kinds and concentrations of defensive chemicals they produce also differ in many other ways.

A powerful method of isolating the costs of producing and maintaining a specific resistance-conferring compound uses plasmids to transfer recombinant DNA into plants (see Chapter 17). The cost associated with resistance to the herbicide chlorosulfuron, conferred by a single allele, was measured in *Arabidopsis thaliana*. This allele, *Csr1-1*, results in the production of an enzyme that is insensitive to chlorosulfuron. Researchers transferred the *Csr1-1* allele to some individuals; other, genetically identical individuals received empty plasmids. Plants with the *Csr1-1* allele produced 34 percent fewer seeds than the nonresistant plants when grown under identical conditions in the absence of the herbicide (Figure 21.15). The reason for the high cost is not fully known, but evidence suggests that the allele results in an accumulation of branched-chain amino acids that interfere with metabolism.

COMPUTER MODELING OF SPIDER WEBS. Many species of spiders construct webs of sticky silk with which they capture flying insects. Because spider webs are relatively simple, two-dimensional structures, they are easy to model with computers. One such computer model, called NetSpinner, builds "webs" on a computer screen using behavioral rules that are actually used by web-building spiders. The model assumes that these behavioral rules are inherited, allowing them to be altered by "mutations." In each generation of the model, six spiders build webs, each using slightly different web-building rules. NetSpinner then shoots "flies" randomly at the "webs" and counts how many are captured. The quality of a web is calculated as the number of "flies" it captures, minus its cost, which is assessed by the length of silk used to make it. A fraction of the population of spiders—those that made the least efficient webs—dies every generation. The remaining spiders mate with one another at random to produce a new generation of spiders.

An example of the webs that emerged from a run of 40 generations of NetSpinner is shown in Figure 21.16. These webs are remarkably similar to those of real web-spinning spiders. Although such computer models do not measure

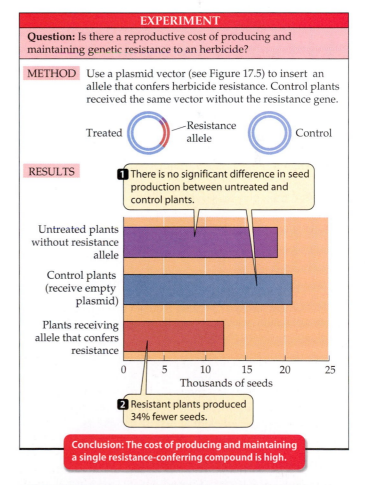

EXPERIMENT

Question: Is there a reproductive cost of producing and maintaining genetic resistance to an herbicide?

METHOD   Use a plasmid vector (see Figure 17.5) to insert an allele that confers herbicide resistance. Control plants received the same vector without the resistance gene.

Treated — Resistance allele   Control

RESULTS

1 There is no significant difference in seed production between untreated and control plants.

Untreated plants without resistance allele

Control plants (receive empty plasmid)

Plants receiving allele that confers resistance

Thousands of seeds

2 Resistant plants produced 34% fewer seeds.

Conclusion: The cost of producing and maintaining a single resistance-conferring compound is high.

21.15 Producing and Maintaining a Chemical Is Costly

Possession of a gene that confers resistance to an herbicide greatly reduced seed production in *Arabidopsis thaliana*.

EXPERIMENT

Question: Have spider webs evolved to be efficient capturers of insects?

METHOD   A computer experiment

1. Write computer rules for web building based on observations of spider behavior.
2. Assign different rules to each of six "virtual" spiders to spin webs.
3. Bombard the webs with the same random distribution of "virtual" flies.

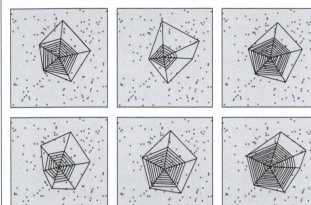

4. Count the number of flies caught.
5. Calculate quality of web as number of flies captured minus cost (length of silk).
6. Eliminate the least successful spiders and mate others to create new generations of spiders.
7. Repeat for many generations.

RESULTS

Virtual web   Actual web

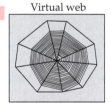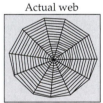

After 40 generations the virtual webs are very similar to real webs.

Conclusion: This computer model suggests that the answer to the question is "yes": Spider webs have evolved to be efficient capturers of insects.

21.16 Computer "Models" Help Us Understand Natural Selection

An "experiment" using a computer program that modeled 40 generations of natural selection on spider webs resulted in webs remarkably similar to real ones.

real natural selection, they simulate the process well enough to lend support to the hypothesis that spider webs have evolved to be efficient at capturing insects.

Maintaining Genetic Variation

Random genetic drift, stabilizing selection, and directional selection all tend to reduce genetic variation within populations. Nevertheless, most populations have considerable

genetic variation. Why isn't the genetic variation of a species lost over time?

Sexual reproduction amplifies existing genetic variation

Recombination in sexually reproducing organisms amplifies existing genetic variation. In asexually reproducing organisms, the cells resulting from a mitotic division normally contain identical genotypes. Each new individual is genetically identical to its parent unless there has been a mutation. When organisms exchange genetic material during sexual reproduction, however, the offspring differ from their parents because chromosomes assort randomly during meiosis, crossing over occurs, and fertilization brings together material from two different cells (see Chapter 9).

Sexual reproduction generates an endless variety of genotypic combinations that increases the evolutionary potential of populations. Because it increases the variation among the offspring produced by an individual, sexual reproduction may improve the chance that at least some of the offspring will be successful in the varying and often unpredictable environments they will encounter. Sexual reproduction does not influence the frequencies of alleles; rather, it generates new combinations of alleles on which natural selection can act. It expands variation in a trait influenced by alleles at many loci by creating new genotypes. That is why selection for bristle number in *Drosophila* (see Figure 21.5) resulted in flies with more bristles than any flies in the initial population had.

Neutral genetic mutations accumulate within species

As we saw in Chapter 12, some mutations do not affect the functioning of the proteins the mutated genes encode. An allele that does not affect the fitness of an organism is called a **neutral allele**. Such alleles, untouched by natural selection, may be lost, or their frequencies may increase with time. Therefore, neutral alleles tend to accumulate in a population over time, providing it with considerable genetic variation.

Much of the variation in those traits we can observe with our unaided senses is not neutral. However, much variation at the molecular level apparently is neutral. Modern molecular techniques enable us to measure variation in neutral traits and provide a means by which we can distinguish adaptive from neutral variation. Chapter 24 will discuss how these techniques enable us to make such discriminations, and how variation in neutral traits can be used to estimate rates of evolution.

Much genetic variation is maintained in geographically distinct subpopulations

Much of the genetic variation in large populations is preserved as differences among subpopulations. Subpopulations often vary genetically because they are subjected to different selective pressures in different environments. Plant subpopulations, for example, may vary geographi-

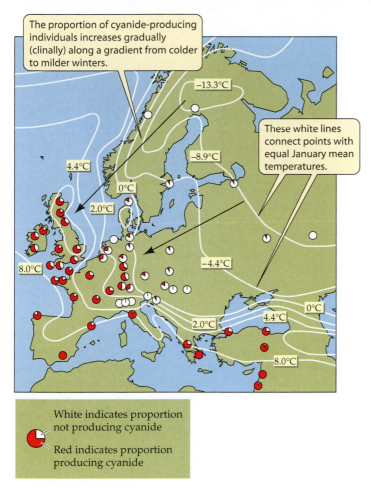

The proportion of cyanide-producing individuals increases gradually (clinally) along a gradient from colder to milder winters.

These white lines connect points with equal January mean temperatures.

-13.3°C
-8.9°C
4.4°C
0°C
2.0°C
8.0°C
-4.4°C
0°C
4.4°C
2.0°C
8.0°C

White indicates proportion not producing cyanide

Red indicates proportion producing cyanide

21.17 Geographic Variation in Poisonous Clovers
The frequency of cyanide-producing individuals in each population of white clover (*Trifolium repens*) is represented by the proportion of the circle that is red.

that of other genotypes (or phenotypes). This process is known as **frequency-dependent selection**.

A small fish that lives in Lake Tanganyika in east central Africa provides an example of frequency-dependent selection. The mouth of this scale-eating fish, *Perissodus microlepis*, opens either to the right or to the left as a result of an asymmetrical jaw joint (Figure 21.18). *Perissodus* approaches its prey (another fish) from behind and dashes in to bite off several scales from its flank. "Right-mouthed" individuals always attack from the victim's left; "left-mouthed" individuals always attack from the victim's right. The distorted mouth enlarges the area of teeth in contact with the prey's flank, but only if the scale eater attacks from the appropriate side.

Prey fish are alert to approaching scale eaters, so attacks are more likely to be successful if the prey must watch both flanks. Guarding by the prey favors equal numbers of right-mouthed and left-mouthed scale eaters, because if one form were more common than the other, prey fish would pay more attention to potential attacks from the corresponding flank. Therefore, success of individuals of the more common morph would be less than that of the less common morph. Over an 11-year period, the polymorphism was found to be stable: the two forms of *Perissodus* remained at about equal frequencies.

cally in the chemicals they synthesize to defend themselves against herbivores. Some individuals of the clover *Trifolium repens* produce the poisonous chemical cyanide. Poisonous individuals are less appealing to herbivores—particularly mice and slugs—than are nonpoisonous individuals. However, clover plants with cyanide are more likely to be killed by frost, because freezing damages cell membranes and releases the toxic cyanide into the plant's own tissues.

In populations of *Trifolium repens*, the frequency of cyanide-producing individuals increases gradually from north to south and from east to west across Europe (Figure 21.17). Poisonous individuals make up a large proportion of clover populations only in areas where the winters are mild. Cyanide-producing individuals are rare where winters are cold, even though herbivores graze them heavily in those areas.

Frequency-dependent selection maintains genetic variation within populations

Natural selection often preserves variation as **polymorphisms**—genetic differences within a population. A polymorphism may be maintained when the fitness of a genotype (or phenotype) varies with its frequency relative to

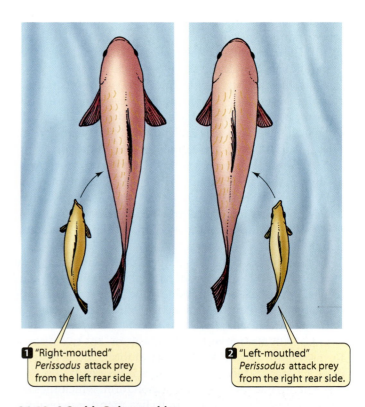

1 "Right-mouthed" *Perissodus* attack prey from the left rear side.

2 "Left-mouthed" *Perissodus* attack prey from the right rear side.

21.18 A Stable Polymorphism
Frequency-dependent selection maintains equal proportions of left-mouthed and right-mouthed individuals of the scale-eating fish *Perissodus microlepis*.

Leaves of a white oak (*Quercus alba*)

Grown in sun   Grown in shade

21.19 Environmentally Induced Variation
Traits may vary among genetically identical individuals or parts of individuals if they are exposed to different environments.

21.20 One Genotype: Two Seasonal Color Forms
The dry-season (left) and wet-season (right) form of the butterfly *Bicyclus anynana* have the same genotype. The environmental conditions experienced by a larva determine the form of the butterfly into which it develops.

How Do Genotypes Determine Phenotypes?

Genotypes do not uniquely determine phenotypes. If one allele is dominant to another, a particular phenotype can be produced by more than one genotype (for example, *AA* and *Aa* individuals may be phenotypically identical).

Similarly, different phenotypes can be produced by a given genotype, depending on the environment encountered during development. For example, the cells of the leaves on a tree or shrub are normally genetically identical. Yet leaves on the same tree often differ in shape and size. Leaves close to the top of an oak tree, where they are exposed to more wind and sunlight, may be more deeply lobed than leaves lower down on the same tree (Figure 21.19). The same differences can be seen between the leaves of individuals growing in sunny and shady sites.

Thus, the phenotype of an organism is the outcome of a complex series of developmental processes that are influenced by both environmental factors and genes. This nearly universal phenomenon is called **phenotypic plasticity**. Although variations in leaf shapes are not passed on to offspring, the ability to produce varied leaf shapes in response to environmental conditions *is* inherited. Phenotypic plasticity of leaf shapes benefits a tree because deeply lobed leaves offer less resistance to wind, absorb less sunlight, lose heat more rapidly by convection, and allow more sunlight to pass to lower leaves.

Because phenotypic plasticity is often adaptive, it may evolve under the influence of natural selection. A particularly thorough demonstration of the adaptive nature of phenotypic plasticity is provided by studies of the tropical African butterfly *Bicyclus anynana*. *B. anynana*, which lives in areas with distinct wet and dry seasons, has two distinct forms (Figure 21.20). The dry-season form, which rests on dried grasses and leaf litter and flies infrequently, has only one small wing spot. The wet-season form, which flies actively in lush, green vegetation, has many conspicuous spots on its wings.

The form of an adult butterfly is determined by the environmental conditions it encounters as a larva. Investigators compared the survival rates of the two forms in both seasons. The dry-season form, which closely matches the brown vegetation on which it rests, survives better during the dry season than does the wet-season form. On the other hand, the more active, conspicuous wet-season form does better during the wet season because its conspicuous spots, which resemble large eyes, deter some predators.

Constraints on Evolution

Thus far we have implicitly assumed that sufficient genetic variation always exists for the evolution of favored traits. A moment's reflection reveals that this assumption cannot be true. As we pointed out in the previous chapter, major evolutionary innovations are rare. Most changes are based on modifications of previously existing traits, even though those traits may come to serve new functions. In addition, evolutionary theory does not allow a population to temporarily become less well adapted. All intermediate forms must work; that is, all modifications must benefit their bearers in every generation.

A striking example that illustrates how natural selection operates by modifying existing states is provided by the evolution of fishes that spend most of their time resting on the sea bottom. One lineage, the bottom-dwelling skates and rays, is beautifully symmetrical (Figure 21.21*a*). These fishes are descended from sharks, which were already somewhat flattened and, therefore, able to lie on their bellies.

Plaice, sole, and flounders, on the other hand, are descendants of deep-bodied ancestors. Unlike sharks, these fishes cannot lie on their bellies; they must flop over on their sides. During development, the eyes of plaice and sole are grotesquely twisted around to bring both eyes to one side of the body (Figure 21.21*b*). No clever designer who was free of constraints would have designed plaice and sole as they are. But small shifts in the position of one eye probably helped ancestral flatfishes see better, resulting in the form found today.

Although constraints on evolution clearly exist, it is difficult to determine whether the absence of certain traits that

(a) Taeniura lymma

(b) Bothus lunatus

21.21 Two Solutions to a Single Problem
(*a*) Stingrays, whose ancestors were dorsally flattened, lie on their bellies. (*b*) Flounders, whose ancestors were laterally flattened, lie on their sides.

would seem to be desirable is due to some constraint or to our having wrongly guessed that the trait would be adaptive. A plausible answer to this question is now available for a puzzling pattern among amphibians. Many salamanders are *neotenic*; that is, individuals become sexually mature while still in their aquatic larval form. Why, then, are there no frogs and toads that reproduce when they are tadpoles? A universal constraint preventing frogs and toads from evolving neoteny may be their need for relatively high levels of thyroid hormones for sex differentiation and reproduction. Neotenic salamanders result only when levels of thyroid hormones are very low—too low to allow frogs and toads to become sexually mature. In many other cases, however, no plausible answer is yet available.

Short-Term versus Long-Term Evolution

Microevolutionary changes within populations are an important focus of study for evolutionary biologists. These changes can be observed directly, they can be manipulated experimentally, and they show the actual processes by which evolution occurs. Studies of these short-term changes identify the genetic bases of evolutionary changes

and demonstrate how natural selection acts. By themselves, however, they do not enable us to predict—or, more properly, "postdict" (because they have already happened)—the macroevolutionary changes we described in Chapter 20.

The reason for this is that patterns of macroevolutionary change can be strongly influenced by events that occur so infrequently or so slowly that they are unlikely to be observed during microevolutionary studies. In addition, the ways in which evolutionary agents act may change with time; even among the descendants of a single ancestral species, different lineages may evolve in different directions. Therefore, additional types of evidence, such as the occurrence of rare and unusual events and trends in the fossil record, must be gathered in order to understand the course of evolution over billions of years.

"Postdiction" problems in science are not unique to evolutionary studies. For example, volcanologists believe they understand the physical theory that explains why Mount St. Helens erupted in 1980, but they lack the detailed information necessary for them to "explain" why the mountain erupted on the exact day it did. Similarly, even though seismologists know the physical principles that govern earthquakes, they cannot predict exactly when or where an earthquake will happen.

In subsequent chapters we will discuss the kinds of information that biologists assemble to study long-term evolutionary changes and infer the processes that led to them.

Chapter Summary

Charles Darwin and Adaptation
▶ Darwin developed his theory of evolution by natural selection by carefully observing nature, especially during his voyage around the world on the *Beagle*.
▶ Darwin based his theory on well-known facts and some key inferences.
▶ Darwin had no examples of the action of natural selection, so he based his arguments on artificial selection by plant and animal breeders.
▶ Modern genetics has elucidated the mechanisms of heredity, which were unknown to Darwin but which have provided the solid base that supports and substantiates his theory.

Genetic Variation within Populations
▶ A single individual has only some of the alleles found in the population of which it is a member. **Review Figure 21.3**
▶ Genetic variation characterizes nearly all natural populations. **Review Figures 21.4, 21.5**
▶ Allele frequencies measure the amount of genetic variation in a population. Genotype frequencies show how a population's genetic variation is distributed among its members.
▶ Biologists estimate allele frequencies by measuring a sample of individuals from a population. The sum of all allele frequencies at a locus is equal to 1. **Review Figure 21.6**
▶ Populations that have the same allele frequencies may nonetheless have different genotype frequencies.

The Hardy–Weinberg Equilibrium
▶ A population that is not changing genetically is said to be at Hardy–Weinberg equilibrium.

▶ The assumptions that underlie the Hardy–Weinberg equilibrium are that the population is large, mating is random, there is no migration, mutation can be ignored, and natural selection is not acting on the population.

▶ In a population at Hardy–Weinberg equilibrium, allele frequencies remain the same from generation to generation. In addition, genotype frequencies remain in the proportions $p^2 + 2pq + q^2 = 1$. **Review Figure 21.7**

▶ Biologists can determine whether an agent of evolution is acting on a population by comparing the genotype frequencies of that population with Hardy–Weinberg equilibrium frequencies.

Microevolution: Changes in the Genetic Structure of Populations

▶ Changes in allele frequencies and genotype frequencies within populations are caused by the actions of several evolutionary agents: mutation, gene flow, random genetic drift, assortative mating, and natural selection.

▶ The origin of genetic variation is mutation. Most mutations are harmful or neutral to their bearers, but some are advantageous, particularly if the environment changes.

▶ Migration of individuals from one population to another, followed by breeding in the new location, produces gene flow. Immigrants may add new alleles to a population or may change the frequencies of alleles already present.

▶ Random genetic drift alters allele frequencies in all populations, but it overrides natural selection only in small populations. Organisms that normally have large populations may pass through occasional periods (population bottlenecks) when only a small number of individuals survive. **Review Figure 21.8**

▶ New populations established by a few founding individuals also have gene frequencies that differ from those in the parent population. **Review Figure 21.10**

▶ If individuals mate more often with individuals that have the same or different genotypes than would be expected on a random basis—that is, when mating is not random—frequencies of homozygous and heterozygous genotypes differ from Hardy–Weinberg expectations. **Review Figure 21.11**

▶ Self-fertilization, an extreme form of nonrandom mating, reduces the frequencies of heterozygous individuals below Hardy–Weinberg expectations without changing allele frequencies.

▶ Natural selection is the only agent of evolution that adapts populations to their environments. Natural selection may preserve allele frequencies or cause them to change with time.

▶ Stabilizing selection, directional selection, and disruptive selection change the distributions of phenotypes governed by more than one locus. **Review Figures 21.12, 21.13, 21.14**

Studying Microevolution

▶ Biologists study microevolution by measuring natural selection in the field, experimentally altering organisms, and building computer models. **Review Figures 21.15, 21.16**

Maintaining Genetic Variation

▶ Random genetic drift, stabilizing selection, and directional selection all tend to reduce genetic variation, but most populations are genetically highly variable.

▶ Sexual reproduction generates an endless variety of genotypic combinations that increases the evolutionary potential of populations, but it does not influence the frequencies of alleles. Rather, it generates new combinations of genetic material on which natural selection can act.

▶ Much genetic variation within many species is maintained in distinct subpopulations. **Review Figure 21.17**

▶ Genetic variation within a population may be maintained by frequency-dependent selection. **Review Figure 21.18**

How Do Genotypes Determine Phenotypes?

▶ Genotypes do not uniquely determine phenotypes. A given phenotype can be produced by more than one genotype.

▶ The phenotype of an organism is the result of a complex series of developmental processes that are influenced by both environmental factors and genes. **Review Figures 21.19, 21.20**

Constraints on Evolution

▶ Natural selection acts by modifying what already exists. A population cannot get temporarily worse in order to achieve some long-term advantage.

Short-Term versus Long-Term Evolution

▶ Patterns of macroevolutionary change can be strongly influenced by events that occur so infrequently or so slowly that they are unlikely to be observed during microevolutionary studies. Additional types of evidence must be gathered to understand why evolution in the long term took the particular course it did.

For Discussion

1. During the past 50 years, more than 200 species of insects that attack crop plants have become highly resistant to DDT and other pesticides. Using your recently acquired knowledge of evolutionary processes, explain the rapid and widespread evolution of resistance. Propose ways of using pesticides that would slow down the rate of evolution of resistance. Now that the use of DDT has been banned in the United States, what do you expect to happen to levels of resistance to DDT among insect populations? Justify your answer.

2. In what ways does artificial selection by humans differ from natural selection in nature? Was Darwin wise to base so much of his argument for natural selection on the results of artificial selection?

3. In nature, mating among individuals in a population is never truly random: Immigration and emigration are common, and natural selection is seldom totally absent. Why, then, does it make sense to use the Hardy–Weinberg model, which is based on assumptions known generally to be false? Can you think of other models in science that are based on false assumptions? How are such models used?

4. As far as we know, natural selection cannot adapt organisms to future events. Yet many organisms appear to respond to natural events before they happen. For example, many mammals go into hibernation while it is still quite warm. Similarly, many birds leave the temperate zone for their southern wintering grounds long before winter arrives. How can such "anticipatory" behaviors evolve?

5. Some people believe that species, like individual organisms, have life cycles. They believe that species are born by a process of speciation, grow and expand, and inevitably die out as a result of "species old age." Could any agents of evolution cause such a species life cycle? If not, how do you explain the high rates of extinction of species in nature?

22 Species and Their Formation

DURING THE 1940S, OFFICIALS IN TRINIDAD launched an intensive campaign to control malaria. Believing that malaria was being transmitted by *Anopheles albimanus*, a swamp-breeding mosquito that is the principal vector of malaria in Latin America, they spent a great deal of money spraying and draining marshes. The campaign failed, however, because the principal vector of malaria in Trinidad was *Anopheles bellator*, a mosquito species that breeds in water held within the leaves of bromeliad plants (relatives of pineapples) growing on tree branches.

Similarly, in Europe, people thought that malaria was transmitted only by mosquitoes of a single species: *Anopheles maculipennis*. European efforts to control malaria sometimes succeeded and sometimes failed, because *A. maculipennis* turned out to be not a single species, but a group of at least 18 species that can be distinguished only by examination of their chromosomes. Some of the species breed in fresh water, others in brackish water. Some enter houses, others do not. Furthermore, which mosquito species is the vector of malaria varies regionally. Control efforts are successful only when directed against the species that actually transmits malaria in that area.

Therefore, to control malaria, we need to know which species of mosquitoes are the vectors of the disease, as well as the details of their life cycles. But how did these many species of mosquitoes arise? What processes keep them cohesive and distinct?

All species, living and extinct, are believed to be descendants of a single ancestral species that lived more than 3 billion years ago. If speciation were a rare event, the biological world would be very different than it is today. Speciation is an essential ingredient of evolutionary diversification, and species are the fundamental units of the biological classification systems we will discuss in Chapter 23. But what are species? How did these millions of species form? How does one species become two? What factors stimulate such splitting? What conditions spur evolutionary radiations? These and related questions are the subject of this chapter.

Trinidad Rainforest
The mosquito that transmits malaria in Trinidad breeds in water held in the bases of leaves of bromeliad plants that grow on the trunks and branches of rainforest trees.

What Are Species?

The word **species** means, literally, "kinds." But what do we mean by "kinds"? Someone who is knowledgeable about a group of organisms, such as orchids or lizards, usually can distinguish the different species of that group found in a particular area simply by examining them superficially. The patterns of similarities and differences that unite groups of organisms and separate them from other groups are familiar to all of us. The standard field guides to birds, mammals, insects, and flowers are possible only because most species are cohesive units that change in appearance only gradually over large geographic distances. We can easily recognize red-winged blackbirds from New York and red-winged blackbirds from California as members of the same species (Figure 22.1).

But not all members of a species look that much alike. For example, males, females, and young individuals may not resemble one another. How do we decide whether similar but easily distinguished individuals should be assigned to different species or regarded as members of the same species? The concept that has guided these decisions for a long time is genetic integration. If individuals within a population mate with one another but not with individuals of

(a) *Agelaius phoeniceus*

(b) *Agelaius phoeniceus*

22.1 Redwings Are Redwings Everywhere
Both of these male birds are obviously red-winged blackbirds, even though (a) lives in the eastern United States and (b) lives in California. In parts of California, males have less yellow in their wings than males elsewhere in the broad range of the species.

other populations, they constitute a distinct group within which genes recombine; that is, they are independent evolutionary units. These independent evolutionary units are usually called species.

More than 200 years ago the Swedish biologist Carolus Linnaeus, who originated the system of naming organisms that we use today, described hundreds of species. Because he knew nothing about the mating patterns of the organisms he was naming, Linnaeus classified them on the basis of their appearances; in other words, he used a *morphological* concept of species. Many species that were classified by their appearances are actually independent evolutionary units. They look alike because they share many alleles that code for body structures. In many groups of organisms for which genetic data are unavailable, species are still recognized by their morphological traits.

A species definition that has been used by many biologists—the "biological" species definition—was proposed by Ernst Mayr in 1940. He stated, "Species are groups of actually or potentially interbreeding natural populations which are reproductively isolated from other such groups." The words "*actually or potentially*" assert that, even if some members of a species are not in the same place and hence are unable to mate, they should not be placed in separate species if they would be likely to mate if they were together. The word "natural" is an important part of the definition because only in nature does the exchange of genes affect evolutionary processes; the interbreeding of two different species in captivity does not. Gene exchange is the main reason why species are cohesive units.

Deciding whether two populations constitute different species can be difficult because speciation is often a gradual process (Figure 22.2). If a barrier divides one population into two populations, the daughter populations may evolve independently long before they become reproductively incompatible—or they may become reproductively incompatible before they evolve any noticeable morphological differences.

 ## How Do New Species Arise?

Speciation is the process by which one species splits into two species, which thereafter evolve as distinct lineages. Not all evolutionary changes result in new species. A single lineage may change through time without giving rise to a new species. Although Charles Darwin entitled his book *The Origin of Species*, he did not discuss how a single species splits into two or more daughter species. Rather, he was concerned principally with demonstrating that species are altered by natural selection over time.

The critical process in the formation of new species is the separation of the gene pool of the ancestral species into two separate gene pools. Subsequently, within each isolated gene pool, allele and gene frequencies may change as a result of the action of evolutionary agents. If sufficient differences accumulate during this period of isolation, the two populations may not exchange genes if they come together again.

Gene flow among populations may be interrupted in several ways, each of which characterizes a mode of speciation. The next three sections focus on these modes of speciation: allopatric speciation, sympatric speciation, and parapatric speciation.

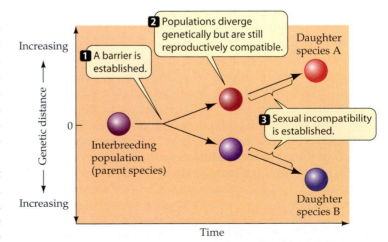

22.2 Speciation May Be a Gradual Process
In this hypothetical example, genetic divergence begins before reproductive incompatibility evolves.

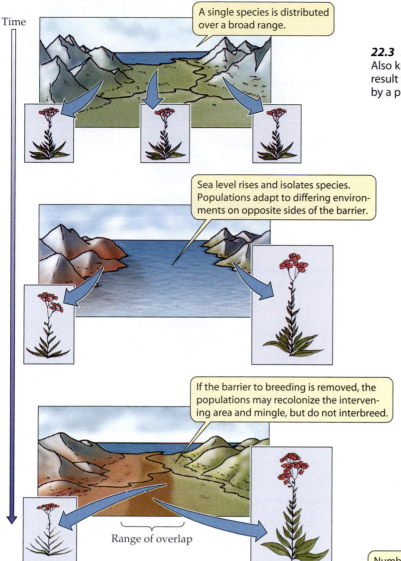

A single species is distributed over a broad range.

Sea level rises and isolates species. Populations adapt to differing environments on opposite sides of the barrier.

If the barrier to breeding is removed, the populations may recolonize the intervening area and mingle, but do not interbreed.

Range of overlap

Time

22.3 Allopatric Speciation
Also known as geographic speciation, allopatric speciation may result when a population is divided into two separate populations by a physical barrier such as rising seas.

sult of new populations founded by individuals dispersing among the islands, because the closest relative of a species on one island is often a species on a neighboring island, rather than a species on the same island. Biologists who have studied the chromosomes of picture-winged *Drosophila* believe that speciation among these flies has resulted from at least 45 such **founder events** (Figure 22.4).

The finches of the Galápagos archipelago, 1,000 km off the coast of Ecuador, demonstrate the importance of geographic isolation for speciation. Darwin's finches (as they are usually called, because Darwin was the first scientist to study them) arose on the Galápagos by speciation from a single South American species that colonized the islands. Today there are 14 species of Galápagos finches, all of which differ strikingly from the blue-black grassquit, their probable mainland ancestor (Figure 22.5).

The islands of the Galápagos archipelago are sufficiently isolated from one another that finches

Allopatric speciation requires total genetic isolation

Speciation that results when a population is divided by a geographic barrier is known as **allopatric speciation** (*allo-*, "different"; *patris*, "country"), or **geographic speciation** (Figure 22.3). Allopatric speciation is thought to be the dominant form of speciation among most groups of organisms. The range of a species may be divided by a barrier such as a water gap for terrestrial organisms, dry land for aquatic organisms, or a mountain range. Barriers can form when continents drift, sea levels rise, or climates change. Populations separated in this way are often large initially. They evolve differences because the places in which they live are, or become, different.

Alternatively, allopatric speciation may result when some members of a population cross an existing barrier and found a new population. Populations established in this way usually differ genetically from their parent populations because a small group of founding individuals has only an incomplete representation of the genes found in its parent population (see Chapter 21). Many of the hundreds of species of the fruit fly *Drosophila* in the Hawaiian Islands are restricted to a single island. They are almost certainly the re-

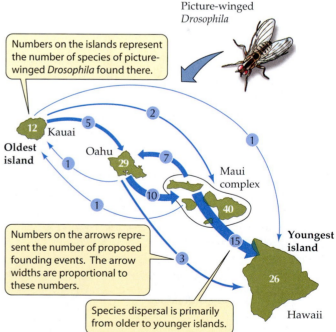

Picture-winged *Drosophila*

Numbers on the islands represent the number of species of picture-winged *Drosophila* found there.

Numbers on the arrows represent the number of proposed founding events. The arrow widths are proportional to these numbers.

Species dispersal is primarily from older to younger islands.

12 Kauai
Oldest island
Oahu
5
2
29
1
7
10
1
40
15
3
1
Youngest island
26
Hawaii
Maui complex

22.4 Founder Events Lead to Allopatric Speciation
The extremely high level of speciation found among picture-winged *Drosophila* in the Hawaiian Islands is almost certainly the result of founder events—new populations founded by individuals dispersing among the islands. The islands, which were formed in sequence as Earth's crust moved over a volcanic "hot spot," vary in age.

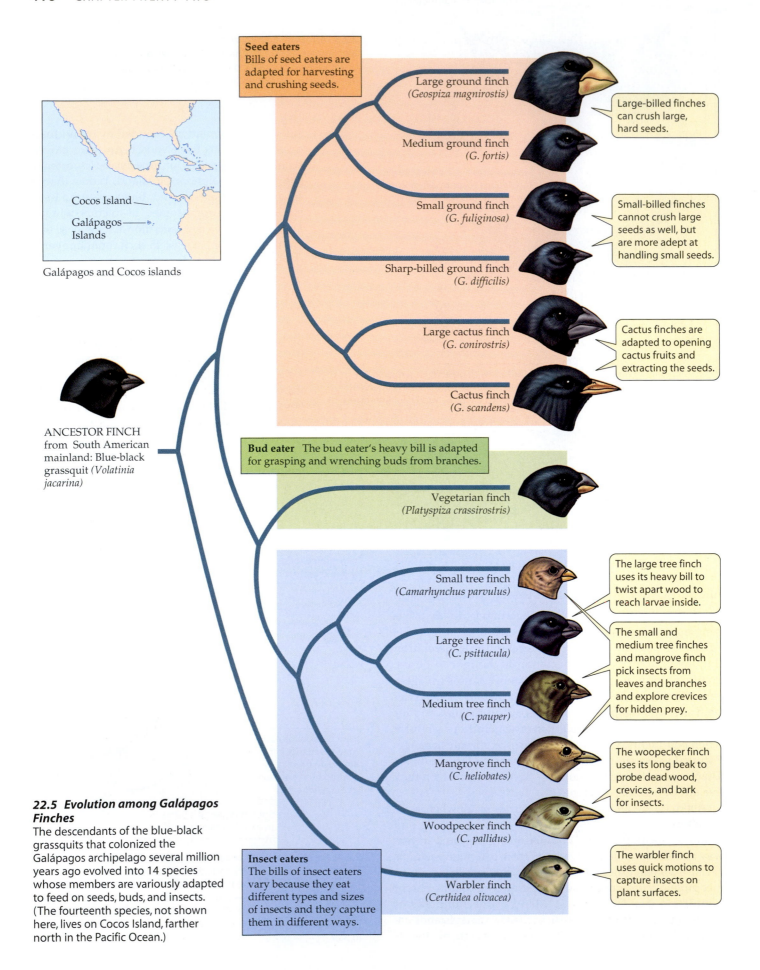

Galápagos and Cocos islands

Seed eaters
Bills of seed eaters are adapted for harvesting and crushing seeds.

Large ground finch
(*Geospiza magnirostis*)

Medium ground finch
(*G. fortis*)

Small ground finch
(*G. fuliginosa*)

Sharp-billed ground finch
(*G. difficilis*)

Large cactus finch
(*G. conirostris*)

Cactus finch
(*G. scandens*)

Large-billed finches can crush large, hard seeds.

Small-billed finches cannot crush large seeds as well, but are more adept at handling small seeds.

Cactus finches are adapted to opening cactus fruits and extracting the seeds.

ANCESTOR FINCH from South American mainland: Blue-black grassquit (*Volatinia jacarina*)

Bud eater The bud eater's heavy bill is adapted for grasping and wrenching buds from branches.

Vegetarian finch
(*Platyspiza crassirostris*)

Small tree finch
(*Camarhynchus parvulus*)

Large tree finch
(*C. psittacula*)

Medium tree finch
(*C. pauper*)

Mangrove finch
(*C. heliobates*)

Woodpecker finch
(*C. pallidus*)

Warbler finch
(*Certhidea olivacea*)

The large tree finch uses its heavy bill to twist apart wood to reach larvae inside.

The small and medium tree finches and mangrove finch pick insects from leaves and branches and explore crevices for hidden prey.

The woopecker finch uses its long beak to probe dead wood, crevices, and bark for insects.

The warbler finch uses quick motions to capture insects on plant surfaces.

Insect eaters
The bills of insect eaters vary because they eat different types and sizes of insects and they capture them in different ways.

22.5 Evolution among Galápagos Finches
The descendants of the blue-black grassquits that colonized the Galápagos archipelago several million years ago evolved into 14 species whose members are variously adapted to feed on seeds, buds, and insects. (The fourteenth species, not shown here, lives on Cocos Island, farther north in the Pacific Ocean.)

seldom migrate between them. Also, environmental conditions differ among the islands. Some are relatively flat and arid; others have forested mountain slopes. Populations of finches on different islands have differentiated enough that when occasional immigrants arrive from other islands, they either do not breed with the residents, or if they do, the resulting offspring do not survive as well as those produced by pairs of island residents. The genetic distinctness and cohesiveness of different populations is thus maintained.

A barrier's effectiveness at preventing gene flow depends on the size and mobility of the species in question. What is an impenetrable barrier to a terrestrial snail may be no barrier at all to a butterfly or a bird. Populations of wind-pollinated plants are isolated at the maximum distance their pollen is blown by the wind, but individual plants are effectively isolated at much shorter distances. Among animal-pollinated plants, the width of the barrier is the distance that animals travel while carrying pollen or seeds. Even animals with great powers of dispersal are often reluctant to cross narrow strips of unsuitable habitat. For animals that cannot swim or fly, narrow water-filled gaps may be effective barriers.

Indirect evidence that most speciation among animals is allopatric is provided by patterns of species distributions. For example, 36 percent of Earth's 20,000 species of bony fishes live in fresh water, even though only 1 percent of Earth's surface is fresh water, and even though fish productivity and populations are higher in some marine environments than in most fresh waters. Because they are highly fragmented, fresh waters have provided abundant opportunities for fishes to form geographically isolated populations. Marine environments provide fewer such opportunities.

Sympatric speciation occurs without physical separation

The subdividing of a gene pool when members of the daughter species are not geographically separated is called **sympatric speciation** (*sym-*, "with"). The most common means of sympatric speciation is **polyploidy**, an increase in the number of chromosomes.

Polyploidy arises in two ways. One way is the accidental production during cell division of cells having four (tetraploid) instead of two (diploid) sets of chromosomes. This process produces an **autopolyploid** individual, one having more than two sets of chromosomes, all derived from a single species. This tetraploid individual cannot produce viable offspring by mating with diploids, but it can do so if it self-fertilizes or mates with other tetraploids.

A polyploid species can also be produced when individuals of two different species interbreed. The resulting offspring are usually sterile, because the chromosomes from one species do not pair properly with those from the other species during meiosis, but they may be able to reproduce asexually. After many generations, some of these individuals may become fertile as a result of further chromosome duplication. Species produced in this way are called **allopolyploids**.

Polyploidy can create new species among plants much more easily than among animals because plants of many species can reproduce by self-fertilization. If polyploidy arises in several offspring of a single parent, the siblings can fertilize one another. Speciation by polyploidy has been very important in the evolution of flowering plants. Botanists estimate that about 70 percent of flowering plant species and 95 percent of all fern species are polyploids. Most of these arose as a result of hybridization between two species, followed by self-fertilization.

The speed with which allopolyploidy can produce new species is illustrated by salsifies (*Tragopogon*), members of the sunflower family. Salsifies are weedy plants that thrive in disturbed areas around towns. People have inadvertently spread them around the world from their ancestral ranges in Eurasia. Three diploid species of salsify were introduced into North America early in the twentieth century: *T. porrifolius*, *T. pratensis*, and *T. dubius*. Two tetraploid hybrids—*T. mirus* and *T. miscellus*—between species of the original three were first reported in 1950. Both hybrids have spread since their discovery and today are more widespread than their diploid parents (Figure 22.6).

Studies of their genetic material have shown that both hybrids have been formed more than once. Some populations of *T. miscellus*—a hybrid of *T. pratensis* and *T. dubius*—have the chloroplast genome of *T. pratensis*, whereas other populations have the chloroplast genome of *T. dubius*. Such differences among local populations of *T. miscellus* show that this allopolyploid has evolved independently at least 21 times; *T. mirus* has formed 12 times! Scientists seldom know the dates and locations of species formation so well.

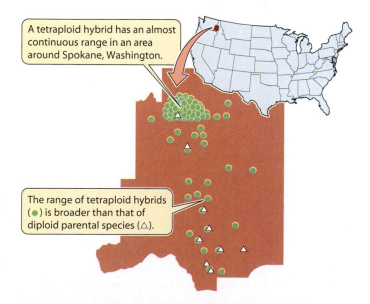

A tetraploid hybrid has an almost continuous range in an area around Spokane, Washington.

The range of tetraploid hybrids (●) is broader than that of diploid parental species (△).

22.6 Polyploids Can Outperform Their Parents
Tragopogon species (salsifies) are members of the sunflower family. The map shows the distribution of the diploid parent species and the tetraploid hybrid species of *Tragopogon* in eastern Washington and adjacent Idaho.

The success of newly formed hybrid species of salsifies illustrates why so many species of flowering plants originated as polyploids.

Among animals, sympatric speciation apparently is rare, but may result from precise selection of habitat and mating sites by individuals. A good example is speciation in a picture-winged fruit fly (*Rhagoletis pomonella*) in New York State. Until the mid-1800s, these fruit flies courted, mated, and deposited their eggs only on hawthorn fruits. The larvae learned the odor of hawthorn as they fed on the fruits, and when they emerged from their pupae, they used this food-based memory to locate other hawthorn plants on which to mate and lay eggs.

About 150 years ago, large commercial apple orchards were planted in the Hudson River Valley. Apple trees are closely related to hawthorns, and a few female *Rhagoletis* laid their eggs on apples, perhaps by mistake. Their larvae did not grow as well as larvae on hawthorn berries, but many did survive. These larvae had learned the odor of apples, so when they emerged as adults they sought out apple trees, where they mated with other flies reared on apples.

Today there are two sympatric species of *Rhagoletis* in the Hudson River Valley. One feeds on hawthorn fruits, the other on apples. The two species are reproductively isolated because they mate only with individuals raised on the same fruit, and because they emerge from their pupae at different times. In addition, apple-feeding flies have evolved so that they now grow more rapidly on apples than they originally did.

Parapatric speciation separates adjacent populations

Sometimes reproductive isolation develops between adjacent populations in the absence of a geographic barrier. This type of speciation, known as **parapatric speciation** (*para-*, "beside"), is in effect allopatric speciation in which the boundary separating species is not a physical barrier, but a difference in conditions. For parapatric speciation to happen, natural selection must be much stronger than gene flow; otherwise, gene flow would prevent differentiation between the two populations. Thus, any factor that reduces gene flow or increases the gradient in selective pressures across small distances can generate conditions favorable for parapatric speciation.

Both kinds of factors are provided by the abrupt changes in soil composition that are created by mining activities that leave rubble (tailings) with high concentrations of heavy metals, such as lead and zinc. Soils that develop on these tailings contain concentrations of heavy metals that are detrimental to the growth of most plants. There is strong selection for heavy metal tolerance in plants growing on the tailings, and within the last several centuries plants able to grow on such soils have evolved in several species of grasses. *Anthoxanthum odoratum* is one of these species. Nearly complete reproductive isolation exists between *A. odoratum* plants growing on tailings and those growing on normal soil because they flower at different times. In addition, metal-tolerant plants self-pollinate more frequently than plants growing on normal soil, further reducing gene flow. Reproductive isolation between metal-tolerant and metal-intolerant plants is almost complete, demonstrating that gene flow can slow or stop even in the absence of a distinct physical barrier.

It is difficult to determine the importance of parapatric speciation in nature because species ranges change over time. Thus, species with adjacent ranges could have arisen parapatrically where their ranges now come into contact, or they could have arisen in geographic isolation and subsequently expanded their ranges. For this reason, parapatric speciation could be more common than it is generally believed to be.

Reproductive Isolating Mechanisms

Once a barrier to gene flow is established, by whatever means, the resulting daughter populations may diverge genetically because of the action of evolutionary agents. Over many generations, differences that reduce the probability of members of the two populations mating and producing viable offspring may accumulate. In this way, reproductive isolation can evolve as an incidental by-product of other genetic changes in allopatric populations. For example, individuals in the two daughter populations may become so different that they are not recognized as suitable mates.

However, geographic isolation does not necessarily lead to reproductive incompatibility. For example, American and European sycamores have been physically isolated from one another for at least 20 million years. Nevertheless, they are morphologically very similar (Figure 22.7), and they can form fertile hybrids. They lack traits that would prevent individuals of the two different populations from producing fertile hybrids. In this section we examine the

(*a*) *Plantanus occidentalis* (American sycamore)

(*b*) *Platanus hispanica* (European sycamore)

22.7 Geographically Separated, Morphologically Similar Although they have been separated on different continents for at least 20 million years, American and European sycamores have diverged very little in appearance.

ways in which such traits—**reproductive isolating mechanisms**—arise. Then we explore what happens when reproductive isolation is incomplete.

Prezygotic barriers operate before mating

Reproductive isolating mechanisms that operate before mating—**prezygotic reproductive barriers**—may prevent individuals of different species from interbreeding.

▶ **Spatial isolation**. Individuals of different species may select different places in the environment in which to live. As a result, they may never come into contact during their respective mating seasons; that is, they are reproductively isolated by location.

▶ **Temporal isolation**. Many organisms have mating periods that are as short as a few hours or days. If the mating periods of two species do not overlap, they will be reproductively isolated by time.

▶ **Mechanical isolation**. Differences in the sizes and shapes of reproductive organs may prevent the union of gametes from different species.

▶ **Gametic isolation**. Sperm of one species may not be attracted to the eggs of another species because the eggs do not release the appropriate attractive chemicals, or the sperm may be unable to penetrate the egg because it is chemically incompatible.

Postzygotic barriers operate after mating

If individuals of two different species still recognize one another and mate, **postzygotic reproductive barriers** may prevent gene exchange. Accumulated genetic differences are likely to reduce the fitness of offspring produced by matings between individuals from the two species.

▶ **Hybrid zygote abnormality**. Hybrid zygotes may fail to mature normally, either dying during development or developing such severe abnormalities that they cannot mate.

▶ **Hybrid infertility**. Hybrids may mature normally, but be infertile when they attempt to reproduce. For example, the offspring of matings between horses and donkeys—mules—are vigorous, but sterile; they produce no descendants (Figure 22.8).

▶ **Low hybrid viability**. Hybrid offspring may survive less well than offspring resulting from matings within each species.

▶ **Absence or sterility of one sex**. In nearly all cases of hybrid sterility and hybrid inviability, it is the sex that is heterozygous for the sex chromosomes (XY, XO, or ZW; see Chapter 10) that is absent or sterile. The reason is that any deleterious recessive alleles on a sex chromosome are fully expressed in hybrids of the sex with only one copy.

If hybrid offspring survive or reproduce poorly, postzygotic barriers may be reinforced by the evolution of more effective prezygotic barriers. More effective prezygotic barriers should evolve if individuals engaging in hybrid matings leave fewer surviving offspring than individuals that mate only within their own species. Reinforcement of prezygotic barriers has been demonstrated in a few laboratory populations, but evidence for it in nature has been slow to accumulate.

22.8 Sturdy but Sterile
Mules are widely used as pack animals because of their stamina. For that purpose, their infertility is unimportant.

Sometimes reproductive isolation is incomplete

If contact is reestablished between two formerly geographically isolated populations before many genetic differences have accumulated, the two populations may interbreed freely with each other, and their hybrid offspring may be as successful as those resulting from matings within each population. If hybrids spread through both populations and reproduce with other individuals, the gene pools combine quickly, and no new species results from the period of isolation. Alternatively, the two populations may interbreed only where they come into contact, resulting in a **hybrid zone**.

Detailed studies are being carried out in a narrow zone in Washington where Townsend's warblers and hermit warblers hybridize (Figure 22.9). Both of these warblers breed in tall conifer forests, and no habitat boundaries exist at the locations of the hybrid zones. The zone is narrow because of natural selection against hybrids. It is shifting southward because Townsend's warblers are replacing hermit warblers. Townsend's males are more aggressive than hermit males toward stuffed males of the other species placed in their territories, and they are better at attracting mates than hermit warbler males.

Species may differ in relatively few genes

If two species hybridize, we know that they are similar genetically. The absence of interbreeding, however, tells us nothing about how dissimilar two species are. Not until modern molecular techniques were developed could biologists measure genetic differences among species.

Molecular studies are now demonstrating that many sympatric species may be genetically very similar to one another. For example, different species of Hawaiian *Drosophila*

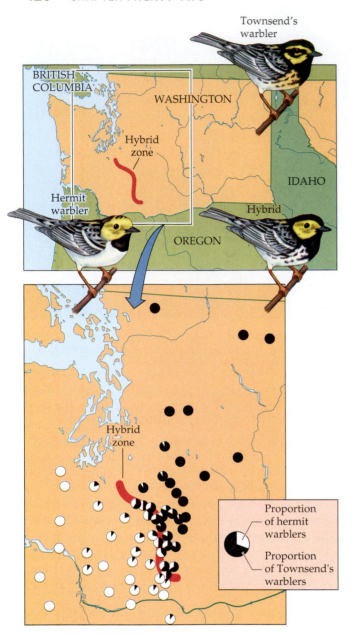

share nearly all of their mitochondrial DNA alleles. However, only a small fraction of the genes of these species have been analyzed, so genetic differences may be greater than we now think. All of the hundreds of species of *Drosophila* that have evolved in Hawaii during the past 32 million years, even those that have diverged morphologically, are relatively similar genetically (Figure 22.10).

Variation in Speciation Rates

Some lineages of organisms contain many species; others have only a few. Hundreds of species of *Drosophila* evolved in the Hawaiian Islands, but there is only one species of horseshoe crab, even though its lineage has survived more than 200 million years. Why do rates of speciation vary so widely among lineages? In the sections that follow we will examine several factors that influence speciation rates: species richness, range size, behavior, environmental changes, and generation times.

Species richness may favor speciation

The larger the number of species there are in a lineage, the larger the number of opportunities for new species to form. This is particularly true of speciation by polyploidy because more species are available to hybridize with one another. It is also partly true of allopatric speciation, because the larger the number of species living in an area, the larger the number of species whose ranges will be bisected by a given barrier.

Drosophila silvestris

Drosophila conspicua

Drosophila balioptera

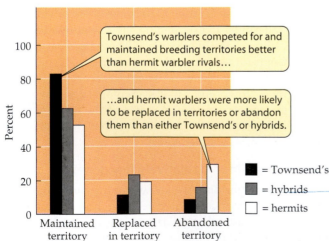

22.9 Hybrid Zones May Shift over Time
The zone where Townsend's and hermit warblers hybridize is shifting to the south because male Townsend's warblers dominate male hermit warblers.

22.10 Morphologically Different, Genetically Similar
Although these fruit flies—a small sample of the hundreds of species found only on the Hawaiian Islands—are extremely variable in appearance, they are genetically similar.

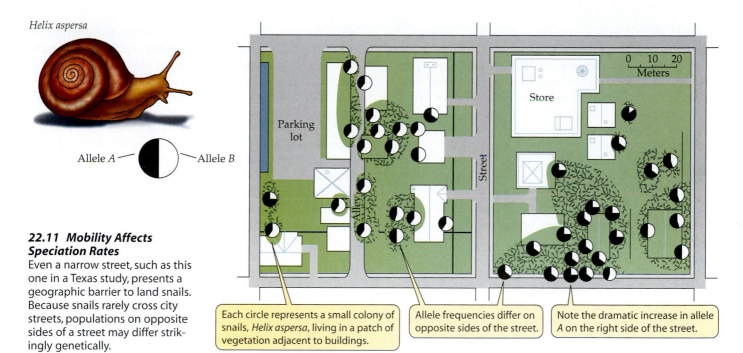

Helix aspersa

Allele *A* — Allele *B*

22.11 Mobility Affects Speciation Rates
Even a narrow street, such as this one in a Texas study, presents a geographic barrier to land snails. Because snails rarely cross city streets, populations on opposite sides of a street may differ strikingly genetically.

Each circle represents a small colony of snails, *Helix aspersa*, living in a patch of vegetation adjacent to buildings.

Allele frequencies differ on opposite sides of the street.

Note the dramatic increase in allele *A* on the right side of the street.

Range size may affect speciation rates

Relationships between range size and speciation rate are not simple, however, because the ranges of individual species tend to be small where there are many species. The larger the range of a species, the more likely a physical barrier is to subdivide it. Also, species with large ranges are more likely than species with small ranges to establish isolated peripheral populations that survive long enough to form new species.

Behavior may influence speciation rates

The mobility of a species may influence how often its range is likely to be divided by barriers. Individuals of species with poor dispersal abilities are unlikely to establish new populations by dispersing across barriers, and even narrow barriers effectively isolate species whose individuals are highly sedentary. Populations of land snails may be separated by barriers as narrow as city streets (Figure 22.11).

Animals with complex behavior are likely to speciate at a high rate because they make sophisticated discriminations among potential mating partners. They distinguish members of their own species from members of other species, and they make subtle discriminations among members of their own species on the basis of size, shape, appearance, and behavior. Such discrimination can greatly influence which individuals are most successful in producing offspring. Therefore, mate selection is probably a major cause of rapid evolution of reproductive isolation between species.

Environmental changes may trigger high speciation rates

African antelopes underwent a burst of speciation and extinction between 2.5 and 2.9 million years ago. During that period, the number of known antelope species doubled,

and 90 percent of all antelope species known to have existed at that time either first appeared or went extinct (Figure 22.12). This burst coincided with a shift in Africa from a warm and wet climate to one that oscillated between warm, wet and cooler, drier conditions. The burst of speciation among antelopes resulted in many more species adapted to grasslands and savannas, environments that increased and decreased, thereby coalescing and separating over much of Africa, as the climate oscillated.

Short generation times enhance speciation

We have been concentrating on factors that influence rates at which the ranges of species are subdivided by barriers.

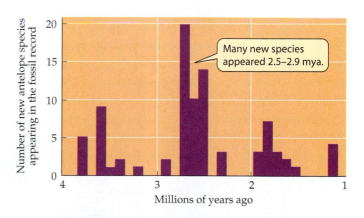

Many new species appeared 2.5–2.9 mya.

Number of new antelope species appearing in the fossil record

Millions of years ago

22.12 Climate Change Drove a Burst of Speciation among Antelopes
The excellent fossil record of African antelopes reveals that there was a sudden burst of speciation between 2.5 and 2.9 million years ago. At that time, the climate of Africa shifted from being consistently warm and wet to oscillating between warm and wet and cool and dry.

But the rate at which new species form also depends on how fast daughter populations diverge. The more rapidly they diverge, the sooner they are likely to evolve reproductive isolating mechanisms, and the less likely they are to hybridize if they again become sympatric. Shorter generation times result in more generations per unit of time and, as a result, generate the potential for more evolutionary changes per unit of time.

Evolutionary Radiations

As we learned in Chapter 20, the fossil record reveals that at certain times in some lineages, speciation rates have been much higher than extinction rates. The result is an *evolutionary radiation* that gives rise to a large number of daughter species. What conditions cause speciation rates to be much higher than extinction rates?

Evolutionary radiations are likely when a population colonizes an environment that has relatively few species. This condition typifies islands because many organisms disperse poorly across large water gaps. Because islands lack many plant and animal groups found on the mainland, ecological opportunities exist that may stimulate rapid evolutionary changes when a new species does reach them. Water barriers also restrict gene flow among islands in an archipelago, so populations on different islands can evolve adaptations to their local environments. Together these two factors make it likely that speciation rates on island archipelagos will exceed extinction rates.

Remarkable evolutionary radiations have occurred in the Hawaiian Islands, the most isolated islands in the world. The Hawaiian Islands lie 4,000 km from the nearest major land mass and 1,600 km from the nearest group of islands. The islands are arranged in a line of decreasing age—the youngest islands to the southeast, the oldest to the northwest. The archipelago is actually much older than the oldest existing islands because even older islands long ago eroded until they no longer rise above the sea surface.

The native biota of the Hawaiian Islands includes 1,000 species of flowering plants, 10,000 species of insects, 1,000 land snails, and more than 100 bird species. However, there were no amphibians, no terrestrial reptiles, and only one native mammal—a bat—until humans introduced additional species. The 10,000 known native species of insects on Hawaii are believed to have evolved from only about 400 immigrant species; only 7 immigrant species are believed to account for all the native Hawaiian land birds.

More than 90 percent of all plant species on the Hawaiian Islands are *endemic*—that is, they are found nowhere else. Several groups of flowering plants have more diverse forms and life histories on the islands and live in a wider variety of habitats than do their close relatives on the mainland. An outstanding example is the group of sunflowers called silverswords (the genera *Argyroxiphium*, *Dubautia*, and *Wilkesia*). Chloroplast DNA data show that these species share a relatively recent common ancestor, which is believed to be a species of tarweed from the Pacific coast of North America. Whereas all mainland tarweeds are small, upright, nonwoody plants (herbs), Hawaiian silversword species include prostrate and upright herbs, shrubs, trees, and vines (Figure 22.13). They occupy nearly all the habitats of the islands, from sea level to above timberline in the

22.13 Rapid Evolution among Hawaiian Plants

Three closely related genera of the sunflower family, two of which are illustrated here, are believed to have descended from a single ancestor, a tarweed (*Madia sativa*) that colonized Hawaii from the Pacific coast of North America. They appear more distantly related than they actually are.

Madia sativa (ancestral tarweed)

Argyoxiphium sandwichense

Dubautia menziesii

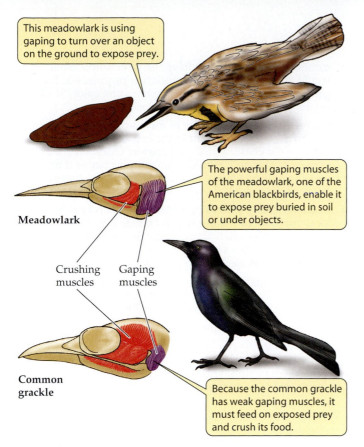

This meadowlark is using gaping to turn over an object on the ground to expose prey.

Meadowlark

The powerful gaping muscles of the meadowlark, one of the American blackbirds, enable it to expose prey buried in soil or under objects.

Crushing muscles   Gaping muscles

Common grackle

Because the common grackle has weak gaping muscles, it must feed on exposed prey and crush its food.

22.14 Blackbirds Expose Food by Gaping
The species of blackbirds that find food by gaping, here illustrated by a meadowlark, differ strikingly in the size and strength of their gaping muscles from blackbird species (such as grackles) that feed on exposed food.

mountains. Despite their extraordinary diversification, however, the silverswords have differentiated very little in their chloroplast genes.

The island silverswords are more diverse in size and shape than the mainland tarweeds because the original colonizers arrived on islands that had very few plant species. In particular, there were few trees and shrubs, because such large-seeded plants rarely disperse to oceanic islands. Many island trees and shrubs have evolved from nonwoody ancestors. On the mainland, however, tarweeds live in ecological communities that contain tree and shrub lineages older than their own—that is, where opportunities to exploit the tree way of life were already preempted.

Evolutionary lineages may also radiate when they acquire a new adaptation that enables them to use the environment in new and varied ways. For example, ancestors of the 95 species of American blackbirds evolved powerful muscles for opening their bills. These muscles enable the birds to obtain food by opening their bills forcibly against objects they wish to move, exposing otherwise hidden prey (Figure 22.14). This behavior is called *gaping*. Birds lacking these powerful muscles can find prey only on exposed surfaces of objects. Blackbirds gape into wood, fruits, leaf clusters, and stems of nonwoody plants; under sticks, stones, and animal droppings; and into the soil. With this feeding method, they have come to occupy nearly all habitat types

in North and South America, and they are among the most abundant birds throughout the region.

The Significance of Speciation

The result of speciation processes operating over billions of years is a world in which life is organized into millions of species, each adapted to live in a particular environment and to use environmental resources in a particular way. Earth would be very different if speciation had been a rare event in the history of life. How the millions of species are distributed over the surface of Earth and organized into ecological communities will be a major focus of Part Seven of this book, "Ecology and Biogeography." There we will also discuss how human activities are causing the extinction of species and what we can do to reduce the rate of species loss.

 ## Chapter Summary

What Are Species?

▶ Species are independent evolutionary units. A commonly accepted definition is that "species are groups of actually or potentially interbreeding natural populations which are reproductively isolated from other such groups."
▶ Because speciation is often a gradual process, it may be difficult to recognize boundaries between species. **Review Figure 22.2**

How Do New Species Arise?

▶ Not all evolutionary changes result in new species.
▶ Allopatric (geographic) speciation is the most important means of speciation among animals and is common in other groups of organisms. **Review Figures 22.3, 22.4, 22.5**
▶ Species may form sympatrically by a multiplication of chromosome numbers because the resulting polyploid organisms cannot interbreed with members of the parent species. Polyploidy has been a major factor in plant speciation, but is rare among animals. **Review Figure 22.6**
▶ Species may form parapatrically where marked environmental differences prevent gene flow among individuals living in adjacent environments.

Reproductive Isolating Mechanisms

▶ When previously allopatric species become sympatric, reproductive isolating mechanisms may prevent the exchange of genes.
▶ Barriers to gene exchange may operate before mating (prezygotic barriers) or after mating (postzygotic barriers).
▶ Hybrid zones may develop if barriers to gene exchange failed to develop during allopatry.
▶ Hybrids may form if separated populations come together again without sufficient genetic differences having accumulated. **Review Figure 22.9**
▶ The existence of hybrids tells us that the two hybridizing species are very similar genetically, but species that do not hybridize may also differ from one another very little genetically.

Variation in Speciation Rates

▶ Rates of speciation differ greatly among lineages of organisms. Speciation rates are influenced by the number of species in a lineage, their range sizes, their behavior, environmental changes, and generation times. **Review Figures 22.11, 22.12**

Evolutionary Radiations

▶ Evolutionary radiations happen when speciation rates exceed extinction rates.

▶ High speciation rates often coincide with low extinction rates when species invade islands that have few other species, or when a new way of exploiting the environment makes a different array of resources available to a species. **Review Figures 22.13, 22.14**

The Significance of Speciation

▶ As a result of speciation, Earth is populated with millions of species, each adapted to live in a particular environment and to use resources in a particular way.

For Discussion

1. Gene exchange between populations is prevented by geographical isolation, by behavioral responses before mating (for example, females may reject courting males of the other species), and by mechanisms that function after mating has occurred (for example, hybrid sterility). All of these are commonly called isolating mechanisms. In what ways are the three types very different? If you were to apply different names to them, which one would you call an isolating mechanism? Why? What names would you give the other types? Why?

2. The blue goose of North America has two distinct color forms, blue and white. Matings between the two color types are common. However, blue individuals pair with blue individuals and white individuals pair with white individuals much more frequently than would be expected by chance. Suppose that 75 percent of all mated pairs consisted of two individuals of the same color. What would you conclude about speciation processes in these geese? If 95 percent of pairs were the same color? If 100 percent of pairs were the same color?

3. Suppose pairs of blue geese of mixed colors were found only in a narrow zone within the broad Arctic breeding range of the geese, would you answer Question 2 the same way you did? Would your answer change if mixed-color pairs were widely distributed across the breeding range of the geese?

4. Although many species of butterflies are divided into local populations among which there is little gene flow, these butterflies often show relatively little geographic variation. Describe studies you would conduct to determine what maintains this morphological similarity?

5. Distinguish among allopatric, parapatric, and sympatric speciation. For each of the three statements below, indicate which type of speciation is implied:

   a. This process in nature is most commonly a result of polyploidy.

   b. The size of national parks and wildlife refuges may be too small to allow this type of speciation among organisms restricted to those areas.

   c. This process usually occurs in species that inhabit areas where sharp environmental contrasts exist.

6. Evolutionary radiations are common and easily studied on oceanic islands. In what types of mainland situations would you expect to find major evolutionary radiations? Why?

7. Fruit flies of the genus *Drosophila* are found worldwide, but most of the species in the genus are found on the Hawaiian Islands. Suggest a hypothesis that might account for this distribution pattern.

23 Reconstructing and Using Phylogenies

SCHISTOSOMIASIS IS A BLOOD INFECTION CAUSED BY a parasitic flatworm, *Schistosoma*. More than 200 million people in South America, Africa, China, Japan, and Southeast Asia have the disease. During part of its life cycle, *Schistosoma* inhabits a freshwater snail. People become infected when they come into contact with water where infested snails live. Larval *Schistosoma* swim from a snail and penetrate the skin. The worm matures and lives in the person's abdominal blood vessels. The disease is progressively debilitating, causing a slow death.

For most of the twentieth century, only one species, *Schistosoma japonicum*, was known to infect humans, and people believed that it was transmitted by a single species of snail in the genus *Oncomelania*. Then, in the 1970s, researchers discovered that a different snail was transmitting *Schistosoma* to humans in the Mekong River in Laos. This discovery stimulated extensive field surveys and anatomical, genetic, and geographic research on the worms and snails of Southeast Asia.

Investigators found that *S. japonicum* was actually a cluster of at least six species. They also discovered that evolutionary relationships among snails influenced which species could host *Schistosoma*. Evolutionary diversification from an ancestral stock of snails produced a group of species of modern snails. Of these, only three can host *Schistosoma*; ten have a genetic trait that allows them to resist invasion by the parasite.

This information is of great value in efforts to combat schistosomiasis. Few of the freshwater snail species in Southeast Asia have been described and named. By using information on evolutionary relationships among snails, scientists can quickly determine whether or not a newly discovered snail is likely to be a host for *Schistosoma*. Control efforts need to be directed toward only the snails that can transmit *Schistosoma* to humans, not all freshwater snails in the region.

How did investigators determine the evolutionary relationships among the snails that are hosts of *Schistosoma*? How could they determine the number of times that genes preventing snails from hosting *Schistosoma* arose? How is knowledge of evolutionary relationships used to help answer other biological questions? How are evolutionary relationships expressed in systems of classification that help guide further studies of organisms?

In this chapter, we discuss systematics, the science that provides answers to these questions. We describe the methods systematists use to infer evolutionary relationships among organisms. Then we illustrate how knowledge of evolutionary relationships is used to solve other biological problems. Finally, we show how evolutionary relationships are incorporated into classification systems.

How Are Phylogenetic Trees Reconstructed?

Ever since its origin nearly 4 billion years ago, life has evolved under the influence of the evolutionary agents we described in Chapter 21. The incredible richness of today's biological world has resulted from millions of speciation events, determined by the processes we discussed in Chapter 22. Biologists have developed methods to trace the history of these processes and make sense of their results.

A **phylogeny** is a history of descent of a group of organisms from their common ancestor. Our understanding of the processes of speciation tells us that lineages of organisms can be represented as branching "trees." These **phylogenetic trees** show the order in which lineages split. A particular tree may portray the evolution of all life, of major

Asian Snails Can Transmit Schistosomiasis
Workers in the rice paddies of tropical Asia are at extreme risk of contracting schistosomiasis (known in some parts of the world as bilharzia). The disease is transmitted to humans via freshwater snails that thrive in the standing water of the paddies.

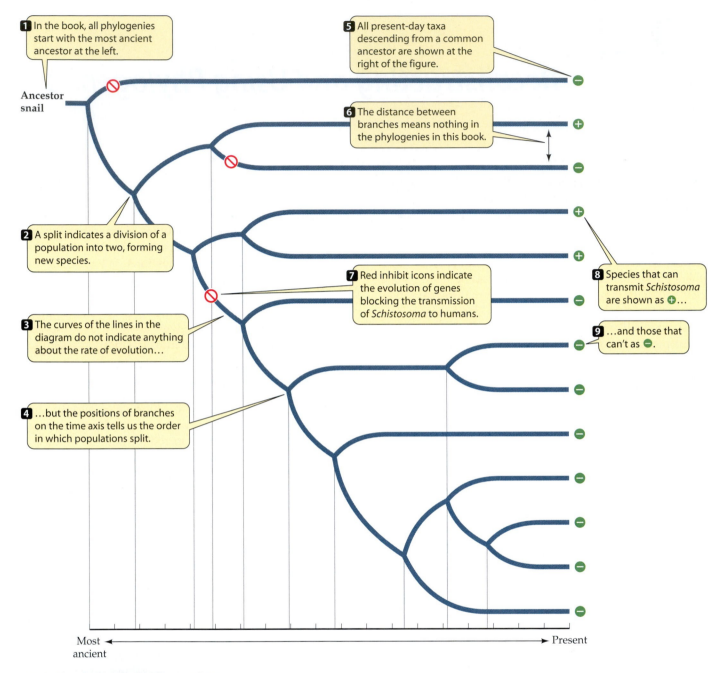

23.1 How to Read a Phylogenetic Tree
A phylogenetic tree displays the order in which lineages split. This example shows the phylogeny of *Oncomelania* snails, the intermediate hosts of the human parasite *Schistosoma*.

evolutionary lineages, or of only a small group of organisms, such as the snail genus *Oncomelania* (Figure 23.1). In the phylogenetic trees in this book, time flows from left (earliest) to right (most recent). It is equally common practice to draw trees with the earliest times at the bottom.

Determining the evolutionary relationships among organisms is intrinsically exciting. We are especially interested in the origin of our own species, but we also care about, for example, the origins of birds and mammals from reptilian ancestors. In addition, phylogenetic information helps us deal with practical problems, such as the control of schistosomiasis. We will return to the uses of phylogeny

after we have described the methods by which systematists reconstruct phylogenetic trees.

Systematists reconstruct phylogenetic trees by analyzing evolutionary changes in the traits of organisms. Phylogenetic trees are rather like pedigrees, except that they are usually constructed with the ancestor at the base rather than on the "twigs." The base of a phylogeny represents the point in the past when the lineage consisted of only the ancestor.

Charles Darwin described evolution as *descent with modification*. He recognized that closely related species—that is, species that share a recent common ancestor—are likely to be very similar. In other words, they should share many traits that they inherited from the common ancestor. Systematists expect traits inherited from an ancestor in the dis-

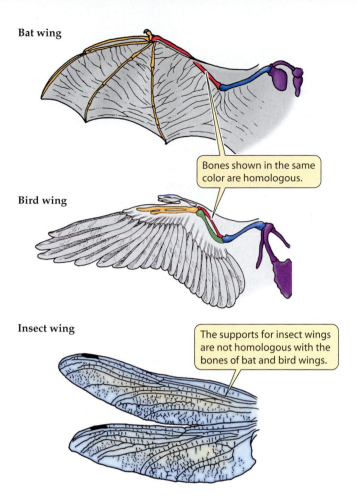

Bat wing

Bird wing

Bones shown in the same color are homologous.

Insect wing

The supports for insect wings are not homologous with the bones of bat and bird wings.

23.2 The Bones of the Wings of Bats and Birds Are Homologous, but the Wings Themselves Are Not
The supporting structures of bat and bird wings are derived from a common tetrapod (four-limbed) ancestor, and are thus homologous. The wings themselves, however, evolved independently in the two groups.

tant past to be shared by a large number of species. Traits that first appeared in a more recent ancestor should be shared by fewer species. But in all cases, the sharing of traits by a group of species indicates that they may be descendants of a common ancestor.

Any two features descended from a common ancestral feature are said to be **homologous**; these features may be anatomical structures, behavior patterns, nucleotides in a DNA sequence, or any other heritable trait. Traits that are shared by most or all organisms in any lineage being studied are likely to have been inherited relatively unchanged from an ancestor that lived very long ago. For example, all living vertebrates have a vertebral column, and all known fossil ancestral vertebrates also had a vertebral column. The vertebral column is therefore judged to be homologous in all vertebrates.

A trait that differs from its ancestral form is called a **derived trait**. In order to identify how traits have changed during evolution, systematists must infer the state of the trait in some ancestor and then determine how it has been

modified in the descendants. Doing so is not easy because real evolutionary patterns are complex. Three processes generate difficulties:

▶ Independently evolved features subjected to similar selective pressures may become superficially similar as a result of **convergent evolution**. For example, although the bones of the wings of bats and birds are homologous, having been inherited from a common ancestor, the wings themselves are not homologous because they evolved independently in bats and in birds from the forelimbs of a nonflying ancestor (Figure 23.2).

▶ Similar developmental processes may result in **parallel evolution** of similar traits in distantly related organisms (Figure 23.3).

▶ Over time, there may be **evolutionary reversals**; that is, a character may revert from a derived state back to an ancestral one. For example, most frogs lack teeth on their lower jaw, but the ancestors of frogs did have such teeth. One frog genus, *Amphignathodon*, has re-evolved teeth in the lower jaw.

Together these processes generate **homoplastic traits**; that is, traits that are similar for some reason other than inheritance from a common ancestor.

Depending on the size of the lineage we are looking at, a given trait may be ancestral or derived. For example, rats and mice (both rodents), but not dogs or other mammals, have long, continuously growing incisor teeth. Continuously growing incisors evidently developed in the common ancestor of rats and mice after their lineage separated from the one leading to dogs and other mammals, because no other mammals have that kind of incisor. Thus, if we were reconstructing a phylogeny of a group of rodents, continuously growing incisors would be an ancestral trait because all rodents have it. However, if we were reconstructing a phylogeny of all mammals, continuously growing incisors would be a derived trait.

The first step in reconstructing a phylogeny is to select the group of organisms whose phylogeny is to be determined. We will refer to these organisms as the *focal group*. The next step is to choose the characters that will be used in the analysis and to identify the possible forms (traits) of

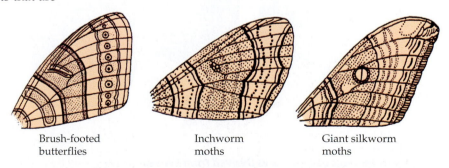

Brush-footed butterflies   Inchworm moths   Giant silkworm moths

23.3 Parallel Evolution in Butterfly Wing Bands
Bands on the wings of these distantly related butterflies and moths conform to a common pattern. Similar processes of wing development in all three species produce these similar patterns.

those characters. Recall from Chapter 10 that a *character* is a feature such as flower color; a *trait* is a particular form of a character, such as white flowers. A trait may be the presence or absence of a character, or the character may exist in more than one form. The next, and usually the most difficult, step is to determine the ancestral and derived traits. Finally, systematists must distinguish homologous from homoplastic traits.

Identifying ancestral traits

Distinguishing derived traits from ancestral traits may be difficult because traits often become so dissimilar that ancestral states are unrecognizable. For example, the leaves of plants have diverged to form many different structures. Several lines of evidence, especially details of their structure and development, indicate that protective spines, tendrils, and brightly colored structures that attract pollinators (Figure 23.4) are all modified leaves; they are *homologs* of one another even though they do not resemble one another closely.

One way to distinguish ancestral traits from derived traits is to assume that an ancestral trait should be found not only among the species of the focal group, but also in outgroups. An **outgroup** is a lineage that is closely related to the focal group , but which branched off from the lineage of the focal group below its base on the evolutionary tree. Traits found only within the focal group, on the other hand, are likely to be derived traits. Species that have a recent common ancestor should share very few homoplastic traits, because little time has been available for convergent evolution to produce them.

The more traits that are measured, the more likely the data will support a single phylogenetic pattern, and the more readily biologists can distinguish between homologies and homoplasies. A few of the traits originally assumed to be homologies may turn out to be homoplasies, but the best way to determine the true status of shared traits is to assume that they are homologous until additional evidence suggests they are not.

 ## Reconstructing a simple phylogeny

To see how a phylogeny is constructed, consider eight vertebrate animals—hagfish, perch, pigeon, chimpanzee, salamander, lizard, mouse, and crocodile. We will assume initially that a given derived trait evolved only once during the evolution of these animals, and that no derived traits were lost from any of the descendant groups. For simplicity, we have selected traits that are either present (+) or absent (–). The traits we will consider are listed in Table 23.1.

As will become evident in Chapter 33, hagfishes are believed to be more distantly related to the other vertebrates than the other vertebrates are to each other. Therefore, we choose hagfishes as the outgroup for our analysis. Derived traits are those that have been acquired by other members of the lineage since they separated from hagfishes.

We begin by noting that the chimpanzee and the mouse share two unique traits, mammary glands and fur. Those traits are absent in both the outgroup and the other species whose relationships we are attempting to determine. Therefore, we infer that mammary glands and fur are derived traits that evolved in a common ancestor of chimpanzees and mice after that lineage separated from the ones leading to the other vertebrates. In other words, we provisionally assume that mammary glands and fur evolved only once among the animals we are classifying.

The pigeon has one unique trait: feathers. As before, we provisionally assume that feathers evolved only once, after the lineage leading to birds separated from that leading to the mouse, chimpanzee, and crocodile. By the same reasoning, we assume that four-chambered hearts evolved only once, after the lineage leading to crocodiles, birds, and mammals separated from the lineage leading to lizards. We assume that claws or nails evolved only once, after the lineage leading to salamanders separated from the lineage leading to those animals that have claws or nails. We make the same assumption for lungs and jaws, continuing to minimize the number of evolutionary events needed to produce the patterns of shared traits among these eight animals.

Using this information, we can reconstruct a provisional phylogeny. The group with no derived traits, the hagfish, is the outgroup, and we assume that the animals that share unique derived traits have a common ancestor not shared with animals lacking those traits. We assume, for exam-

The spines of the barrel cactus and the bracts of *Heliconia* are both modified leaves.

Cheiridopsis tuberculata   *Heliconia* sp.

23.4 Homologous Structures Derived from Leaves
The leaves of plants have diverged during their evolution to form many different structures, some of which bear very little resemblance to each other. *Heliconia* bracts support flowers and attract pollinators.

23.1 Eight Vertebrates Ordered According to Unique Shared Derived Traits

TAXON	JAWS	LUNGS	CLAWS OR NAILS	FEATHERS	FUR	MAMMARY GLANDS	FOUR-CHAMBERED HEART
Hagfish (outgroup)	–	–	–	–	–	–	–
Perch	+	–	–	–	–	–	–
Salamander	+	+	–	–	–	–	–
Lizard	+	+	+	–	–	–	–
Crocodile	+	+	+	–	–	–	+
Pigeon	+	+	+	+	–	–	+
Mouse	+	+	+	–	+	+	+
Chimpanzee	+	+	+	–	+	+	+

^a A plus sign indicates the trait is present, a minus sign that it is absent.

ple, that mice and chimpanzees, the only two animals that share fur and mammary glands, share a more recent common ancestor with each other than they do with birds and crocodiles. Otherwise we would need to assume that the ancestors of birds and crocodiles also had fur and mammary glands, but that those traits were subsequently lost—unnecessary additional assumptions.

A phylogeny for these eight vertebrates, based on the traits we used and the assumption that each derived trait evolved only once, is shown in Figure 23.5. Notice that the phylogeny does not describe the ancestors or date the splits

between lineages. It shows only the sequential order of the splits: The oldest splits are to the left, and the more recent ones are to the right. Notice also that the *y* axis has no scale. In this and all other phylogenies in this book, vertical distances between groups do not correlate with degree of similarity or difference between them.

The phylogeny of these eight vertebrates was easy to construct because the traits we chose fulfilled the assumptions that derived traits appeared only once in the lineage and were never lost after they appeared. If we had included a snake in the group, however, our second assumption

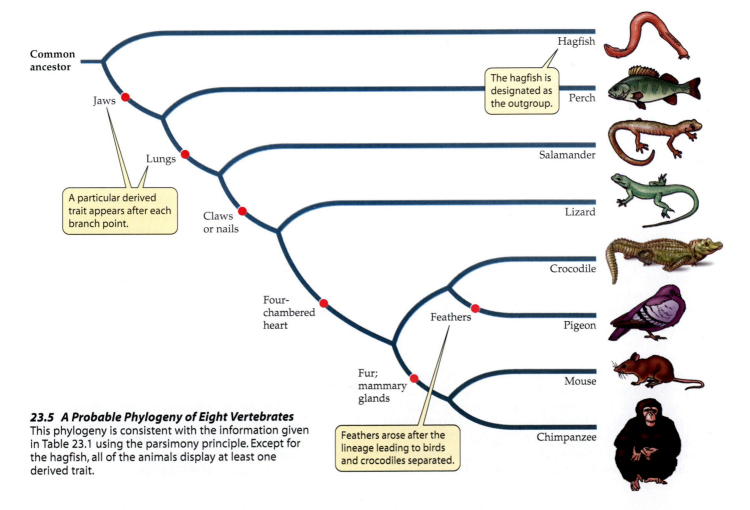

23.5 A Probable Phylogeny of Eight Vertebrates
This phylogeny is consistent with the information given in Table 23.1 using the parsimony principle. Except for the hagfish, all of the animals display at least one derived trait.

would have been violated, because the lizard ancestors of snakes had limbs, which were subsequently lost, along with their claws. We would need to examine additional traits to determine that the lineage leading to snakes separated from the one leading to lizards long after the lineage leading to lizards separated from the others. In fact, the analysis of many traits shows that snakes evolved from burrowing lizards that lost their limbs during a long period of subterranean existence.

Many traits must be analyzed to reconstruct a phylogeny, and systematists use various methods to combine information from the different traits. The simple method we used in our vertebrate example does not work in the vast majority of cases because we know from fossil and other evidence that traits can change more than once, or even undergo reversal. How do systematists deal with these complexities when they reconstruct phylogenies?

The most widely used methods of reconstructing phylogenetic trees employ the **parsimony principle**. In its most general form, the parsimony principle states that one should prefer the simplest hypothesis that is capable of explaining the known facts. Its application to the reconstruction of phylogenies means minimizing the number of evolutionary changes that need to be assumed over all characters in all groups in the tree—that is, the best hypothesis is the one that requires the fewest homoplasies.

Parsimony works best for morphological traits, whose evolutionary rates are generally slow enough that similarities due to homoplasies are uncommon relative to the number of traits retained because they were inherited from the common ancestor.

Another method, called the **maximum likelihood method**, is used primarily for the reconstruction of phylogenies based on molecular data. The computer programs employed in this method are complicated. They are designed to deal with the fact that mutations that result in substitutions of nucleotides are common, but that their frequencies can be estimated independently from other genetic information (see Chapter 24).

Using the parsimony principle is helpful not because evolutionary changes are necessarily parsimonious, but because it is generally wiser not to adopt complicated explanations when simpler ones explain the known facts. More complicated explanations are accepted only when evidence requires them. Phylogenetic trees are hypotheses about evolutionary relationships that are repeatedly tested and modified as additional traits are measured and as new fossil evidence becomes available.

Whatever method is employed, determining the most likely phylogeny for any group of organisms is difficult. For example, there are 34,459,425 possible phylogenetic trees for a lineage with only 11 species! Computer programs using the parsimony principle employ various search routines that calculate the shortest possible phylogenetic tree—that is, with the fewest homoplasies—for a given data set and then compare other possible phylogenies with the shortest one. If, as is usually the case, several

trees are of approximately equal length, they can be merged into a **consensus tree** that retains only those lineage splits that are found in all the most parsimonious trees. In a consensus tree, groups whose relationships differ among the trees form nodes with more than two branches. These nodes are considered "unresolved" because during speciation, a lineage typically splits into only two daughter species.

Traits Used in Reconstructing Phylogenies

Because organisms differ in many ways, systematists use many traits to reconstruct phylogenies. Some of these traits are readily preserved in fossils; others, such as behavior and molecular structure, rarely survive fossilization processes. Systematists take into consideration behavioral and molecular traits as well as structural traits in both living and fossil organisms. The more traits that are measured, the more inferred phylogenies should converge on one another and on the actual evolutionary pattern.

Morphology and development

An important source of information for systematists is **morphology**—that is, the sizes and shapes of body parts. Because living organisms have been studied for centuries, we have a wealth of morphological data, as well as extensive museum and herbarium collections of organisms whose traits can be measured. Sophisticated methods are now available for measuring and analyzing morphology and for estimating the amount of morphological variation among individuals, populations, and species.

The fossil record, which reveals when lineages diverged and began their independent evolutionary histories, can tell us the timing of evolutionary events. Fossils provide important evidence that helps us distinguish ancestral from derived traits. They provide the only available information about where and when organisms lived in the past and what they looked like. When available, this information is valuable, but sometimes few or no fossils have been found for a group whose phylogeny we wish to determine.

The early developmental stages of many organisms reveal similarities to other organisms that are lost by the time of adulthood. For example, the larvae of the marine creatures called sea squirts have a rod in the back—the *notochord*—that disappears as they develop into adults. Many other animals—all the animals called *vertebrates*—also have this structure at some time during their development. This shared structure is one of the reasons for believing that sea squirts are more closely related to vertebrates than would be suspected by examination of the adults only (Figure 23.6).

Molecular traits

Like the sizes and shapes of their body parts, the molecules that make up organisms are heritable characteristics that may diverge among lineages over evolutionary time. Molecular evolution will be discussed in detail in Chapter 24. The molecular traits most useful for constructing phyloge-

Sea squirt
(seen in section)

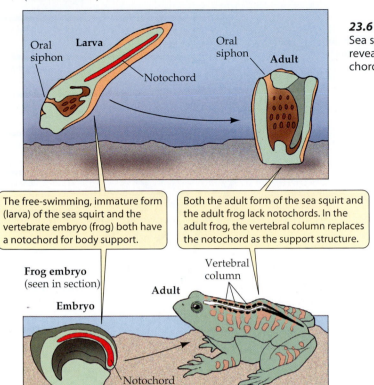

23.6 A Larva Reveals Evolutionary Relationships
Sea squirt larvae, but not adults, have a well-developed notochord that reveals their relationship with vertebrates, all of which have a notochord at some time during their life cycle.

The free-swimming, immature form (larva) of the sea squirt and the vertebrate embryo (frog) both have a notochord for body support.

Both the adult form of the sea squirt and the adult frog lack notochords. In the adult frog, the vertebral column replaces the notochord as the support structure.

moglobin pseudogene (a nonfunctional DNA sequence derived early in primate evolution by duplication of a hemoglobin gene). The outgroup in the analysis was the genus *Ateles*, the New World spider monkeys. The DNA data strongly indicate that chimpanzees and humans share a more recent ancestor with each other than they do with gorillas (Figure 23.7), a conclusion supported by other types of molecular data.

Phylogenetic Trees Have Many Uses

Phylogenetic trees contain information that is useful to scientists investigating a wide variety of biological questions. Here we illustrate how phylogenetic trees are being used to determine how many times a particular trait may have arisen during evolution, and to assess when lineages may have split.

nies are the structures of proteins and nucleic acids (DNA and RNA).

PROTEIN STRUCTURE. Relatively precise information about phylogenies can be obtained by comparison of the molecular structure of proteins. We can estimate genetic differences between two lineages by obtaining homologous proteins from both and determining the number of amino acids that have changed since the lineages diverged from a common ancestor.

DNA BASE SEQUENCES. The base sequences of DNA provide excellent evidence of evolutionary relationships among organisms. The cells of eukaryotes have genes in their mitochondria as well as in their nuclei; plant cells also have genes in their chloroplasts. The chloroplast genome (cpDNA), which is used extensively in phylogenetic studies of plants, has changed little over evolutionary time. Mitochondrial DNA (mtDNA), which evolved much more rapidly than cpDNA, has been used extensively for evolutionary studies of animals.

Relationships among apes and humans were investigated by sequencing more than 10,000 base pairs making up a segment of nuclear DNA that includes a he-

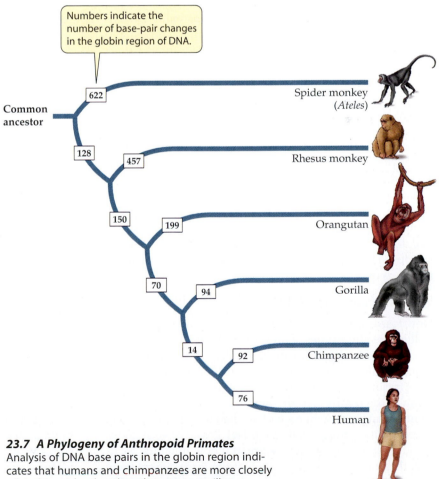

Numbers indicate the number of base-pair changes in the globin region of DNA.

23.7 A Phylogeny of Anthropoid Primates
Analysis of DNA base pairs in the globin region indicates that humans and chimpanzees are more closely related to each other than they are to gorillas.

How Often Have Traits Evolved?

Most flowering plants reproduce by mating with another individual, or *outcrossing*, and have mechanisms to prevent self-fertilization. Many species, however, can fertilize themselves with their own pollen—they are *self-compatible*. How can we tell how often self-compatiblity has evolved in a lineage? We can do so by plotting on a phylogenetic tree which species are outcrossing and which are selfing.

The evolution of fertilization methods was examined in *Linanthus* (a genus in the phlox family), a lineage of plants with a diversity of breeding systems and pollination mechanisms. The outcrossing (self-incompatible) species of *Linanthus* have flowers with long tubes and are pollinated by long-tongued flies. The self-fertilizing (self-compatible) species all have short-tubed flowers.

Investigators reconstructed a phylogeny for 12 species in a section of the genus using the internal-transcriber-spacer (ITS) region of nuclear ribosomal DNA (Figure 23.8). This region was known to be useful for reconstructing species-level phylogenies in other plant groups and had already been used for constructing a phylogeny of the phlox family. The investigators determined whether each species was self-compatible by artificially pollinating flowers with their own or outcrossed pollen and observing the results.

Several lines of evidence suggested that self-incompatibility is the ancestral state in *Linanthus*. First, multiple origins of self-incompatibility are not known in any other flowering plant family. Second, self-incompatibility systems involve physiological mechanisms in both the pollen and the stigma and require the presence of at least three distinct alleles. Therefore, a change from self-incompatibility to self-compatibility is easier than the reverse change. Third, in all self-incompatible species of *Linanthus*, the site of pollen rejection is the stigma, even though sites of pollen rejection vary greatly among other plant groups.

Assuming that self-incompatibility is the ancestral state, the phylogeny suggests that self-compatibility has evolved three times in this *Linanthus* lineage (Figure 23.8). The change to self-compatibility has been accompanied by the evolution of reduced flower size. Interestingly, the striking similarity in flower form among the self-compatible groups had led to their classification as members of a single species. The phylogenetic analysis showed them to be members of three distinct lineages!

When Did Lineages Split?

How fossils can help us determine evolutionary pathways is illustrated by studies of lungfishes. A phylogenetic tree of the three extant genera of lungfishes, all of which are strictly limited to fresh water (Figure 23.9), indicates that the African *Protopterus* and the South American *Lepidosiren* (both in the family Lepidosirenidae) share a more recent common ancestor with each other than with the Australian *Neoceradotus* (family Ceratodontidae). Fossils of each genus are known only from the continent it now inhabits. By itself, this information suggests that the three genera were isolated by the breakup of Gondwana (see Chapter 20).

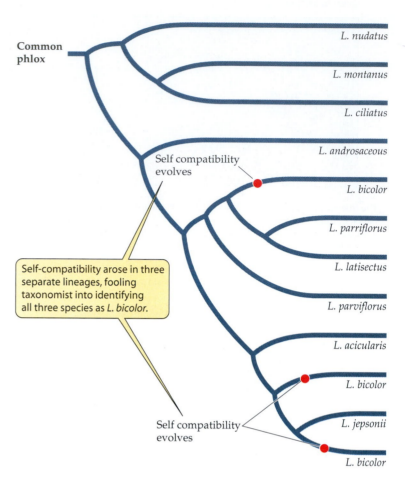

Common phlox

L. nudatus
L. montanus
L. ciliatus
L. androsaceous

Self compatibility evolves

L. bicolor
L. parriflorus
L. latisectus

Self-compatibility arose in three separate lineages, fooling taxonomist into identifying all three species as *L. bicolor*.

L. parviflorus
L. acicularis
L. bicolor

Self compatibility evolves

L. jepsonii
L. bicolor

23.8 Phylogeny of a Section of the Phlox Genus Linanthus
Self-compatibility apparently evolved three times in this lineage. Because the form of flowers converged in the selfing lineages, taxonomists mistakenly thought that they were all members of a single species.

However, fossils of other members of the family Ceratodontidae have been found in all continents except South America and Antarctica, and fossil lepidosirenids have been found in both Europe and North America. Thus, the ancestors of both families probably ranged over much of Pangaea. The combination of the phylogenetic tree and fossil evidence informs us that their divergence happened long before the breakup of Gondwana.

Why Classify Organisms?

Classification systems are important for several reasons. They improve our ability to explain relationships among things. They are also an aid to memory. It is impossible to remember the characteristics of many different things unless we can group them into categories based on shared characteristics. They are also useful as predictors. For example, the discovery of biochemical precursors of the drug cortisone in certain species of yams (genus *Dioscorea*) stimulated a successful search for higher concentrations of the drug in other *Dioscorea* species. And, as we saw at the be-

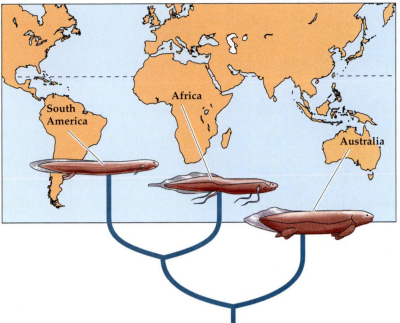

Linnaeus gave each species two names, one identifying the species itself and the other the genus to which it belongs. A **genus** (plural genera; adjectival form, generic) is a group of closely related species. In many cases the name of the taxonomist who first proposed the species name is added at the end. Thus, *Homo sapiens* Linnaeus is the name of the modern human species. *Homo* is the genus to which the species belongs, and *sapiens* identifies the species; Linnaeus proposed the species name *sapiens*. You can think of the generic name *Homo* as equivalent to your surname and the specific name *sapiens* as equivalent to your first name. The generic name is always capitalized; the species name is not. Both names are always italicized, whereas common names are not.

23.9 Evolutionary Pathways in Lungfish Species
In this phylogeny, the ancestor is at the bottom of the figure.

ginning of this chapter, a phylogeny of *Oncomelania* snails is helping to devise methods to control schistosomiasis.

Biological classification systems provide unique names for organisms. If the names are changed, the systems provide a means of tracing the changes. Common names, even if they exist (most organisms have none), are very unreliable and often confusing. For example, plants called bluebells are found in England, Scotland, Texas, and the Rocky Mountains—but none of the bluebells in any of those places is closely related evolutionarily to the bluebells in any of the other places (Figure 23.10).

Recognizing and interpreting similarities and differences among organisms is easier if the organisms are classified into groups that are ordered and ranked. Any group of organisms that is treated as a unit in a biological classification system is called a **taxon** (plural taxa). **Taxonomy** is the theory and practice of classifying organisms.

The Hierarchical Classification of Species

The biological classification system that is used today was developed by the Swedish biologist Carolus Linnaeus in 1758. His two-name system, referred to as *binomial nomenclature*, replaced the cumbersome descriptions biologists had previously used. For example, the honeybee, which had been named *Apis pubescens, thorace subgriseo, abdomine fusco, pedibus posticis glabris utrinque margine ciliatis*, became simply *Apis mellifera*. Binomial nomenclature is universally employed in biology today. Using this system, scientists throughout the world refer to the same organisms by the same names.

(a) *Campanula* sp.

(b) *Endymion nonscriptus*

(c) *Mertensia virginica*

23.10 Many Different Plants Are Called Bluebells
(a) These flowers from the plains of North Dakota are often called bluebells. (b) This English bluebell is a member of the lily family. (c) These are known as Virginia bluebells. None of these plants is closely related to the others.

Kingdom	Plantae (plants)	
± 275,000 species		
Phylum	Angiospermae (flowering plants)	
± 250,000 species		
Class	Eudicotyledonae (true dicots)	
± 235,000 species		
Order	Rosales (roses and their allies)	
± 18,000 species		
Family	Rosaceae	
± 3,500 species		
Genus	*Rosa*	
± 500 species		
Species	*Rosa gallica*	
	Moss rose	

Less specific ↑ More specific ↓

23.11 Hierarchy in the Linnaean System
The moss rose and the Blackburnian warbler as
they are classified under the Linnaean system.

When referring to more than one species in a genus without naming each one, we use the abbreviation "spp." after the generic name (for example, "*Drosophila* spp." means more than one species in the genus *Drosophila*). The abbreviation "sp." is used after a generic name if the identity of the species is uncertain. Rather than repeating a generic name when it is used several times in the same discussion, biologists often spell it out only once and abbreviate it to the initial letter thereafter (for example, *E. coli* is the abbreviated form of *Escherichia coli*).

In the Linnaean system, species and genera are grouped into higher taxonomic categories. The category (taxon) above genus in the Linnaean system is **family**. The names of animal families end in the suffix "-idae." Thus Formicidae is the family that contains all ant species, and the family Hominidae contains humans, a few of our fossil relatives, and chimpanzees and gorillas. Family names are based on the name of a member genus. Formicidae is based on *Formica*, and Hominidae is based on *Homo*. Plant classification follows the same procedures except that the suffix

"-aceae" is used with family names instead of "-idae." Thus Rosaceae is the family that includes the genus of roses (*Rosa*) and its close relatives.

Families, in turn, are grouped into **orders**, and orders into **classes**. Classes are grouped into **phyla** (singular phylum), and phyla into **kingdoms**. The hierarchical units of this classification system, as applied to an animal species, the Blackburnian warbler (*Dendroica fusca*), and a plant species, the moss rose (*Rosa gallica*), are shown in Figure 23.11.

It should be obvious from this discussion that although the species category has real meaning and can be defined fairly rigorously, higher taxonomic categories are only mental constructs. They help us understand the diversity of life and its evolution, but they have only relative meaning. A family is always less inclusive than an order, but more inclusive than a genus. However, there are no rigorous criteria by means of which to decide whether a particular lineage should be given the status of a family or an order. Therefore, an avian family may have a more recent common ancestor than a family of flowering plants, or vice versa.

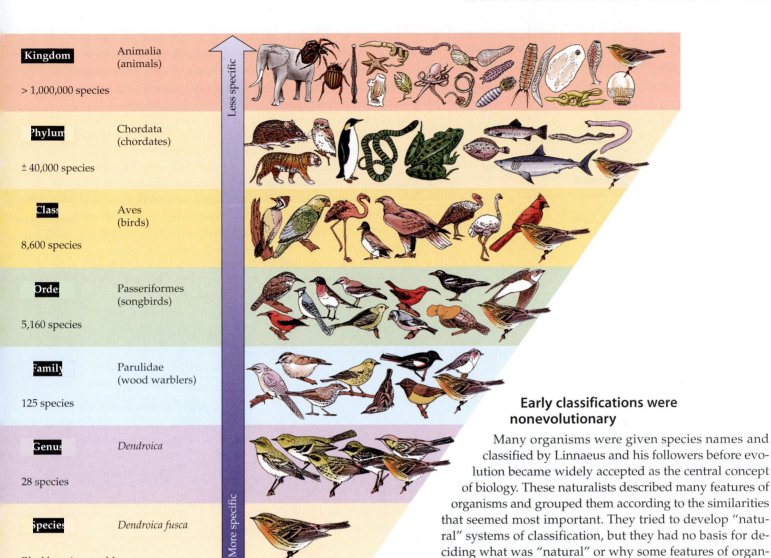

Kingdom	Animalia (animals)
> 1,000,000 species	
Phylum	Chordata (chordates)
± 40,000 species	
Class	Aves (birds)
8,600 species	
Order	Passeriformes (songbirds)
5,160 species	
Family	Parulidae (wood warblers)
125 species	
Genus	*Dendroica*
28 species	
Species	*Dendroica fusca*
Blackburnian warbler	

Less specific → More specific

Biological Classification and Evolutionary Relationships

Biological classification systems are designed to express relationships among organisms. The kind of relationship we wish to express influences which features we use to classify organisms. If, for instance, we were interested in a system that would help us decide what plants and animals were desirable as food, we might devise a classification system based on tastiness, ease of capture, and the type of edible parts each organism possessed. Early Hindu classifications of plants were designed according to these criteria. Biologists do not use such systems today, but they served the needs of the people who developed them.

Classification systems should be judged only in terms of their utility and consistency with their stated goals. To evaluate any classification system, we must first ask, What relationships is it trying to express? Then, How well does it express those relationships?

Early classifications were nonevolutionary

Many organisms were given species names and classified by Linnaeus and his followers before evolution became widely accepted as the central concept of biology. These naturalists described many features of organisms and grouped them according to the similarities that seemed most important. They tried to develop "natural" systems of classification, but they had no basis for deciding what was "natural" or why some features of organisms were more important than others.

Current biological classifications reflect evolutionary relationships

Most taxonomists today believe that classification systems should reflect the evolutionary relationships of organisms—that is, that taxonomic groups should be **monophyletic**. A monophyletic group (also called a **clade**) contains all the descendants of a particular ancestor and no other organisms. In other words, a monophyletic group is one that can be removed from a phylogenetic tree by one "cut" in the tree. A taxon consisting of members that do not share the same common ancestor is **polyphyletic**. A group that contains some, but not all, of the descendants of a particular ancestor is said to be **paraphyletic** (Figure 23.12).

Taxonomists agree that polyphyletic groups are inappropriate as taxonomic units. The classifications used today still contain many polyphyletic groups because many organisms have not been studied enough to distinguish between homologies and homoplasies. However, as soon as they detect homoplasies, systematists change their classifications to eliminate polyphyletic taxa. Thus the three lineages of self-compatible *Linanthus* shown in Figure 23.8 are now treated as distinct species.

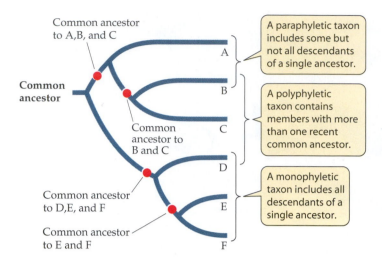

Common ancestor to A,B, and C

A paraphyletic taxon includes some but not all descendants of a single ancestor.

A polyphyletic taxon contains members with more than one recent common ancestor.

A monophyletic taxon includes all descendants of a single ancestor.

Common ancestor

Common ancestor to B and C

Common ancestor to D,E, and F

Common ancestor to E and F

23.12 Monophyletic, Polyphyletic, and Paraphyletic Taxa
Taxa are classified in terms of their evolutionary relationships. Polyphyletic groups are considered inappropriate as taxonomic units, but systematists sometimes use paraphyletic taxa.

In phylogenetic classification systems, formal taxonomic names are given only to monophyletic groups. But this does not mean that every monophyletic group should have a name. For example, it would be very cumbersome to put every pair of species into its own genus and every pair of genera into its own family. In addition, such a classification system would need to be changed every time a new species was described. Therefore, many monophyletic groups have no formal names. Systematists generally name only groups linked by many shared derived traits or by the presence of a major, obvious character that can be used to identify members of the group. These informal practices give stability to the classification system and aid in identifying organisms and their traits.

Although most systematists favor phylogenetic classifications, some believe that classification systems should also reflect degrees of difference among organisms, not only their evolutionary pedigree. According to this view, names should be retained for paraphyletic groups that have undergone rapid evolutionary change and diversification. The perspective of these taxonomists can be illustrated by birds, crocodilians, and their relatives.

We now know, from both fossil and anatomical evidence, that birds, turtles, and crocodilians (a group that includes crocodiles and alligators) share a more recent common ancestor than crocodilians and turtles share with snakes and lizards (Figure 23.13a). Traditionally, crocodilians were grouped with snakes, lizards, and turtles in the class Reptilia. Birds were placed in a separate class, Aves (Figure 23.13b). This classification came about because, since the time the two lineages separated, crocodilians have evolved more slowly than birds. As a result, crocodilians are more similar in many features to snakes and lizards than they are to birds. They look like very large lizards.

Figure 23.13b shows that the traditional class Reptilia is paraphyletic because it does not include all the descendants of its common ancestor; that is, birds are not included. If only monophyletic taxa were permitted, birds would be included with crocodilians, turtles, and their ancestors in a single taxon separate from snakes and lizards (Figure 23.13c). Retaining birds as a separate class (that is, retaining reptiles as a paraphyletic group) emphasizes that birds have undergone rapid evolution since they separated from reptiles and have developed major, unique derived traits.

The current tendency is to change classifications to eliminate paraphyletic groups, but some of the most familiar taxonomic categories—gymnosperms and reptiles, for example—are paraphyletic. Because of their familiarity and the extensive literature devoted to them, these categories are likely to remain in use for some time, even after their formal taxonomic designations change.

(a) **The evolutionary relationships**

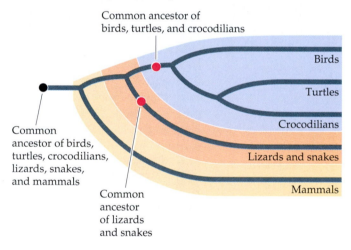

Common ancestor of birds, turtles, and crocodilians

Birds

Turtles

Crocodilians

Lizards and snakes

Mammals

Common ancestor of birds, turtles, crocodilians, lizards, snakes, and mammals

Common ancestor of lizards and snakes

(b) **The traditional classification**

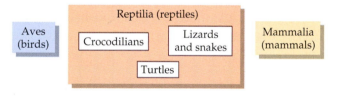

Reptilia (reptiles)

Aves (birds)

Crocodilians

Lizards and snakes

Turtles

Mammalia (mammals)

(c) **A phylogenetic classification**

Birds   Turtles

Crocodilians

Lizards and snakes

Mammals

23.13 Phylogeny and Classification
A phylogenetic classification based on their evolutionary relationships would group crocodilians and turtles together with birds. The traditional classification unites crocodilians and turtles with lizards and snakes in the paraphyletic taxon Reptilia because these animals share many morphological traits.

The Future of Systematics

The development of molecular methods and powerful computers has ushered in a new era of taxonomy. Computers enable systematists to analyze many characters and to compare many possible phylogenetic trees. Many phylogenies are being reconstructed, and classifications are being revised. Information from many sources continues to be used in constructing phylogenies. The range of data used in classification is likely to increase rather than decrease in the future because modern chemical, biochemical, and microscopic methods allow systematists to measure more traits of organisms than they could previously.

Often phylogenies are reconstructed as part of efforts to determine evolutionary relationships among organisms. In addition, as we have just seen, phylogenies are increasingly being used to answer many other types of biological questions. Many biological statements are phylogenetic statements. Any statement claiming an association between a trait and a group of organisms is a claim about when during a lineage the trait first arose and about the fate of the trait since its first appearance. For example, the statement that possession of a cytoskeleton is a trait possessed by all eukaryotes is a statement that the cytoskeleton is an ancestral, homologous trait that has been maintained during the subsequent evolution of all surviving eukaryote lineages.

Chapter Summary

How Are Phylogenetic Trees Reconstructed?

▶ Phylogenetic trees display the patterns of evolution of life on Earth. In addition, they help biologists deal with a wide variety of practical problems. Review Figure 23.1

▶ Traits that are inherited from a common ancestor are said to be homologous. A derived trait is one that differs from its form in the ancestor of a lineage. **Review Figure 23.2**

▶ Traits that are similar as a result of convergent or parallel evolution or evolutionary reversals are said to be homoplastic. **Review Figure 23.3**

▶ To determine true evolutionary relationships, systematists must distinguish between ancestral and derived traits within a lineage, as well as between homologous and homoplastic traits. This task is often difficult because divergent evolution may make homologous traits appear dissimilar and convergent evolution may make homoplastic traits appear similar.

▶ Systematists often employ the principle of parsimony to reconstruct phylogenetic trees. **Review Figure 23.5**

Traits Used in Reconstructing Phylogenies

▶ Systematists use data from fossils and the rich array of morphological and molecular data available from living organisms to determine evolutionary relationships.

▶ Structures in early developmental stages sometimes show evolutionary relationships that are not evident in adults. **Review Figure 23.6**

▶ The structures of proteins and the base sequences of nucleic acids are important taxonomic data. **Review Figure 23.7**

Phylogenetic Trees Have Many Uses

▶ Phylogenetic trees help biologists determine how many times evolutionary traits have arisen and when lineages diverged. **Review Figures 23.8, 23.9**

Why Classify Organisms?

▶ Classification systems improve our ability to explain relationships among things, aid our memory, and provide unique, universally used names for organisms.

The Hierarchical Classification of Species

▶ Biological nomenclature assigns to each organism a unique combination of a generic and a specific name.

▶ In the Linnaean classification system, species are grouped into higher-level units called genera, families, orders, classes, phyla, and kingdoms. **Review Figure 23.11**

Biological Classification and Evolutionary Relationships

▶ Taxonomists agree that taxa should share a common ancestor and that polyphyletic groups should not be used. **Review Figure 23.12**

▶ Paraphyletic taxa may be retained because of their familiarity and to highlight the fact that members of some lineages evolved especially rapidly. **Review Figure 23.13**

The Future of Systematics

▶ Molecular methods and powerful computers have ushered in a new era of systematics.

For Discussion

1. The great blue heron, *Ardea herodias*, is found over most of North America. The very similar gray heron, *Ardea cinerea*, ranges over most of Europe and Asia. These two herons currently are treated as different species, but a colleague argues that they should be treated as a single species. What facts should you consider in evaluating your colleague's suggestion? What taxonomic theories would be relevant to your evaluation?

2. Why are systematists so concerned with identifying lineages that share a single common ancestor?

3. How are fossils used to identify ancestral and derived forms of traits of organisms?

4. Taxonomists regularly use the "parsimony principle" when reconstructing phylogenetic trees. Given that nature is not always parsimonious, why is parsimony used as a guiding principle?

5. A student of the evolution of frogs has proposed a strikingly new classification of frogs based on an analysis of a few mitochondrial genes from about 25 percent of frog species. Should frog taxonomists immediately accept the new classification? Why or why not?

6. Linnaeus developed his system of classification before Darwin proposed his theory of evolution by natural selection, and most early classifications of organisms were developed by non-evolutionists. Yet many of these classifications are still used today, with minor modifications, by most evolutionary taxonomists. Why?

24 Molecular and Genomic Evolution

NEANDERTHALS ARE A GROUP OF EXTINCT HUMAN relatives that lived in Europe and eastern Asia from about 300,000 to 30,000 years ago. During part of that time they coexisted with *Homo sapiens*. Some researchers have identified Neanderthals as direct ancestors of modern humans. Others believe that Neanderthals contributed only a few genes to the human gene pool. Still others think that they contributed no genes.

In an attempt to discover which of these three hypotheses is correct, scientists extracted mitochondrial DNA from a section of a leg bone of a Neanderthal fossil between 30,000 and 100,000 years old. The base sequences of the Neanderthal mtDNA fell well outside the variation found in modern human mtDNA. From these results, investigators judged that Neanderthal mtDNA and modern human mtDNA have been evolving separately for at least 500,000 years. This finding provided evidence that Neanderthals contributed few or no genes to the human gene pool.

How can investigators compare the mtDNA of humans and Neanderthals? In this chapter we review how molecular biologists determine the structures of nucleic acids and proteins and use those structures to infer both the patterns and the causes of molecular evolution. With these insights, we explore how the functions of molecules change, where new genes come from, and the evolution of the genomes of organisms. Finally, we show how knowledge of the patterns of molecular evolution helps us solve other biological problems, including inferring phylogenetic relationships among organisms and determining how humans spread over Earth.

What Is Molecular Evolution?

The molecules of interest to molecular evolutionists are nucleotides, nucleic acids, amino acids, and proteins. Nucleic acids evolve by means of nucleotide base substitutions, which in turn result in changes in the amino acids they encode. Alterations in the structure and functioning of proteins result from changes in the ordering of the amino acids of which they are composed. Molecular evolutionists investigate the evolution of these macromolecules to determine how rapidly they have changed and why they have

changed. To do so, they must be able to characterize the precise structures of these macromolecules.

Molecular evolutionists also try to reconstruct the evolutionary histories of genes and organisms, a field known as **molecular phylogenetics**. These two components of the study of molecular evolution are intimately related because phylogenetic information is essential for determining the order of changes in molecular characters, and knowing the order of such changes is usually the first step in inferring their causes. Conversely, knowledge of the pattern and rate of change in a given molecule is crucial for attempts to reconstruct the evolutionary history of a group of organisms.

For most of its history, evolutionary biology depended on the study of the obvious morphological features of organisms. During his 5-year voyage aboard the *Beagle*, Charles Darwin observed morphological differences among species found in different geographic areas. He later synthesized these observations into descriptions of how species change over time. He was able to hypothesize *why*

Neanderthal Bones
DNA recovered from bones of Neanderthals can be used to infer whether Neanderthals contributed many or no genes to the modern human genome. This skeleton was unearthed in 1908 from a cave in France.

many of these morphological changes had happened, but he could not determine *how* they occurred. Understanding the mechanisms of morphological change had to await discoveries in biochemistry a century later.

Even though genetic differences underlie all components of the adaptive evolution of organisms, molecular evolution differs from phenotypic evolution in one important way. In addition to natural selection, random genetic drift and mutation exert important influences on the rates and directions of molecular evolution.

A *mutation*, as you know from Chapter 12, is a change in the sequence of a single copy of a gene (see pages 234–235). A **substitution** is the partial or complete replacement of a nucleotide base or longer sequence by another throughout an entire population or species. It is substitutions that are of interest to molecular evolutionists.

Many mutations in sequences of genes do not alter the proteins encoded by those genes. The reason is that most amino acids are specified by more than one codon. Leucine, for example, is specified by six different codons: UUA, UUG, CUU, CUC, CUA, and CUG (see Figure 12.5; in this and all other cases, most of the redundancy is in the third codon position).

When it occurs throughout a population, a nucleotide substitution that does not change the amino acid specified—UUA to UUG, for example—is known as a **synonymous** or **silent substitution**. Synonymous substitutions are unlikely to affect the functioning of the protein (and hence the organism) and are therefore unlikely to be influenced by natural selection.

Because they are unlikely to be influenced by natural selection, synonymous substitutions are free to accumulate in a population over evolutionary time at rates determined by rates of mutation and genetic drift. Because modern molecular techniques enable us to detect substitutions at the level of nucleotides, molecular evolutionists can measure even these nonfunctional changes.

The occurrence in a population of nucleotide substitutions that *do* change the amino acid that is specified—UUA to UCA, for example, which would result in serine rather than leucine—is known as **nonsynonymous substitution**. In general, nonsynonymous mutations are likely to be deleterious to the individual organism. But even an amino acid change does not necessarily change a protein's shape and, hence, its functional properties. Therefore, a nonsynonymous substitution may be selectively neutral, or nearly so.

Most natural populations of organisms harbor much more genetic variation than we would expect if genetic variation were influenced primarily by natural selection. This discovery, combined with the knowledge that many substitutions do not change molecular function, stimulated the development of the neutral theory of molecular evolution.

The neutral theory, first articulated by Motoo Kimura in 1968, postulates that, at the molecular level, the majority of mutations are selectively neutral: they confer neither an advantage nor a disadvantage on their bearers. If so, the majority of evolutionary changes in macromolecules, and much of the genetic variation within species, results from

neither positive selection of advantageous alleles nor stabilizing selection, but from random genetic drift.

To see why this is so, consider a population with a size of N and a rate of neutral mutation at a particular locus of μ per gamete per generation. The number of new mutations would on average be $\mu \times 2N$, because $2N$ gene copies are available to mutate. According to genetic drift theory (see Chapter 21), the probability that a mutation will be fixed by genetic drift is its frequency, p, which equals $1/(2N)$ for a newly arisen (and hence very rare) mutation. Therefore, the number of neutral mutations that arise per generation that are likely to become fixed is $2N\mu \times 1/(2N) = \mu$, which equals the mutation rate.

In other words, the rate of fixation of mutations is theoretically constant and is equal to the neutral mutation rate. This is the theoretical basis of the concept of the **molecular clock**, which states that macromolecules should diverge from one another over time at a constant rate. We will discuss molecular clocks later in this chapter and show how, with care, the concept can be used to study many features of molecular evolution.

According to the neutral theory of molecular evolution, most polymorphisms at specific genetic loci are transitory rather than stable, because the frequency of neutral alleles in a population should change slowly over time (Figure 24.1a). In contrast, advantageous mutations are rapidly fixed in a population, and deleterious mutations are quickly lost (Figure 24.1b). The neutral theory and the theory of natural selection agree that most mutations are deleterious, but the neutral theory asserts that the selective ad-

(a) **Neutral mutations**

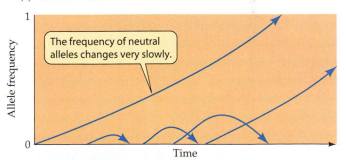

The frequency of neutral alleles changes very slowly.

(b) **Advantageous and deleterious mutations**

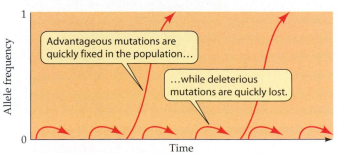

Advantageous mutations are quickly fixed in the population…

…while deleterious mutations are quickly lost.

24.1 Polymorphisms May be Transitory or Stable
(a) The frequencies of neutral alleles change slowly. Much polymorphism in these alleles is transitory. (b) Most polymorphisms in advantageous or disadvantageous alleles are stable, not transitory.

vantages or disadvantages of most molecular mutations are so small that selection on them is too weak to offset the influences of genetic drift.

Determining and Comparing the Structure of Macromolecules

To reveal patterns of molecular evolution, biologists may determine the precise structure of biological molecules. An investigator attempting to determine the structure of a nucleic acid or a protein begins by extracting and purifying it from a natural source. The molecule can then be analyzed by X ray crystallography. The molecule is crystallized, and the crystal is bombarded with a beam of X rays. The regularly spaced atoms in the crystal deflect the X rays into an orderly array of spots on a photographic film. With data from successive cross-sections through the crystal, a computer can generate a three-dimensional electron density map of the molecule. Graphics software enables the computer to create a picture showing the position of each atom in the molecule (Figure 24.2).

The base sequences of nucleic acids also provide important information about evolutionary histories. The invention of the polymerase chain reaction (PCR) technique (see Chapter 11) allowed biologists to determine the sequence of regions of DNA not only from living tissues, but also from fossilized remains, mummified tissues, dried skins in museums, and pressed plants in herbaria, even though these objects contain only tiny amounts of DNA. DNA has been extracted and amplified from human fossils more than

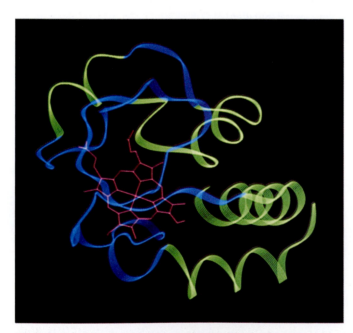

24.2 Computer Graphic Shows the Positions of Atoms in Molecules
The positions of atoms and the three-dimensional structure of tuna cytochrome *c* were computed from data generated by cross-sections through the crystallized molecule.

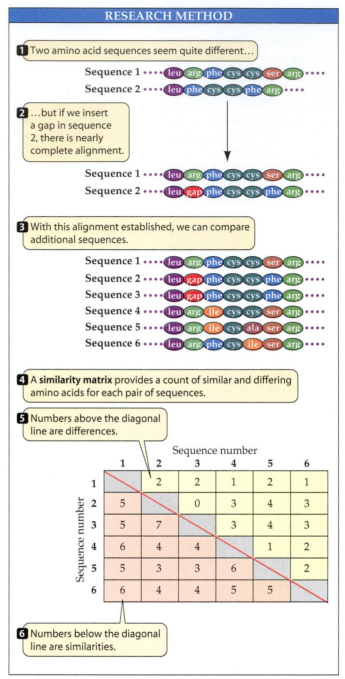

24.3 Amino Acid Sequence Alignment
Inserting a gap allows us to align two sequences so that we can compare homologous amino acids. Once the alignment is established, more sequences can be added and compared. The larger the number of similarities, the more recent the presumed common ancestor of the species.

30,000 years old, plant leaf fossils 40,000 years old, and insects fossilized in amber 135,000,000 years ago.

Once the sequences of amino acids in molecules from different organisms have been determined, they must be compared. A simple example illustrates how this is done. In Figure 24.3, two amino acid sequences (1 and 2) are compared. The two sequences come from homologous proteins in different organisms, and they differ in number and identity of amino acid residues. Our goal is to align these sequences so that we can compare homologous portions of the protein. To do so, we first observe that, although the sequences appear quite different, they would become similar if we were to insert a gap after the first amino acid in sequence 2 (after the leucine residue). In fact, these sequences then differ by only one amino acid at position 6 (serine or phenylalanine). A single insertion aligns the sequences in this case, but longer sequences and those that have diverged more extensively require more elaborate adjustments.

After we have aligned the sequences, we can compare them in several ways. First, we can simply count the number of nucleotides or amino acids that differ between the sequences. Let's add some more sequences to our previous example and compare them with our original two sequences. By adding up the number of similar and different amino acids in the sequences, we can construct a *similarity matrix* (see Figure 24.3). The assumption is that the longer the molecules have been evolving separately, the more differences they will have.

Enough analyses of mammalian genes have been performed to show that the rate of nonsynonymous nucleotide substitution in mammals varies from nearly zero to about 3×10^{-9} substitutions per site per year. Synonymous substitutions in the protein-coding regions of nuclear genes have occurred about 5 times more rapidly than nonsynonymous substitutions; in other words, substitution rates are highest at codon sites that do not change the amino acid being expressed (Figure 24.4). The rate of substitution is even higher in pseudogenes—duplicate copies of genes that have undergone one or more mutations that eliminate their ability to be expressed.

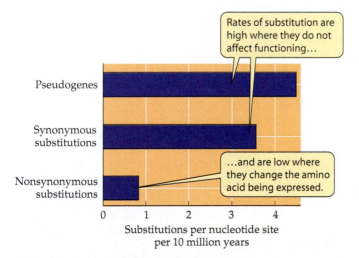

24.4 Rates of Base Substitution Differ
Rates of nonsynonymous substitutions in mammals are much slower than rates of synonymous substitutions and substitutions in pseudogenes.

Why do rates of nucleotide substitution vary so greatly?

The fact that rates of nucleotide substitution are highest at sites and in molecules where they have no functional significance is consistent with the hypothesis that substitution rates at these sites are driven primarily by a combination of mutation and genetic drift. The much slower rates of substitution at sites that *do* affect molecular function is consistent with the view that most nonsynonymous mutations are disadvantageous and are eliminated from the population by natural selection. An interesting consequence of these processes is that, in general, *the more essential a molecule is for cell functioning, the slower the rate of its evolution.*

A molecule that illustrates this principle is the enzyme cytochrome *c*, one component of the respiratory chain of mitochondria. Together with other proteins of the citric acid cycle and respiratory chain, cytochrome *c* is found in all eukaryotes. The amino acid sequences of cytochrome *c* are known for more than 100 species of organisms, including microbial eukaryotes, plants, fungi, and mammals. Within these cytochromes *c* are regions that accumulated changes relatively quickly; for example, positions 44, 89, and 100 differ among many of the organisms compared (Figure 24.5 on pages 442–443).

There are also invariant positions, such as 14, 17, 18, and 80. This particular set of invariant residues is known to interact with the iron-containing heme group that is essential for the functioning of the enzyme. Presumably, because any mutations that changed these amino acids diminished the functioning of the heme group, they were removed by natural selection when they arose.

Using biological molecules as molecular clocks

Earlier in this chapter, we stated the theoretical basis for expecting macromolecules to evolve at constant rates. But do they actually behave as the theory says they should? For example, if we plot the time since the divergence of certain organisms, as determined by the fossil record, against the number of amino acids by which their cytochromes *c* differ, we find that differences in cytochrome *c* sequences have evolved at a relatively constant rate (Figure 24.6).

Many other proteins show constancy in the rate at which they have accumulated changes over time. It would be convenient if the rates of change were the same for all protein molecules. Unfortunately, different molecular clocks tick at different rates. These differences exist because proteins differ in the nature of functional constraints on their evolution.

Despite these differences, the rates at which many molecular clocks tick appear to be relatively constant. This is especially true for nucleotide or amino acid substitutions that do not affect the functioning of the molecule and, hence, the fitness of the organism. Even if the rate of ticking of a molecular clock changes slightly over time, the variations may not be great enough to seriously affect our estimates of the dates of divergences of gene and organism

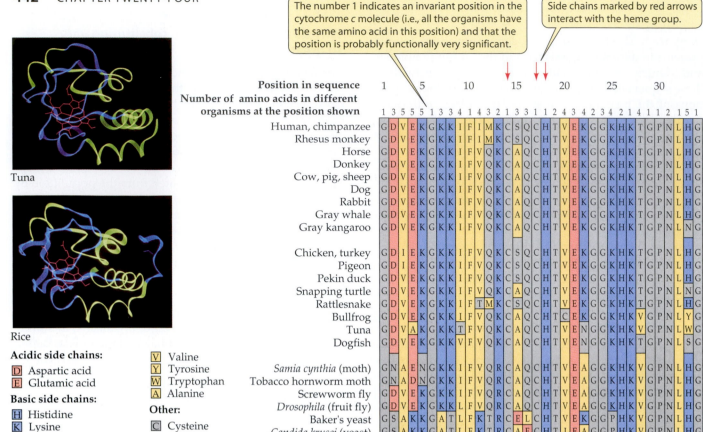

24.5 Amino Acid Sequences of Cytochrome c
The two computer graphics show how similar the three-dimensional structure of tuna and rice cytochrome c are. The amino acid sequences shown here were obtained from analyses of cytochromes c from 33 species of plants, fungi, and animals.

lineages. By comparing the rates of a variety of molecular clocks, further insights can be gained into why different protein molecules have evolved at such different rates.

Where Do New Genes Come From?

The earliest forms of life must have had very few organized nucleic acid sequences. Because we believe that life is monophyletic—that all living organisms arose from a single ancestor—the many thousands of different functional genes in modern organisms must have arisen from these few ancestral genes. How has this happened? By far the most important process appears to be gene duplication.

Gene duplication may involve part of a gene, a single gene, parts of a chromosome, an entire chromosome, or the whole genome (see Chapter 14). We saw in Chapter 22 that duplication of the entire genome (polyploidy) has been important in speciation. Polyploid individuals are usually vi-

able because all of their chromosomes are duplicated, so that they avoid imbalances in gene expression. As we have already discussed, polyploidy is widespread among plants. Genome duplication was probably widespread among animals before the sex chromosomes became differentiated. Among organisms with differentiated sex chromosomes, however, genome duplication disrupts the mechanisms of sex determination.

Duplications of part or all of a chromosome are probably unimportant as sources of new genes because they typically result in severe imbalances in gene products. *Drosophila* in which more than half of one arm of a chromosome is present in three doses (trisomy) do not survive. In humans, trisomies larger than one chromosome are lethal; even smaller ones result in sterility. For example, individuals having three copies of chromosome 21 have Down syndrome, and are usually sterile. Therefore, duplications of whole chromosomes or parts of chromosomes generally are not passed along to any offspring.

Duplication of genes can lead to new gene families

The two identical copies of a gene produced by gene duplication may retain their original function, with the result that the organism produces larger quantities of their RNA

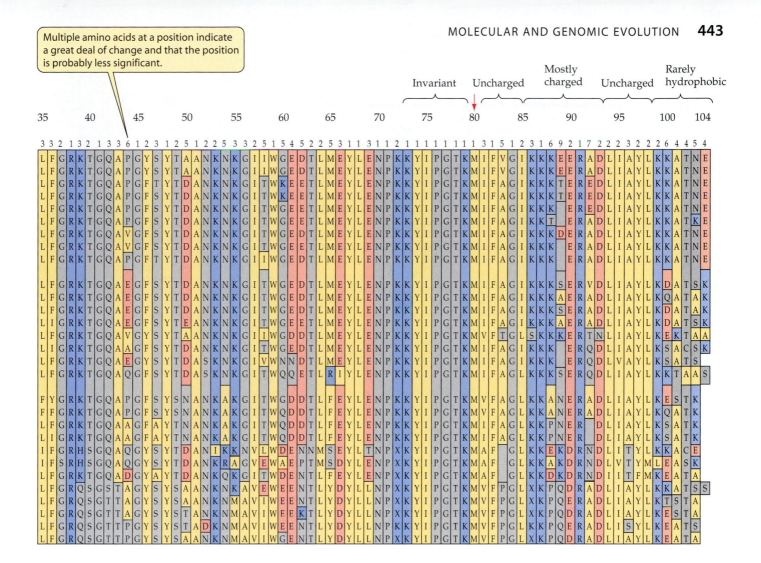

Multiple amino acids at a position indicate a great deal of change and that the position is probably less significant.

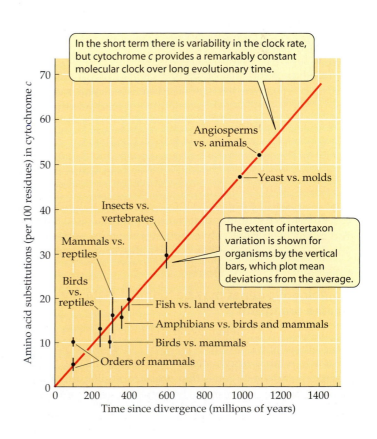

In the short term there is variability in the clock rate, but cytochrome *c* provides a remarkably constant molecular clock over long evolutionary time.

Angiosperms vs. animals

Yeast vs. molds

Insects vs. vertebrates

Mammals vs. reptiles

The extent of intertaxon variation is shown for organisms by the vertical bars, which plot mean deviations from the average.

Birds vs. reptiles

Fish vs. land vertebrates

Amphibians vs. birds and mammals

Birds vs. mammals

Orders of mammals

or protein product. Alternatively, one copy may be incapacitated by the accumulation of deleterious mutations and become a functionless pseudogene. More importantly for evolution, one copy may retain its original function while the other accumulates enough mutations that it can perform a different task. Several successive rounds of duplication may result in a **gene family**, a group of homologous genes with related functions. Members of a gene family are often arrayed in tandem along a chromosome.

Molecular evolution by gene duplication has been well studied in the globin gene family (see Chapter 14). Globins were among the first proteins to be sequenced and their amino acid sequences compared. Humans have three families of globin genes: the myoglobin family, whose single member is located on chromosome 22; the α-globin family, on chromosome 16; and the β-globin family, on chromosome 11 (see Figure 14.9).

Two types of proteins are produced by these three gene families: myoglobin and hemoglobin. Comparisons of their

24.6 Cytochrome c Molecules Evolved at a Constant Rate
Rates of substitution in cytochrome *c* are constant enough that this molecule can be used as a molecular clock.

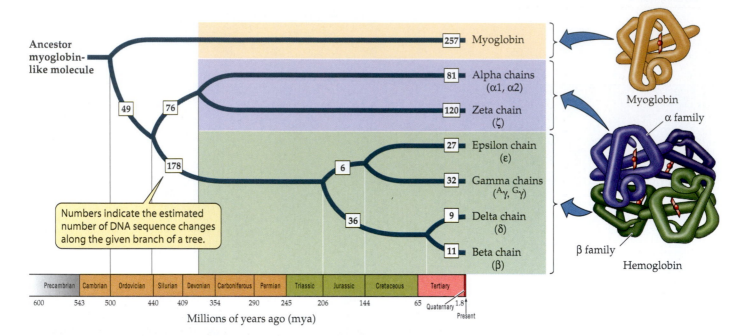

24.7 A Globin Gene Tree
The globin family gene tree suggests that myoglobin diverged from modern hemoglobin precursors about 500 mya, at about the time of the origin of vertebrates.

amino acid sequences strongly suggest that rather than arising from different genes that independently converged on similar functions, the different forms of globins arose through gene duplications. How long the globins have been evolving separately can be inferred by comparing their amino acid sequences. The greater the number of amino acid differences between two globins, the farther back in time was their most recent common ancestor.

To estimate the time of the first globin gene duplication, we can create a *gene tree*, similar to a phylogenetic tree. Our tree is based on the estimated number of base substitutions necessary to account for the observed amino acid differences between the globins. Based on this tree, the earliest organisms known to have both myoglobin and hemoglobin must have lived about 500 million years ago. Thus, the initial duplication event by which myoglobins diverged from all other globins probably happened at least that long ago (Figure 24.7). Assuming that the rate of amino acid substitution has been relatively constant since then—about 100 substitutions per 500 million years—the α and β hemoglobins are estimated to have split about 450 mya.

Homologous genes may be found in distantly related organisms

How can we tell whether genes in different species are really homologous? One way to detect homologous genes in distantly related organisms is to find identical or nearly identical families of genes that produce similar effects in a wide variety of organisms. The homeotic gene complex is one such family.

As we saw in Chapter 16, some mutations in *Drosophila* cause appendages that are appropriate to one body segment to appear in another. Thus, leglike appendages may grow where there should be antennae, or a body segment may be duplicated (Figure 24.8). These unusual changes are caused by genes that occur in two tightly linked clusters that together constitute the homeotic gene complex.

All homeotic genes contain a region called the homeobox, which in *Drosophila* specifies a sequence of 60 amino acids. **Homeobox genes** (Hox genes for short) are responsible for turning many other genes on or off. They are crucial in that they specify body segments and, in the absence of mutations, are responsible for the appropriate development of those segments.

This pattern of developmental control is not unique to fruit flies (see Figure 16.17). More than 350 homeobox elements have been identified in cnidarians, fungi, plants, and animals. Sponges, the simplest of multicellular animals, have only one homeobox-like gene; sea anemones (cnidarians) have up to seven. Vertebrate Hox gene clusters can have as many as thirteen genes. Hox genes occur in the same order along the chromosomes in all animals.

Hox genes carry out similar functions in all animals. A gene found in a simple cnidarian, the hydra, coordinates development of the animal's tentacles. This finding suggests that early in evolution, before animals had strongly developed head, body, and tail regions, Hox genes may have worked to specify the axis of development of the body. Structural and functional similarities strongly suggest that all homeobox genes have a common evolutionary origin,

Normal *Drosophila*

Second thoracic segment

Third thoracic segment

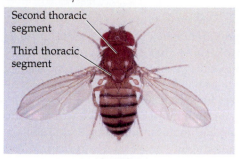

The third thoracic segment is mutated to produce an extra second thoracic segment.

bithorax mutation

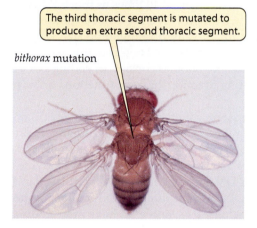

24.8 The bithorax *Mutation in* Drosophila
Mutations of one of the homeotic genes, *bithorax*, transform the third thoracic segment into a second copy of the second thoracic segment. The result is a fly with two pairs of wings.

and that the mechanisms that determine the differentiation of the major morphological regions (body, head, trunk, and tail) may have arisen only once in animal evolution.

How Do Proteins Acquire New Functions?

Evolution as we know it would not have been possible if proteins were unable to change their functional roles. Gene duplication frees one copy of a gene from having to perform its original function. The copy is redundant because the original protein is still encoded by the original gene. Therefore, duplication allows the evolution of entirely novel functions.

Gene families provide evidence of functional diversification

For an example of how gene duplication permits the genes, and the proteins they encode, to evolve different functions, let's look again at the globin families. Hemoglobin, a tetramer consisting of two α chains and two β chains, carries oxygen in the blood. Myoglobin, a monomer, is the primary oxygen storage protein in muscle. It has evolved an affinity for oxygen that is much higher than that of hemoglobin.

In contrast to myoglobin, hemoglobin evolved to be much more refined and diversified in its role as the blood oxygen carrier. Hemoglobin binds oxygen from the lungs or gills, where the oxygen concentration is relatively high,

transports it to regions of low oxygen concentration, and releases it in those areas. With its more complex, tetrameric structure (see Figure 3.7), hemoglobin also is able to transfer hydrogen ions and carbon dioxide in the blood and bind together four molecules of oxygen.

In humans, the α-globin family has four functional genes and three pseudogenes. The four functional genes diversified in function, while the three pseudogenes lost all function. Thus, duplication events may result in increased genomic complexity (as seen with the alternate genes for α-globin) as well as nonfunctional DNA (pseudogenes).

The globin genes show that molecular functions may change after gene duplication, but how do these functional changes happen? We will explore this interesting component of molecular evolution by using lysozyme as an example.

Lysozyme evolved a novel function

Lysozyme is an enzyme found in almost all animals. It is produced in the tears, saliva, and milk of mammals and in the whites of bird eggs. Lysozyme digests the cell walls of bacteria, rupturing and killing them. As a result, lysozyme plays an important role as a first line of defense against invading bacteria. All animals defend themselves against bacteria by digesting them, which is probably why all animals have lysozyme. Some animals, however, also use lysozyme in the digestion of food.

Among mammals, a novel mode of digestion called *foregut fermentation* has evolved twice. The anterior part of the stomach (the foregut) has been converted into a chamber in which bacteria break down ingested plant matter by fermentation. Mammals with this adaptation can obtain nutrients from the otherwise indigestible cellulose of plant material.

Foregut fermentation evolved independently in ruminants, such as cows, and certain leaf-eating monkeys, such as langurs. We know that these evolutionary events were independent because close relatives of langurs and ruminants do not ferment their food in the foregut. In both foregut-fermenting lineages, lysozyme has been modified to play a new, nondefensive role in the foregut. Lysozyme ruptures some of the bacteria that live in the foregut, releasing nutrients, which the mammal absorbs.

How many changes were incorporated into the lysozyme molecule to allow it to function amid the digestive enzymes and acidic conditions of the mammalian foregut? To answer this question, molecular evolutionists compared the amino acid sequences of lysozyme in foregut fermenters and in several of their nonfermenting relatives. They then determined which amino acids differed and which were shared among the species (Table 24.1). Finally, they compared the patterns of these changes with the known phylogenetic relationships among the species.

The most striking finding is that amino acid changes have occurred about twice as rapidly in the lineage leading to langur lysozyme as in any other primate lineage. This high rate of substitution shows that lysozyme went

24.1 Comparisons of Lysozyme Amino Acid Sequences of Different Species

SPECIES	LANGUR	BABOON	HUMAN	RAT	COW	HORSE
Langur*		14	18	38	32	65
Baboon	0		14	33	39	65
Human	0	1		37	41	64
Rat	0	1	0		55	64
Cow*	5	0	0	0		71
Horse	0	0	0	0	1	

Shown above the diagonal line is the number of amino acid sequence *differences* between the two species being compared; below the line are the number of sequences uniquely *shared* by the two species. Asterisks (*) indicate foregut-fermenting species.

through a period of rapid adaptation to the stomachs of langurs. The lysozymes of langurs and cows share five amino acid substitutions, all of which lie on the surface of the lysozyme molecule, well away from the active site. Several of the shared substitutions involve changes from arginine to lysine, which makes the lysozymes more resistant to attack by the pancreatic enzyme trypsin. By understanding the functional significance of amino acid substitutions, molecular evolutionists can explain observed changes in amino acid sequences in terms of the changing function of the protein.

A large body of fossil, morphological, and physiological evidence shows that langurs and cows do not share a recent common ancestor. However, langur and ruminant lysozymes share many amino acid residues that neither animal shares with the lysozymes of their own closer relatives. The lysozymes have converged on a similar sequence despite having very different ancestry; in other words, they are homoplasies. The amino acid residues they share give these lysozymes the ability to lyse the bacteria that ferment leaves in the foregut.

An even more remarkable story emerges if we look at lysozyme in the crop of the hoatzin, a leaf-eating South American cuckoo, the only known avian foregut fermenter. Hoatzins have an enlarged crop that contains resident bacteria and acts as a fermenting chamber. Many of the amino acid changes that occurred in the adaptation of hoatzin crop lysozyme are identical to the changes that evolved in ruminants and langurs. Thus, even though these three groups have evolved independently from one another for more than 300 million years, they have each evolved a similar molecule that enables them to recover nutrients from their fermenting bacteria in a highly acidic environment. The lysozyme story also illustrates why using single molecules to infer phylogenetic histories can be very misleading.

Genome Organization and Evolution

Investigations of the sizes and compositions of the genomes of many species have revealed tremendous variation. Multicellular organisms have more DNA than do single-celled organisms. The genome of *Mycoplasma genitalium*, the sim-

plest free-living prokaryote, has only 470 genes. *Rickettsia prowazekii*, the prokaryote that causes typhus, has 634 genes. The genome of the yeast *Saccharomyces cerevisiae*—a eukaryote—has about 6,000 genes; that of the nematode *Caenorhabditis elegans* has about 20,000 genes (Figure 24.9).

It is not surprising that more complex instructions are needed for building and maintaining a large, complex organism than a small, simpler one. What is surprising is that lungfishes and lilies have about 40 times as much DNA as humans do. Clearly, a lungfish or a lily is not 40 times more complex than a human. Why does genome size vary so enormously among organisms? How did this variation arise? What fraction of genomes consists of coding DNA? Does the noncoding fraction have a function, or is it "junk?"

Some of the apparent differences in genome size disappear when we compare the portion of DNA that actually encodes functional RNA's or proteins. The size of the coding genome varies in a way that makes sense. Eukaryotes have more coding DNA than prokaryotes; vascular plants have more coding DNA than single-celled organisms; invertebrates with wings, legs, and eyes have more coding DNA than roundworms; and vertebrates have more DNA than invertebrates. The species with the largest amount of nuclear DNA has 80,000 times as much as the simplest organisms, but the species with the largest number of genes has only 20 times as many genes as a bacterium. Therefore, most of the variation in genome size is not due to differences in the number of functional genes, but in the amount of noncoding DNA (Figure 24.10).

What maintains such large quantities of noncoding DNA in the cells of most organisms? Most of this DNA appears to be nonfunctional. Much of it may consist of pseudogenes that are simply carried in the genome because

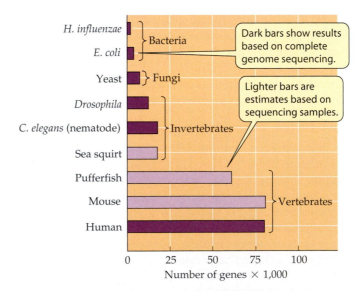

24.9 Complex Organisms Have More Genes than Simpler Organisms
Genome sizes have been measured or estimated in a variety of organisms, ranging from single-celled prokaryotes to vertebrates.

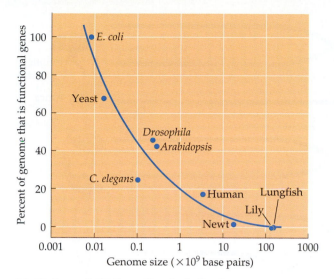

24.10 Some DNA Does Not Code for Genes
Most of the DNA of bacteria and yeast codes for genes, but most of the DNA of more complex organisms is noncoding. We do not know how much of this noncoding DNA is nonfunctional.

the cost of doing so is small. Some of the DNA may be parasitic transposable elements (see Chapter 14) that spread through populations because they reproduce faster than the host's genome. Nonetheless, it is still possible that some of this DNA has undetected functions.

Using Biological Molecules to Reconstruct Phylogenetic Trees

By comparing the structures of molecules from different species, we both gain insights into how the molecules function and acquire a tool for inferring phylogenies. Molecules that have evolved slowly can be used to estimate relationships among organisms that diverged long ago. Molecules that have evolved rapidly are useful for studying organisms that share more recent common ancestors.

As we have seen, there is much evidence to suggest that sequences of amino acids in proteins or base sequences in RNA and DNA that have changed very little during evolution probably have the same function in all species. Sequences that have changed rapidly during evolution either have less important functions in the cell or have undergone major changes in function, as happened to lysozyme in foregut fermenters.

If you are interested in determining the evolutionary relationships of all existing organisms, you must choose a molecule that all organisms possess, such as ribosomal RNA. Equally important, rRNA experiences strong functional constraints; that is, even minor changes in the rRNA sequence prevent ribosomes from functioning properly. As a result, rRNA has evolved so slowly that comparisons of differences among the rRNA's of living organisms can be used to estimate lineage splits that may have happened billions of years ago.

Although molecular data are often the only data available with which to estimate the timing of ancient lineage divisions or reconstruct the phylogenies of prokaryotes,

molecular data are also regularly used in combination with morphological and fossil data. Why do we use molecules when morphology is available? The answer is simple: The more characters that are used (morphological, molecular, fossil, and so on) to reconstruct a phylogeny, the less likely we are to be misled by losses of traits or convergent evolution, as we discussed in Chapter 23.

The more types of molecules we use, the better we can detect homoplasies. For example, if we were to infer a phylogeny only from lysozyme sequences, we might falsely conclude that langurs, cows, and hoatzins are all closely related. However, if we compared the structures of many molecules in those animals, even in the absence of morphological and fossil data, it is clear that these species do not share a recent common ancestor.

No fossils exist to document the most ancient splits in the lineages of life. Molecular evolutionists have used small-subunit rRNA molecules, which are found in all organisms and evolve very slowly, to infer the times of these lineage separations. These rRNA's strongly support the division of living organisms into three major branches, or domains: the Bacteria, the Archaea, and the Eukarya.

The structure of DNA extracted from extinct organisms is being used to determine the evolutionary relationships between those organisms and their surviving relatives. For example, DNA was obtained from the bones and mummified soft tissues of moas—large, flightless birds (weighing up to 200 kg) that lived in New Zealand until humans arrived a thousand years ago and hunted them to extinction. Comparison of their DNA with the DNA of other groups of flightless birds, such as kiwis and rheas, suggests that although kiwis and moas both lived in New Zealand, they are not each other's closest relatives (Figure 24.11). The closest relatives of the moas are unknown, extinct flightless birds that also gave rise to flightless descendants in Australia. Kiwis came to New Zealand more recently; their closest relatives, emus and cassowaries, live in their ancestral home, Australia and New Guinea.

Molecular Studies of Human Evolution

Molecular data have influenced our understanding of our own evolution. Fossil evidence suggests that the hominoid lineage leading to modern humans diverged from a chimpanzee-like lineage about 5 million years ago in Africa. About 2 mya, the human ancestor known as *Homo erectus* arose in Africa, then spread to other continents. Fossil remains of *Homo erectus* have been found in Africa, Indonesia, China, the Middle East, and Europe.

The transition from *Homo erectus* to *Homo sapiens* probably occurred about 400,000 years ago, but there is considerable controversy about the place of origin of modern humans. The "out of Africa" hypothesis suggests a single origin in Africa followed by several dispersals. The "multiple regions" hypothesis, in contrast, proposes several earlier origins of *Homo sapiens* from *Homo erectus* in different regions of Europe, Africa, and Asia (Figure 24.12).

Common ancestor

Common ancestor of all species except rhea

Common ancestor of kiwi, emu, and cassowary

The branching patterns on this tree suggest that kiwis evolved on Australia and then moved to New Zealand at a later date.

Rhea (2 species) — South America

Ostrich — Africa

Cassowary — Australia and New Guinea

Emu — Australia

Kiwi (3 species) — New Zealand

Moa (4 species) — New Zealand

24.11 A Flightless Bird Phylogeny
A phylogeny of living and extinct flightless birds based on DNA sequences suggests that although moas and kiwis both lived in New Zealand, they are not each other's closest relatives. Kiwis are more closely related to the emus and cassowaries of Australia and New Guinea.

The limited number of human fossils and their patchy distribution do not allow us to choose between these two hypotheses. However, DNA sequences of several mitochondrial genes from individuals from more than 100 ethnically distinct modern human populations have provided valuable evidence. Mitochondrial DNA (mtDNA) is useful for studying the recent evolution of closely related species and populations because it accumulates mutations rapidly and because it is maternally inherited. The Y chromosome serves the same role for following male lineages.

The mtDNA sequences of modern humans imply a common ancestry of all mtDNA's about 200,000 years ago. This date of shared ancestry was calculated using the number of nucleotide differences among existing humans; the rate of mtDNA sequence divergence was calibrated using mammals with better fossil records.

The multiple-origins hypothesis requires at least 1 million years of divergence since the last common ancestor. Thus, the mtDNA analysis lends support to the "out of Africa" hypothesis, suggesting that all modern human populations share a recent mitochondrial ancestor. Studies of 26 nuclear genes and a family of repeated sequences (called the *Alu* family), as well as analysis of sequences on the Y chromosome, also support a recent African ancestry. However, the issue is not completely settled, because some types of genetic data can be interpreted in different ways under different assumptions of population structure and of the strength of natural selection acting on the traits. These differences illustrate the importance of gathering data on many different molecules, just as data need to be gathered on many morphological traits, when constructing phylogenies.

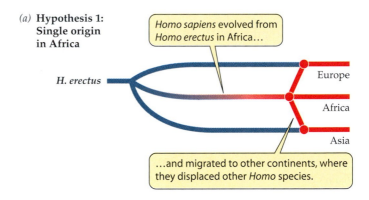

(a) **Hypothesis 1: Single origin in Africa**

Homo sapiens evolved from *Homo erectus* in Africa...

H. erectus

Europe

Africa

Asia

...and migrated to other continents, where they displaced other *Homo* species.

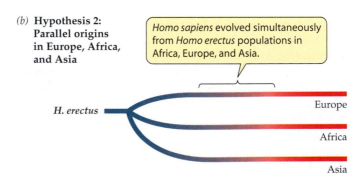

(b) **Hypothesis 2: Parallel origins in Europe, Africa, and Asia**

Homo sapiens evolved simultaneously from *Homo erectus* populations in Africa, Europe, and Asia.

H. erectus

Europe

Africa

Asia

24.12 Two Models for the Origin of Modern Humans
There is considerable controversy among scientists as to whether the transition to *Homo sapiens* (red lineage) took place (a) only in Africa (hypothesis 1), or (b) occurred simultaneously on three continents (hypothesis 2). Current evidence from mtDNA and nuclear genes supports hypothesis 1.

Chapter Summary

What Is Molecular Evolution?

▶ Molecular evolution differs from phenotypic evolution in that mutations and genetic drift are much more important determinants of rates of molecular evolution.

▶ The goals of the study of molecular evolution are to determine the patterns of evolutionary change in the molecules of which organisms are composed, to determine the processes that caused those changes, and to use those insights to help solve other biological problems.

▶ Neutral alleles are fixed slowly, whereas advantageous and disadvantageous alleles are fixed rapidly. **Review Figure 24.1**

Determining and Comparing the Structure of Macromolecules

▶ The polymerase chain reaction method allows biologists to determine the nucleotide base sequences of organisms from their fossilized remains.

▶ Biological molecules can be compared by aligning their sequences. **Review Figure 24.3**

▶ Changes evolve slowly in regions of molecules that are functionally significant, but more rapidly in regions where base substitutions do not affect the functioning of the molecules. **Review Figures 24.4, 24.5**

▶ Rates of amino acid substitutions in some molecules are relatively constant over evolutionary time. **Review Figure 24.6**

Where Do New Genes Come From?

▶ Most new genes arise from gene duplication. The most important types of duplication are genome duplication (polyploidy) and domain duplication.

▶ Globin diversity evolved via gene duplication. **Review Figure 24.7**

▶ Groups of genes that are aligned in the same order on chromosomes of distantly related species are likely to be homologs of one another.

How Do Proteins Acquire New Functions?

▶ Changes in the functions performed by molecules may result from gene duplication if one gene retains the original function and the other evolves a new one.

▶ Homeotic genes have acquired varied functions in development.

Genome Organization and Evolution

▶ The genome sizes of organisms vary more than a hundred-fold, but the amount of coding DNA varies much less. In general, eukaryotes have more coding DNA than do prokaryotes, vascular plants and invertebrate animals have more coding DNA than do single-celled organisms, and vertebrates have more coding DNA than do invertebrates. **Review Figures 24.9, 24.10**

Using Biological Molecules to Reconstruct Phylogenetic Trees

▶ Biological molecules are an important source of data that can be used to infer phylogenetic relationships among organisms. For ancient splits and phylogenies of prokaryotes, molecular data are the only source of information about phylogenetic relationships.

▶ Molecules that have evolved slowly are useful for determining ancient lineage splits. Molecules that have evolved rapidly are useful for determining more recent lineage splits. **Review Figure 24.11**

Molecular Studies of Human Evolution

▶ Comparisons of mtDNA from more than 100 ethnically distinct modern human populations strongly suggest that all modern humans shared a common African ancestor no more than 200,000 years ago. **Review Figure 24.12**

For Dicussion

1. If you were interested in reconstructing the phylogeny of a subgenus of fruit flies using molecular data, what kinds of molecule(s) would you choose to examine? Why? If you wanted to reconstruct the phylogeny of all vertebrates, would you use the same molecule(s)? Why or why not?

2. How have our views about organismal evolution been affected by recent applications of molecular methods to the study of evolution?

3. Discuss the relative importance of molecular characters versus morphological characters in reconstructing the phylogeny of a group of organisms.

4. Existing evidence suggests that for some molecules, a molecular clock ticks at a fairly constant rate, but that rates of change differ widely among molecules. How does this variation limit how and in what ways we can use the concept of a molecular clock to help us answer questions about the evolution of both molecules and organisms?

5. One hypothesis for the existence of large amounts of noncoding ("junk") DNA is that the cost of maintaining all that DNA is so small that natural selection is too weak to reduce it. What other hypotheses might account for the existence of so much noncoding DNA?

6. Why do reconstructions of the phylogenies of genes and phylogenies of the organisms that contain them often differ?

7. We are, by nature, interested in our own evolution. This chapter presented a brief introduction to the application of molecular methods to studying questions about human evolution. Make a short list of additional questions about human evolution and develop a rough outline of the molecules and methods you might bring to bear in addressing these questions.

25 *The Origin of Life on Earth*

SCIENTISTS BELIEVE THAT BETWEEN 10 AND 20 billion years ago there was a mighty explosion. The matter of the universe, which had been highly concentrated, began to spread apart rapidly. Eventually clouds of matter collapsed through gravitational attraction, forming the galaxies—great clusters of hundreds of billions of stars.

Somewhat less than 5 billion years ago, toward the outer edge of our galaxy (the Milky Way), our solar system (the sun, Earth, and our sister planets) took form. Earth probably formed about 4.5 billion years ago by gravitational attraction of rocks of various sizes. As Earth grew by this process, the weight of the outer layers compressed the interior of the planet. The resulting pressures, combined with energy from radioactive decay, heated the interior until it melted.

Within this viscous liquid, the heavier elements settled to produce a fluid iron and nickel core with a radius of approximately 3,700 km that persists to this day. Around the core lies a mantle of dense silicate material, called *magma*, that is 3,000 km thick. Over the mantle is a lighter crust, more than 40 km thick under the continents but as little as 5 km thick in some places under the oceans.

During the first half-billion years of its existence, Earth was bombarded by hundreds of rock bodies left over from the formation of the solar system. Collision with one of these bodies, which was at least as large as Mars, dislodged the material that became the moon. Many of these collisions were large enough to create a superheated atmosphere of vaporized rock that would have vaporized water and sterilized Earth's surface and subsurface.

Before the evolution of life, Earth's mantle and crust released carbon dioxide, nitrogen, and other heavier gases. These gases were held by Earth's gravitational field, and gradually, over several hundred million years, formed a new atmosphere consisting mostly of methane (CH_4), carbon dioxide (CO_2), ammonia (NH_3), hydrogen (H_2), nitrogen (N_2), and water vapor (H_2O). Eventually Earth cooled enough that the water vapor escaping from inside the planet condensed to liquid water and formed the oceans.

After Earth cooled enough for oceans to form and the bombardment was reduced to a very low level, life evolved on Earth. Fossils of complex unicellular life have been found in geological formations dated to 3.5 billion years ago (bya) (Figure 25.1). Scientists now believe that life first appeared on Earth about 4 billion years ago.

The first life must have come from nonliving matter. How did this happen? Under what conditions did life originate on Earth? This chapter describes how scientists try to answer these questions.

How Can We Study a Unique Event that Happened Several Billion Years Ago?

For the most part, scientists seek generalizations about nature. The most powerful scientific theories explain processes that occur repeatedly. Indeed, reproducibility is a key element of the hypothetico-deductive method. The scientific study of life's beginnings is different. The events leading to the origin of life may have happened only once, and we have no direct observational evidence of them. Since

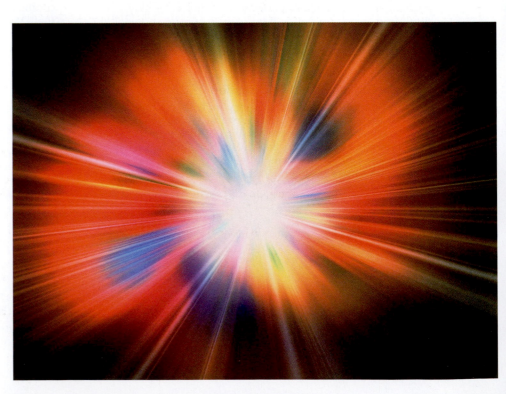

The Big Bang
This computer-generated illustration of the Big Bang helps us visualize what that huge explosion might have looked like.

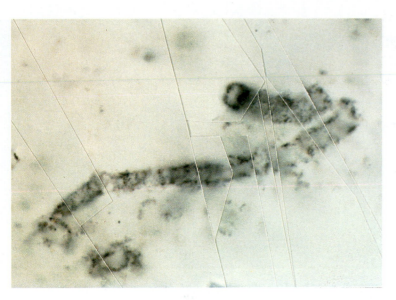

25.1 The Oldest Traces of Life
Some of the oldest known fossils of microscopic unicellular life have been found in Australia. This 3.5 billion-year-old bacterium was found in Western Australia.

then, the state of Earth has been so altered that most of the traces of those early events have vanished. For this reason, the study of life's origins shares many characteristics with the study of history. It is inevitably more speculative than most components of scientific inquiry.

Nevertheless, three scientific principles can guide the study of the origin of life.

▶ The **principle of continuity** states that, because life probably evolved from nonlife by a continuous, gradual process, any stage in life's evolution that we propose should be derivable from preexisting states. In other words, we should not expect to find sudden major changes.

▶ The **signature principle** states that because of this historical continuity, prebiotic processes should leave some signatures—traces—in contemporary biochemistry.

▶ The third principle, which we may call the **no-free-lunch principle**, states that all living organisms require some form of energy for growth. More specifically, they must oxidize some material and obtain energy from that oxidation.

Using these three principles, scientists can focus their attention on hypotheses that are plausible, testable, and worthy of serious consideration.

To remind ourselves what it is that we must explain, let's briefly summarize the essential characteristics of life:

▶ All life is cellular.

▶ Life is based on aqueous solutions.

▶ The major atoms in all cells are carbon, hydrogen, nitrogen, oxygen, phosphorus, and sulfur.

▶ Biochemical reactions take place inside cells.

▶ All proteins are made from the same group of amino acids, all RNA from the same group of ribonucleotides, and all DNA from the same group of deoxyribonucleotides.

▶ All carbohydrates are formed from a small group of sugars, and all phospholipids from a limited group of fatty acids.

▶ The flow of energy in the living world is accompanied by the formation and hydrolysis of phosphate bonds, usually those of ATP.

▶ All cells have an osmotically active barrier composed of lipids and associated proteins.

▶ The genome of every replicating cell is composed of DNA or RNA that is translated into polypeptides.

▶ All cells have ribosomes, and ribosomes are the sites of protein synthesis.

▶ All reactions that proceed rapidly in cells are catalyzed by proteins.

▶ Reproducing biological systems give rise to altered phenotypes as a result of mutated genotypes.

These basic chemical properties are the potential signatures of early life that can guide our study of life's origins. Their ubiquity shows that biochemical evolution has been remarkably conservative.

This conservatism helps us focus the scientific study of how life may have evolved from nonlife. For example, the earliest known molecular fossils were recently found in exceptionally well preserved shales from northwestern Australia, dated at 2.7 bya. These molecules are a type of lipid that is found today in the cell membranes of some photosynthetic cyanobacteria.

Necessary Conditions for the Origin of Life

Living organisms are complex nonequilibrium systems that are maintained by the flow of usable energy—free energy (see Chapter 6). Disordered energy—entropy—cannot do work. Only two possible long-term sources of free energy for metabolism exist: radiation from the sun, and the chemical potential of reduced compounds in Earth's magma that are released when it flows to the surface. Both sources are used by organisms today. Either one, or both, could have powered the origin of life.

Conditions on early Earth differed from those of today

Free oxygen (O_2) probably was not present in Earth's early atmosphere. Any oxygen that was present reacted with hydrogen to form water, and with components of Earth's crust and atmosphere to form iron oxides, silicates, carbon dioxide, and carbon monoxide. Because oxygen was bound up with other elements, Earth had a **reducing** (electron-adding) **atmosphere**. As a setting for chemical reactions, then, early Earth differed fundamentally from present-day Earth, which has an **oxidizing atmosphere** containing large quantities of O_2.

What sort of chemical reactions could have occurred in a reducing environment? Could such reactions have been the

EXPERIMENT

Question: Can organic compounds be generated under conditions similar to those that existed on primeval Earth?

METHOD

A solution of simple chemicals is heated, producing a reducing "atmosphere" of methane, ammonia, hydrogen, and water vapor.

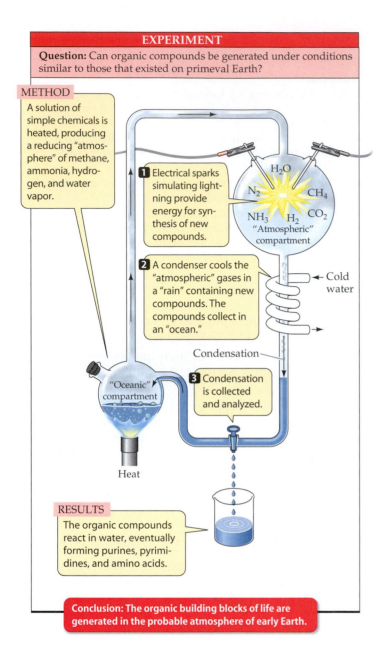

1 Electrical sparks simulating lightning provide energy for synthesis of new compounds.

H_2O

N_2 CH_4

NH_3 H_2 CO_2
"Atmospheric" compartment

2 A condenser cools the "atmospheric" gases in a "rain" containing new compounds. The compounds collect in an "ocean."

Cold water

Condensation

"Oceanic" compartment

3 Condensation is collected and analyzed.

Heat

RESULTS

The organic compounds react in water, eventually forming purines, pyrimidines, and amino acids.

Conclusion: The organic building blocks of life are generated in the probable atmosphere of early Earth.

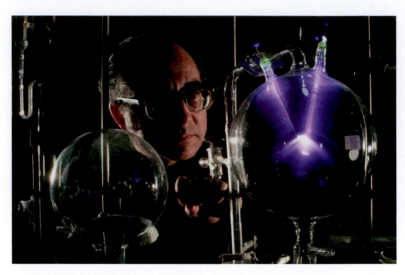

25.2 Synthesis of Prebiotic Molecules in an Experimental Atmosphere
Stanley Miller used an apparatus similar to this one to determine which molecules could be produced in a reducing atmosphere such as existed on early Earth.

absent. These findings suggest that once Earth cooled enough for water to condense and form oceans, molecules of many kinds probably formed. Over millions of years, these organic molecules would have accumulated in the oceans. They would have reached even higher concentrations in drying ponds or on the surfaces of clays.

Polymerization provided diverse macromolecules

The next stage in the sequence leading to life was the generation of large molecules by polymerization of small molecules. Polysaccharides, proteins, and nucleic acids are all polymers formed by the combination of subunits called monomers. As we saw in Chapter 3, polymers are assembled through repeated condensations of monomers. Each of these condensation reactions requires energy. Polymers that formed faster or were more stable would have come to predominate. High concentrations of polymers, in turn, would have stimulated further polymerization by shifting chemical equilibria from unstable monomers to more stable polymers.

Protobionts: Enclosing Prebiotic Systems

The experiments showing that a rich array of prebiotic molecules can be formed under the conditions likely to have existed on early Earth are highly informative, but a prebiotic soup of small molecules does not lead to life. For life to evolve, three additional conditions must be met:

▸ There must be a supply of **replicators**—molecules that are self-reproducing.

▸ The copying of these replicators must be subject to error via *mutation*.

first step toward the origin of life? The first person to investigate these questions was Stanley Miller. In the 1950s, he established an experimental reducing atmosphere of hydrogen, ammonia, methane gas, and water vapor. Through these gases, he passed a spark to simulate lightning, then cooled the system so the gases would condense and collect in an aqueous solution, or "ocean" (Figure 25.2). Within hours, the system contained numerous simple organic compounds (compounds containing carbon, nitrogen, and hydrogen), including, for example, hydrogen cyanide and formaldehyde. Such compounds eventually reacted in water to form amino acids, purines, and pyrimidines—some of the building blocks of life.

The same or similar compounds can be produced under a variety of conditions, including ones that simulate conditions in aquatic environments, provided that free oxygen is

▶ The system of replicators requires a perpetual supply of *free energy* and *partial isolation* from the general environment.

We will describe the source of replicators in the following sections. The requirement for mutation would have been easy to fulfill, because at the high temperatures found on early Earth, prebiotic molecules would have been continually altered as a result of thermal motion.

The evolution of membranes provided partial isolation

Partial isolation from the general environment can be achieved within aggregates of artificially produced prebiotic molecules. Called **protobionts**, these aggregates cannot reproduce, but they can maintain internal chemical environments that differ from their surroundings.

In the 1920s the Russian scientist Alexander Oparin observed that if he shook a mixture of a large protein and a polysaccharide, protobionts formed. Their interiors, which were primarily protein and polysaccharide, with some water, were separated from the surrounding aqueous solution, which had much lower concentrations of proteins and polysaccharides. Such protobionts, known as **coacervates**, are quite stable. They can be formed in solutions of many different types of polymers.

Oparin's coacervates also exhibited a simple form of metabolism. They absorbed substrates, catalyzed reactions, and let the products diffuse back into the aqueous solution (Figure 25.3). However, because these coacervates lacked lipid outer membranes, they differed from the probable precursors of life.

Other protobionts, called **microspheres**, form when mixtures of a variety of artificially produced organic compounds are mixed with cool water. If the mixture of compounds includes lipids, the surface of a microsphere consists of a lipid bilayer, similar to the lipid bilayer of cell membranes.

Membrane components became energy-transducing devices

Molecules that absorb visible or near-ultraviolet light—called **chromophores**—are likely to have been components of the lipid membranes of some protobionts. When light shines on protobionts that have chromophores in their membranes, electric potentials develop across the membranes. Such protobionts can become energy-transducing devices.

Given a continuous flux of light, oxidation–reduction reactions are possible if electrons can be conducted across the membrane. Acid–base-driven reactions are also possible if protons can be conducted across the membrane. These two types of reactions can be coupled because both are driven by the same electric potential.

Today, the principal route of biological energy flow is from solar radiation to an oxidized and a reduced compound and the formation of some type of phosphate. The universality of this process suggests that it may have char-

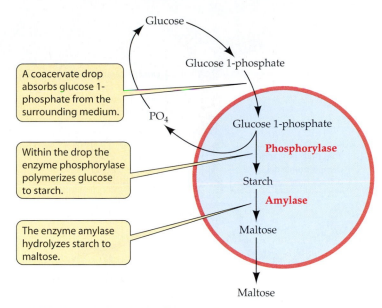

25.3 "Metabolism" of a Coacervate
The properties of artificial coacervates are similar to some of the properties of living cells. They are held intact by a nonlipid membranelike coating. Chemical reactions take place in the interior in the presence of enzymes.

acterized the earliest life and that protolife may have been driven primarily by solar energy.

RNA was probably the first biological catalyst

Some polymers can direct the synthesis of molecules identical to themselves. Which of the molecules on prebiotic Earth were most likely to reproduce themselves? The nucleic acids—the basis of today's genetic code—are good candidates. They are clearly capable of self-copying, and the purine and pyrimidine constituents of nucleotides were formed in Miller's experiment, under conditions similar to those believed to have prevailed on early Earth.

However, there is a problem with this idea. The enzymes that control the types and rates of reactions within organisms are proteins. As you learned in Chapter 12, proteins are synthesized by a process that begins with the transcription of information from DNA to mRNA. The information in mRNA is eventually used to synthesize a polypeptide from amino acids using another kind of RNA, tRNA. This system of protein synthesis (DNA → RNA → protein) probably evolved gradually from much simpler processes. How could such a system have evolved if proteins needed nucleic acids to form, but nucleic acids needed proteins to catalyze their replication? Which came first?

Inability to solve this dilemma held up research on the origin of life for several decades. The first clue came from experiments in the late 1970s. When RNA molecules were added to solutions containing purines and pyrimidines, sequences of 5 to 10 nucleotides were formed. If a simple inorganic ion such as zinc was added, much longer sequences were copied.

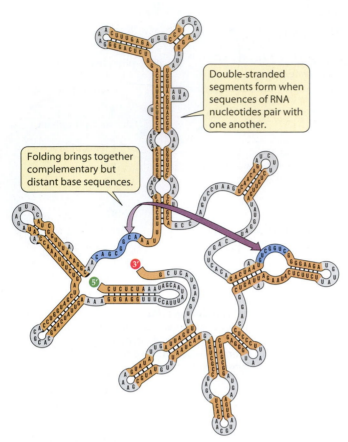

Double-stranded segments form when sequences of RNA nucleotides pair with one another.

Folding brings together complementary but distant base sequences.

25.4 A Ribozyme from a Protist
The folded three-dimensional structure of this catalytic RNA, or ribozyme, enables it to catalyze chemical reactions during protein synthesis. RNA catalysis may have preceded enzyme catalysis in the evolution of life.

The next discovery that provided a solution to the dilemma came in 1981 from scientists studying the excision of introns and the splicing together of exons. They found—entirely contrary to expectations—that these processes took place in the absence of enzymes! The intron itself—a 400-nucleotide sequence of RNA—catalyzed the excision and splicing.

In addition, it was discovered that ribosomes, which contain several molecules of RNA and a variety of proteins, have a catalytic RNA that operates in protein synthesis (see Chapter 12). RNA's that catalyze chemical reactions are called *ribozymes* (Figure 25.4).

Taken together, these discoveries suggest that the first genetic code was based on RNA that catalyzed its own replication as well as catalyzing other chemical reactions. A high concentration of RNA would have been needed so that it could participate in many different chemical reactions. The accumulated products of RNA-catalyzed reactions could then participate in other reactions and form structures. For example, RNAs could have catalyzed the formation of lipidlike molecules that could form plasma membranes, and of proteins that could catalyze the synthesis of other proteins. However, after proteins evolved, they

eventually took over most enzymatic functions because they are better catalysts than RNA and are capable of more diverse specific activities.

To replicate, different RNA's would have competed with one another for monomers. Some RNA molecules would have been better at replicating in certain environments because their base sequences produced the most stable folded structures under the conditions of temperature and salinity they encountered. With their higher rates of replication and greater stability, these RNA molecules would have come to dominate the populations of RNA in their environments. Investigators have simulated the "evolution" of RNA molecules by selecting in test tubes for ribozymes with high catalytic ability. By this method they have produced ribozymes with reaction rates 7 million times faster than the uncatalyzed reaction rate, showing how highly catalytic RNA's might have evolved.

DNA evolved from an RNA template

If the first cells used RNA as their hereditary molecule, then RNA must have provided the template for the synthesis of DNA. In solution, DNA is less stable than RNA. Therefore, DNA probably did not evolve as a hereditary molecule until RNA-based life became enclosed in membranes within which water concentrations were lower than in the surrounding environment. In such cellular environments, DNA is a more stable storage molecule for genetic information than RNA. Therefore, once cells evolved, DNA probably rapidly replaced RNA as the genetic code for most organisms. But by then RNA's had assumed their current roles as intermediaries in the translation of genetic information into proteins.

Photosynthesis Is the Source of Atmospheric O$_2$

The evolution of noncyclic photophosphorylation slightly more than 2 billion years ago changed the course of evolution and changed Earth. The key change was the ability of living organisms to use water as their source of hydrogen:

$$2 \; H_2O \rightarrow 4 \; H^+ + O_2 + 4 \; e^-$$

By chemically splitting H$_2$O, they generated O$_2$ as a waste product and made electrons available for reducing CO$_2$ to form organic compounds.

The ability to split water molecules appeared first in certain sulfur bacteria that evolved into cyanobacteria. Remains of these bacteria are abundantly fossilized in concentrations called **stromatolites**. Cyanobacteria are still forming stromatolites in a few very salty places on Earth (Figure 25.5).

Their water-splitting ability was doubtless the cause of the extraordinary success of cyanobacteria. The O$_2$ they liberated opened the way for the evolution of aerobic oxidation reactions as the energy source for the synthesis of ATP. Aerobic metabolism was much more rapid and efficient than the anaerobic metabolism that had dominated life

(a)

(b)

25.5 Stromatolites
(a) A vertical section through a fossil stromatolite. (b) These rock-like structures are living stromatolites that thrive in the very salty waters of Shark Bay, Western Australia. Layers of cyanobacteria are found in the uppermost parts of the structures.

until then. The success of the cyanobacteria made possible the evolution of the full respiratory chain of reactions now carried out by all aerobic cells.

The evolution of life irrevocably changed the nature of our planet. When it first appeared, oxygen was poisonous to the anaerobic organisms that were living on Earth at the time. Those prokaryotes that evolved a tolerance to O_2 were able to successfully colonize environments empty of other organisms and proliferate in great abundance. Life created the O_2 of our atmosphere, and it removed most of the carbon dioxide from the atmosphere by incorporating it into organic compounds and subsequently transferring it to ocean sediments.

Is Life Evolving from Nonlife Today?

Scientists have gathered information that provides many insights into the origin of life on Earth. Taken together, this information suggests that the evolution of life as we know it was highly probable under the conditions that prevailed on Earth 4 billion years ago. The molecules on which life is based form readily under such conditions, and those molecules readily organize themselves into larger units. Thus, the origin of life may have been almost inevitable.

However, new life apparently is not being assembled from nonliving matter on Earth today. Until the mid-1800s, people believed that life could arise by *spontaneous generation* from nonliving substances—for example, that frogs could arise from moist soil. The experiments that finally disproved the theory of spontaneous generation were performed in 1862 by the great French scientist Louis Pasteur. His experiments showed that microorganisms come only from other microorganisms and that a genuinely sterile solution remains lifeless indefinitely unless contaminated by living creatures (Figure 25.6). As a result of Pasteur's experiments and similar ones by other scientists, most people now accept that all life comes from existing life.

Why is it that new life is not being assembled from nonliving matter on today's Earth? The reason is that simple biological molecules released into today's environment are quickly consumed by existing life. They cannot accumulate to the densities that characterized the "primordial soup," even in anaerobic environments. In aerobic environments, these molecules are quickly oxidized to other forms. Thus, they could not accumulate even if they were not consumed. Generation of life from nonlife on Earth did happen, but it was an event of the remote past. Once life had evolved, it prevented other life from arising from nonlife.

Does Life Exist Elsewhere in the Universe?

People have long speculated about the possibility of life on other planets in our solar system or in other solar systems. Recent evidence suggesting that life may exist on Mars or on one of the moons of Jupiter has fueled these speculations. Whether life exists or has existed on other planets in our solar system may be determined by future explorations of those planets. For other solar systems, we must rely on indirect methods. One approach is to identify the conditions on Earth that enabled life to evolve and that maintain life today.

Life was able to evolve on Earth because a set of conditions existed here that were suitable for the origin of single-celled organisms. For a planet to be able to support simple life, it must be associated with a star that has a relatively constant energy output and be far enough from it to be sufficiently cool for liquid water to form on its surface. These conditions are likely to be found in many places in the universe.

Multicellular organisms, on the other hand, are more exacting in their requirements. A more stable environment is

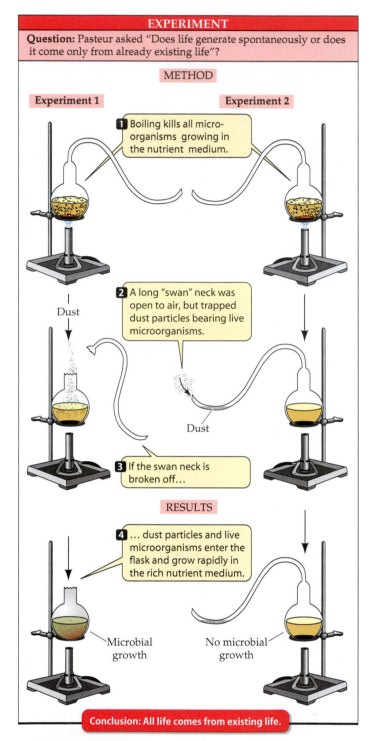

EXPERIMENT

Question: Pasteur asked "Does life generate spontaneously or does it come only from already existing life"?

METHOD

Experiment 1   Experiment 2

1 Boiling kills all micro-organisms growing in the nutrient medium.

Dust

2 A long "swan" neck was open to air, but trapped dust particles bearing live microorganisms.

Dust

3 If the swan neck is broken off…

RESULTS

4 … dust particles and live microorganisms enter the flask and grow rapidly in the rich nutrient medium.

Microbial growth   No microbial growth

Conclusion: All life comes from existing life.

25.6 Experiments Disproved the Spontaneous Generation of Life
Louis Pasteur's classic experiments showed that, under today's conditions, a genuinely sterile solution remains lifeless indefinitely. Only after being "contaminated" by living organisms did life appear in the flasks.

required for their evolution and long-term survival. For multicellular life to evolve and survive, a planet must have other nearby planets large enough to intercept most of the comets and large meteors that would otherwise regularly strike it and obliterate complex organisms. The planet also must have a nearly circular orbit and a rate of spin fast enough that extreme cold does not develop during the long night nor extreme heat during the long day. It also must have a moon large enough to dampen the planet's rotational irregularity.

These conditions result in a water-bathed environment in which temperature fluctuations, both diurnally and seasonally, are relatively small and the rate of occurrence of major perturbations that result in mass extinctions is low. The combination of these conditions is probably extremely rare in the universe. Therefore, even though microbial life may be widespread in the universe, Earth may be the only place, or one of only a few places, where multicellular life evolved and exists today. But the search for other planets that meet these stringent conditions is still in its infancy, and the universe of possibilities is very large.

Archaea and Bacteria had the planet to themselves for almost 3 billion years, and thousands of highly successful species of these prokaryotes exist today. The fossil record we discussed in Chapter 20 shows that although periods of mass extinction have occurred, conditions on Earth have been suitable for multicellular life for nearly a billion years. The result is that today Earth supports a rich and diverse array of species of both unicellular and multicellular organisms. The next section of this book is a brief overview of the many diverse forms life on Earth takes.

Chapter Summary

How Can We Study a Unique Event that Happened Several Billion Years Ago?

▶ Life originated from nonliving matter nearly 4 billion years ago. Even though the origin of life was a unique event, it can be studied scientifically by following three principles—the principle of continuity, the signature principle, and the "no-free-lunch" principle.

Necessary Conditions for the Origin of Life

▶ Conditions on Earth at the time of life's origin differed from those of today because Earth had a reducing atmosphere. Under conditions that resemble Earth's early atmosphere, small molecules essential to living systems form and polymerize. **Review Figure 25.2**

▶ Before life appeared, polymerization reactions generated the carbohydrates, lipids, amino acids, and nucleic acids of which organisms are composed. These molecules accumulated in the oceans.

Protobionts: Enclosing Prebiotic Systems

▶ The earliest protobionts probably had lipid-based membranes. **Review Figure 25.3**

▶ The first genetic material may have been RNA that had a catalytic function as well as an information transfer function. Some RNA's—called ribozymes—have catalytic functions today. **Review Figure 25.4**

▶ DNA probably evolved after RNA-based life became surrounded by membranes that provided an environment in which DNA was stable.

Photosynthesis Is the Source of Atmospheric O_2

▶ Cyanobacteria, which evolved the ability to split water into hydrogen ions and O_2, proliferated and created atmospheric O_2. The accumulation of free O_2 in Earth's atmosphere made possible the evolution of aerobic metabolism.

Is Life Evolving from Nonlife Today?

▶ Because most of the chemical reactions that gave rise to life occur readily under the conditions that prevailed on the early Earth, life's evolution was probably nearly inevitable.

▶ Experiments by Louis Pasteur and others convinced scientists that life does not come from nonlife on Earth today. **Review Figure 25.6**

▶ New life is no longer being assembled from nonliving matter today because simple biological molecules that form in today's environment are quickly oxidized or consumed by existing life.

Does Life Exist Elsewhere in the Universe?

▶ The conditions that permit the evolution and maintenance of simple prokaryotic life may be widespread in the universe, but multicellular life has more stringent requirements, including a planet with a relatively circular orbit, a rapid rate of spin, nearby planets that intercept impacts, and a large moon that stabilizes the planet's orbit. Such conditions may be very rare.

▶ Although conditions on Earth have fluctuated greatly, they have been suitable for multicellular organisms for nearly a billion years.

For Discussion

1. Why is determining the composition of Earth's early atmosphere a key component of inferring how life arose?

2. Why is the ability of ribozymes to catalyze both their own synthesis and the synthesis of proteins so important for understanding the origin of life?

3. Scientists are confident that life no longer arises from non-living matter under current conditions on Earth. Yet, biologists believe that life did arise on this planet, nearly 4 billion years ago, from nonliving matter. How can scientists hold both of these beliefs?

4. Why do biologists believe that the evolution of life was highly probable on early Earth?

5. How might each of the following have been involved in the evolution of coacervates?

   a. Coating coacervate boundaries with lipids

   b. Wave action in bodies of water

   c. Catalysts within coacervates

6. Some people think that intelligent life exists on many planets in the universe. Others think that Earth may be the only planet where complex, multicellular life evolved. Which view do you support? Why?

Part Four

THE EVOLUTION OF DIVERSITY

26 Bacteria and Archaea: The Prokaryotic Domains

A TEAM OF GERMAN SCIENTISTS HAD FOUND some organisms living in oceanic sediments off the coast of Chile. These organisms, too small to be seen except with a microscope, were able to grow and reproduce using sulfur present in the ooze of the ocean floor for their energy supply. Each was a bacterium, well known as being among the smallest of cells.

Now the oceanographers were looking for similar sulfur-using bacteria in sediments off the coast of Namibia in southwestern Africa. And they found them—but this time they didn't need a microscope. Strings of white bacteria, each the size of the period at the end of this sentence, were plainly visible to the naked eye. At up to 0.75 mm in diameter, these cells were the largest bacteria ever found. In comparison to typical bacteria, they were as large as a blue whale—the largest animal in the world—would be compared with a mouse.

How can single-celled bacteria be so different in size, yet carry out the same functions? A key in this case is that most of the huge bacterial cell (named *Thiomargarita namibienses* by scientists) is filled with stored nitrate, which the cell uses to oxidize sulfur. But the "working chemistry" of *Thiomargarita* is remarkably similar to that of microscopic bacterial species.

Bacteria were first identified by the early microscopists some 300 years ago. Bacteria are prokaryotes, but they are not the only prokaryotes. The Archaea is a superficially similar group of microscopic, unicellular prokaryotes. Both the biochemistry and the genetics of bacteria differ in numerous ways from those of archaea. Not until the 1970s did biologists discover how radically different bacteria and archaea really are. And only with the sequencing of an archaean genome in 1996 did we realize just how extensively archaea differ from both bacteria and eukaryotes.

Many biologists acknowledge the antiquity of these lineages and the importance of their differences by recognizing three domains of living things: Bacteria, Archaea, and Eukarya. The domain Bacteria comprises the "true bacteria"; the domain Archaea (from the Greek *archaios*, "ancient") comprises other prokaryotes once called, inaccurately,

"ancient bacteria." The domain Eukarya comprises all other living things on Earth. Dividing the living world in this way, with two prokaryotic domains and a single domain for all the eukaryotes, fits with the current trend toward reflecting evolutionary relationships in classification systems.

In the eight chapters of Part Four, we celebrate and describe the diversity of the living world—the products of evolution. This chapter focuses on the two prokaryotic domains. Chapters 27–33 deal with the protists and the kingdoms Plantae, Fungi, and Animalia.

In this chapter, we will pay close attention to the ways in which the two domains of prokaryotic organisms resemble each other, and how they differ. We will describe the impediments to the resolution of evolutionary relationships among the prokaryotes. Then we will survey the surprising diversity of organisms within each of the two domains, relating the characteristics of different prokaryotic groups to their roles in the biosphere and in our lives.

Why Three Domains?

What does it mean to be *different*? You and the person nearest you look very different—certainly you appear more dif-

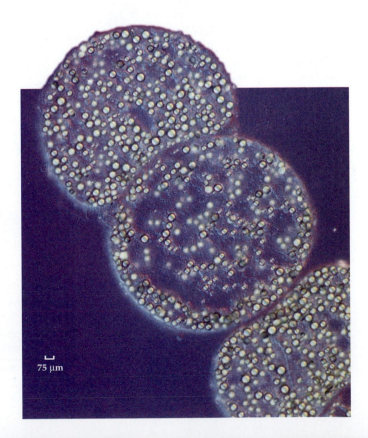

The Largest Known Bacterium
Three cells of the bacterium *Thiomargarita namibiensis*. The middle cell is about 0.2 mm in diameter; the scale bar at the left represents the size of a large "typical" bacterium, giving a sense of the size of this giant among its kind. The light dots are globules of sulfur.

75 μm

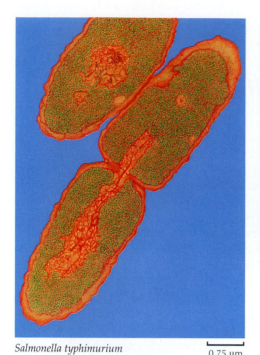

Salmonella typhimurium   0.75 μm

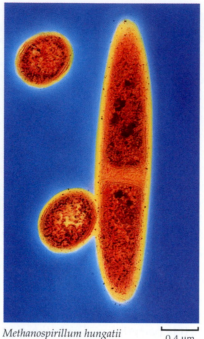

Methanospirillum hungatii   0.4 μm

26.1 Very Different Prokaryotes
In each image, one of the cells has nearly finished dividing. On the left are bacteria; on the right are archaea, which are more closely related to eukaryotic organisms than they are to the bacteria.

mon ancestor and that the present-day archaea share a more recent common ancestor with eukaryotes than they do with bacteria (Figure 26.2). Because of the ancient time at which these three lineages diverged, the major differences among the three kinds of organisms, and especially the fact that the archaea are more closely related to the eukaryotes than are either of those groups to the bacteria, many biologists agree that it makes sense to treat these three groups as *domains*—a higher taxonomic category than *kingdoms*. To treat all the prokaryotes as a single kingdom within a five-kingdom classification of organisms would result in a kingdom that is paraphyletic. That is, a single kingdom "Prokaryotes" would not include all the descendants of their common ancestor. (See Chapter 23, especially Figure 23.11, for a discussion of paraphyletic groups.) The domain concept is still controversial, and it may have to be abandoned if new data fail to support it. In this book, however, we will use the domain concept.

ferent than the two cells shown in Figure 26.1. But the two of you are members of the same species, and these two tiny organisms are classified in entirely separate domains. You (in the domain Eukarya) and those two prokaryotes (in the domains Bacteria and Archaea) have a lot in common. Members of all three domains conduct glycolysis, and they replicate their DNA semiconservatively. In all three, the DNA encodes polypeptides that are produced by transcription and translation, and the cells have plasma membranes and ribosomes in abundance.

As a member of the domain Eukarya, you have cells with nuclei, membrane-enclosed organelles, and a cytoskeleton—things that no prokaryote has. However, a glance at Table 26.1 will show you that there are also major differences, most of which cannot be seen even under the microscope, between the two prokaryotic domains. In some ways the archaea are more like us; in other ways they are more like bacteria.

Genetic studies have led many biologists to conclude that all three domains had a single com-

The common ancestor of all three domains was prokaryotic. Its genetic material was DNA; common machinery for transcription and translation produced RNA's and proteins, respectively. It probably had a circular chromosome, and many of its structural genes were grouped into operons.

26.1   The Three Domains of Life on Earth

CHARACTERISTIC	BACTERIA	ARCHAEA	EUKARYA
		DOMAIN	
Membrane-enclosed nucleus	Absent	Absent	*Present*
Membrane-enclosed organelles	Absent	Absent	*Present*
Peptidoglycan in cell wall	*Present*	Absent	Absent
Membrane lipids	Ester-linked	*Ether-linked*	Ester-linked
	Unbranched	*Branched*	Unbranched
Ribosomes[a]	70S	70S	*80S*
Initiator tRNA	*Formylmethionine*	Methionine	Methionine
Operons	Yes	Yes	*No*
Plasmids	Yes	Yes	*Rare*
RNA polymerases	One	Several	Three
Sensitive to chloramphenicol and streptomycin	*Yes*	No	No
Ribosomes sensitive to diphtheria toxin	*No*	Yes	Yes
Some are methanogens	No	*Yes*	No
Some fix nitrogen	Yes	Yes	*No*
Some conduct chlorophyll-based photosynthesis	Yes	*No*	Yes

[a] 70S ribosomes are smaller than 80S ribosomes.

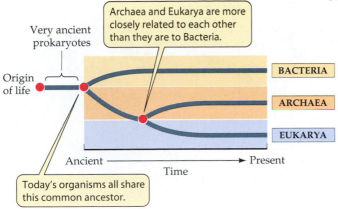

Very ancient prokaryotes

Archaea and Eukarya are more closely related to each other than they are to Bacteria.

Origin of life

BACTERIA

ARCHAEA

EUKARYA

Today's organisms all share this common ancestor.

Ancient —————→ Present

Time

26.2 The Three Domains of the Living World
Biologists believe that the three domains share a common pro-karyotic ancestor. The relationships shown here are controversial.

The Archaea, Bacteria, and Eukarya of today are all the products of billions of years of natural selection and genetic drift, and they are all highly adapted to present-day environments. None are "primitive." The common ancestor of the Archaea and the Eukarya probably lived more than 2 billion years ago, and the common ancestor of the Archaea, the Eukarya, and the Bacteria probably lived more than 3 billion years ago.

The earliest prokaryotic fossils date back at least 3.5 billion years, as we saw in Chapter 25, and these ancient fossils indicate that there was considerable diversity among the prokaryotes even during the earliest days of life. The prokaryotes were alone on Earth for a very long time, adapting to new environments and to changes in existing environments.

General Biology of the Prokaryotes

There are many, many prokaryotes around us—everywhere. Although most are so small that we cannot see them with the naked eye, the prokaryotes are the most successful of all creatures on Earth, if success is measured by numbers of individuals. The bacteria in one person's intestinal tract, for example, outnumber all the humans who have ever lived, and even the total number of *human* cells in that person's body. Some of these bacteria form a thick lining along the intestinal wall.

Although small, prokaryotes play many critical roles in the biosphere, interacting in one way or another with every other living thing. In this section on the general biology of the prokaryotes, we'll see that some perform key steps in the cycling of nitrogen, sulfur, and carbon. Other prokaryotes trap energy from the sun or from inorganic chemical sources, and some help animals digest their food. The members of the two prokaryotic domains outdo all other groups in metabolic diversity. Eukaryotes, in contrast, are much more diverse in size and shape, but their metabolism is much less diverse. In fact, much of the energy metabolism of eukaryotes is carried out in organelles—mitochondria and chloroplasts—that are descended from bacteria.

Prokaryotes are found in every conceivable habitat on the planet, from the coldest to the hottest, from the most acidic to the most alkaline, and to the saltiest. Some live where oxygen is abundant and others where there is no oxygen at all. They have established themselves at the bottom of the seas, in rocks more than 2 km into Earth's solid crust, and even inside other organisms, large and small. Their effects on our environment are diverse and profound.

Prokaryotes and their associations take a few characteristic forms

Three shapes are particularly common among the prokaryotes: spheres, rods, and curved or spiral forms (Figure 26.3). A spherical prokaryote is called a **coccus** (plural cocci). Cocci may live singly or may associate in two- or three-dimensional arrays as chains, plates, or blocks of cells. A rod-shaped prokaryote is called a **bacillus** (plural bacilli). Bacilli and spiral forms, the third main prokaryotic shape, may be single or may form chains.

Prokaryotes are almost all unicellular, although some multicellular ones are known. Associations such as chains do not signify multicellularity, because each cell is fully viable and independent. Associations arise as cells adhere to one another after reproducing by fission. Some bacteria associate in chains that become enclosed within delicate tubular sheaths. These associations are called **filaments**. All the cells of a filament divide simultaneously.

26.3 Shapes of Prokaryotic Cells
(a) These spherical cocci of an acid-producing bacterium grow in the mammalian gut. (b) Rod-shaped E. coli are the most thoroughly studied of all bacteria—indeed, of almost any organism on Earth. (c) A freshwater spiral bacteria species. The cells move by means of the tufts of flagella at each pole.

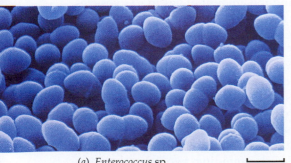

(a) *Enterococcus* sp.   1 µm

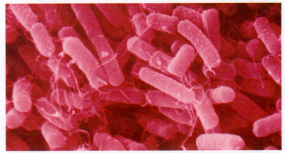

(b) *Escherichia coli*   1 µm

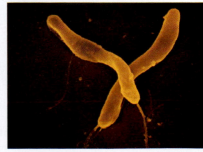

(c) *Aquaspirillum sinosum*   1 µm

Prokaryotes lack nuclei, organelles, and a cytoskeleton

The architectures of prokaryotic and eukaryotic cells were compared in Chapter 4. The basic unit of archaea and bacteria is the prokaryotic cell (see Figure 4.5), which contains a full complement of genetic and protein-synthesizing systems, including DNA, RNA, and all the enzymes needed to transcribe and translate the genetic information into proteins. The prokaryotic cell also contains at least one system for generating the ATP it needs.

In what follows, bear in mind that most of what we know about the structure of prokaryotes comes from studies of bacteria. We still know relatively little about the diversity of archaea, although the pace of research on archaea is accelerating.

The prokaryotic cell differs from the eukaryotic cell in three important ways. First, the organization and replication of the genetic material differs. The DNA of the prokaryotic cell is not organized within a membrane-enclosed nucleus. DNA molecules in prokaryotes are usually circular; in the best-studied prokaryotes, there is a single chromosome, but there are often plasmids as well (see Chapter 13).

Second, prokaryotes have none of the membrane-enclosed cytoplasmic organelles that modern eukaryotes have—mitochondria, Golgi apparatus, and others. However, the cytoplasm of a prokaryotic cell may contain a variety of infoldings of the plasma membrane (see Figure 4.6) and photosynthetic membrane systems not found in eukaryotes. Membranous infoldings frequently associate with new cell walls during cell division. In electron micrographs, the DNA of a bacterial cell is often seen attached to such an infolding, called a *mesosome* (Figure 26.4).

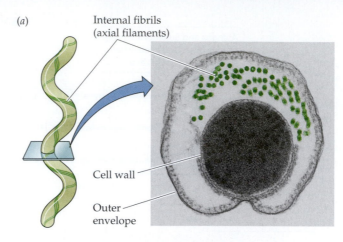

(a) Internal fibrils (axial filaments)

Cell wall

Outer envelope

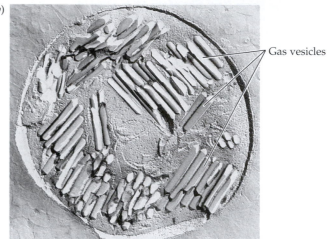

Gas vesicles

26.5 Structures Associated with Prokaryote Motility
(a) A spirochete from the gut of a termite, seen in cross section, shows the fibrils used to produce a rolling motion. (b) Gas vesicles in a cyanobacterium, visualized by the freeze-etch technique.

Third, prokaryotic cells lack a cytoskeleton, and, without the cytoskeletal proteins, they lack mitosis. Prokaryotic cells divide by their own elaborate method, **fission**, after replicating their DNA.

Prokaryotes have distinctive modes of locomotion

Although many prokaryotes are not motile, others can move by one of several means. Some spiral bacteria called spirochetes use a rolling motion made possible by internal fibrils (Figure 26.5a). Many cyanobacteria and some other bacteria use various poorly understood gliding mechanisms, including rolling. Some aquatic prokaryotes, including some cyanobacteria, can move slowly up and down in the water by adjusting the amount of gas in gas vesicles (Figure 26.5b). By far the most common type of locomotion in prokaryotes is that driven by flagella.

Bacterial flagella are whiplike filaments that extend singly or in tufts from one or both ends of the cell (see Figure 26.3c and 26.8a), or all around it (Figure 26.6). A bacterial flagellum consists of a single fibril made of the protein *flagellin*, projecting from the cell surface (see Figure 4.7). In contrast, the flagellum of eukaryotes is enclosed by the plasma membrane and usually contains a circle of nine pairs of microtubules surrounding two central micro-

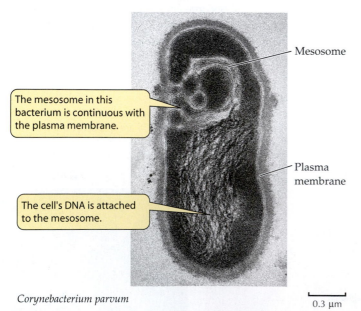

Mesosome

The mesosome in this bacterium is continuous with the plasma membrane.

Plasma membrane

The cell's DNA is attached to the mesosome.

Corynebacterium parvum

0.3 μm

26.4 Some Prokaryotes Have Internal Membranes
Unlike eukaryotic organelles, the mesosome in this bacterial cell is not a separate, membrane-enclosed compartment.

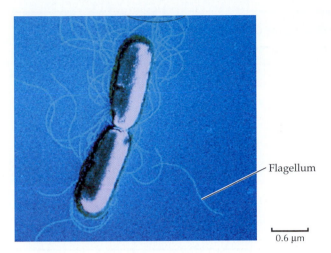

26.6 Some Bacteria Use Flagella for Locomotion
Flagella surround the rod-shaped cells of this *Bacillus* species.

tubules, all containing the protein tubulin, along with many other, associated proteins. The prokaryotic flagellum rotates about its base, rather than beating, as a eukaryotic flagellum or cilium does.

Prokaryotes have distinctive cell walls

Most prokaryotes have a thick and relatively stiff cell wall. This wall is quite different from the cell walls of plants and algae, which contain cellulose and other polysaccharides, and of fungi, which contain chitin. Almost all bacteria have cell walls containing *peptidoglycan* (a polymer of amino sugars). Archaean cell walls are of differing types, but most contain significant amounts of protein. One group of archaea has pseudopeptidoglycan in its wall; as you have probably already guessed from the prefix *pseudo-*, pseudopeptidoglycan is similar to, but distinct from, the peptidoglycan of bacteria. Peptidoglycan is a substance unique to

bacteria; its absence from the walls of archaea indicates a key difference between the two prokaryotic domains.

In 1884 Hans Christian Gram, a Danish physician, developed a simple staining process that has lasted into our high-technology era as the single most common tool in the identification of bacteria. The **Gram stain** separates most types of bacteria into two distinct groups, Gram-positive and Gram-negative, on the basis of their cell wall structure (Figure 26.7). A smear of cells on a microscope slide is soaked in a violet dye and treated with iodine; it is then washed with alcohol and counterstained with safranine (a red dye). **Gram-positive** bacteria retain the violet dye and appear blue to purple (Figure 26.7*a*). The alcohol washes the violet stain out of **Gram-negative** cells; these cells then pick up the safranine counterstain and appear pink to red (Figure 26.7*b*). Gram-staining characteristics are a crucial consideration in classifying some kinds of bacteria and are important in determining the identity of bacteria in an unknown sample. Mycoplasmas, which lack cell walls, are not stained at all by the Gram stain.

For the majority of the bacteria, the Gram-staining results correlate with the structure of the cell wall. Peptidoglycan forms a thick layer outside the plasma membrane of Gram-positive bacteria. The Gram-negative cell wall usually has only one-fifth as much peptidoglycan, and outside the peptidoglycan layer the cell is surrounded by a second, outer membrane quite distinct in chemical makeup from the plasma membrane (see Figure 26.7*b*). The space between the inner (plasma) and outer membranes of Gram-negative bacteria is called the *periplasmic space*. The periplasmic space contains enzymes that are important in

26.7 The Gram Stain and the Cell Wall
When treated with a Gram stain, the cell wall components of different bacteria react in one of two ways. (*a*) Gram-positive bacteria retain the violet dye and appear deep blue or purple; the pink counterstain surrounds the cells in this micrograph. (*b*) Gram-negative bacteria do not retain the violet dye but are made visible on the slide by the counterstain and appear pink-red.

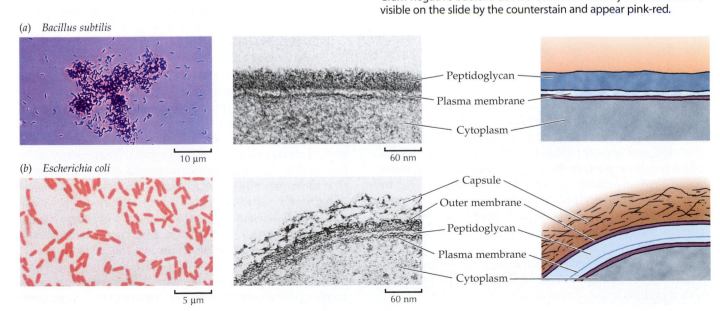

(a) *Bacillus subtilis*

Peptidoglycan
Plasma membrane
Cytoplasm

10 μm   60 nm

(b) *Escherichia coli*

Capsule
Outer membrane
Peptidoglycan
Plasma membrane
Cytoplasm

5 μm   60 nm

digesting some materials, transporting others, and detecting chemical gradients in the environment.

The consequences of the different features of prokaryotic cell walls are numerous and relate to the disease-causing characteristics of some prokaryotes. Indeed, the cell wall is a favorite target in medical combat against diseases that are caused by prokaryotes, because it has no counterpart in eukaryotic cells. Antibiotics and other agents that specifically interfere with the synthesis of peptidoglycan-containing cell walls tend to have little, if any, effect on the cells of humans and other eukaryotes.

Prokaryotes reproduce asexually, but genetic recombination does occur

Prokaryotes reproduce by fission, an asexual process. Recall, however, that there are also processes—transformation, conjugation, and transduction—that allow the exchange of genetic information between some prokaryotes quite apart from either sex or reproduction (see Chapter 13).

Many prokaryotes multiply very rapidly. One of the fastest is the bacterium *Escherichia coli*, which under optimal conditions has a generation time of about 20 minutes. The shortest known prokaryote generation times are about 10 minutes. Values of 1 to 3 hours are common; some extend to days. Bacteria living in rock deep in Earth's crust may suspend their growth for more than a century without dividing and then grow for a few days before suspending growth again.

Prokaryotes have exploited many metabolic possibilities

The long evolutionary history of the bacteria and archaea, including their explorations of new environments, has led to the extraordinary diversity of their metabolic "lifestyles"—their use or nonuse of oxygen, their energy sources, the sources of their carbon atoms, and the materials they secrete.

ANAEROBIC VERSUS AEROBIC METABOLISM. Some prokaryotes can live only by anaerobic metabolism because oxygen gas is poisonous to them. These oxygen-sensitive organisms are called **obligate anaerobes**.

Other organisms can shift their metabolism between anaerobic and aerobic modes (see Chapter 7) and thus are called **facultative anaerobes**. Some facultative anaerobes cannot conduct cellular respiration, but are not damaged by oxygen when it is present. Many prokaryotes are facultative anaerobes that alternate between anaerobic metabolism (such as fermentation) and cellular respiration as conditions dictate.

At the other extreme from the obligate anaerobes, some prokaryotes are **obligate aerobes**, unable to survive for extended periods in the *absence* of oxygen.

26.2	How Organisms Obtain Their Energy and Carbon	
NUTRITIONAL CATEGORY	**ENERGY SOURCE**	**CARBON SOURCE**
Photoautotrophs (some Bacteria, some Eukarya)	Light	Carbon dioxide
Photoheterotrophs (some Bacteria)	Light	Organic compounds
Chemoautotrophs (some Bacteria, a few Archaea)	Inorganic substances	Carbon dioxide
Chemoheterotrophs (found in all three domains)	Organic compounds	Organic compounds

NUTRITIONAL CATEGORIES. Biologists recognize four broad nutritional categories of organisms: photoautotrophs, photoheterotrophs, chemoautotrophs, and chemoheterotrophs. Prokaryotes are represented in all four groups (Table 26.2).

Photoautotrophs are photosynthetic. They use light as their source of energy and carbon dioxide as their source of carbon. Like the photosynthetic eukaryotes, the cyanobacteria, one group of photoautotrophic bacteria, use chlorophyll *a* as their key photosynthetic pigment and produce oxygen as a by-product of noncyclic photophosphorylation (see Chapter 8).

By contrast, the other photosynthetic bacteria use bacteriochlorophyll as their key photosynthetic pigment, and they do not release oxygen gas. Some of these photosynthesizers produce particles of pure sulfur instead because hydrogen sulfide (H_2S), rather than H_2O, is their electron donor for photophosphorylation (Figure 26.8*a*). Bacteriochlorophyll absorbs light of longer wavelengths than the chlorophyll used by all other photosynthesizing organisms does. As a result, bacteria using this pigment can grow in water beneath fairly dense layers of algae, using light of wavelengths that are not appreciably absorbed by the algae (Figure 26.8*b*).

Photoheterotrophs use light as their source of energy, but must obtain their carbon atoms from organic compounds made by other organisms. They use compounds such as carbohydrates, fatty acids, and alcohols as their organic "food." The purple nonsulfur bacteria, among others, are photoheterotrophs.

Chemoautotrophs obtain their energy by oxidizing inorganic substances, and they use some of that energy to fix carbon dioxide. Some chemoautotrophs use reactions identical to those of the photosynthetic carbon reduction cycle (see Chapter 8), but others use other pathways to fix carbon dioxide. Some bacteria oxidize ammonia or nitrite ions to form nitrate ions. Others oxidize hydrogen gas, hydrogen sulfide, sulfur, and other materials. Some archaea are chemoautotrophs (Figure 26.9).

Some deep-sea ecosystems are based on chemoautotrophic prokaryotes that are incorporated into large com-

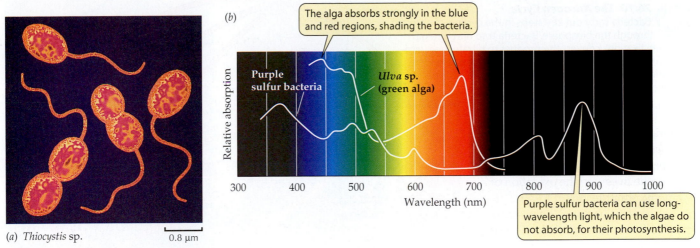

(b)

The alga absorbs strongly in the blue and red regions, shading the bacteria.

Purple sulfur bacteria

Ulva sp. (green alga)

Relative absorption

Wavelength (nm)

Purple sulfur bacteria can use long-wavelength light, which the algae do not absorb, for their photosynthesis.

(a) Thiocystis sp. 0.8 μm

26.8 Some Bacteria are Photosynthetic
(*a*) Cells of purple sulfur bacteria store granules of sulfur that they produce via anaerobic photosynthesis. (*b*) *Ulva*, a green alga, absorbs no light of wavelengths longer than 750 nm. Purple sulfur bacteria can conduct photosynthesis using the longer wavelengths that pass through the algae.

munities of crabs, mollusks, and giant worms, all living in near-boiling water at a depth of 2,500 meters, below any hint of light from the sun, but in the immediate neighborhood of volcanic vents in the ocean floor. These bacteria obtain energy by oxidizing hydrogen sulfide and other substances released from the vents.

Finally, **chemoheterotrophs** obtain both energy and carbon atoms from one or more complex organic compounds.

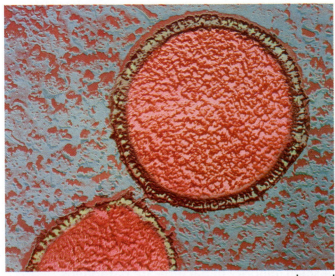

Staphylothermus marinus 0.2 μm

26.9 Chemoautotrophs in Hot Water
These archaea are chemoautotrophs that live in the extremely hot water surrounding deep-sea volcanic vents. Chemoautotrophic archaea and bacteria fix carbon and support the nutritional needs of entire communities of hydrothermal vent organisms that thrive far below the reach of sunlight.

Most known bacteria and archaea are chemoheterotrophs—as are all animals and fungi, and many protists.

NITROGEN AND SULFUR METABOLISM. Some bacteria carry out respiratory electron transport without using oxygen as an electron acceptor. These forms use oxidized inorganic ions such as nitrate, nitrite, or sulfate as electron acceptors. Among these organisms are the ocean-dwelling bacteria mentioned at the beginning of this chapter. Other examples include the denitrifiers, bacteria that return nitrogen to the atmosphere as nitrogen gas (N_2), completing the cycle of nitrogen in nature. These normally aerobic bacteria, mostly species of the genera *Bacillus* and *Pseudomonas*, use nitrate (NO_3^-) in place of oxygen if they are kept under anaerobic conditions:

$$2\ NO_3^- + 10\ e^- + 12\ H^+ \rightarrow N_2 + 6\ H_2O$$

Nitrogen fixers convert atmospheric nitrogen gas into chemical forms usable by the nitrogen fixers themselves and by other living things. Some, for example, convert nitrogen gas to ammonia:

$$N_2 + 6\ H \rightarrow 2\ NH_3$$

All organisms require nitrogen for their proteins, nucleic acids, and other important compounds. The vital process of nitrogen fixation is carried out by a wide variety of bacteria, including cyanobacteria, but by no other organisms. We'll discuss this process in detail in Chapter 36.

Ammonia is oxidized to nitrate by the process of **nitrification**. This process is carried out in the soil by chemoautotrophic bacteria called nitrifiers. Bacteria of two genera, *Nitrosomonas* and *Nitrosococcus*, convert ammonia to nitrite ions (NO_2^-), and *Nitrobacter* oxidizes nitrite to nitrate (NO_3^-). What do the nitrifiers get out of these reactions? Their chemosynthesis is powered by the energy released by oxidation of ammonia or nitrite. For example, by passing the electrons from nitrite through an electron transport chain, *Nitrobacter* can make ATP, and using some of this ATP, it can also make NADH. With the ATP and NADH, the

26.10 The Nitrogen Cycle
Bacteria carry out key steps in the cycling of nitrogen through the biosphere. Bacteria trap nitrogen gas (nitrogen fixation), convert the product to nitrate ions (nitrification), and return nitrogen gas to the atmosphere (denitrification) in the final step. Plants provide the nitrate reduction steps.

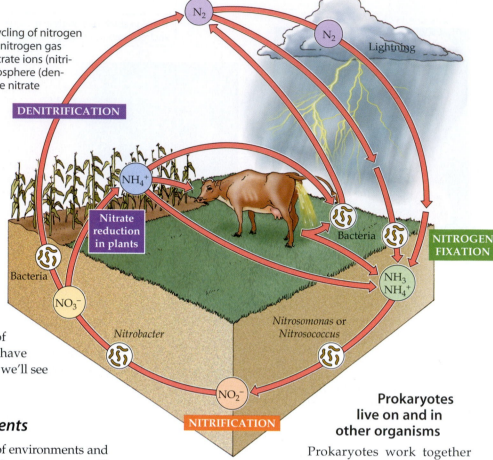

bacterium can convert CO_2 and H_2O to glucose and other foods. The nitrifiers base their entire biochemistry—their entire lives—on the oxidation of ammonia or nitrite ions.

Numerous bacteria base their metabolism on the modification of sulfur-containing ions and compounds in their environments. As examples, we have already mentioned the photoautotrophic bacteria and chemoautotrophic archaea that use H_2S as an electron donor in place of H_2O. Such uses of nitrogen and sulfur have obvious environmental implications, as we'll see in the next section.

Prokaryotes in Their Environments

Prokaryotes live in and exploit all sorts of environments and are parts of many ecosystems. In the following pages, we'll examine prokaryotes living in soils, in water, and even in other living things, where they may exist in neutral, benevolent, or parasitic relationship with the host's tissues.

Prokaryotes are important players in element cycling

Animals depend on photosynthetic plants for their food, directly or indirectly. But plants depend on other organisms—prokaryotes—for their own nutrition. The extent and diversity of life on Earth would not be possible without biological nitrogen fixation (Figure 26.10). Nitrifiers are also crucial to the biosphere, because they convert the products of nitrogen fixation into nitrate ions, the form of nitrogen most easily used by many plants. Plants, in turn, are the source of nitrogen compounds for animals and fungi. Denitrifiers also play a key role in keeping the nitrogen cycle going. Without denitrifiers, which convert nitrate ions back into nitrogen gas, all forms of nitrogen would leach from the soil and end up in lakes and oceans, making life on land impossible. Other prokaryotes contribute to a similar cycle for sulfur.

In the ancient past, the cyanobacteria had an equally dramatic impact on life: Their photosynthesis generated oxygen, converting Earth from an anaerobic to an aerobic environment. The result was the wholesale loss of species that couldn't tolerate the O_2 generated by the cyanobacteria, but this transformation made possible the evolution of cellular respiration and the subsequent explosion of eukaryotic life.

Prokaryotes live on and in other organisms

Prokaryotes work together with eukaryotes in many ways. In fact, mitochondria and chloroplasts are descended from what were once free-living bacteria. Much later in evolutionary history, some plants formed associations with bacteria to form cooperative nitrogen-fixing nodules on their roots (see Chapter 36).

Many animals, including humans, harbor a variety of bacteria and archaea in their digestive tracts. Cows depend on prokaryotes to perform important steps in digestion. Like most animals, cows cannot produce cellulase, the enzyme needed to start the digestion of the cellulose that makes up the bulk of their plant food. However, bacteria living in a special section of the gut called the rumen produce enough cellulase to process the cow's daily diet. Humans use some of the metabolic products—especially vitamins B_{12} and K—of bacteria living in our large intestine.

We are heavily populated, inside and out, by bacteria. Although very few of them are agents of disease, popular notions of bacteria as "germs" arouse our curiosity about those few. Let's briefly consider the roles of some bacteria as pathogens.

A small minority of bacteria are pathogens

The late nineteenth century was a productive era in the history of medicine—a time during which bacteriologists, chemists, and physicians proved that many diseases are caused by microbial agents. During this time the German physician Robert Koch laid down a set of rules for establishing that a particular microorganism causes a particular disease:

▶ The microorganism must always be found in individuals with the disease.

▶ The microorganism can be taken from the host and grown in pure culture.

▶ A sample of the culture produces the disease when injected into a new, healthy host.

▶ The newly infected host yields a new, pure culture of microorganisms identical to those obtained in the second step.

These rules—called **Koch's postulates**—were very important in a time when it was not widely accepted that microorganisms cause disease. Today, medical science makes use of other, more powerful diagnostic tools.

Only a tiny percentage of all prokaryotes are **pathogens** (disease-producing organisms), and of those that are known, all are bacteria. For an organism to be a successful pathogen, it must overcome several hurdles:

▶ It must arrive at the body surface of a potential host.

▶ It must enter the host's body.

▶ It must evade the host's defenses.

▶ It must multiply inside the host.

▶ It must infect a new host.

Failure to overcome any of these hurdles ends the reproductive career of a pathogenic organism. However, in spite of the many defenses available to potential hosts that we considered in Chapter 19, some bacteria are very successful pathogens.

For the host, the consequences of a bacterial infection depend on several factors. One is the **invasiveness** of the pathogen—its ability to multiply within the body of the host. Another is its **toxigenicity**—its ability to produce chemical substances (toxins) harmful to the tissues of the host. *Corynebacterium diphtheriae*, the agent that causes diphtheria, has low invasiveness and multiplies only in the throat, but its toxigenicity is so great that the entire body is affected. In contrast, *Bacillus anthracis*, which causes anthrax (a disease primarily of cattle and sheep), has low toxigenicity but an invasiveness so great that the entire bloodstream ultimately teems with the bacteria.

There are two general types of bacterial toxins: exotoxins and endotoxins. **Endotoxins** are released when certain Gram-negative bacteria lyse (burst). These toxins are lipopolysaccharides that form part of the outer bacterial membrane. Endotoxins are rarely fatal; they normally cause fever, vomiting, and diarrhea. Among the endotoxin producers are some strains of *Salmonella* and *Escherichia*.

Exotoxins are proteins released by living, multiplying bacteria, and they may travel throughout the host's body. They are highly toxic—often fatal—to the host, but do not produce fevers. Exotoxin-induced human diseases include tetanus (from *Clostridium tetani*), botulism (from *Clostridium botulinum*), cholera (from *Vibrio cholerae*) and plague (from *Yersinia pestis*).

Remember that in spite of our frequent mention of human pathogens, only a small minority of the known prokaryotic species are pathogenic. Many more species play positive roles in our lives and in the biosphere. We make direct use of many bacteria and a few archaea in such diverse applications as cheese production, sewage treatment, and the industrial production of an amazing variety of antibiotics, vitamins, organic solvents, and other chemicals.

Prokaryote Phylogeny and Diversity

The prokaryotes comprise a diverse array of microscopic organisms. To explore this diversity, let's first consider how they are classified, and some of the difficulties involved in doing so; then we'll look at some specific examples.

Nucleotide sequences of prokaryotes reveal their evolutionary relationships

There are three primary motivations for classification schemes: to help identify unknown organisms, to reveal evolutionary relationships, and to provide universal names (see Chapter 23). Many scientists and medical technologists must be able to identify bacteria quickly and accurately—when the bacteria are pathogenic, lives may depend on it.

Until recently, taxonomists based their classification schemes for the prokaryotes on readily observable phenotypic characters such as color, motility, nutritional requirements, antibiotic sensitivity, and reaction to the Gram stain. Although such schemes have facilitated the identification of prokaryotes, they have not provided insights into how these organisms evolved—a question of great interest to microbiologists and to all students of evolution. The prokaryotes and the protists (see Chapter 27) have long been major challenges to those who attempted phylogenetic classifications. Only recently have systematists had the right tools for tackling this task.

Analyses of the nucleotide sequences of ribosomal RNA's provided us with the first apparently reliable measures of evolutionary distance among taxonomic groups. Ribosomal RNA (rRNA) is particularly useful for evolutionary studies of living organisms for several reasons:

▶ rRNA is evolutionarily ancient.

▶ No living organism lacks rRNA.

▶ rRNA plays the same role in translation in all organisms.

▶ rRNA has evolved slowly enough that sequence similarities between groups of organisms are easily found.

Let's look at just one of the approaches to the use of rRNA for studying evolutionary relationships.

Comparisons of rRNA's from a great many organisms showed that there are recognizable short base sequences characteristic of particular taxonomic groups. These *signature sequences*, approximately 6 to 14 bases long, appear at the same approximate positions in rRNA's from related groups. For example, the signature sequence AAACUUAAAG occurs about 910 bases from one end of the RNA of the light subunit of ribosomes in 100 percent of the Archaea and Eukarya tested, but in *none* of the Bacteria tested. Several signature sequences distinguish each of the three domains. Similarly, the major groups within the bacteria and archaea possess unique signature sequences.

These data sound promising, but things aren't as easy as we might wish. When biologists examined other genes and RNA's, contradictions began to appear. Analyses of different nucleotide sequences suggested different phylogenetic patterns. How could such a situation have arisen?

 ### Lateral gene transfer muddied the phylogenetic waters

It is now clear that, from early in evolution to the present day, genes have been moving among prokaryotic species by **lateral gene transfer**. As we have seen, a gene from one species can become incorporated into the genome of another. Mechanisms of lateral gene transfer include transfer by plasmids and viruses and uptake of DNA by transformation. Such transfers are well documented, not just between bacterial species or archaean species, but also across the boundary between bacteria and archaea.

A gene that has been transferred will be inherited by the recipient's progeny and in time will be recognized as part of the normal genome of the descendants. Biologists are still assessing the extent of lateral gene transfer among prokaryotes and its implications for phylogeny, especially at the early stages of evolution.

There is great controversy now over prokaryotic phylogeny. Figure 26.11 is an overview of some major groups in the domains Bacteria and Archaea that we will discuss further in this chapter. Keep in mind that a new picture will likely emerge within the next decade, based on the addition of more nucleotide sequence data and on new information about the currently understudied archaea.

Mutations are the most important source of prokaryotic variation

Assuming that the prokaryote groups we are about to describe do indeed represent monophyletic groups, we discover that they are amazingly complex. A single group of bacteria or archaea may contain the most extraordinarily diverse species, and a species in one group may be phenotypically almost indistinguishable from one or many species in another group. What are the sources of this diversity?

Although prokaryotes can acquire new alleles by transformation, transduction, or conjugation, the most important source of genetic variation in populations of prokaryotes is probably mutation. Mutations, especially recessive mutations, are slow to make their presence felt in populations of humans and other diploid organisms. In contrast, a mutation in a prokaryote, which is haploid, has immediate consequences for that organism. If it is not lethal, it will be transmitted to and expressed in the organism's daughter cells—and in their daughter cells, and so forth. Thus, a beneficial mutant allele spreads rapidly.

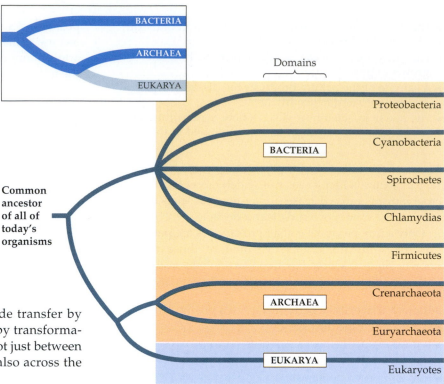

26.11 Two Domains: A Brief Overview
An abridged summary classification of the domains Bacteria and Archaea shows their relationships to each other and to the Eukarya. The relationships among the many lineages of bacteria, not all of which are listed here, are unresolved at this time.

The rapid multiplication of many prokaryotes, coupled with mutation, selection, and genetic drift, allows rapid phenotypic changes within their populations. Important changes, such as loss of sensitivity to an antibiotic, can occur over broad geographic areas in just a few years. Think how many significant metabolic changes could have occurred over even modest time spans in relation to the history of life on Earth. When we introduce the Proteobacteria, the largest group of bacteria, you will see that its different subgroups easily and rapidly adopted and abandoned metabolic pathways under selective pressure from their environments.

The Bacteria

The great majority of known prokaryotes are bacteria. Here we will describe bacterial diversity using a currently popular classification scheme that enjoys considerable support from nucleotide sequence data. More than a dozen monophyletic groups have been proposed under this scheme; we will describe just of a few of them here. The higher-order relationships among these groups of prokaryotes are not known. Some biologists describe them as kingdoms, some as subkingdoms, others as phyla. Here we call them groups. We'll pay the closest attention to the Proteobacteria, Cyanobacteria, Spirochetes, Chlamydias, and Firmicutes (see Figure 26.11), but first we mention one property that is shared by members of three other groups.

Some bacteria are heat lovers

Three of the bacterial groups that may have branched out earliest during bacterial evolution are all **thermophiles** (heat lovers), as are the most ancient of the archaea. This observation supports the hypothesis that the first living organisms were thermophiles that appeared in an environment much hotter than those that predominate today.

The Proteobacteria are a large and diverse group

By far the largest group of bacteria, in terms of number of described species, is the **Proteobacteria**, sometimes referred to as the *purple bacteria*. Among the proteobacteria are many species of Gram-negative, bacteriochlorophyll-containing, sulfur-using photoautotrophs. However, this group also includes a dramatically diverse group of bacteria that bear no resemblance to the purple bacteria in phenotype. The mitochondria of eukaryotes were derived from proteobacteria by endosymbiosis.

No characteristic demonstrates the diversity of the proteobacteria more clearly than their metabolic pathways (Figure 26.12). The common ancestor of all the proteobacteria was probably a photoautotroph. Early in evolution, two groups of proteobacteria lost their ability to photosynthesize and have been chemoheterotrophs ever since. The other three groups still have photoautotrophic members, but in *each* group, some evolutionary lines have abandoned photoautotrophy and taken up other modes of nutrition. There are chemoautotrophs and chemoheterotrophs in all three groups. Why? We can view each of the trends in Figure 26.12 as an evolutionary response to selective pressures encountered as these bacteria encountered new habitats that presented new challenges and opportunities.

Among the proteobacteria are some nitrogen-fixing genera such as *Rhizobium* (see Figure 34.10) and other bacteria that contribute to the global nitrogen and sulfur cycles. *E. coli*, one of the most studied organisms on Earth, is a proteobacterium. So, too, are many of the most famous human pathogens, such as *Yersinia pestis*, *Vibrio cholerae*, and *Salmonella typhimurium*, all mentioned in our discussion of pathogens.

Fungi cause most plant diseases, and viruses cause others, but about 200 plant diseases are of bacterial origin. *Crown gall*, with its characteristic tumors (Figure 26.13), is one of the most striking. The causal agent of crown gall is *Agrobacterium tumefaciens*, which harbors a plasmid used in recombinant DNA studies as a vehicle for inserting genes into new plant hosts (see Chapter 17).

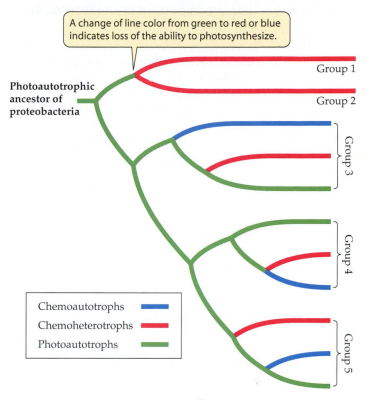

A change of line color from green to red or blue indicates loss of the ability to photosynthesize.

Photoautotrophic ancestor of proteobacteria

Group 1
Group 2
Group 3
Group 4
Group 5

Chemoautotrophs
Chemoheterotrophs
Photoautotrophs

26.12 The Evolution of Metabolism in the Proteobacteria
The common ancestor of all proteobacteria was probably a photoautotroph. As they encountered new environments, groups 1 and 2 lost the ability to photosynthesize; in the other three groups, some evolutionary lines became chemoautotrophs or chemoheterotrophs.

Proteobacteria
Cyanobacteria
Spirochetes
Chlamydias
Firmicutes
Crenarchaeota
Euryarchaeota
Eukaryotes

Crown gall is a plant disease caused by the Gram-negative rod *Agrobacterium tumefaciens*.

26.13 Crown Gall
This colorful tumor is a crown gall growing on the stem of a geranium plant.

The Cyanobacteria are important photoautotrophs

The **Cyanobacteria** (blue-green bacteria) require only water, nitrogen gas, oxygen, a few mineral elements, light, and carbon dioxide to survive. They use chlorophyll *a* for photosynthesis and liberate oxygen gas; many species also fix nitrogen. Their photosynthesis was the basis of the "oxygen revolution" that transformed Earth's atmosphere.

Cyanobacteria carry out the same type of photosynthesis that is characteristic of eukaryotic photosynthesizers. They contain elaborate and highly organized internal membrane systems called photosynthetic lamellae, or *thylakoids* (Figure 26.14). The chloroplasts of photosynthetic eukaryotes are derived from an endosymbiotic cyanobacterium.

Cyanobacteria may associate in colonies or live free as single cells. Depending on the species and on growth conditions, colonies of cyanobacteria may range from flat sheets one cell thick to spherical balls of cells.

Some filamentous colonies differentiate into three cell types: vegetative cells, spores, and heterocysts (Figure 26.15). Vegetative cells photosynthesize, **spores** are resting cells that can eventually develop into new filaments, and **heterocysts** are cells specialized for nitrogen fixation. All of the known cyanobacteria with heterocysts fix nitrogen. Heterocysts also have a role in reproduction: When filaments break apart to reproduce, the heterocyst may serve as a breaking point.

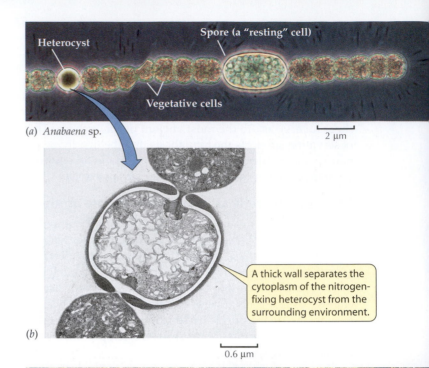

(a) *Anabaena* sp.

Heterocyst

Spore (a "resting" cell)

Vegetative cells

2 μm

A thick wall separates the cytoplasm of the nitrogen-fixing heterocyst from the surrounding environment.

(b)

0.6 μm

Proteobacteria
Cyanobacteria
Spirochetes
Chlamydias
Firmicutes
Crenarchaeota
Euryarchaeota
Eukaryotes

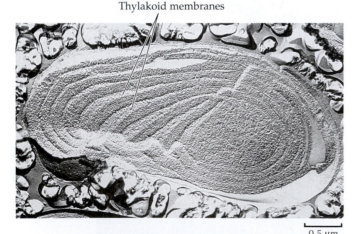

Thylakoid membranes

0.5 μm

26.14 Thylakoids in Cyanobacteria

This cyanobacterium was prepared by the freeze-etch technique to emphasize the extensive system of internal membranes. These photosynthetic thylakoid membranes are present through most of the cytoplasm and clearly identify the specimen as a cyanobacterium, even though the exact species is not identified here.

(c)

26.15 Cyanobacteria

(a) *Anabaena* is a genus of colonial, filamentous cyanobacteria. The vegetative cells are photosynthetic. (b) A thin neck attaches a heterocyst to each of two other cells in a colony. (c) Cyanobacteria appear in enormous numbers in some environments. This California pond has experienced eutrophication: Phosphorus and other nutrients generated by human activity have accumulated in the pond, feeding an immense green mat—commonly referred to as "pond scum"—made up of several species of unicellular cyanobacteria.

Spirochetes look like corkscrews

Spirochetes are Gram-negative bacteria characterized by unique structures called **axial filaments**, fibrils running through the periplasmic space (see Figure 26.5*a*). The cell body is a long cylinder coiled into a spiral (Figure 26.16). The axial filaments begin at either end of the cell and overlap in

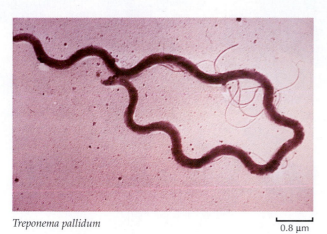

Treponema pallidum   0.8 µm

26.16 A Spirochete
This corkscrew-shaped spirochete causes syphilis in humans.

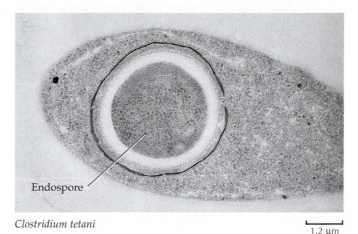

Endospore

Clostridium tetani   1.2 µm

26.18 The Endospore: A Structure for Waiting Out Bad Times
This firmicute, which causes tetanus, produces endospores as resistant resting structures.

the middle, and there are typical basal rings where they are attached to the cell wall. Many spirochetes live in humans as parasites. Others live free in mud or water.

Chlamydias are extremely small

Chlamydias are among the smallest bacteria (0.2–1.5 µm in diameter). They can live only as parasites within the cells of other organisms. These tiny spheres are unique prokaryotes because of their complex life cycle, which involves two different forms of cells (Figure 26.17). In humans, various strains of chlamydias cause eye infections (especially trachoma), sexually transmitted disease, and some forms of pneumonia.

Most Firmicutes are Gram-positive

The **Firmicutes** are sometimes referred to as the *Gram-positive bacteria*, but some firmicutes are Gram-negative, and some have no cell wall at all. Nonetheless, the firmicutes constitute a monophyletic group.

Some firmicutes produce **endospores** (Figure 26.18)—heat-resistant resting structures—when nutrients become scarce. The bacterium replicates its DNA and encapsulates one copy, along with some of its cytoplasm, in a tough cell wall heavily thickened with peptidoglycan and surrounded by a spore coat. The parent cell then breaks down, releasing the endospore. *Endospore production is not a reproductive process*; the endospore merely replaces the parent cell. The endospore can survive harsh environmental conditions, such as high or low temperatures or drought, because it is *dormant*—its normal activity is suspended. Later, if it encounters favorable conditions, the endospore becomes metabolically active and divides, forming new cells like the parent. Some endospores apparently can be reactivated even after more than a thousand years of dormancy.

Members of this endospore-forming group include the many species of *Bacillus* and *Clostridium*. The toxins produced by *C. botulinum* are among the most poisonous ever discovered; the lethal dose for humans is about one-millionth of a gram (1 µg).

The genus *Staphylococcus*—the staphylococci—includes firmicutes that are abundant on the human body surface; they are responsible for boils and

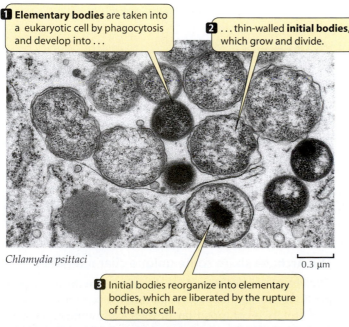

1 **Elementary bodies** are taken into a eukaryotic cell by phagocytosis and develop into . . .

2 . . . thin-walled **initial bodies**, which grow and divide.

Chlamydia psittaci   0.3 µm

3 Initial bodies reorganize into elementary bodies, which are liberated by the rupture of the host cell.

26.17 Chlamydias Change Form during Their Life Cycle
Elementary bodies and initial bodies are the two major phases of the life cycle of a chlamydia.

Proteobacteria

Cyanobacteria

Spirochetes

Chlamydias

Firmicutes

Crenarchaeota

Euryarchaeota

Eukaryotes

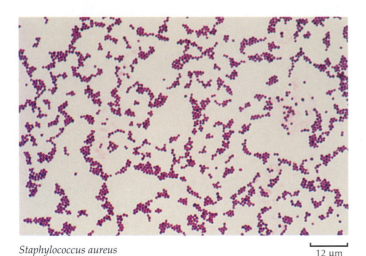

Staphylococcus aureus   12 μm

26.19 Gram-Positive Firmicutes
"Grape clusters" are the usual arrangement of Gram-positive staphylococci.

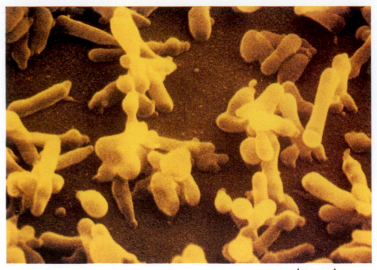

Mycoplasma gallisepticum   0.4 μm

26.21 The Tiniest Living Cells
Containing only about one-fifth as much DNA as *E. coli*, mycoplasmas are the smallest known bacteria.

many other skin problems (Figure 26.19). *S. aureus* is the best-known human pathogen; it is found in 20 to 40 percent of normal adults (and in 50 to 70 percent of hospitalized adults). It can cause respiratory, intestinal, and wound infections, in addition to skin diseases.

Actinomycetes are firmicutes that develop an elaborately branched system of filaments (Figure 26.20). These bacteria closely resemble the filamentous bodies of fungi. Some actinomycetes reproduce by forming chains of spores at the tips of the filaments. In the species that do not form spores, the branched, filamentous growth ceases and the structure breaks up into typical cocci or rods, which then reproduce by fission.

The actinomycetes include several medically important bacteria. *Mycobacterium tuberculosis* causes tuberculosis. *Streptomyces* produces streptomycin, as well as hundreds of

other antibiotics, including several dozen in general use. We derive most of our antibiotics from members of the actinomycetes.

Another interesting group of firmicutes, the **mycoplasmas**, lack cell walls, although some have a stiffening material outside the plasma membrane. Some of them are the smallest cellular creatures ever discovered—they are even smaller than chlamydias (Figure 26.21). The smallest mycoplasmas capable of growth have a diameter of about 0.2 μm, and they are small in another crucial sense: They have less than half as much DNA as do most other prokaryotes. It has been speculated that the amount of DNA in a mycoplasma may be the minimum amount required to code for the essential properties of a living cell.

The Archaea

The domain Archaea consists mainly of prokaryotic genera that live in habitats notable for characteristics such as extreme salinity (salt content), low oxygen concentration, high temperature, or high or low pH. On the face of it, the Archaea do not seem to belong together as a group; in fact, some evidence suggests that the domain Archaea is paraphyletic. One current classification scheme treats the domain as two kingdoms: **Euryarchaeota** and **Crenarchaeota**. In fact, we know very little about the phylogeny of archaea, in part because the study of archaea is still in its early stages. We do know that archaea share certain characteristics.

The Archaea share some unique characteristics

Two characteristics shared by all archaea are the absence of peptidoglycan in their cell walls and the presence of lipids of distinctive composition in their cell membranes (see Table 26.1). The base sequences of their ribosomal RNA's support a close evolutionary relationship among them. Their separation from the Bacteria and Eukarya was clari-

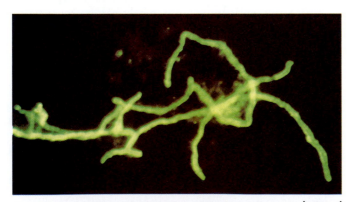

Actinomyces israelii   10 μm

26.20 Filaments of an Actinomycete
These branching filaments are visualized with a fluorescent stain. This species is part of the normal flora in the human tonsils, mouth, intestinal tract, and lungs, but will invade body tissues and cause severe abscesses when afforded the opportunity.

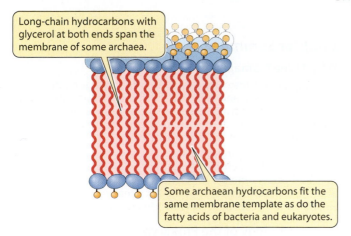

Long-chain hydrocarbons with glycerol at both ends span the membrane of some archaea.

Some archaean hydrocarbons fit the same membrane template as do the fatty acids of bacteria and eukaryotes.

26.22 Membrane Architecture in Archaea
The long-chain hydrocarbons of archaean membranes are branched, and may have glycerol at both ends. This structure still fits into a biological membrane, however; in fact, all three domains have similar membrane structures.

fied when biologists sequenced the first archaean genome: It consisted of 1,738 genes, *more than half of which* were unlike any genes ever found in the other two domains.

The unusual lipids in the membranes of archaea deserve some description. They are found in all archaea, and in no bacteria or eukaryotes. Most membrane lipids of bacteria and eukaryotes contain unbranched long-chain fatty acids connected to glycerol by **ester linkages**:

$$\begin{array}{ccc} O & & H \\ \parallel & & | \\ -C-O- & & C- \\ & & | \\ & & H \end{array}$$

(see also Figure 3.19). In contrast, archaean membrane lipids contain long-chain hydrocarbons connected to glycerol by **ether linkages**:

$$\begin{array}{ccc} H & & H \\ | & & | \\ -C-O- & & C- \\ | & & | \\ H & & H \end{array}$$

In addition, the long-chain hydrocarbons are branched in the archaea. One class of these lipids contains glycerol at *both* ends of the hydrocarbons. This structure still fits in a biological membrane, as shown in Figure 26.22. In spite of the striking difference in membrane lipids, all three domains have lipid bilayer membranes with similar overall structures, dimensions, and functions.

Proteobacteria
Cyanobacteria
Spirochetes
Chlamydias
Firmicutes
Crenarchaeota
Euryarchaeota
Eukaryotes

Most Crenarchaeota live in hot, acidic places

Most known Crenarchaeota are both thermophilic (heat-loving) and *acidophilic* (acid-loving). Members of the genus *Sulfolobus* live in hot sulfur springs at temperatures of 70–75°C. They die of "cold" at 55°C (131°F). Hot sulfur springs are also extremely acidic. *Sulfolobus* grows best in the pH range from 2 to 3, but it readily tolerates pH values as low as 0.9. Some acidophilic hyperthermophiles maintain an internal pH near 7 (neutral) in spite of the acidity of their environment. These and other hyperthermophiles thus thrive where very few other organisms can even survive (Figure 26.23).

The Euryarchaeota live in many amazing places

Some species of Euryarchaeota, once assigned to unrelated bacterial groups, share the property of producing methane (CH_4) by reducing carbon dioxide. All of these *methanogens* are obligate anaerobes, and methane production is the key step in their energy metabolism. Comparison of rRNA nucleotide sequences revealed a close evolutionary relationship among all these methanogens.

Methanogens release approximately 2 billion tons of methane gas into Earth's atmosphere each year, accounting for all the methane in our air, including that associated with mammalian belching. Approximately a third of this methane comes from methanogens in the guts of grazing herbivores such as cows.

26.23 Some Would Call It Hell; Archaea Call It Home
Masses of heat- and acid-loving archaea form an orange mat inside a volcanic vent on the island of Kyushu, Japan. Sulfurous residue is visible at the edges of the archaean mat.

26.24 Extreme Halophiles
Commercial seawater evaporating ponds, such as these in San Francisco Bay, are attractive homes for salt-loving archaea.

One methanogen, *Methanopyrus*, lives on the ocean bottom near blazing volcanic vents. *Methanopyrus* can survive and grow at 110°C. It grows best at 98°C and not at all at temperatures below 84°C.

Another group of Euryarchaeota, the *extreme halophiles* (salt lovers), lives exclusively in very salty environments. Because they contain pink carotenoids, they can be seen easily under some circumstances (Figure 26.24). Halophiles grow in the Dead Sea and in brines of all types: Pickled fish may sometimes show reddish pink spots that are colonies of halophilic archaea. Few other organisms can live in the saltiest of the homes that the strict halophiles occupy; most would "dry" to death, losing too much water to the hypertonic environment. Strict halophiles have been found in lakes with pH values as high as 11.5—the most alkaline environment inhabited by living organisms, almost as alkaline as household ammonia.

Some of the extreme halophiles have a unique system for trapping light energy and using it to form ATP—without using any form of chlorophyll—when oxygen is in short supply. They use the pigment *retinal* (also found in the vertebrate eye) combined with a protein to form **bacteriorhodopsin**, and form ATP by a chemiosmotic mechanism of the sort described in Figure 7.12.

Another member of the Euryarchaeota, *Thermoplasma*, has no cell wall. It is thermophilic and acidophilic, its metabolism is aerobic, and it lives in coal deposits. It has the smallest genome among the archaea, and perhaps the smallest (along with the mycoplasmas) of any free-living organisms—1,100,000 base pairs.

This chapter has provided a brief summary of two of the three domains of the living world. The world of the eukaryotes, both unicellular and multicellular, will be the subject of the next seven chapters.

Chapter Summary

Why Three Domains?

▶ Living organisms can be divided into three domains: Bacteria, Archaea, and Eukarya. Both the Archaea and the Bacteria are prokaryotic, but they differ from each other more radically than do the Archaea from the Eukarya, which constitute the rest of the living world.

▶ The evolutionary relationships of the three domains were first revealed by their rRNA sequences. The common ancestor of all three domains lived more than 3 billion years ago, and the common ancestor of the Archaea and Eukarya at least 2 billion years ago. **Review Figure 26.2 and Table 26.1**

General Biology of the Prokaryotes

▶ The prokaryotes are the most numerous organisms on Earth, and they occupy an enormous variety of habitats.

▶ Most prokaryotes are cocci, bacilli, or spiral forms. Some link together to form associations, but very few are truly multicellular. **Review Figure 26.3**

▶ Prokaryotes lack nuclei, membrane-enclosed organelles, and cytoskeletons. Their chromosomes are circular. They often contain plasmids. Some prokaryotes contain internal membrane systems. **Review Figure 26.4**

▶ Many prokaryotes move by means of flagella, gas vesicles, or gliding mechanisms. Prokaryotic flagella rotate rather than beat.

▶ Prokaryotic cell walls differ from those of eukaryotes. Bacterial cell walls generally contain peptidoglycan. Differences in peptidoglycan content result in different reactions to the Gram stain. **Review Figure 26.7**

▶ Prokaryotes reproduce asexually by fission, but also exchange genetic information.

▶ Prokaryotes have diverse metabolic pathways and nutritional modes. They include obligate anaerobes, facultative anaerobes, and obligate aerobes. The major nutritional types are photoautotrophs, photoheterotrophs, chemoautotrophs, and chemoheterotrophs. Some prokaryotes base their energy metabolism on nitrogen- or sulfur-containing ions. **Review Figure 26.8 and Table 26.2**

Prokaryotes in Their Environments

▶ Some prokaryotes play key roles in global nitrogen and sulfur cycles. Important players in the nitrogen cycle are the nitrogen fixers, nitrifiers, and denitrifiers. **Review Figure 26.10**

▶ Photosynthesis by cyanobacteria generated the oxygen gas that permitted the evolution of aerobic respiration and the appearance of present-day eukaryotes.

▶ Many prokaryotes live in or on other organisms, with neutral, beneficial, or harmful effects.

▶ A small minority of bacteria are pathogens. Pathogens vary with respect to their invasiveness and toxigenicity. Some produce endotoxins, which are rarely fatal; others produce exotoxins, which tend to be highly toxic.

Prokaryote Phylogeny and Diversity

▶ Phylogenetic classification of prokaryotes is now based on rRNA sequences and other molecular evidence.

▶ Lateral gene transfer among prokaryotes, which has occurred throughout evolutionary history, makes it difficult to infer prokaryote phylogeny.

▶ Evolution, powered by mutation, natural selection, and genetic drift, can proceed rapidly in prokaryotes because they are haploid and can multiply rapidly.

The Bacteria

▶ There are far more known bacteria than known archaea. One phylogenetic classification of the domain Bacteria groups them into more than a dozen groups.

▶ The most ancient bacteria, like the most ancient archaea, may be thermophiles, suggesting that life originated in a hot environment.

▶ All four nutritional types occur in the largest bacterial group, the Proteobacteria. Metabolism in different groups of proteobacteria has evolved along different lines. **Review Figure 26.12**

▶ Cyanobacteria, unlike other bacteria, photosynthesize using the same pathways plants use. Many cyanobacteria fix nitrogen.

▶ Spirochetes move by means of axial filaments.

▶ Chlamydias are tiny parasites that live within the cells of other organisms.

▶ Firmicutes are diverse; some of them produce endospores as resting structures that resist harsh conditions. Actinomycetes, some of which produce important antibiotics, grow as branching filaments.

▶ Mycoplasmas, the tiniest living things, lack conventional cell walls. They have very small genomes.

The Archaea

▶ Archaea have cell walls lacking peptidoglycan, and their membrane lipids differ from those of bacteria and eukaryotes, containing branched long-chain hydrocarbons connected to glycerol by ether linkages. **Review Figure 26.22**

▶ The domain Archaea can be divided into two kingdoms: Crenarchaeota and Euryarchaeota.

▶ Crenarchaeota are heat-loving and often acid-loving archaea.

▶ Methanogens produce methane by reducing carbon dioxide. Some methanogens live in the guts of herbivorous animals; others occupy high-temperature environments on the ocean floor.

▶ Extreme halophiles are salt lovers that often lend a pinkish color to salty environments; some halophiles also grow in extremely alkaline environments.

▶ Archaea of the genus *Thermoplasma* lack cell walls, are thermophilic and acidophilic, and have a tiny genome (1,100,000 base pairs).

For Discussion

1. Why do systematic biologists find rRNA sequence data more useful than data on metabolism or cell structure for classifying prokaryotes?

2. Why does lateral gene transfer make it so difficult to arrive at agreement on phylogeny?

3. Differentiate among the members of the following sets of related terms:

   a. prokaryotic/eukaryotic

   b. obligate anaerobe/facultative anaerobe/obligate aerobe

   c. photoautotroph/photoheterotroph/chemoautotroph/ chemoheterotroph

   d. Gram-positive/Gram-negative

4. Why are the endospores of firmicutes not considered to be reproductive structures?

5. Until fairly recently, the cyanobacteria were called blue-green algae and were not grouped with the bacteria. Suggest several reasons for this (abandoned) tendency to separate the bacteria and cyanobacteria. Why are the cyanobacteria now grouped with the other bacteria?

6. The actinomycetes are of great commercial interest. Why?

7. Hyperthermophiles are of great interest to molecular biologists and biochemists. Why? What practical concerns might motivate that interest?

27 Protists and the Dawn of the Eukarya

THE BACTERIA AND THE ARCHAEA HAD the living world to themselves for more than a billion years. As we saw in Chapter 26, members of these two domains differ sharply in several important ways—but neither of these prokaryotes is like the single-celled organism shown here. What strikes you the most about this amoeba? Probably the most obvious visible difference between it and the prokaryotes is that the amoeba has numerous compartments—membrane-enclosed organelles.

What are compartments useful for? For one thing, they keep items separate—like keeping greasy tools away from clean socks by storing them in different cabinets. Compartments also keep things together when that makes sense, such as keeping all your socks in one drawer, or keeping all the files for your term paper in a single directory on your computer. Rooms are another example of compartments, and they can be specialized for different activities: One room in the Biological Sciences Building might be set up as a laboratory, while another room serves as a lecture hall, and still another contains special protective seals that prevent radiation or pathogens from escaping.

The single-celled amoeba is an example of an organism that compartmentalizes: It has a cytoskeleton, a nucleus enclosed by a nuclear envelope, and several kinds of organelles. Amoebas are members of the domain Eukarya, and they differ from members of the two prokaryotic domains in other important ways as well.

The flexibility and options that arose once the eukaryotic cell had evolved resulted in a profusion of body forms and myriad specialized functions. Eukaryotic evolution has produced great diversity, especially among the multicellular lineages, but even among the unicellular protists. In both multicellular and unicellular forms, however, there are also many cases of convergent evolution; for example, organisms with an amoeba-like body form arose several times.

Protists Defined

Many modern members of the Eukarya are familiar to us—trees, dogs, and mushrooms, not to mention ourselves. These members of the kingdoms Plantae, Animalia, and Fungi are not strange to us. However, amoebas and a dazzling assortment of other eukaryotes, mostly microscopic organisms, don't fit into these three kingdoms. We call all those eukaryotes that are neither plants, animals, nor fungi **protists** (Figure 27.1; Table 27.1). *The protists are not a monophyletic group.* Some protists are more closely related to the animals than they are to other protists. Some protists are motile, while others are stationary; some are photosynthetic, while others are heterotrophic; most are unicellular, while some giant kelps are not only multicellular but also huge, sometimes achieving lengths greater than that of a football field.

The origin of the eukaryotic cell was one of the pivotal events in evolutionary history. In this chapter on the protists we describe and celebrate the origin and early diversification of the eukaryotes and the complexity achieved by some single cells. We'll explore some of the diversity of protist body forms, and we'll try to give a sense of developing current views of the evolutionary relationships of some of the protists.

The Origin of the Eukaryotic Cell

The eukaryotic cell differs in many ways from prokaryotic cells. How did it originate? Given the nature of evolutionary processes, the differences cannot all have arisen simultaneously. We think we can make some reasonable guesses

An Amoeba
Amoebas have several kinds of organelles (seen here as bubble-like "compartments"). Their flowing pseudopods are constantly changing shape as the amoeba moves and feeds.

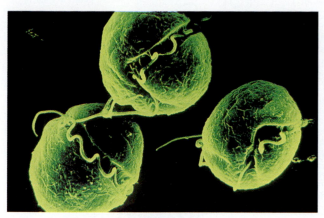

(a) *Gonyaulax* sp.

27.1 Three Eukaryote Protists
(a) Dinoflagellates are photosynthetic unicellular protists. (b) *Giardia* is a unicellular parasite of humans. (c) Giant kelps are some of the world's longest organisms.

(b) *Giardia* sp.   (c) *Macrocystis* sp.

about the important events, bearing in mind that the global environment underwent an enormous change—from anaerobic to aerobic—during the course of these events. As you read this chapter, keep in mind that the steps we suggest are just that: guesses. This version of the story is one of a few under current consideration. We present it as a framework for thinking about this challenging problem, not as a set of facts.

27.1 Major Monophyletic Protist Groups

GROUP	COMMON NAME	ATTRIBUTES	EXAMPLES
Euglenozoa		Unicellular, with flagella	
Euglenoids		Mostly photoautotrophic	*Euglena*
Kinetoplastids		Have a single, large mitochondrion	*Trypanosoma*
Alveolata		Unicellular; cavities (alveoli) below cell surface	
Pyrrophyta	Dinoflagellates	Pigments give golden-brown color	*Gonyaulax*
Apicomplexa		Apical complex for penetration of host	*Plasmodium*
Ciliophora	Ciliates	Cilia; two types of nuclei	*Paramecium*
Stramenopila		Two unequal flagella, one with hairs	
Bacillariophyta	Diatoms	Unicellular; photoautotrophic; two-part walls	
Phaeophyta	Brown algae	Multicellular; marine; photoautotrophic	*Fucus, Macrocystis*
Oomycota	Water molds, powdery mildews	Mostly coenocytic; heterotrophic	*Saprolegnia*
Rhodophyta	Red algae	No flagella; photoautrophic; phycocyanin	*Chondrus*
Chlorophyta	Green algae[a]	Photoautotrophic	*Chlamydomonas, Ulothrix*
Choanoflagellida		Resemble sponge cells; heterotrophic	

[a] The green algae do not constitute a monophyletic group. The Chlorophyta are one lineage of green algae that qualifies as a monophyletic group; another lineage gave rise to the plant kingdom.

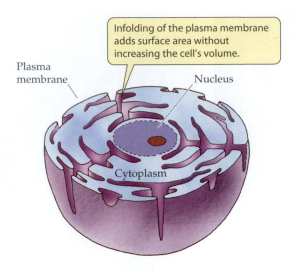

27.2 *Membrane Infolding*
The loss of the rigid prokaryotic cell wall allowed the plasma membrane to elaborate inward and create more surface area.

The modern eukaryotic cell arose in several steps

The essential steps in the origin of the eukaryotic cell include

▸ The origin of a flexible cell surface
▸ The origin of a cytoskeleton
▸ The origin of a nuclear envelope
▸ The appearance of digestive vesicles
▸ The endosymbiotic acquisition of certain organelles

WHAT A FLEXIBLE CELL SURFACE ALLOWS.   Many ancient fossil prokaryotes look like rods, and we presume that they, like most present-day prokaryotic cells, had firm cell walls. The first step toward the eukaryotic condition may have been the loss of the cell wall by an ancestral prokaryotic cell. This may not seem like an obvious first step, but consider the possibilities open to a flexible cell without a wall.

First, think of cell size. As a cell grows, its surface area-to-volume ratio decreases (see Figure 4.2). Unless the surface is flexible and can fold inward and elaborate itself, creating more surface area for gas and nutrient exchange (Figure 27.2), the cell volume will reach an upper limit. With a surface flexible enough to allow infolding, the cell can exchange materials with its environment rapidly enough to sustain a larger volume and more rapid metabolism. Further, a flexible surface can pinch off bits of the environment, bringing them into the cell by endocytosis (Figure 27.3).

Also recall that the chromosome of a prokaryotic cell is attached to a site on its plasma membrane (see Figure 26.4). If that region of the plasma membrane were to fold into the cell, the first step would be taken

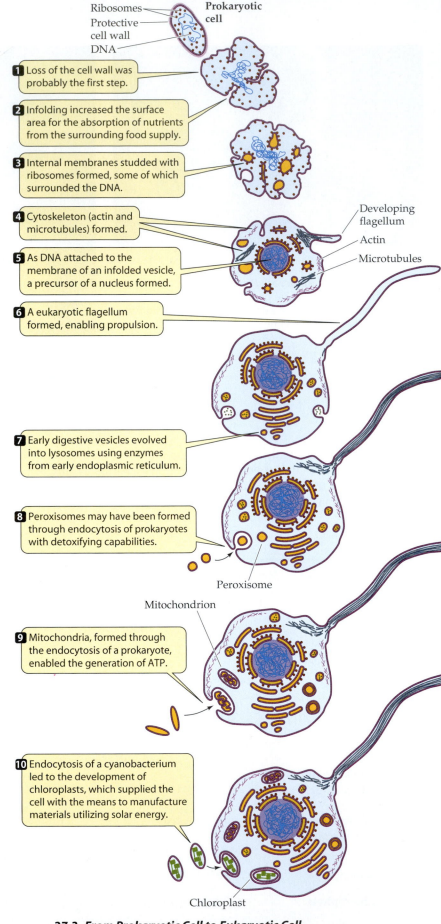

27.3 *From Prokaryotic Cell to Eukaryotic Cell*
One possible evolutionary sequence is shown here. The exact sequence, of course, is not known.

toward the evolution of a nucleus, the key feature of the eu-karyotic cell.

CHANGES IN CELL STRUCTURE AND FUNCTION. Early steps in the evolution of the eukaryotic cell are likely to have included three advances: the formation of ribosome-studded internal membranes, some of which surrounded the DNA (see Figure 27.3); the appearance of a cytoskeleton and the evolution of digestive vesicles.

A cytoskeleton made up of actin fibers and microtubules would allow the cell to manage changes in shape, to distribute daughter chromosomes, and to move materials from one part of the now much larger cell to other parts. The origin of the cytoskeleton remains a mystery, heightened by the fact that the genes that encode the cytoskeleton are present in neither bacteria nor archaea. An intriguing and controversial suggestion is that a fourth domain of life, now long extinct, originated these genes and transferred them laterally to an ancestor of the early eukaryotes.

From an intermediate kind of cell, the next advance was probably to a cell that we could call a *phagocyte*—a motile cell that could prey on other cells by engulfing and digesting them. The first true eukaryote possessed a cytoskeleton and a nuclear envelope. It may have had an associated endoplasmic reticulum and Golgi apparatus, and perhaps one or more flagella of the eukaryotic type. Notice how much of the progress to this point was made possible by the loss of the cell wall and the elaboration of what was originally the plasma membrane.

ENDOSYMBIOSIS AND ORGANELLES. While the processes already outlined were taking place, the cyanobacteria were very busy, generating oxygen gas as a product of photosynthesis. The increasing O_2 levels in the atmosphere had disastrous consequences for most other living things, because most living things of the time (archaea and bacteria) were unable to tolerate the newly aerobic, oxidizing environment. But some prokaryotes managed to cope, and—fortunately for us—so did some of the ancient phagocytes.

According to one hypothesis, the key to the survival of early phagocytes was the ingestion and incorporation of a prokaryote that became symbiotic within the phagocyte and evolved into the peroxisomes of today (see Figure 27.3). These organelles were able to disarm the toxic products of oxygen action, such as hydrogen peroxide. This association may have been the first important endosymbiosis in the evolution of the eukaryotic cell.

In Chapter 4 we introduced the concept of endosymbiosis (organisms living together, one inside the other). A crucial endosymbiotic event in the history of the Eukarya was the incorporation of a proteobacterium that evolved into the mitochondrion. Upon completion of this step, the basic modern eukaryotic cell was complete. Some very important eukaryotes are the result of yet another endosymbiotic step, the incorporation of prokaryotes related to today's cyanobacteria, which became chloroplasts. We'll see how this happened later in the chapter.

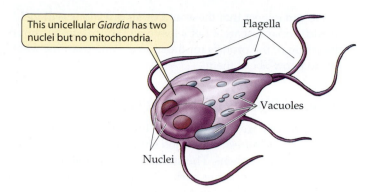

This unicellular *Giardia* has two nuclei but no mitochondria.

Flagella

Vacuoles

Nuclei

27.4 *Giardia*: A Protist without Mitochondria
Current evidence indicates that *Giardia* is descended from an ancestor that possessed mitochondria.

"Archaezoa": The little kingdom that was

The hypothesis that the eukaryotic nucleus evolved before the mitochondrion gained early support from the existence of a few unicellular eukaryotes, such as *Giardia*, that lack mitochondria. *Giardia lamblia* is a familiar parasite that contaminates water supplies and causes the intestinal disease giardiasis (Figure 27.4). This tiny organism has no mitochondria, chloroplasts, or other membrane-enclosed organelles, but it contains two nuclei bounded by nuclear envelopes, and it has a cytoskeleton. Some biologists treated such eukaryotes without mitochondria as the modern descendants of a hypothetical ancient group, which they called "archaezoans." It was later learned that at least some archaezoans may have descended from eukaryotes that lost their mitochondria. Research sometimes takes surprising twists and turns! We no longer speak of a *kingdom* of archaezoans. However, the existence of such organisms today shows that eukaryotic life is feasible without mitochondria, and the eukaryotes that lack mitochondria are the focus of much attention.

Many uncertainties remain

Several uncertainties cloud our current understanding of the origins of eukaryotic cells. Lateral gene transfer complicates the study of eukaryotic origins just as it complicates the study of relationships among the prokaryote lineages. At the same time, it may not have been extensive enough to account for the fact that, as genetic studies advance, more and more genes of bacterial origin are being found in eukaryotes.

An endosymbiotic origin of mitochondria and chloroplasts accounts for the presence of bacterial genes encoding enzymes for energy metabolism (respiration and photosynthesis), but it does not explain the presence of many other bacterial genes. The eukaryotic genome clearly is a mixture of genes with two distinct origins. A recent suggestion is that the Eukarya might have arisen from the mutualistic fusion (not endosymbiosis) of a Gram-negative bacterium and an archaean. There are many interesting ideas about eukaryotic origins awaiting additional data and analysis.

We can expect that these and other questions will yield to additional research. Let's leave our speculations about the origin of the protists for the moment and examine what we do know about them.

General Biology of the Protists

Most protists are aquatic. Some live in marine environments, others in fresh water, and still others in the body fluids of other organisms. The slime molds inhabit damp soil and the moist, decaying bark of rotting trees. Many other protists also live in soil water, some of them contributing to the global nitrogen cycle by preying on soil bacteria and recycling their nitrogen compounds to nitrates.

Protists are strikingly diverse in their structure, but not so diverse in their metabolism as the prokaryotes—in fact, some of the eukaryotes' most important metabolic pathways were "borrowed" from bacteria through endosymbiosis. However, protists do display a number of nutritional modes. Some are autotrophs, some are heterotrophs, and some switch with ease between the autotrophic and heterotrophic modes of nutrition.

Some protists, formerly classified as animals, are sometimes referred to as **protozoans**, although biologists increasingly regard this term as inappropriate because it lumps together protist groups that are phylogenetically distant from one another. Most protozoans are ingestive heterotrophs. There are several kinds of photosynthetic protists that some biologists still refer to as **algae** (singular alga). Although these two terms—protozoans and algae—are useful in some contexts, they do not correspond with natural phy-

logeny, and we generally avoid them in this book except as parts of descriptive names such as "brown algae."

EXPERIMENT

Question: Where—and how—does *Paramecium* digest its food?

METHOD   Yeast cells are stained with Congo red, a pH indicator.

RESULTS

Stained yeast cells

1 A food vacuole forms around yeast cells.

2 The change in color shows that the vacuole has become acidic, like your stomach; acid helps digest the yeast cells.

3 Digestion continues.

4 As products of digestion move into the cytosol, the pH increases in the vacuole. The dye becomes red again.

5 Waste material is expelled.

Conclusion: Digestion, assisted by low pH, took place in the food vacuole.

27.6 Food Vacuoles Handle Digestion and Excretion
An experiment with *Paramecium* demonstrates the function of food vacuoles. *Paramecium* ingests food by way of the oral groove at the left. The dye Congo red turns green at acidic pH and red at neutral or basic pH.

Protists have diverse means of locomotion

Although a few protist groups consist entirely of nonmotile organisms, most groups include cells that move, either by amoeboid motion, by ciliary action, or by means of flagella.

In amoeboid motion, the cell forms **pseudopods** ("false feet") that are extensions of its constantly changing body mass. Cells such as the amoeba on page 476 simply extend a pseudopod and then flow into it. Cilia are tiny, hairlike organelles that beat in a coordinated fashion to move the cell forward or backward (see Figure 4.24). A eukaryotic flagellum moves like a whip; some flagella *push* the cell forward, others *pull* the cell forward.

Vesicles perform a variety of functions

Unicellular organisms tend to be of microscopic size. As we noted above, an important reason that cells are small is that they need enough membrane surface area in relation to their volume to support the exchange of materials required for their existence. Many relatively large unicellular protists minimize this problem by having membrane-enclosed **vesicles** of various types that increase their effective surface area.

As we saw in Chapter 5, organisms living in fresh water are hypertonic to their environments. Many freshwater pro-

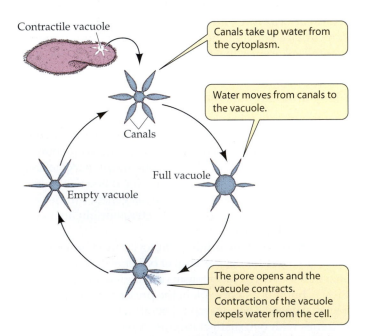

Contractile vacuole

Canals take up water from the cytoplasm.

Canals

Water moves from canals to the vacuole.

Full vacuole

Empty vacuole

The pore opens and the vacuole contracts. Contraction of the vacuole expels water from the cell.

27.5 Contractile Vacuoles Bail Out Excess Water
Water constantly enters freshwater protists by osmosis. A pore in the cell surface allows the contractile vacuole to expel the water it accumulates.

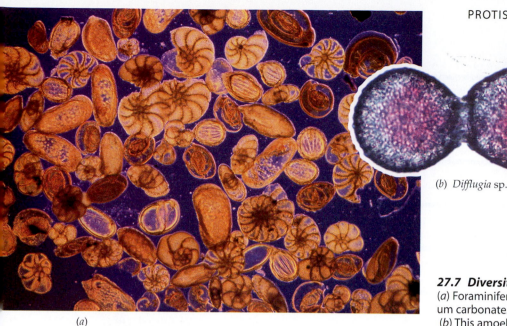

(a)

(b) *Difflugia* sp.

(c) *Paramecium caudatum*

27.7 Diversity among Protist Cell Surfaces
(*a*) Foraminiferan shells are made of protein hardened with calcium carbonate. Several species are shown in this photograph.
(*b*) This amoeba constructed its shell by cementing sand grains together. (*c*) Spirals of protein make this *Paramecium*'s surface—known as its pellicle—flexible but resilient.

tists address this problem by means of vesicles that contract to excrete excess water. Members of several protist groups have such **contractile vacuoles**. Because these organisms have a higher concentration of solutes than their freshwater environment does, they constantly take in water by osmosis. The excess water collects in the contractile vacuole and is then pushed out (Figure 27.5).

It is easy to confirm that bailing out water is the principal function of the contractile vacuole. First, we can observe some protists under a light microscope and note the rate at which the vacuoles are contracting—they look like little eyes winking. Then we can place other protists of the same species in solutions of differing osmotic potential. The less negative the osmotic potential of the surrounding solution, the more hypertonic the cells are, and the faster water rushes into them, causing the contractile vacuoles to pump more rapidly. Conversely, the contractile vacuoles will stop pumping if the solute concentration of the medium is increased so that it is equal to that of the cells.

A second important type of vesicle found in many protists is the **food vacuole**. Protists such as *Paramecium* engulf solid food by endocytosis, forming a food vacuole within which the food is digested (Figure 27.6). Smaller vesicles containing digested food pinch away from the food vesicle and enter the cytoplasm. These tiny vesicles provide a large surface area across which the products of digestion may be absorbed by the rest of the cell.

The cell surfaces of protists are diverse

A few protists, such as some amoebas, are surrounded by only a plasma membrane, but most have stiffer surfaces that maintain the structural integrity of the cell. Many protists have cell walls, which are often complex in structure. Other protists that lack cell walls have a variety of ways of strengthening their surfaces. Some have internal "shells," which the organism either produces itself, as foraminiferans do, or makes from bits of sand and thickenings immediately beneath the plasma membrane, as some amoebas do (Figure 27.7).

Many protists contain endosymbionts

Endosymbiosis is very common among the protists, and in some instances both the host and the endosymbiont are protists. Many radiolarians, for example, harbor photosynthetic protists (Figure 27.8). As a result, these radiolarians appear greenish or yellowish, depending on the type of endosymbiont they contain. This arrangement is beneficial to the radiolarian, for it can make use of the food produced by its photosynthetic guest. The guest, in turn, may make use of metabolites made by the host, or it may simply receive physical protection. In other cases, the guest may be a victim, exploited for its photosynthetic products while receiving no benefit itself.

27.8 Protists within Protists
Photosynthetic organisms living as endosymbionts within these radiolarians provide food for the radiolarians, as well as part of the pigmentation seen through their glassy skeletons. Both the endosymbionts and the radiolarians are protists.

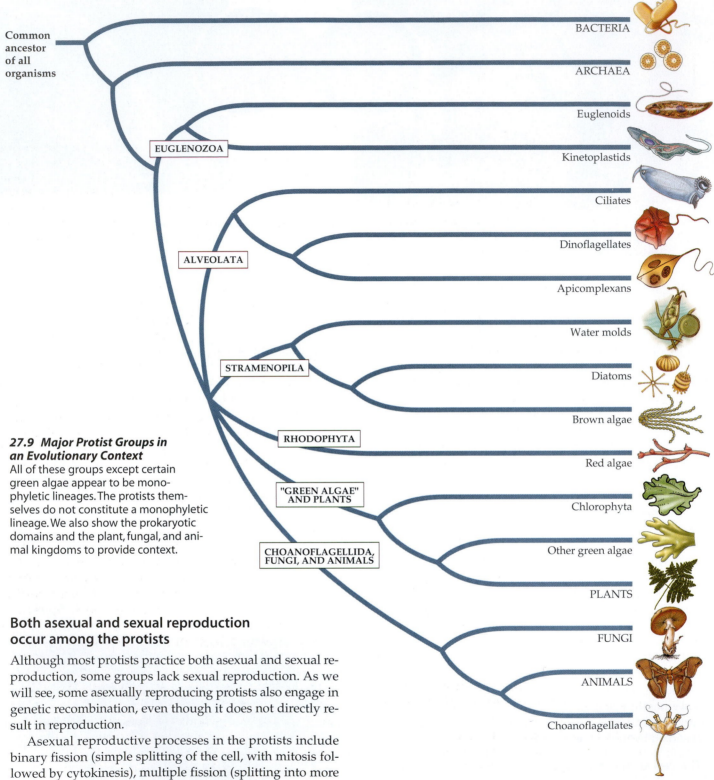

BACTERIA

ARCHAEA

Euglenoids

EUGLENOZOA

Kinetoplastids

Ciliates

ALVEOLATA

Dinoflagellates

Apicomplexans

Water molds

STRAMENOPILA

Diatoms

Brown algae

RHODOPHYTA

Red algae

"GREEN ALGAE" AND PLANTS

Chlorophyta

CHOANOFLAGELLIDA, FUNGI, AND ANIMALS

Other green algae

PLANTS

FUNGI

ANIMALS

Choanoflagellates

27.9 Major Protist Groups in an Evolutionary Context
All of these groups except certain green algae appear to be monophyletic lineages. The protists themselves do not constitute a monophyletic lineage. We also show the prokaryotic domains and the plant, fungal, and animal kingdoms to provide context.

Both asexual and sexual reproduction occur among the protists

Although most protists practice both asexual and sexual reproduction, some groups lack sexual reproduction. As we will see, some asexually reproducing protists also engage in genetic recombination, even though it does not directly result in reproduction.

Asexual reproductive processes in the protists include binary fission (simple splitting of the cell, with mitosis followed by cytokinesis), multiple fission (splitting into more than two cells), budding (the outgrowth of a new cell from the surface of an old one), and the formation of spores (cells that are capable of developing into new organisms). Sexual reproduction also takes various forms. In some protists, as in animals, the gametes are the only haploid cells. In some other protists, by contrast, both diploid and haploid cells undergo mitosis, giving rise to alternation of generations, which will be described later in the chapter.

The diversity of form, habitat, metabolism, locomotion, reproduction, and life cycles found among the protists reflects the diversity of avenues pursued during the early evolution of eukaryotes. Many of these avenues led to great success, judging from the abundance and diversity of today's protists and other eukaryotes.

Protist Diversity

As we have seen, the phylogeny of protists is an area of exciting, challenging research. The marvelous diversity of protist body forms and metabolic lifestyles seems reason enough for a fascination with these organisms, but questions about how the multicellular eukaryotic kingdoms originated from the protists stimulate further interest. Fortunately, the tools of molecular biology, such as rRNA sequencing, make it possible to explore evolutionary relationships among the protists in greater detail and with somewhat greater confidence than previously (see Chapters 24 and 26).

We will discuss several apparently monophyletic groups of protists, as well as a few other groups of more uncertain phylogenetic status. Some biologists refer to many of these monophyletic groups as kingdoms; others refer to them as subkingdoms; still others refer to them as phyla. This choice of words is not of immediate concern to us here, so we'll just call them "groups." We'll describe the Euglenozoa, Alveolata, Stramenopila, Rhodophyta, Chlorophyta, and Choanoflagellida (Figure 27.9).

As we shall see, some of the monophyletic protist groups consist of organisms with very diverse body plans. On the other hand, certain body plans, such as those of amoebas and those of slime molds, have arisen again and again during evolution, in groups only distantly related to one another.

Euglenozoa

The **Euglenozoa** are a monophyletic group of *flagellates:* unicellular organisms with flagella. They reproduce asexually by binary fission. There are two subgroups of Euglenozoa: euglenoids and kinetoplastids.

Euglenoids have anterior flagella

The **euglenoids** possess flagella arising from a pocket at the anterior end of the cell. Euglenoids used to be claimed by the zoologists as animals and by the botanists as plants. They are unicellular flagellates, but many members of the group are photosynthetic.

Figure 27.10 depicts a cell of the genus *Euglena*. Like most other euglenoids, this common freshwater organism has a complex cell structure. It propels itself through the water with one of its two flagella, which may also serve as an anchor to hold the organism in place. The flagellum provides power by means of a wavy motion that spreads from base to tip. The second flagellum is often rudimentary.

Euglena has very flexible nutritional requirements. Many species are always heterotrophic. Other species are fully autotrophic in sunlight, using chloroplasts to synthe-

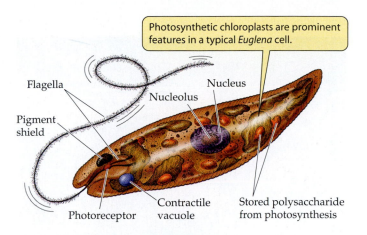

Photosynthetic chloroplasts are prominent features in a typical *Euglena* cell.

Flagella · Nucleolus · Nucleus · Pigment shield · Photoreceptor · Contractile vacuole · Stored polysaccharide from photosynthesis

27.10 A Photosynthetic Euglenoid
Several *Euglena* species are among the best-known flagellates. In this species, the second flagellum is rudimentary.

size organic compounds through photosynthesis. The chloroplasts of euglenas are surrounded by three membranes (unlike plant chloroplasts, which have only two). When kept in the dark, these euglenas lose their photosynthetic pigment and begin to feed exclusively on dead organic material floating in the water around them. Such a "bleached" *Euglena* resynthesizes its photosynthetic pigment when it is returned to the light and becomes autotrophic again. But *Euglena* cells treated with certain antibiotics or mutagens lose their photosynthetic pigment completely; neither they nor their descendants are ever autotrophs again. However, those descendants function well as heterotrophs.

Kinetoplastids have mitochondria that edit their own RNA

The **kinetoplastids** are unicellular, parasitic flagellates with a single, large mitochondrion. That mitochondrion contains a *kinetoplast*—a unique structure housing DNA and associated proteins. The kinetoplast DNA is of two types, called minicircles and maxicircles. The maxicircles encode enzymes associated with oxidative metabolism, and the minicircles encode "guides" that accomplish a remarkable type of RNA editing within the mitochondrion.

Some kinetoplastids are human pathogens. Sleeping sickness, one of the most dreaded diseases of Africa, is caused by the parasitic kinetoplastid *Trypanosoma* (Figure 27.11). The vector (intermediate host) for *Trypanosoma* is an insect, the tsetse fly. Carrying its deadly cargo, the tsetse fly bites livestock, wild animals, and even humans, infecting them with the parasite. *Trypanosoma* then multiplies in the mammalian bloodstream and produces toxins. When these parasites invade the nervous system, the neurological symptoms of sleeping sickness appear and are followed by death. Other trypanosomes cause leishmaniasis, Chagas' disease, and East Coast fever; all are major diseases in the tropics.

Undulating membrane of trypanosome

Trypanosoma gambiense

25 μm

27.11 A Parasitic Kinetoplastid
Trypanosomes, shown here among human red blood cells, cause sleeping sickness in mammals. A flagellum runs along one edge of the cell as part of a structure called the undulating membrane.

Alveolata

The **Alveolata** are a monophyletic group of unicellular organisms characterized by the possession of cavities called *alveoli* just below their plasma membranes. They are diverse in body form. The alveolate groups we'll consider here are the dinoflagellates, apicomplexans, and ciliates.

Dinoflagellates are unicellular marine organisms with two flagella

The **dinoflagellates** are all unicellular, and most are marine organisms. A distinctive mixture of photosynthetic and accessory pigments gives their chloroplasts a golden-brown color. The dinoflagellates are of great ecological, evolutionary, and morphological interest. They are among the most important primary photosynthetic producers of organic matter in the oceans.

Euglenozoa

Alveolata

Stramenopila

Rhodophyta

Chlorophyta

Many dinoflagellates are endosymbionts, living within the cells of other organisms, including various invertebrates and even other marine protists. Dinoflagellates are particularly common endosymbionts in corals, to whose growth they contribute by photosynthesis. Some dinoflagellates are nonphotosynthetic and live as parasites within other marine organisms.

Dinoflagellates have a distinctive appearance (see Figure 27.1*a*). They have two flagella, one in an equatorial groove around the cell, the other starting at the same point as the first and passing down a longitudinal groove before extending into the surrounding medium.

Some dinoflagellates reproduce in enormous numbers in warm and somewhat stagnant waters. The result can be a "red tide," so called because of the reddish color of the sea that results from pigments in the dinoflagellates (Figure 27.12). During a red tide, the concentration of dinoflagellates may reach 60 million cells per liter of ocean water. Certain red tide species produce a potent nerve toxin that can kill tons of fish. The genus *Gonyaulax* produces a toxin that can accumulate in shellfish in amounts that, although not fatal to the shellfish, may kill a person who eats the shellfish.

Many dinoflagellates are bioluminescent. In complete darkness, cultures of these organisms emit a faint glow. If you suddenly stir or bubble air through the culture, the organisms each emit numerous bright flashes. A ship passing through a tropical ocean that contains a rich growth of these species produces a bow wave and wake that glow eerily as billions of these dinoflagellates discharge their light systems.

Apicomplexans are parasites with unusual spores

Exclusively parasitic organisms, the **apicomplexans** derive their name from the *apical complex*, a mass of organelles contained within the apical end of their spores. These organelles help the apicomplexan spore invade its host tissue. Unlike many other protists, apicomplexans lack contractile vacuoles.

Apicomplexans generally have an amorphous body form like that of an amoeba. This body form has evolved over and over again in parasitic protists. It appears even among parasitic dinoflagellates, a group of organisms whose nonparasitic relatives have highly distinctive, complex body forms.

Like many obligate parasites, apicomplexans have elaborate life cycles featuring asexual and sexual reproduction

A red tide along the coast of Baja California

27.12 A Red Tide of Dinoflagellates
By reproducing in astronomical numbers, the dinoflagellate *Gonyaulax tamarensis* can cause a toxic red tide.

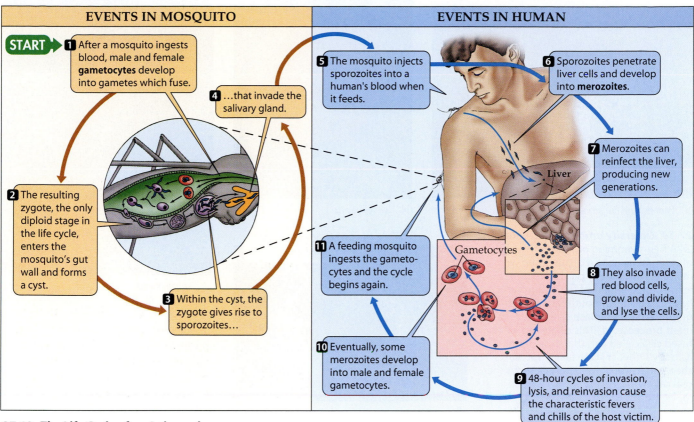

EVENTS IN MOSQUITO	EVENTS IN HUMAN

START ➤ **1** After a mosquito ingests blood, male and female **gametocytes** develop into gametes which fuse.

4 …that invade the salivary gland.

5 The mosquito injects sporozoites into a human's blood when it feeds.

6 Sporozoites penetrate liver cells and develop into **merozoites**.

2 The resulting zygote, the only diploid stage in the life cycle, enters the mosquito's gut wall and forms a cyst.

7 Merozoites can reinfect the liver, producing new generations.

Liver

3 Within the cyst, the zygote gives rise to sporozoites…

11 A feeding mosquito ingests the gametocytes and the cycle begins again.

Gametocytes

8 They also invade red blood cells, grow and divide, and lyse the cells.

10 Eventually, some merozoites develop into male and female gametocytes.

9 48-hour cycles of invasion, lysis, and reinvasion cause the characteristic fevers and chills of the host victim.

27.13 The Life Cycle of an Apicomplexan
Malaria-causing *Plasmodium* species spend part of their life cycle in humans and part in mosquitoes. Sporozoites and merozoites are spores with apical complexes.

by a series of very dissimilar life stages. Often these stages are associated with two different types of host organism.

The best-known apicomplexans are the malarial parasites of the genus *Plasmodium*, a highly specialized group of organisms that spend part of the life cycle within human red blood cells (Figure 27.13). Although it has been almost eliminated from the United States, malaria continues to be a major problem in many tropical countries. In terms of the number of people infected, malaria is one of the world's most serious diseases.

Female mosquitoes of the genus *Anopheles* transmit *Plasmodium* to humans. The parasite enters the human circulatory system when an infected *Anopheles* mosquito penetrates the human skin in search of blood. The parasites find their way to cells in the liver and the lymphatic system, change their form, multiply, and reenter the bloodstream, attacking red blood cells. The apical complex enables *Plasmodium* to enter human liver cells and red blood cells.

The parasites multiply inside red blood cells, which then burst, releasing new swarms of parasites. If another *Anopheles* bites the victim, the mosquito takes in some of the parasitic *Plasmodium* cells along with blood. The infecting cells develop into gametes, which unite to form zygotes that lodge in the mosquito's gut, divide several times, and

move into its salivary glands, from which they can be passed on to another human host. Thus, *Plasmodium* is an extracellular parasite in the mosquito vector and an intracellular parasite in the human host.

Malaria kills more than a million people each year, and *Plasmodium* has proved to be a singularly difficult pathogen to attack. The *Plasmodium* life cycle is best broken by the removal of stagnant water, in which mosquitoes breed. The use of insecticides to reduce the *Anopheles* population can be effective, but their benefits must be weighed against the possible ecological, economic, and health risks posed by the insecticides themselves. However, there is now new hope in the form of a genome-sequencing project that targets the common form of *Plasmodium*. Scheduled to be completed by the year 2002, this project may provide the information needed to end the epidemic.

Ciliates have two types of nuclei

The **ciliates** are so named because they characteristically have hairlike cilia. This group is noteworthy for its diversity and ecological importance (Figure 27.14). Almost all ciliates are heterotrophic (a few contain photosynthetic endosymbionts), and they are much more specialized in body form than are most flagellates and other protists.

The definitive characteristic of ciliates is the possession of two types of nuclei, from one to as many as a thousand large **macronuclei** and, within the same cell, from one to eighty **micronuclei**. The micronuclei, which are typical eukaryotic nuclei, are essential for genetic recombination. The

(a) *Paramecium bursaria*

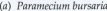
10 μm

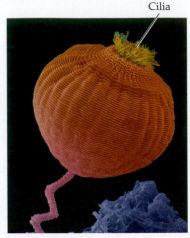

Cilia

(b) *Epistylis* sp.

27.14 Diversity among the Ciliates
(a) A free-swimming organism, this paramecium belongs to a ciliate group whose members have many cilia of uniform length. (b) Members of this subgroup have cilia on their mouthparts. (c) In this group, tentacles replace cilia as development proceeds. (d) This ciliate "walks" on fused cilia called cirri that project from its body. Other cilia are fused into flat sheets that sweep food particles into the oral cavity.

Cirri

This individual has ingested green algae.

(d) *Euplotes* sp.

25 μm

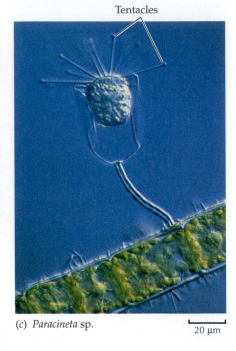

Tentacles

(c) *Paracineta* sp.

20 μm

macronuclei are derived from micronuclei. Each macronucleus contains many copies of the genetic information, packaged in units containing very few genes each; the macronuclear DNA is transcribed and translated to regulate the life of the cell. Although we do not know how this system of macro- and micronuclei came into being, we do know something about the behavior of these nuclei, which we will discuss after describing the body plan of one important ciliate, *Paramecium*.

A CLOSER LOOK AT ONE CILIATE. *Paramecium*, a frequently studied ciliate genus, exemplifies the complex structure and behavior of ciliates (Figure 27.15*a*). The slipper-shaped cell is covered by an elaborate **pellicle**, a structure composed principally of an outer membrane and an inner layer of closely packed, membrane-enclosed sacs (the alveoli) that surround the bases of the cilia. Defensive organelles called *trichocysts* are also present in the pellicle. In response to a threat, a microscopic explosion expels the trichocysts in a few milliseconds, and they emerge as sharp darts, driven forward at the tip of a long, expanding filament (Figure 27.15*b*).

The cilia provide a form of locomotion that is generally more precise than locomotion by flagella or pseudopods. A

paramecium can direct the beat of its cilia to propel itself either forward or backward in a spiraling manner (Figure 27.16). It can also back off swiftly when it encounters a barrier or a negative stimulus. The coordination of ciliary beating is probably the result of a differential distribution of ion channels in the plasma membrane near the two ends of the cell.

REPRODUCTION WITHOUT SEX, AND SEX WITHOUT REPRODUCTION. Paramecia reproduce asexually by binary fission. The micro-

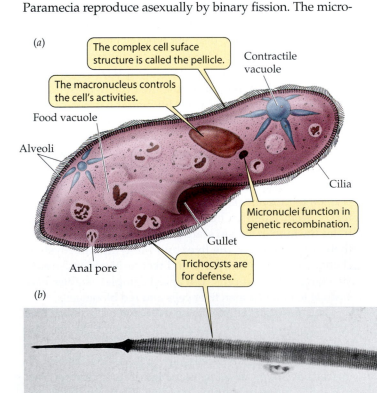

(a)

The complex cell suface structure is called the pellicle.

Contractile vacuole

The macronucleus controls the cell's activities.

Food vacuole

Alveoli

Cilia

Micronuclei function in genetic recombination.

Gullet

Anal pore

Trichocysts are for defense.

(b)

27.15 Anatomy of Paramecium
(a) The major structures of a typical paramecium. (b) A trichocyst discharged from beneath the pellicle of a paramecium has a sharp point and a straight filament.

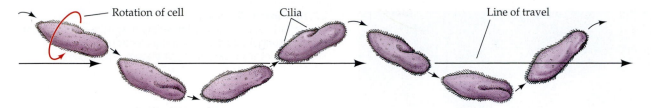

27.16 "Swimming" with Cilia
Beating its cilia in coordinated waves that progress from one end of the cell to the other, a paramecium can move in either direction with respect to the long axis of the cell. The cell rotates in a spiral as it travels.

nuclei divide mitotically. The macronuclei divide by a still unknown mechanism following a round of DNA replication.

Paramecia also have an elaborate sexual behavior called **conjugation**. Two paramecia line up tightly against each other and fuse in the oral region of the body. Nuclear material is extensively reorganized and exchanged over the next several hours (Figure 27.17). As a result of this process, each cell ends up with two haploid micronuclei, one from itself and one from the other cell, which fuse to form a new diploid micronucleus. New macronuclei develop from the micronuclei through a series of dramatic chromosomal rearrangements. The exchange of nuclei is fully reciprocal—each of the two paramecia gives and receives an equal amount of DNA. The two organisms then separate and go their own ways, each equipped with new combinations of alleles.

Conjugation in *Paramecium* is a *sexual* process of genetic recombination, but it is not a *reproductive* process. The same two cells that begin the process are there at the end, and no new cells are created. As a rule, each clone of paramecia must periodically conjugate. Experiments have shown that if some species are not permitted to conjugate, the asexual clones can live through no more than approximately 350 cell divisions before they die out.

Stramenopila

The **stramenopiles** include three prominent groups, two of which are photosynthetic. The two flagella of a stramenopile cell are typically unequal in length. The longer of the two bears rows of tubular hairs. Some stramenopiles lack flagella, but they are presumed to be descended from ancestors that possessed typical stramenopile flagella. The stramenopiles include the diatoms and the brown algae, which are photosynthetic, and the oomycetes, which are not. Other, smaller stramenopile groups include some that are nonphotosynthetic. Some botanists prefer to call the stramenopiles the "brown plant kingdom."

Diatoms are everywhere in the marine environment

Diatoms (Bacillariophyta) are single-celled organisms, although some species associate in filaments. Many have sufficient carotenoids in their chloroplasts to give them a yellow or brownish color. All make chrysolaminarin (a carbohydrate) and oils as photosynthetic storage products. They lack flagella.

Architectural magnificence on a microscopic scale is the hallmark of the diatoms (Figure 27.18a). Many diatoms deposit silicon in their cell walls. The cell wall of some species

Euglenozoa
Alveolata
Stramenopila
Rhodophyta
Chlorophyta

27.17 Paramecia Achieve Genetic Recombination by Conjugating
Conjugating *Paramecium* individuals exchange micronuclei, thereby permitting genetic recombination. After conjugation, the cells separate and continue their lives as two individuals.

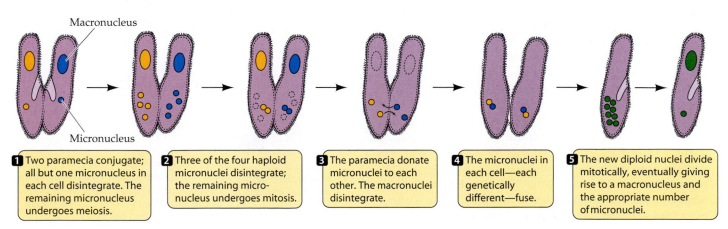

Macronucleus

Micronucleus

1 Two paramecia conjugate; all but one micronucleus in each cell disintegrate. The remaining micronucleus undergoes meiosis.

2 Three of the four haploid micronuclei disintegrate; the remaining micronucleus undergoes mitosis.

3 The paramecia donate micronuclei to each other. The macronuclei disintegrate.

4 The micronuclei in each cell—each genetically different—fuse.

5 The new diploid nuclei divide mitotically, eventually giving rise to a macronucleus and the appropriate number of micronuclei.

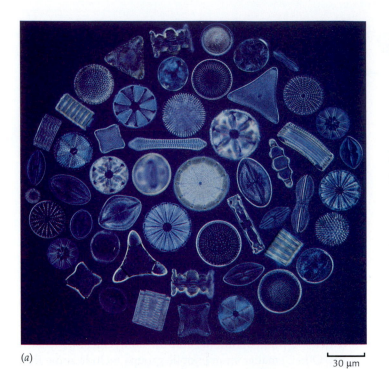

(a)

30 μm

27.18 Diatom Diversity
(a) Diatoms exhibit a splendid variety of species-specific forms. (b) This artificially colored scanning electron micrograph shows the intricate patterning of diatom cell walls.

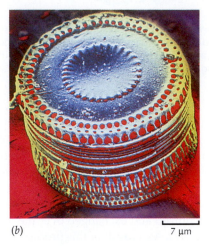

(b)   7 μm

is constructed in two pieces, with the wall of the top overlapping the wall of the bottom like the top and bottom of a petri plate. The silicon-impregnated walls have intricate, unique patterns (Figure 27.18b). Despite their remarkable morphological diversity, however, all diatoms are symmetrical—either bilaterally (with "right" and "left" halves) or radially (with the type of symmetry possessed by a circle).

Diatoms reproduce both sexually and asexually. Asexual reproduction is by cell division and is somewhat constrained by the stiff, silica-containing cell wall. Both the top and the bottom of the "petri plate" become tops of new "plates" without changing appreciably in size; as a result, the new cells made from former bottoms are smaller than the parent cells (Figure 27.19). If this process continued indefinitely, one cell line would simply vanish, but sexual reproduction largely solves this potential problem. Gametes are formed, shed their cell walls, and fuse. The resulting zygote then increases substantially in size before a new cell wall is laid down.

Diatoms are everywhere in the marine environment and are frequently present in great numbers, making them major photosynthetic producers in coastal waters. Diatoms are also common in fresh water. Because the silicon-containing walls of dead diatom cells resist decomposition, certain sedimentary rocks are composed almost entirely of diatom skeletons that sank to the seafloor over time. Diatomaceous earth, which is obtained from such rocks, has many industrial uses, such as insulation, filtration, and metal polishing. It has also been used as an "Earth-friendly" insecticide that clogs the tracheae (breathing structures) of insects.

The brown algae include the largest protists

All the **brown algae** (Phaeophyta) are multicellular and composed either of branched filaments (Figure 27.20) or of leaflike growths called **thalli** (singular thallus) (Figure 27.21a). The brown algae obtain their namesake color from the carotenoid *fucoxanthin*, which is abundant in their chloro-

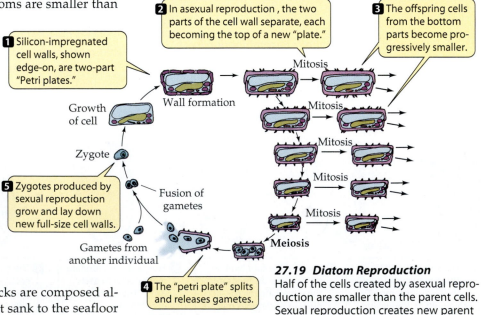

2 In asexual reproduction, the two parts of the cell wall separate, each becoming the top of a new "plate."

3 The offspring cells from the bottom parts become progressively smaller.

1 Silicon-impregnated cell walls, shown edge-on, are two-part "Petri plates."

Growth of cell

Wall formation

Mitosis

Mitosis

Mitosis

Mitosis

Mitosis

Meiosis

Zygote

5 Zygotes produced by sexual reproduction grow and lay down new full-size cell walls.

Fusion of gametes

Gametes from another individual

4 The "petri plate" splits and releases gametes.

27.19 Diatom Reproduction
Half of the cells created by asexual reproduction are smaller than the parent cells. Sexual reproduction creates new parent cells with full-sized cell walls.

(a) *Hormosira banksii*

(b) *Ectocarpus* sp.

27.20 Brown Algae
(a) A filamentous brown alga growing in Australia. This species is sometimes called "Neptune's necklace." (b) A filamentous brown alga seen through a light microscope.

plasts. The combination of this yellow-orange pigment with the green of chlorophylls *a* and *c* yields a brownish tinge.

The brown algae include the largest of the protists. Giant kelps, such as those of the genus *Macrocystis*, may be up to 60 meters long (see Figure 27.1c). The brown algae are almost exclusively marine. Some float in the open ocean; the most famous example is the genus *Sargassum*, which forms dense mats of vegetation in the Sargasso Sea in the mid-Atlantic. Most brown algae, however, are attached to rocks near the shore. A few thrive only where they are regularly exposed to heavy surf; a notable example is the sea palm *Postelsia palmaeformis* of the Pacific coast (Figure 27.21a). All

of the attached forms develop a specialized structure, called a **holdfast**, that literally glues them to the rocks (Figure 27.21b).

Some brown algae differentiate extensively into stemlike stalks and leaflike blades, and some develop gas-filled cavities or bladders. For biochemical reasons that are only poorly understood, these gas cavities often contain as much as 5 percent carbon monoxide—a concentration high enough to kill a human. In addition to organ differentiation, the larger brown algae also exhibit considerable tissue differentiation. Most of the giant kelps have photosynthetic filaments only in the outermost regions of their stalks and blades. Within the photosynthetic region lie filaments of long cells that closely resemble the food-conducting tissue of plants. Called *trumpet cells* because they have flaring ends, these tubes rapidly conduct the products of photosynthesis through the body of the organism.

The cell walls of brown algae may contain as much as 25 percent *alginic acid*, a gummy polymer of sugar acids. Alginic acid cements cells and filaments together and provides good holdfast glue. It is used commercially as an emulsifier in ice cream, cosmetics, and other products.

Many protist and all plant life cycles feature alternation of generations

Brown algae, like many photosynthetic protists and all plants, exhibit a type of life cycle known as **alternation of generations**, in which a multicellular, diploid, spore-producing organism gives rise to a multicellular, haploid, gamete-producing organism. When two gametes fuse (a process called *syngamy*), a diploid organism is formed (Figure 27.22). The haploid organism, the diploid organism, or both may also reproduce asexually.

(a) *Postelsia palmaeformis*

(b)

The leaflike structures are the thalli of sea palm.

27.21 Brown Algae in a Turbulent Environment
Brown algae growing in the intertidal zone on an exposed rocky shore take a tremendous pounding by the surf. (a) Sea palm growing along the California coast. (b) The tough, branched holdfast that anchors the sea palm.

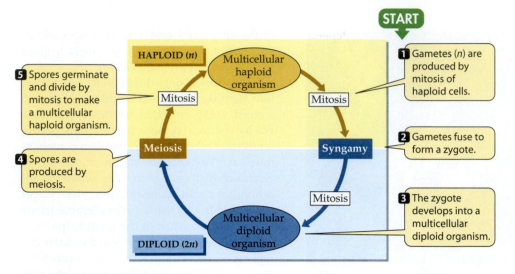

START

HAPLOID (n)

Multicellular haploid organism

5 Spores germinate and divide by mitosis to make a multicellular haploid organism.

Mitosis

Mitosis

1 Gametes (*n*) are produced by mitosis of haploid cells.

Meiosis

Syngamy

2 Gametes fuse to form a zygote.

4 Spores are produced by meiosis.

Mitosis

3 The zygote develops into a multicellular diploid organism.

Multicellular diploid organism

DIPLOID (2n)

27.22 Alternation of Generations
In many multicellular photosynthetic protists and all plants, a diploid generation that produces spores alternates with a haploid generation that produces gametes.

The two generations (spore-producing and gamete-producing) differ genetically (one has haploid cells and the other has diploid cells), but they may or may not differ morphologically. In **heteromorphic** alternation of generations, the two generations differ morphologically; in **isomorphic** alternation of generations they do not, despite their genetic difference. We will see examples of both heteromorphic and isomorphic alternation of generations in some representative brown and green algae. In discussing the life cycles of plants and multicellular photosynthetic protists, we will use the terms **sporophyte** ("spore plant") and **gametophyte** ("gamete plant") to refer to the multicellular diploid and haploid generations, respectively.

Gametes are not produced by meiosis because the gamete-producing generation is already haploid. Instead, specialized cells of the diploid sporophyte, called **sporocytes**, divide meiotically to produce four haploid spores. The spores may eventually germinate and divide mitotically to produce multicellular haploid gametophytes, which produce gametes by mitosis and cytokinesis.

Gametes, unlike spores, can produce new organisms only by fusing with other gametes. The fusion of two gametes produces a diploid zygote, which then undergoes mitotic divisions to produce a diploid organism: the sporophyte generation. The sporocytes of the sporophyte generation then undergo meiosis and produce haploid spores, starting the cycle anew.

The brown algae exemplify the extraordinary diversity found among the photosynthetic protists. One genus of simple brown algae is *Ectocarpus* (see Figure 27.20*b*). Its branched filaments, a few centimeters long, commonly grow on shells and stones. The gametophyte and sporophyte generations of *Ectocarpus* can be distinguished only by chromosome number or reproductive products (spores or gametes). Thus the generations are isomorphic.

By contrast, some kelps of the genus *Laminaria* and some other brown algae show a more complex heteromorphic alternation of generations. The larger and more obvious generation of these species is the sporophyte. Meiosis in special fertile regions of the leaf-like fronds produces haploid **zoospores**—motile spores that are propelled by flagella. These germinate to form a tiny, filamentous gametophyte that produces either eggs or sperm. The eggs and sperm of brown algae typically have flagella.

The oomycetes include water molds and their relatives

A nonphotosynthetic stramenopile group called **oomycetes** consists in large part of the water molds and their terrestrial relatives, such as the downy mildews. Water molds are filamentous and stationary, and they feed by absorption. If you have seen a whitish, cottony mold growing on dead fish or dead insects in water, it was probably a water mold of the common genus *Saprolegnia* (Figure 27.23).

The oomycetes are **coenocytes**: They have many nuclei enclosed in a single plasma membrane. Their filaments have no cross-walls to separate the many nuclei into discrete cells. Their cytoplasm is continuous throughout the body of the mold, and there is no single structural unit with a single nucleus, except in certain reproductive stages. A distinguishing feature of the oomycetes is their flagellated reproductive cells. Oomycetes are diploid throughout most of their life cycle and have cellulose in their cell walls.

The water molds, such as *Saprolegnia*, are all aquatic and **saprobic** (they feed on dead organic matter). Some other

Saprolegnia sp.

27.23 An Oomycete
The filaments of a water mold radiate from the carcass of an insect.

oomycetes are terrestrial. Although most terrestrial oomycetes are harmless or helpful decomposers of dead matter, a few are serious plant parasites that attack crops such as avocados, grapes, and potatoes. The mold *Phytophthora infestans*, for example, is the causal agent of late blight of potatoes, which brought about the great Irish potato famine of 1845–1847. *P. infestans* destroyed the entire Irish potato crop in a matter of days in 1846. Among the consequences of the famine were a million deaths from starvation and the emigration of about 2 million people, mostly to the United States.

Rhodophyta

Almost all **red algae** (Rhodophyta) are multicellular (Figure 27.24). Some botanists now refer to the red algae as the "red plant kingdom." Their characteristic color is a result of the photosynthetic pigment phycoerythrin, which is found in relatively large amounts in the chloroplasts of many species. In addition to phycoerythrin, red algae contain phycocyanin, carotenoids, and chlorophyll.

The red algae include species that grow in the shallowest tide pools, as well as the algae found deepest in the ocean (as deep as 260 meters if nutrient conditions are right and the water is clear enough to permit the penetration of light). Very few red algae inhabit fresh water. Most grow attached to a substrate by a holdfast.

```
                    ┌─ Euglenozoa
                ┌───┤
            ┌───┤   └─ Alveolata
            │   │
        ────┤   └───── Stramenopila
            │
            │   ┌────── Rhodophyta
            └───┤
                └────── Chlorophyta
```

In a sense the red algae, like several other groups of algae, are misnamed. They have the capacity to change the relative amounts of their various photosynthetic pigments depending on the light conditions where they are growing. Thus the leaflike *Chondrus crispus*, a common North Atlantic red alga, may appear bright green when it is growing at or near the surface of the water and deep red when growing at greater depths. The ratio of pigments present depends to a remarkable degree on the intensity of the light that reaches the alga. In deep water, where the light is dimmest, the alga accumulates large amounts of phycoerythrin, an accessory photosynthetic pigment (see Figure 8.7). Algae in deep water have as much chlorophyll as the green ones near the surface, but the accumulated phycoerythrin makes them look red.

In addition to being the only photosynthetic protists with phycoerythrin and phycocyanin among their pigments, the red algae have two other unique characteristics: They store the products of photosynthesis in the form of *floridean starch*, which is composed of very small, branched chains of approximately 15 glucose units. And they produce no motile, flagellated cells at any stage in their life cycle. The male gametes lack cell walls and are slightly amoeboid; the female gametes are completely immobile.

Some red algal species enhance the formation of coral reefs. Like the coral animals, they possess the biochemical machinery for depositing calcium carbonate both in and around their cell walls. After the death of the corals and algae, the calcium carbonate persists, sometimes forming substantial rocky masses.

Some red algae produce large amounts of mucilaginous polysaccharide substances, which contain the sugar galactose with a sulfate group attached. This material readily forms solid gels and is the source of agar, a substance widely used in the laboratory for making a solid aqueous medium on which tissue cultures and many microorganisms can be grown.

Certain red algae became endosymbionts, long ago, within the cells of other, nonphotosynthetic protists, eventually giving rise to chloroplasts. This was the evolutionary origin of the distinctive chloroplasts of the photosynthetic stramenopiles (the brown algae and the diatoms).

(a) *Palmaria palmata*

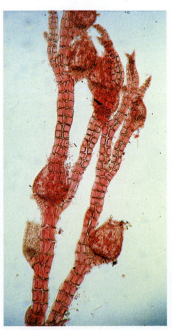

(b) *Polysiphonia* sp.

27.24 Red Algae
(a) Dulse, a large, edible red alga, is growing here on rocks in New Brunswick, Canada. (b) Both vegetative and reproductive structures of this alga can be seen under the light microscope.

Chlorophyta

The "green algae" do not form a monophyletic group, but include at least two multicellular lineages. A major lineage constitutes the **Chlorophyta**, a monophyletic group. A sister lineage to the Chlorophyta consists of other green algal lineages and the plant kingdom. There are more than 17,000 species of chlorophytes. The green algal lineages share characters that distinguish them from other protists: Like the plants, they contain chlorophylls *a* and *b*, and their reserve of photosynthetic products is stored as starch in plastids. Most chlorophytes are aquatic—some are marine, but more are freshwater forms—but others are terrestrial, living in moist environments. The chlorophytes range in size from microscopic unicellular forms to multicellular forms many centimeters in length.

Chlorophytes vary in shape and cellular organization

We find in the Chlorophyta an incredible variety in shape and construction of the algal body. *Chlorella* is an example of the simplest type: unicellular and flagellated.

Surprisingly large and well-formed colonies of cells are found in such freshwater groups as the genus *Volvox*. These cells are not differentiated into tissues and organs, as in plants and animals, but the colonies show vividly how the preliminary step of this great evolutionary development might have been taken. In *Volvox*, the origins of cell specialization can be seen as certain cells within the colony (Figure 27.25*a*) are specialized for reproduction.

While *Volvox* is colonial and spherical, *Oedogonium* is multicellular and filamentous, and each of its cells has only one nucleus. *Cladophora* is multicellular, but each cell is multinucleate. *Bryopsis* is tubular and coenocytic, forming cross-walls only when reproductive structures form. *Acetabularia* is a single, giant uninuclear cell a few centimeters long that becomes multinucleate only at the end of the reproductive stage. *Ulva lactuca* is a membranous sheet two cells thick; its unusual appearance justifies its common name: sea lettuce (Figure 27.25*b*).

Chlorophyte life cycles are diverse

The life cycles of chlorophytes show great diversity. Let's examine two chlorophyte life cycles in detail, beginning with that of the sea lettuce *Ulva lactuca* (Figure 27.26). The diploid sporophyte of this common seashore organism is a thin cellular sheet a few centimeters in diameter. Some of its cells (sporocytes) differentiate and undergo meiosis and cytokinesis, producing motile haploid spores (zoospores). These swim away, each propelled by four flagella, and some eventually find a suitable place to settle. The spores then lose their fla-

(*a*) *Volvox* sp.   Parent colony   Somatic cells

Specialized reproductive cells produce and release daughter colonies.

18 μm

(*b*) *Ulva lactuca*

(*c*) *Micrasterias* sp.   Isthmus

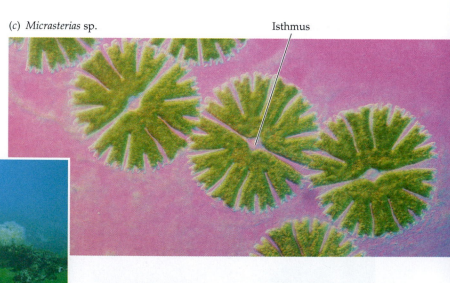

27.25 Chlorophytes
(*a*) *Volvox* colonies are precisely spaced arrangements of cells. Several daughter colonies can be seen within the parent colonies. (*b*) A stand of sea lettuce, submerged in a tidal pool. (*c*) A microscopic desmid. A narrow isthmus containing the nucleus joins two elaborate semicells—halves—of this unicellular organism. A single large, ornate chloroplast fills much of the volume of each semicell.

Euglenozoa
Alveolata
Stramenopila
Rhodophyta
Chlorophyta

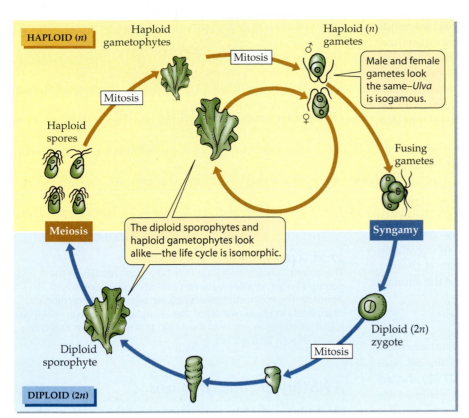

27.26 An Isomorphic Life Cycle
The life cycle of *Ulva lactuca* is an example of isomorphic alternation of generations.

flagella, undergo mitosis, and produce a new gametophyte directly; in other words, the gametes can also function as zoospores. Few chlorophytes other than *Ulva* have motile gametes that can also function as zoospores.

In contrast to the isomorphic life cycle of *Ulva*, many other chlorophytes have a heteromorphic life cycle: Sporophyte and gametophyte generations differ in structure. In one variation of the heteromorphic life cycle—the **haplontic** life cycle (Figure 27.27)—a multicellular haploid individual produces gametes that fuse to form a zygote. The zygote functions directly as a sporocyte, undergoing meiosis to produce spores, which in turn produce a new haploid individual. In the entire haplontic life cycle, only one cell—the zygote—is diploid. The filamentous organisms of the genus *Ulothrix* are examples of haplontic chlorophytes.

Other chlorophytes have a **diplontic** life cycle like that of many animals. In a diplontic life cycle, meiosis of sporocytes

gella and begin to divide mitotically, producing a thin filament that develops into a broad sheet only two cells thick. The gametophyte thus produced looks just like the sporophyte—in other words, *Ulva* has an isomorphic life cycle.

An individual gametophyte can produce only male or female gametes—never both. The gametes arise mitotically within single cells (called *gametangia*), rather than within a specialized multicellular structure, as in plants. Both types of gametes bear two flagella (in contrast to the four flagella of a haploid spore) and hence are motile.

In most species of *Ulva* the female and male gametes are indistinguishable structurally, making those species **isogamous**—having gametes of identical appearance. Other chlorophytes, including some other species of *Ulva*, are **anisogamous**—having female gametes that are distinctly larger than the male gametes.

Female and male gametes come together and unite, losing their flagella as the zygote forms and settles. After resting briefly, the zygote begins mitotic division, producing a multicellular sporophyte. Any gametes that fail to find partners can settle down on a favorable substrate, lose their

27.27 A Haplontic Life Cycle
In the life cycle of *Ulothrix*, a filamentous, multicellular gametophyte generation alternates with a sporophyte generation consisting of a single cell.

produces gametes directly; the gametes fuse, and the resulting zygote divides mitotically to form a new multicellular sporophyte. In such organisms, every cell except the gametes is diploid. Between these two extremes are chlorophytes whose gametophyte and sporophyte generations are both multicellular, but that have one phase (usually the sporophyte) that is much larger and more prominent than the other.

There are green algae other than chlorophytes

As we mentioned above, the Chlorophyta are the largest lineage of green algae, but there are other green algal lineages as well. Those lineages are branches of a lineage that also includes the plant kingdom. The green algal lineage that is sister to the plant kingdom, a group of organisms called charophytes, will be described in the next chapter. But now let's consider a close protist relative of the animals.

Choanoflagellida

One group of protists with flagella, the **Choanoflagellida**, is thought to comprise the closest relatives of the animals. Members of this group are colonial (Figure 27.28) and are thought to be closely related to the sponges, the most ancient of the surviving phyla of animals. Sponges are also colonial rather than truly multicellular, in that they lack organized tissues and their cells can be separated and recombined. Choanoflagellates bear a striking resemblance to the most characteristic type of cell found in the sponges (compare Figures 27.28 and 31.4).

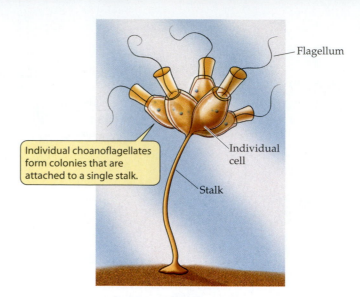

27.28 A Link to the Animal Kingdom
This colonial choanoflagellate may be a close relative of the sponges and thus a link between protists and the kingdom Animalia. The connection of unicellular organisms into colonies often leads to the evolution of specialized cells and true multicellularity; another example of a colonial protist was seen earlier in *Volvox* (Figure 27.25a).

A History of Endosymbiosis

As we have already seen, many protists possess chloroplasts. Groups with chloroplasts appear in several distantly related protist lineages. Some of these groups differ from others in the photosynthetic pigments their chloroplasts contain. And we've seen that not all chloroplasts have a pair of surrounding membranes—in some protists, they

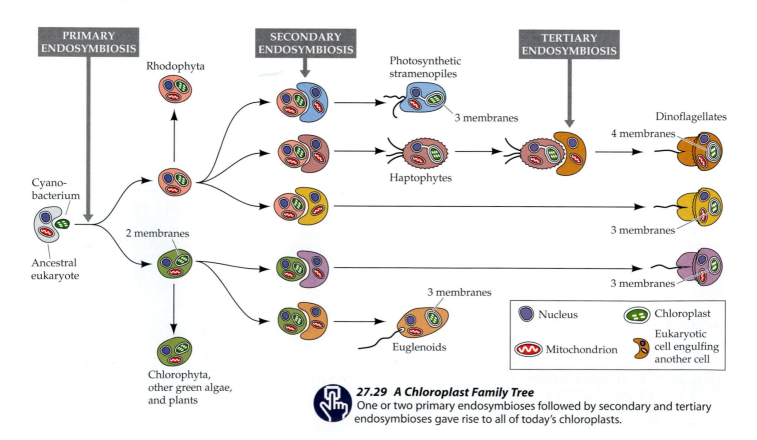

27.29 A Chloroplast Family Tree
One or two primary endosymbioses followed by secondary and tertiary endosymbioses gave rise to all of today's chloroplasts.

have three membranes. We now understand these observations in terms of a remarkable series of endosymbioses.

All chloroplasts trace back to the engulfment of an ancestral cyanobacterium by a larger eukaryotic cell (Figure 27.29). This event constituted *primary endosymbiosis*. The cyanobacterium had a single membrane—its plasma membrane—and that membrane was surrounded by part of the eukaryote's plasma membrane that wrapped around the bacterium as it was taken up. Thus, the original chloroplasts had two surrounding membranes.

Primary endosymbiosis gave rise to the chloroplasts of the green algae and the red algae. We do not yet know whether both trace back to a single primary endosymbiosis, with later divergence, or whether they resulted from independent occurrences of primary endosymbiosis. In either case, each line participated in further endosymbioses.

The photosynthetic euglenoids derived their chloroplasts from *secondary endosymbiosis*. Their ancestor took up a unicellular chlorophyte, retaining the endosymbiont's chloroplast and eventually losing the rest of its constituents. This history accounts for the fact that the photosynthetic euglenoids have the same photosynthetic pigments as the chlorophytes and plants. It also accounts for the third membrane of the euglenoid chloroplast, which is derived from the euglenoid's plasma membrane.

Other photosynthetic protist groups derived their chloroplasts by endosymbiosis with unicellular red algae. Both the green lineage and the red lineage of chloroplasts appear to have given rise to more than one secondary endosymbiosis. At least one secondary endosymbiosis produced a unicellular protist that became, itself, a partner in a *tertiary endosymbiosis*!

Although euglenoid chloroplasts are descendants of a chlorophyte, and stramenopile chloroplasts are descendants of a red alga, this does not mean that euglenoids themselves are descendants of a chlorophyte, nor are stramenopiles themselves descendants of a red alga. The ancestors that took up green or red algae in secondary endosymbiosis had their own evolutionary histories. It has taken much research to piece together the lineages as we now understand them.

The monophyletic groups of protists that we have discussed are summarized in Table 27.1 and Figure 27.9. Now let's consider two major types of protist body forms—amoebas and other organisms with pseudopods, and slime molds—that are not monophyletic. Rather, they have reappeared in various branches of the eukaryote family tree.

Some Recurrent Body Forms

Amoebas used to be classified together in a single protist group. However, the amoeba body plan has popped up again and again in the course of the evolution of the Eukarya (Figure 27.30). Similarly, three kinds of organisms called slime molds, once classified together, may be quite different phylogenetically.

Amoebas form pseudopods

The pseudopods used by **amoebas** for locomotion are a hallmark of these protists. This body plan has appeared by convergent evolution in various protist groups. The mechanism of amoeboid motion will be discussed in Chapter 47.

Amoebas have often been portrayed in popular writing as blobs—the simplest form of "animal" life imaginable. Superficial examination of a typical amoeba shows how such an impression might have been obtained. An amoeba consists of a single cell. It feeds on small organisms and particles of organic matter by phagocytosis, engulfing them with its pseudopods.

But amoebas are specialized protists. Many are adapted for life on the bottoms of lakes, ponds, and other bodies of water. Their creeping locomotion and their manner of engulfing food particles fit them for life close to a relatively rich supply of sedentary organisms or organic particles. Most amoebas exist as predators, parasites, or scavengers. A few are photosynthetic.

Amoebas of the free-living genus *Naegleria*, some of which can enter humans and cause a fatal disease of the nervous system, have a two-stage life cycle, one stage having amoeboid cells and the other flagellated cells. Some

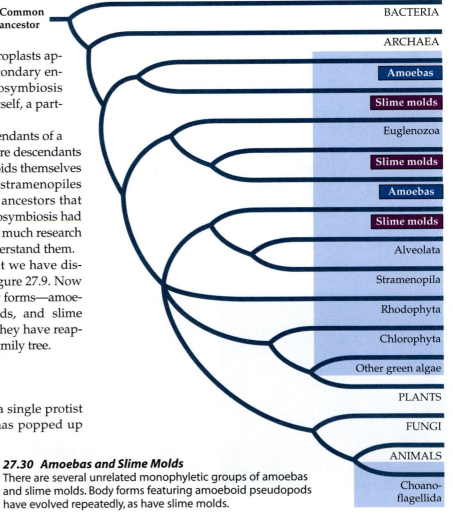

27.30 Amoebas and Slime Molds
There are several unrelated monophyletic groups of amoebas and slime molds. Body forms featuring amoeboid pseudopods have evolved repeatedly, as have slime molds.

amoebas are shelled, living in casings of sand grains glued together (see Figure 27.7*b*) or in shells secreted by the organism itself.

Actinopods have thin, stiff pseudopods

The **actinopods** are recognizable by their thin, stiff pseudopods, which are reinforced by microtubules. The pseudopods play at least four roles:

▶ They greatly increase the surface area of the cell for exchange of materials with the environment.
▶ They help the cell float in its marine or freshwater environment.
▶ They provide locomotion in some species.
▶ They are the cell's feeding organs, trapping smaller organisms and often taking them up by endocytosis.

Radiolarians, actinopods that are exclusively marine, are perhaps the most beautiful of all microorganisms (Figure 27.31*a*). Almost all radiolarian species secrete glassy *endoskeletons* (internal skeletons) from which needlelike pseudopods project. Part of the skeleton is a central capsule within the cytoplasm. The skeletons of the different species are as varied as snowflakes, and many have elaborate geometric designs. A few radiolarians are among the largest of the unicellular protists, with skeletons measuring several millimeters across. Innumerable radiolarian skeletons, some as old as 700 million years, form the sediments under some tropical seas.

Heliozoans lack an endoskeleton (Figure 27.31*b*). Most heliozoans live in fresh water. They roll along the substrate by shortening and elongating their pseudopods.

Foraminiferans have created vast limestone deposits

Foraminiferans are marine protists that secrete shells of calcium carbonate (see Figure 27.7*a*). Some foraminiferans live as **plankton** (free-floating microscopic organisms), and many others live at the bottom of the sea. Their long, threadlike, branched pseudopods reach out through numerous microscopic pores in the shell and interconnect to create a sticky net, which the foraminiferan uses to catch smaller plankton.

After foraminiferans reproduce (by mitosis and cytokinesis), the daughter cells abandon the parent shell and make new shells of their own. The discarded skeletons of ancient foraminiferans make up extensive limestone deposits in various parts of the world, forming a layer hundreds to thousands of meters deep over millions of square kilometers of ocean bottom. Foraminiferan skeletons also make up the sand of some beaches. A single gram of such sand may contain as many as 50,000 foraminiferan shells.

The shells of individual foraminiferan species have distinctive shapes and are easily preserved as fossils in marine sediments. Each geological period has distinctive foraminiferan species. For this reason, and because they are so abundant, the remains of foraminiferans are especially valuable as indicators in the classification and dating of sedimentary rocks, as well as in oil prospecting.

Slime molds release spores from erect fruiting bodies

The three groups of **slime molds** seem so similar at first glance that they were once grouped in a single phylum. However, the slime molds are actually so different that some biologists now classify them in different *kingdoms*. We will consider two of these groups, called acellular slime molds and cellular slime molds.

The slime molds share only general characteristics. All are motile, all ingest particulate food by endocytosis, and all form spores on erect fruiting bodies. They undergo striking changes in organization during their life cycles, and one stage consists of isolated cells that engage in absorptive nutrition. Some slime molds may cover areas of 1 meter or more in diameter while in their less aggregated stage. Such a large slime mold may weigh more than 50 grams. Slime molds of both types favor cool, moist habitats, primarily in forests. They range from colorless to brilliantly yellow and orange.

ACELLULAR SLIME MOLDS FORM MULTINUCLEATE MASSES. If the nucleus of an amoeba began rapid mitotic division, accompanied by a tremendous increase in cytoplasm and organelles, the resulting organism might resemble the **acel-**

(*a*) Radiolarian (species not identified)

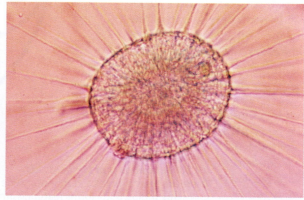

(*b*) *Actinosphaerium eichorni*

150 µm

27.31 Actinopods
(*a*) A radiolarian displays its intricate glassy skeleton. (*b*) A heliozoan with long pseudopods.

(a) *Physarum polycephalum*

27.32 Acellular Slime Molds
(a) Plasmodia of yellow slime mold cover a rock in Nova Scotia.
(b) The fruiting structures—sporangiophores (yellow) and sporangia (black)—of *Physarum*.

(b) *Physarum* sp.

1 mm

lular slime molds (Myxomycota). During its *vegetative* (feeding) phase, an acellular slime mold is a wall-less mass of cytoplasm with numerous diploid nuclei. This mass streams very slowly over its substrate in a remarkable network of strands called a **plasmodium*** (Figure 27.32a). The plasmodium of an acellular slime mold is an example of a coenocyte, a body in which many nuclei are enclosed in a single plasma membrane. The outer cytoplasm of the plasmodium (closest to the environment) is normally less fluid than the interior cytoplasm and thus provides some structural rigidity.

Acellular slime molds such as *Physarum* (a popular research subject) provide a dramatic example of movement by **cytoplasmic streaming**. The outer cytoplasmic region becomes more fluid in places, and cytoplasm rushes into those areas, stretching the plasmodium. This streaming somehow reverses its direction every few minutes as cytoplasm rushes into a new area and drains away from an older one, moving the plasmodium over its substrate in search of food. Sometimes an entire wave of plasmodium moves across the substrate, leaving strands behind. Actin filaments and a contractile protein called myxomyosin interact to produce the streaming movement. As it moves, the plasmodium engulfs food particles—predominantly bacteria, yeasts, spores of fungi, and other small organisms, as well as decaying animal and plant remains.

An acellular slime mold can grow almost indefinitely in its plasmodial stage, as long as the food supply is adequate and other conditions, such as moisture and pH, are favorable. However, one of two things can happen if conditions become unfavorable. First, the plasmodium can form an irregular mass of hardened cell-like components called a **sclerotium**. This resting structure rapidly becomes a plasmodium again when favorable conditions are restored.

Alternatively, the plasmodium can transform itself into

*Do not confuse the plasmodium of an acellular slime mold with the genus *Plasmodium*, the apicomplexan that is the cause of malaria.

spore-bearing fruiting structures (Figure 27.32b). These stalked or branched structures, called **sporangiophores**, rise from heaped masses of plasmodium. They derive their rigidity from walls that form and thicken between their nuclei. The nuclei of the plasmodium are diploid, and they divide by meiosis as the sporangiophore develops. One or more knobs, called **sporangia**, develop on the end of the stalk. Within a sporangium, haploid nuclei become surrounded by walls and form spores. Eventually, as the sporangiophore dries, it sheds its spores.

The spores germinate into wall-less, flagellated, haploid cells called **swarm cells**, which can either divide mitotically to produce more haploid swarm cells or function as gametes. Swarm cells can live as separate individual cells, and can become walled and resistant resting cysts when conditions are unfavorable. When conditions improve again, the cysts release flagellated swarm cells. Two swarm cells can also fuse to form a diploid zygote, which divides by mitosis (but without a wall forming between the nuclei) and thus forms a new, coenocytic plasmodium.

CELLS RETAIN THEIR IDENTITY IN THE CELLULAR SLIME MOLDS. Whereas the plasmodium is the basic vegetative unit of the acellular slime molds, an amoeboid cell is the vegetative unit of the cellular slime molds. Large numbers of cells called **myxamoebas**, which have single haploid nuclei, engulf bacteria and other food particles by endocytosis and reproduce by mitosis and fission. This simple life cycle stage, consisting of swarms of independent, isolated cells, can persist indefinitely as long as food and moisture are available.

When conditions become unfavorable, however, the cellular slime molds aggregate and form fruiting structures, as do their acellular counterparts. The apparently independent myxamoebas aggregate into a mass called a **slug** or *pseudoplasmodium* (Figure 27.33a). Unlike the true plasmodium of the acellular slime molds (see Figure 27.32a), this structure is not simply a giant sheet of cytoplasm with many nuclei; the individual myxamoebas retain their plasma membranes and, therefore, their identity.

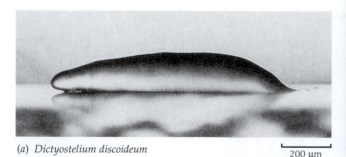

(a) *Dictyostelium discoideum*

200 µm

27.33 A Cellular Slime Mold
(a) A pseudoplasmodium migrates over its substrate. (b) Fruiting structures in various stages of development.

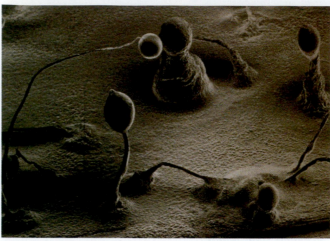

(b) *Dictyostelium discoideum*

500 µm

The chemical signal that causes the myxamoebas of cellular slime molds to aggregate into a slug is 3′,5′-cyclic adenosine monophosphate (cAMP), a compound that plays many important roles in chemical signaling in animals (see Chapter 15). A slug may migrate over its substrate for several hours before becoming motionless and reorganizing to construct a delicate, stalked fruiting structure (Figure 27.33b). Cells at the top of the fruiting structure develop into thick-walled spores, which are eventually released. Later, under favorable conditions, the spores germinate, releasing myxamoebas.

The cycle from myxamoebas through slug and spores to new myxamoebas is asexual. Cellular slime molds also have a sexual cycle, in which two myxamoebas fuse. The product of this fusion develops into a spherical structure that ultimately germinates, releasing new haploid myxamoebas.

In the remaining chapters of Part Four, we will explore the three classic kingdoms of multicellular eukaryotes. Chapters 28 and 29 deal with the kingdom Plantae (which, combined with the Chlorophyta and other green algae, is called by some botanists the "green plant kingdom"). Chapter 30 presents the kingdom Fungi, and Chapters 31–33 describe the kingdom Animalia. These kingdoms all arose from protist ancestors.

Chapter Summary

Protists Defined

▶ In this book we define the protists simply as all eukaryotes that are not plants, fungi, or animals. The protists are not a monophyletic group.

The Origin of the Eukaryotic Cell

▶ The modern eukaryotic cell arose from an ancestral prokaryote in several steps. Probable steps included loss of the cell wall and inward folding of the plasma membrane. **Review Figure 27.2**

▶ In subsequent steps, an infolded plasma membrane attached to the chromosome may have led to the formation of a nuclear envelope. A primitive cytoskeleton evolved. **Review Figure 27.3**

▶ The first truly eukaryotic cell was larger than its prokaryote ancestor, and it may have possessed one or more flagella of the eukaryotic type.

▶ The incorporation of prokaryotic cells as endosymbionts gave rise to eukaryotic organelles. Peroxisomes, which protected the host cell from an oxygen-rich atmosphere, may have been the first organelles of endosymbiotic origin. Mitochondria evolved from once free-living proteobacteria, and chloroplasts evolved from once free-living cyanobacteria. Cells with nuclei probably appeared before the first cells with mitochondria. **Review Figure 27.3**

General Biology of the Protists

▶ Most protists are aquatic; some live within other organisms. The great majority are unicellular and microscopic, but many are multicellular and a few are enormous.

▶ "Protozoan" is an outdated term sometimes applied to protists, mostly ingestive heterotrophs, that were once classified as animals. "Alga" is an outdated term sometimes applied to photosynthetic protists.

▶ Protists vary widely in their modes of nutrition, metabolism, and locomotion. Some protist cells contain contractile vacuoles, and some digest their food in food vacuoles. **Review Figures 27.5, 27.6**

▶ Protists have a variety of cell surfaces, some of them protective. **Review Figure 27.7**

▶ Many protists contain endosymbiotic prokaryotes. Some protists are endosymbionts in other cells, including other protists. Some endosymbiotic protists perform photosynthesis, to the advantage of their hosts.

▶ Most protists reproduce both asexually and sexually.

Protist Diversity

▶ Molecular and other techniques are enabling biologists to identify many monophyletic groups of protists. **Review Figure 27.9 and Table 27.1**

Euglenozoa

▶ The Euglenozoa are a monophyletic group of unicellular protists with flagella.

▶ Euglenoids are Euglenozoa, such as *Euglena*, that are often photosynthetic and have anterior flagella.

▶ Kinetoplastids are Euglenozoa, such as *Trypanosoma*, that have a single, large mitochondrion, in which RNA is edited.

Alveolata

▶ The Alveolata are a monophyletic group of unicellular organisms with alveoli (cavities) beneath their plasma membranes.

▶ Dinoflagellates are marine alveolates with a golden-brown color that results from their photosynthetic and accessory pigments. They are major contributors to world photosynthesis. Many are endosymbionts; in that role they are important contributors to coral reef growth. Dinoflagellates are responsible for toxic "red tides."

▶ Apicomplexans are parasitic alveolates with an amoeba-like body form. Their spores, containing a mass of organelles at the apical end, are adapted to the invasion of host tissue. The apicomplexan *Plasmodium*, which causes malaria, uses two alternate hosts (humans and *Anopheles* mosquitoes). **Review Figure 27.13**

▶ Ciliates are alveolates such as *Paramecium* that move rapidly by means of cilia and have two kinds of nuclei. The macronuclei control the cell by means of transcription and translation. The micronuclei are responsible for genetic recombination, accomplished by conjugation, which is sexual but not reproductive. Some ciliates have a remarkably complex internal structure. **Review Figures 27.15, 27.16, 27.17**

Stramenopila

▶ Stramenopiles typically have two flagella of unequal length, the longer bearing rows of tubular hairs. Some stramenopile groups are photosynthetic.

▶ Diatoms are unicellular stramenopiles, many of which have complex, two-part, glassy cell walls. They contribute extensively to world photosynthesis. **Review Figure 27.19**

▶ The brown algae are predominantly multicellular, photosynthetic stramenopiles. They include the largest of all protists, and some show considerable tissue differentiation.

▶ In many multicellular photosynthetic protists and in all plants, both haploid and diploid cells undergo mitosis, leading to an alternation of generations. The diploid sporophyte generation forms spores by meiosis, and the spores develop into haploid organisms. This haploid gametophyte generation forms gametes by mitosis, and their fusion yields zygotes that develop into the next generation of sporophytes. **Review Figure 27.22**

▶ Oomycetes are a group of nonphotosynthetic stramenopiles including water molds and downy mildews. The oomycetes are coenocytic. They are diploid for most of their life cycle.

Rhodophyta

▶ Red algae (Rhodophyta) are multicellular, photosynthetic protists. They differ from the other photosynthetic protist groups in having a characteristic storage product (floridean starch) and lacking flagellated reproductive cells.

Chlorophyta

▶ The Chlorophyta, a monophyletic group of green algae, are often multicellular. Like plants, they contain chlorophylls *a* and *b* and use starch as a storage product. The chlorophytes have diverse life cycles; among these are the isomorphic alternation of generations of *Ulva* and the haplontic life cycle of *Ulothrix*. **Review Figures 27.26, 27.27**

▶ The chlorophytes are sister to a lineage that includes other green algae and the plant kingdom.

Choanoflagellida

▶ The Choanoflagellida are protists with flagella and a body type similar to the most characteristic type of cell found in sponges. The Choanoflagellida are sister to the animal kingdom.

A History of Endosymbiosis

▶ Primary endosymbiosis of a cyanobacterium and a eukaryote gave rise to the chloroplasts of green algae, plants, and red algae. **Review Figure 27.29**

▶ Secondary endosymbiosis of eukaryotes with unicellular green or red algae gave rise to the chloroplasts of euglenoids, stramenopiles, and other groups. A cell of one of those groups, in tertiary endosymbiosis, has given rise to another type of chloroplast.

Some Recurrent Body Forms

▶ Some similar body forms are found in several different, unrelated protist groups. **Review Figure 27.30**

▶ Amoebas, which appear in many protist groups, move by means of pseudopods.

▶ Actinopods have thin, stiff pseudopods that serve various functions, including food capture.

▶ Foraminiferans also use pseudopods for feeding, and secrete shells of calcium carbonate.

▶ Acellular slime molds and cellular slime molds are superficially very similar, moving as slimy masses and producing stalked fruiting structures. However, they differ at the cellular level. Acellular slime molds are coenocytes with diploid nuclei. Cellular slime molds consist of individual haploid cells that aggregate into masses consisting of distinct cells.

For Discussion

1. For each type of organism below, give a single characteristic that may be used to differentiate it from the other, related organism(s) in parentheses.
   a. foramineferans (radiolarians)
   b. *Euglena* (*Volvox*)
   c. *Trypanosoma* (*Giardia*)
   d. *Amoeba* (flagellate)
   e. *Physarum* (*Dictyostelium*)

2. For each of the following groups, give at least two characteristics used to distinguish the group from other groups.
   a. Ciliophora
   b. Apicomplexa
   c. Phaeophyta
   d. Rhodophyta

3. In what sense are sex and reproduction independent of each other in the ciliates? What does that suggest as to the most important role of sex in biology?

4. Why are dinoflagellates and apicomplexans placed in one group of protists and brown algae and oomycetes in another?

5. Giant seaweed (mostly brown algae) have "floats" that aid in keeping their fronds suspended at or near the surface of the water. Why is it important that the fronds be suspended?

6. Why are algal pigments so much more diverse than those of plants?

7. For each of the groups Chlorophyta, Euglenozoa, and Rhodophyta, indicate how many membranes surround their chloroplasts, and offer a reasonable explanation in each case. Why do some dinoflagellates have more membranes around their chloroplasts than other dinoflagellates?

28 *Plants without Seeds: From Sea to Land*

HOW DO WE RUN OUR ENGINES, HEAT OUR homes, smelt our metals, and generate much of our electricity? We do these things by burning plant-based fuels. The great majority of such fuels—petroleum and natural gas, the so-called fossil fuels—comes from the remains of plants without seeds that grew in great forests hundreds of millions of years ago.

In some parts of the world, people derive the majority of their fuel from peat bogs. They harvest and burn peat, another substance produced largely by a nonseed plant.

Peat consists of partially decomposed plant material. It forms as rapidly growing upper layers of moss, primarily the genus *Sphagnum*, along with some other plants, compress the deeper-lying layers. Peatlands cover an area approximately half as large as the United States—more than 1 percent of Earth's total surface.

Sphagnum is one of the most abundant plants on Earth, yet it and its mossy neighbors at first glance seem to lack adaptations to life on land. Mosses have no internal "plumbing system" to move water and nutrients within their bodies, and their leafy photosynthetic organs are only one cell thick. They require liquid water in order to reproduce, and indeed seem at first glance to be highly dependent on external moisture. They can dry out to the point of becoming brittle, yet they snap back as soon as a bit of water is available. How do they manage to survive on land?

That mosses and their relatives do have effective adaptations for life in terrestrial environments is obvious from their wide distribution. Most live in moist habitats, but a few mosses even live in deserts.

Earth did not take on a green tint until about half a billion years ago, long after the ancestors of today's plants invaded the land sometime during the Paleozoic era (see Table 20.1). The earliest land plants were tiny, but their metabolic activities helped convert native rock into soil that could support the needs of their successors. Larger and larger plants evolved rapidly (in geological terms),

and during the Carboniferous period (354 to 290 million years ago) great forests were widespread. However, few of the trees in those forests were like those we know today. During the tens of millions of years since the Carboniferous, these early trees have been replaced by the modern trees whose adaptations and appearance are familiar to us.

In this chapter, we will see how members of the plant kingdom conquered the land and evolved. We will see what made early plants different and made Plantae a unique kingdom, and will survey the diverse products of plant evolution. In the next chapter we will complete our survey of the plant kingdom by considering the seed plants, which dominate the terrestrial scene today.

The Plant Kingdom

As we use the term, a plant is a photosynthetic eukaryote that uses chlorophylls *a* and *b*, stores carbohydrates, usually as starch, and develops from an embryo protected by tissues of the parent plant. Most plants have, or had and then lost in the course of evolution, two whiplash flagella at the anterior end of their motile cells. Thus defined, the kingdom Plantae is monophyletic—it forms a single branch of the evolutionary tree. Because of their development from embryos, plants are sometimes referred to as *embryophytes*.

Fuel from Plants
Peat, formed from layers of plant matter including moss of the genus *Sphagnum*, is the major source of fuel for residents of the Falkland Islands off the coast of Argentina.

28.1 Classification of Plants[a]

PHYLUM	COMMON NAME	CHARACTERISTICS
NONTRACHEOPHYTES		
Hepatophyta	Liverworts	No filamentous stage; gametophyte flat
Anthocerophyta	Hornworts	Embedded archegonia; sporophyte grows basally
Bryophyta	Mosses	Filamentous stage; sporophyte grows apically (from the tip)
TRACHEOPHYTES		
NONSEED TRACHEOPHYTES		
Lycophyta	Club mosses	Simple leaves in spirals; sporangia in leaf axils
Sphenophyta	Horsetails	Simple leaves in whorls; stems jointed
Psilotophyta	Whisk ferns	No true leaves; roots absent
Pterophyta	Ferns	Complex leaves; sporangia on underside of leaves
SEED PLANTS		
Gymnosperms		
Cycadophyta	Cycads	Compound leaves; swimming sperm; seeds on modified leaves
Ginkgophyta	Ginkgo	Deciduous; fan-shaped leaves; swimming sperm
Gnetophyta	Gnetophytes	Vessels in vascular tissue; opposite, simple leaves
Coniferophyta	Conifers	Seeds in cones; needlelike or scalelike leaves
Angiosperms		
Angiospermae	Flowering plants	Endosperm; carpels; much reduced gametophytes; seeds in fruit

[a] No extinct groups are included in this classification.

Some botanists refer to a group consisting of the Plantae plus the green algae as the "green plant kingdom," to the Stramenopila as the "brown plant kingdom," and the Rhodophyta as the "red plant kingdom" (see Figure 27.29).

There are twelve surviving phyla of plants

The surviving members of the kingdom Plantae fall naturally into twelve phyla (Table 28.1). All members of nine of the phyla possess well-developed vascular systems that transport materials throughout the plant body. We call these nine phyla, collectively, the **tracheophytes** because they all possess conducting cells called tracheids.

The remaining three phyla (liverworts, hornworts, and mosses), which lack tracheids, were once considered classes of a single larger phylum, of which the most familiar examples are mosses. Now we use the term **nontracheophytes** to refer collectively to these three phyla. The nontracheophytes are sometimes collectively called bryophytes, but in this text we reserve that term for the mosses. Collectively, the nontracheophytes are not a monophyletic group.

Life cycles of plants feature alternation of generations

A universal feature of the life cycles of plants is the alternation of generations (see Figure 9.13b). If we begin looking at the plant life cycle at the single-cell stage—the diploid zygote—then the first phase of the cycle features the formation, by mitosis and cytokinesis, of a multicellular embryo and eventually the mature diploid plant (see Figure 27.22).

This multicellular, diploid plant is the **sporophyte** ("spore plant"). Cells contained in **sporangia** (singular sporangium, "spore reservoir") on the sporophyte undergo meiosis to produce haploid, unicellular spores. By mitosis and cytokinesis a spore forms a haploid plant. This multicellular, haploid plant is the **gametophyte** ("gamete plant") and produces haploid gametes. The fusion of two gametes (syngamy, or fertilization) results in the formation of a diploid cell, the zygote, and the cycle repeats.

The *sporophyte generation* extends from the zygote through the adult, multicellular, diploid plant; the *gametophyte generation* extends from the spore through the adult, multicellular, haploid plant to the gamete. The transitions between the phases are accomplished by fertilization and meiosis. In all plants, the sporophyte and gametophyte differ genetically: The former has diploid cells, the latter haploid cells.

The Plantae arose from a green algal lineage

Much evidence indicates that the closest living relatives of the plants are a group of green algae called *charophytes*. The charophytes, along with some other green algae and the plants, are in a lineage that is sister to the Chlorophyta (see Figure 27.9), but we don't yet know which charophyte lineage is the true sister group to the plants. Stoneworts of the genus *Chara* are charophytes that resemble plants in terms of their rRNA and DNA sequences, peroxisome contents, mechanics of mitosis and cytokinesis, and chloroplast structure (Figure 28.1a). On the other hand, strong evidence from morphology-based cladistic analysis suggests that the

(a) *Chara* sp. (stonewort)

28.1 The Closest Relatives of Land Plants
The plant kingdom probably evolved from a common ancestor shared with the charophytes, a green algal group. (a) Molecular evidence seems to favor stoneworts of the genus *Chara*. (b) Evidence from morphology indicates that the group including this coleochaete alga may be the ancestor of land plants.

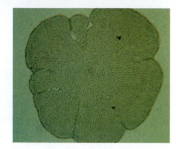

(b) *Coleochaete* sp.

sister group of the plants is a group of charophytes that includes the genus *Coleochaete* (Figure 28.1b). *Coleochaete*-like algae have features found in plants, such as plasmodesmata and a tendency to protect the young sporophyte.

Whether more similar to stoneworts or to *Coleochaete*, the ancestral green algae lived at the margins of ponds or marshes, ringing them with a green mat. From these margin habitats, which were sometimes wet and sometimes dry, early plants made the move onto land.

The Conquest of the Land

Plants or their immediate ancestors in the green mat pioneered and modified the terrestrial environment. That environment differs dramatically from the aquatic environment. The most obvious difference is the availability of the water that is essential for life: It is everywhere in the aquatic environment, but hard to find and to retain in the terrestrial environment. Water also provides aquatic organisms with support against gravity; a plant on land, however, must either have some other support system or sprawl unsupported on the ground. A land plant must also use different mechanisms for dispersing its gametes and progeny than its aquatic relatives use. How did terrestrial organisms arise from aquatic ancestors to thrive in such a challenging environment?

Adaptations to life on land distinguish plants from green algae

Most of the characteristics that distinguish plants from green algae are evolutionary adaptations to life on land (Figure 28.2). Many of the characteristics that proved adaptive to land plants probably evolved before the appearance of any of the plant groups we will discuss in this chapter. These characteristics include:

▶ The *cuticle*, a waxy covering that retards desiccation (drying).
▶ *Gametangia*, cases that enclose plant gametes and prevent them from drying. Eggs are housed in archegonia, sperm in antheridia.
▶ *Embryos*, which are young sporophytes contained within a protective structure.
▶ Certain *pigments* that afford protection against the mutagenic ultraviolet radiation that bathes the terrestrial environment.
▶ *Thick spore walls* that prevent desiccation and resist decay.

All these characteristics were probably shared by a plant ancestral to today's plants. Further adaptations to the terrestrial environment appeared as plant evolution continued. We will identify the most important ones in this and the next chapter. For now, let's look at one of the key later adaptations: the appearance of vascular tissues.

Most present-day plants have vascular tissue

The first plants were truly nonvascular, lacking both water-conducting and food-conducting cells. Although the term "nonvascular plants" is a time-honored name, it is misleading to apply it to the entire nontracheophyte lineage, because some mosses (unlike liverworts and hornworts) do have a limited amount of vascular-like tissue. Thus the more unwieldy name nontracheophyte is more descriptive. The first true tracheophytes—possessing specialized conducting cells called tracheids—arose later.

The nontracheophytes (the liverworts, hornworts, and mosses) have never been large plants. Except for some of the mosses, they have no water-transporting tissue, yet some are found in dry environments. Many grow in dense masses (see Figure 28.7a), through which water can move by capillary action. Nontracheophytes have leaflike structures that readily catch and hold any water that splashes onto them. These plants are small enough that minerals can be distributed internally by diffusion. They lack the leaves, stems, and roots that characterize tracheophytes, although they have structures analogous to each.

Familiar tracheophytes include the ferns, conifers, and flowering plants. Tracheophytes differ from liverworts, hornworts, and mosses in crucial ways, one of which is the possession of a well-developed **vascular system** consisting of specialized tissues for the transport of materials from one part of the plant to another. One such tissue, the *phloem*, conducts the products of photosynthesis from sites where they are produced or released to sites where they are used

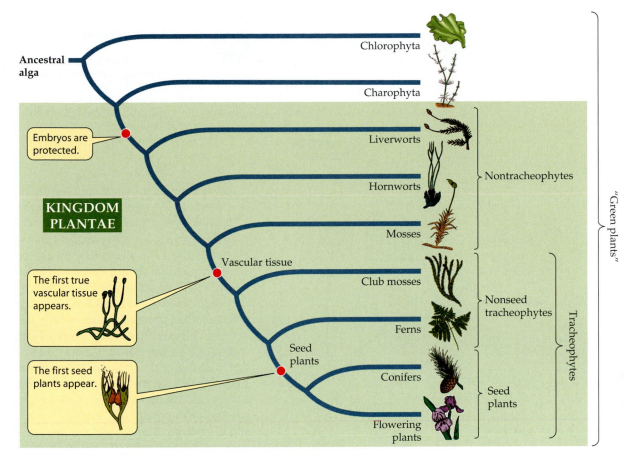

28.2 From Green Algae to Plants
Green algae called charophytes are sister to the plants; green
algae called chlorophytes are sister to the lineage that includes
the charophytes and plants.

or stored. The other vascular tissue, the *xylem*, conducts
water and minerals from the soil to aerial parts of the plant;
because some of its cell walls are stiffened by a substance
called *lignin*, xylem also provides support in the terrestrial
environment.

Nontracheophyte plants evolved tens of millions of
years before the earliest tracheophytes, even though tracheophytes appear earlier in the fossil record. The oldest
tracheophyte fossils date back more than 410 million years,
whereas the oldest nontracheophyte fossils are only about
350 million years old, dating from a time when tracheophytes were already widely distributed. This simply means
that, given their different structures and the chemical
makeup of their cell walls, tracheophytes are more likely to
form fossils than nontracheophytes are.

We will examine the adaptations of the tracheophytes
later in the chapter, concentrating first on the nontracheophytes.

The Nontracheophytes:
Liverworts, Hornworts, and Mosses

Most liverworts, hornworts, and mosses grow in dense
mats, usually in moist habitats (see Figure 28.7*a*). The

largest of these plants are only about 1 meter tall, and most
are only a few centimeters tall or long. Why have no large
nontracheophytes ever evolved? The probable answer is
that they lack an
efficient system
for   conducting
water and minerals
from the soil to dis-
tant parts of the plant
body. However, to limit
water loss, layers of maternal tissue protect the embryos of all nontracheophytes.
All nontracheophyte lineages
also have a cuticle, although it is
often very thin (or even absent in
some species) and thus not highly effective in retarding
water loss.

Most nontracheophytes live on the soil or on other
plants, but some grow on bare rock, dead and fallen tree
trunks, and even on buildings. Nontracheophytes are
widely distributed over six continents and exist very locally
on the coast of the seventh (Antarctica). They are very successful plants, well-adapted to their environments. Most
are terrestrial. Some live in wetlands. Although a few nontracheophyte species live in fresh water, these aquatic forms
are descended from terrestrial ones. There are no marine
nontracheophytes.

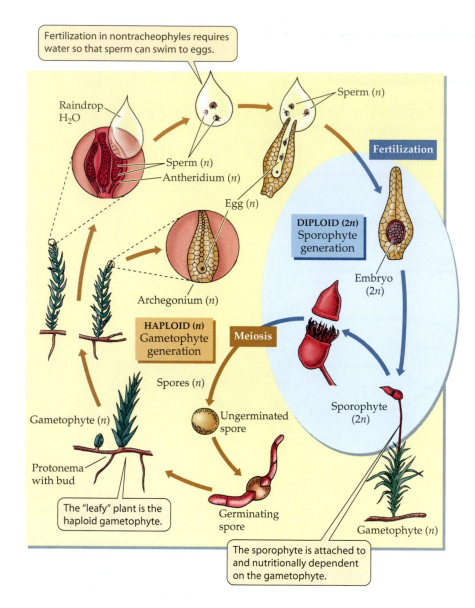

Fertilization in nontracheophyles requires water so that sperm can swim to eggs.

Raindrop H₂O

Sperm (*n*)

Sperm (*n*)
Antheridium (*n*)

Egg (*n*)

Fertilization

DIPLOID (2*n*) Sporophyte generation

Embryo (2*n*)

Archegonium (*n*)

HAPLOID (*n*) Gametophyte generation

Meiosis

Spores (*n*)

Ungerminated spore

Sporophyte (2*n*)

Gametophyte (*n*)

Protonema with bud

The "leafy" plant is the haploid gametophyte.

Germinating spore

Gametophyte (*n*)

The sporophyte is attached to and nutritionally dependent on the gametophyte.

28.3 A Nontracheophyte Life Cycle
The life cycle of nontracheophytes, illustrated here by a moss, is dependent on an external source of liquid water. The visible green structure of nontracheophytes is the gametophyte; in nontracheophyte plants, the "leafy" structures are sporophytes.

Once released, the sperm must swim or be splashed by raindrops to a nearby archegonium on the same or a neighboring plant. The sperm are aided in this task by chemical attractants released by the egg or the archegonium. Before sperm can enter the archegonium, certain cells in the neck of the archegonium must break down, leaving a water-filled canal through which the sperm swim to complete their journey. Note that all of these events require liquid water.

On arrival at the egg, one of the sperm nuclei fuses with the egg nucleus to form the zygote. Mitotic divisions of the zygote produce a multicellular, diploid sporophyte embryo. The base of the archegonium grows to protect the embryo during its early growth. Eventually the developing sporophyte elongates sufficiently to break out of the archegonium, but it remains connected to the gametophyte by a "foot" that is embedded in the parent tissue and absorbs water and nutrients from it (see Figure 28.3). The sporophyte remains attached to the gametophyte throughout its life. The sporophyte produces a sporangium, or **capsule**, within which meiotic divisions produce spores and thus the next gametophyte generation.

The structure and pattern of elongation of the sporophyte differ among the three phyla of nontracheophytes—the liverworts (Hepatophyta), hornworts (Anthocerophyta), and mosses (Bryophyta). The evolutionary relationships of the three phyla and the tracheophytes can be seen in Figure 28.2.

Nontracheophyte sporophytes are dependent on gametophytes

In nontracheophytes, the conspicuous green structure visible to the naked eye is the gametophyte (Figure 28.3). In contrast, the familiar forms of tracheophytes such as ferns and seed plants are sporophytes. The gametophyte of nontracheophytes is photosynthetic and therefore nutritionally independent, whereas the sporophyte may or may not be photosynthetic but is *always* dependent on the gametophyte and remains permanently attached to it.

A sporophyte produces unicellular, haploid spores as products of meiosis. A spore germinates, giving rise to a multicellular, haploid gametophyte whose cells contain chloroplasts and are thus photosynthetic. Eventually gametes form within specialized sex organs, the **gametangia**. The **archegonium** is a multicellular, flask-shaped female sex organ with a long neck and a swollen base (Figure 28.4*a*). The base contains a single egg. The **antheridium** is a male sex organ in which sperm, each bearing two flagella, are produced in large numbers (Figure 28.4*b*).

Liverworts are the most ancient surviving plant lineage

The gametophytes of some liverworts (phylum Hepatophyta) are green, leaflike layers that lie flat on the ground (Figure 28.5*a*). The simplest liverwort gametophytes, however, are flat plates of cells, a centimeter or so long, that produce antheridia or archegonia on their upper surfaces and water-absorbing filaments called **rhizoids** on the lower. Liverwort sporophytes are shorter than those of mosses and hornworts, rarely exceeding a few millimeters.

The sporophyte has a stalk that connects capsule and foot. The stalk elongates and thus raises the capsule above

(a)

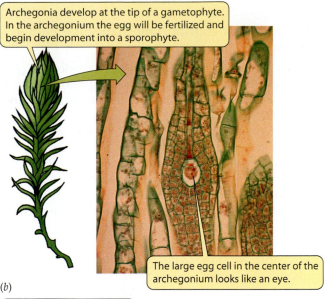

Archegonia develop at the tip of a gametophyte. In the archegonium the egg will be fertilized and begin development into a sporophyte.

The large egg cell in the center of the archegonium looks like an eye.

(b)

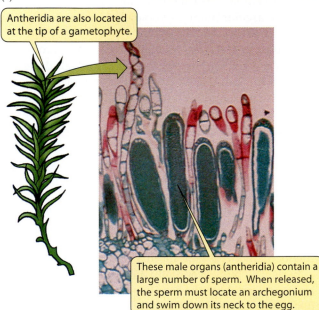

Antheridia are also located at the tip of a gametophyte.

These male organs (antheridia) contain a large number of sperm. When released, the sperm must locate an archegonium and swim down its neck to the egg.

28.4 Sex Organs in Plants
Archegonia (*a*) and antheridia (*b*) of the moss *Mnium* (phylum Bryophyta). Gametophytes of all plants have archegonia and antheridia, but they are much reduced in seed plants.

ground level, favoring dispersal of spores when they are released. The capsules of liverworts are simple: a globular capsule wall surrounding a mass of spores. In some species of liverworts, spores are not released by the sporophyte until the surrounding capsule wall rots.

In other liverworts, however, the spores are disseminated by structures called **elaters** located within the capsule. Elaters are long cells that have a helical thickening of the cell wall. As an elater loses water, the whole cell shrinks to a fraction of its former length, thus compressing the helical thickening like a spring. When the stress becomes sufficient, the compressed "spring" snaps back to its resting position, throwing spores in all directions.

Among the most familiar liverworts are species of the genus *Marchantia* (Figure 28.5*a*). *Marchantia* is easily recognized by the characteristic structures on which its male and female gametophytes bear their antheridia and archegonia (Figure 28.5*b*). Like most liverworts, *Marchantia* also reproduces vegetatively by simple fragmentation of the gametophyte. Along with sexual reproduction, *Marchantia* and some other liverworts and mosses also reproduce vegetatively by means of **gemmae** (singular gemma), which are lens-shaped clumps of cells. In a few liverworts the gemmae are loosely held in structures called gemma cups, which promote dispersal by raindrops (Figure 28.5*c*).

Hornworts evolved stomata as an adaptation to terrestrial life

The phylum Anthocerophyta comprises the hornworts, so named because their sporophytes look like little horns (Figure 28.6). Hornworts appear at first glance to be liverworts with very simple gametophytes. These gametophytes consist of flat plates of cells, a few cells thick.

However, the hornworts, along with the mosses and tracheophytes, share an advance over the liverwort lineage in their adaptation to life on land. They have *stomata*—pores

28.5 Liverwort Structures
Members of the phylum Hepatophyta display various characteristic structures. (*a*) Gametophytes. (*b*) Structures bearing antheridia and archegonia. (*c*) Gemmae.

The finger-headed structures bear archegonia.

The disc-headed structures bear antheridia.

These cups are filled with gemmae—small, lens-shaped outgrowths of the body, each capable of developing into a new plant.

(a) *Marchantia* sp.

(b) *Marchantia* sp.

(c) *Lunularia* sp.

The sporophytes of hornworts can reach 20 cm in height.

Gametophytes are flat plates a few cells thick.

Anthoceros dieteret

28.6 A Hornwort
The sporophytes of hornworts can resemble little horns.

that, when open, allow the uptake of CO_2 for photosynthesis and the release of O_2, but that can close to prevent excessive water loss.

Hornworts have two characteristics that distinguish them from both liverworts and mosses. First, the cells of hornworts each contain a single large, platelike chloroplast, whereas the other nontracheophytes contain numerous small, lens-shaped chloroplasts. Second, of all the nontracheophyte sporophytes, those of the hornworts come closest to being capable of indefinite growth (without a set limit).

The stalk of either the liverwort or moss sporophyte stops growing as the capsule matures, so elongation of the sporophyte is strictly limited. In a hornwort such as *Anthoceros,* however, there is no stalk, but a basal region of the capsule remains capable of indefinite cell division, continuously producing new spore-bearing tissue above.

Sporophytes of some hornworts growing in mild and continuously moist conditions can become as tall as 20 centimeters. Eventually the sporophyte's growth is limited by the lack of a transport system. To support their growth, the hornworts need access to nitrogen. Hornworts have internal cavities filled with a mucilage; these cavities are often populated by cyanobacteria that fix atmospheric nitrogen gas into a nutrient form usable by the host plant.

We have presented the hornworts as sister to the lineage consisting of mosses and tracheophytes, but this is only one possible interpretation of the current data. The exact evolutionary status of the hornworts is still in doubt.

Water- and sugar-transport mechanisms emerged in the mosses

The most familiar nontracheophytes are the mosses (phylum Bryophyta). There are more species of mosses than of liverworts and hornworts combined, and these hardy little plants are found in almost every terrestrial environment. They often are found on damp, cool ground, where they form thick mats (Figure 28.7a). The mosses are sister to the tracheophytes (see Figure 28.2).

Many mosses contain a type of cell called a *hydroid*, which dies and leaves a tiny channel through which water may travel. The hydroid likely is a progenitor of the tracheid, the characteristic water-conducting cell of the tracheophytes, but it lacks lignin (a waterproofing substance) and the wall structure found in tracheids. The possession of hydroids and of a limited system for transport of sucrose by some mosses (via cells called *leptoids*) shows that the old term "nonvascular plant" is somewhat misleading when applied to mosses.

In contrast to liverworts and hornworts, the sporophytes of mosses and tracheophytes grow by **apical cell division**. A region at the growing tip provides an organized pattern of cell division, elongation, and differentiation. This allows extensive and sturdy vertical growth of sporophytes.

The moss gametophyte that develops following spore germination is a branched, filamentous structure, or **protonema** (see Figure 28.3). Although the protonema looks much like a filamentous green alga, it is unique to the mosses. Some of the filaments contain chloroplasts and are photosynthetic; others, called rhizoids, are nonphotosynthetic and anchor the protonema to the substrate. After a period of linear growth, cells close to the tips of the photosynthetic filaments divide rapidly in three dimensions to form buds. The buds eventually differentiate a distinct tip, or apex, and produce the familiar leafy moss shoot with leaflike structures arranged spirally.

These leafy shoots produce antheridia or archegonia (see Figure 28.4). The antheridia release sperm that travel through liquid water to the archegonia, where they fertilize the eggs. Sporophyte development in most mosses follows a precise pattern, resulting ultimately in the formation of an absorptive foot, a stalk, and, at the tip, a swollen capsule. In contrast to hornworts, which grow from the base, the moss sporophyte stalk grows at its apical end, as tracheophytes do. Cells at the tip of the stalk divide, supporting elongation of the structure and giving rise to the capsule. For a while, archegonial tissue grows rapidly as the stalk elongates, but eventually the archegonium is outgrown and is torn apart by the expanding sporophyte.

The top of the capsule is shed after completing meiosis and spore development. Groups of cells just below the lid form a series of toothlike structures surrounding the opening. Highly responsive to humidity, these structures dig into the mass of spores when the atmosphere is dry; then, when the atmosphere becomes moist, they fling out, scooping out the spores as they go (Figure 28.7b). The spores are thus dispersed when the surrounding air is

(a)

(b)

The teeth of this moss capsule help expel the thousands of spores.

28.7 The Mosses
(a) Dense moss forms hummocks in a valley on New Zealand's South Island. (b) The moss capsule, from which spores are dispersed, grows at the tip of the plant.

moist—that is, when conditions favor their subsequent germination.

Only a few mosses depart from this pattern of capsule development. A familiar exception is the genus *Sphagnum*, which we discussed at the beginning of this chapter. Species in this genus have a simple capsule with an air chamber in it. Air pressure builds up in this chamber, eventually causing the capsule lid to pop open, dispersing the spores with an audible explosion.

With their simple system of internal transport, the mosses are in a sense vascular plants. However, they are not tracheophytes, because they lack true xylem and phloem.

Introducing the Tracheophytes

Although an extraordinarily large and diverse group, the tracheophytes can be said to have been launched by a single evolutionary event. Sometime during the Paleozoic era, probably well before the Silurian period (440 mya), the sporophyte generation of a now long-extinct organism produced a new cell type, the **tracheid**. The tracheid is the principal water-conducting element of the xylem in all tracheophytes except the angiosperms (flowering plants); and even in angiosperms the tracheid persists alongside a more specialized and efficient system of vessels and fibers derived from tracheids.

The evolutionary appearance of a tissue composed of tracheids had two important consequences. First, it provided a pathway for long-distance transport of water and mineral nutrients from a source of supply to regions of need. Second, it provided something almost completely lacking—and unnecessary—in the largely aquatic green algae: rigid structural support. Support is important in a terrestrial environment because plants tend to grow upward as they compete for sunlight to power photosynthesis. Thus the tracheid set the stage for the complete and permanent invasion of land by plants.

The tracheophytes feature a further evolutionary novelty: *a branching, independent sporophyte*. A branching sporophyte can produce more spores than an unbranched body, and it can develop in complex ways. The sporophyte of a tracheophyte is nutritionally independent of the gametophyte.

The present-day evolutionary descendants of the early tracheophytes belong to nine distinct phyla (Figure 28.8). We can sort these phyla into two groups: those that produce seeds and those that do not. The nonseed tracheophytes include ferns, horsetails, club mosses, and whisk ferns. In the nonseed tracheophytes, the haploid and diploid generations are independent at maturity. The sporophyte is the large and obvious plant that one normally notices in nature (in contrast to the nontracheophyte sporophyte, which is attached to, dependent on, and usually much smaller than the gametophyte). Gametophytes of the nonseed tracheophytes are rarely more than 1 or 2 centimeters long and are short-lived, whereas their sporophytes are often highly visible; the sporophyte of a tree fern, for example, may be 15 or 20 meters tall and may live for many years.

The most prominent resting stage in the life cycle of a nonseed tracheophyte is the single-celled spore. This feature makes this life cycle similar to those of the fungi, the green algae, and the nontracheophytes but not, as we will see in the next chapter, to that of the seed plants. Nonseed tracheophytes must have an aqueous environment for at least one stage of their life cycle because fertilization is accomplished by a motile, flagellated sperm.

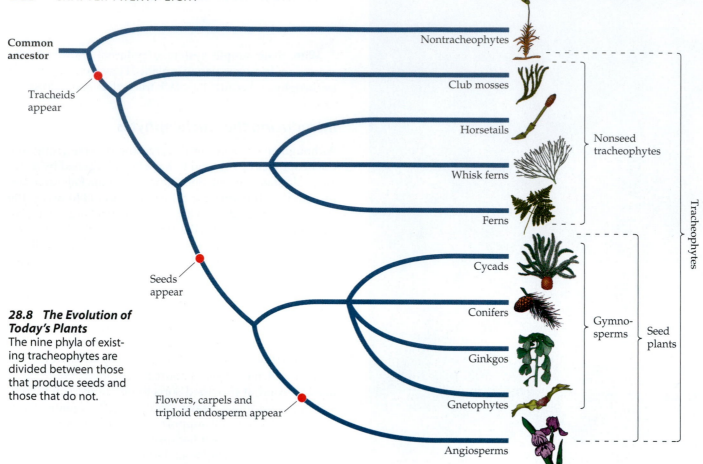

Common ancestor

Tracheids appear

Seeds appear

28.8 The Evolution of Today's Plants
The nine phyla of existing tracheophytes are divided between those that produce seeds and those that do not.

Flowers, carpels and triploid endosperm appear

Nontracheophytes

Club mosses

Horsetails

Whisk ferns

Ferns

Cycads

Conifers

Ginkgos

Gnetophytes

Angiosperms

Nonseed tracheophytes

Gymnosperms

Seed plants

Tracheophytes

We now turn to a more detailed account of the evolution of the nonseed tracheophytes.

Tracheophytes have been evolving for almost half a billion years

The plant kingdom successfully invaded the terrestrial environment between 400 and 500 million years ago. The evolution of a water-impermeable cuticle and of protective layers for the gamete-bearing structures (archegonia and antheridia) helped make the invasion successful, as did the initial absence of herbivores (plant-eating animals).

By the late Silurian period, tracheophytes were being preserved as fossils that we can study today. Several remarkable developments arose during the Devonian period, 409 to 354 million years ago. Three groups of nonseed tracheophytes that still exist made their first appearances during that period: the lycopods (club mosses), horsetails, and ferns. Their proliferation made the terrestrial environment more hospitable to animals: Amphibians and insects arrived soon after the plants became established.

Trees of various kinds appeared in the Devonian period, and dominated the landscape of the Carboniferous. Mighty forests of lycopods up to 40 meters tall, horsetails, and tree ferns flourished in the tropical swamps of what would become North America and Europe (Figure 28.9). In the subsequent Permian period the continents came ponderously together to form a single gigantic land mass, called Pangaea. The continental interior became warmer and drier, but late in the period glaciation was extensive. The 200-million-year reign of the lycopod–fern forests came to an

end as they were replaced by forests of seed plants (gymnosperms) that ruled until other seed plants (angiosperms) became dominant less than 80 million years ago.

The earliest tracheophytes lacked roots and leaves

The first tracheophytes belonged to the now-extinct phylum Rhyniophyta. The rhyniophytes appear to have been the only tracheophytes in the Silurian period. The landscape at that time probably consisted of bare ground, with stands of rhyniophytes in low-lying moist areas. Early versions of the structural features of all the other tracheophyte phyla appeared in the rhyniophytes of that time. These shared features strengthen the case for the origin of all tracheophytes from a common nontracheophyte ancestor.

In 1917, the British paleobotanists Robert Kidston and William H. Lang reported well-preserved fossils of tracheophytes embedded in Devonian rocks near Rhynie, Scotland. The preservation of these plants was remarkable, considering that the rocks were more than 395 million years old. These fossil plants had a simple vascular system of phloem and xylem. Flattened scales on the stems of some of the plants lacked vascular tissue and thus were not comparable with the true leaves of any other tracheophytes.

These plants lacked roots. They were apparently anchored in the soil by horizontal portions of stem, called **rhizomes**, that bore water-absorbing rhizoids. These rhizomes also bore aerial branches, and sporangia—homologous with the nontracheophyte capsule—were found at

28.9 An Ancient Forest
A little more than 300 million years ago, a forest grew in a setting similar to tropical river delta habitats of today. Most of the plants depicted here were nonseed tracheophytes 10 to 20 m tall. Far in the distance, early seed plants—giants up to 40 m tall—towered over the forest. This artist's impression is based on evidence from fossils.

the tips of these branches. Branching was dichotomous; that is, the shoot apex divided to produce two equivalent new branches, each pair diverging at approximately the same angle from the original stem (Figure 28.10). Scattered fragments of such plants had been found earlier, but never in such profusion or so well preserved as those discovered near Rhynie by Kidston and Lang.

The presence of xylem indicated that these plants, named *Rhynia* after the site of their discovery, were tracheophytes. But were they sporophytes or gametophytes? Close inspection of thin sections of fossil sporangia revealed that the spores were in groups of four. In almost all living nonseed tracheophytes (with no evidence to the contrary from fossil forms), the four products of a meiotic division and cytokinesis remain attached to one another during their development into spores. The spores separate only when they are mature, and even after separation their walls reveal the exact geometry of how they were attached. Therefore, a group of four closely packed spores is found only immediately after meiosis, and a plant that produces such a group of four must be a diploid sporophyte—and so the Rhynie fossils must have been sporophytes. The gametophytes of the Rhyniophyta also were branched, and depressions at the apices of the branches contained archegonia and antheridia.

Although apparently ancestral to the other tracheophyte phyla, the rhyniophytes themselves are long gone. None of their fossils appear anywhere after the Devonian period.

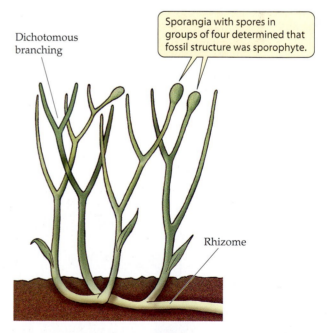

Sporangia with spores in groups of four determined that fossil structure was sporophyte.

Dichotomous branching

Rhizome

28.10 A Very Ancient Tracheophyte
This extinct plant in the genus *Rhynia* (phylum Rhyniophyta) lacked roots and leaves. The rhizome is a horizontal underground stem, not a root. The aerial shoots were less than 50 cm tall, and some were topped by sporangia.

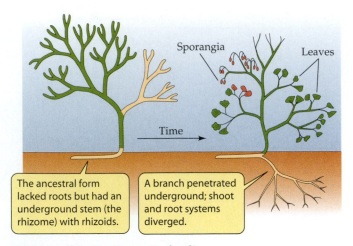

The ancestral form lacked roots but had an underground stem (the rhizome) with rhizoids.

A branch penetrated underground; shoot and root systems diverged.

28.11 Is This How Roots Evolved?
According to Lignier's hypothesis, branches from ancestral root-less plants could have penetrated the soil, where they gradually evolved into a root system.

Early tracheophytes added new features

Within a few tens of millions of years, during the Devonian period, three new phyla of tracheophytes—Lycophyta, Sphenophyta, and Pterophyta—appeared on the scene, arising from rhyniophyte-like ancestors. These new groups featured specializations not found in the rhyniophytes, including one or more of the following: true roots, true leaves, and a differentiation between two types of spores.

THE ORIGIN OF ROOTS.   *Rhynia* and its close relatives lacked true roots. They had only rhizoids arising from a rhizome (Figure 28.11, left) with which to gather water and minerals. How, then, did subsequent groups of tracheophytes come to have the complex roots we see today?

In 1903, a French botanist, E. A. O. Lignier, proposed an attractive hypothesis that is still widely accepted today. Lignier argued that the ancestors of the first tracheophytes grew by branching dichotomously. This explanation is supported by the dichotomous branching observed in the rhyniophytes. Lignier suggested that such a branch could bend, penetrate the soil, and branch there (Figure 28.11,

right). The underground portion could anchor the plant firmly, and even in this primitive condition it could absorb water and minerals. The subsequent discovery of fossil plants from the Devonian period, all having horizontal stems (rhizomes) with both underground and aerial branches, supported Lignier's hypothesis.

Underground and aboveground branches, growing in sharply different environments, were subjected to very different selection during the succeeding millions of years. Thus the two parts of the plant axis (the shoot and root systems) diverged in structure and evolved distinct internal and external anatomies. In spite of these differences, scientists believe that the root and shoot systems of tracheophytes are homologous—that they were once part of the same organ.

THE ORIGIN OF TRUE LEAVES.   Thus far we have used the term "leaf" rather loosely. We spoke of "leafy" mosses and commented on the absence of "true leaves" in rhyniophytes. In the strictest sense, a **leaf** is a flattened photosynthetic structure emerging laterally from a main axis or stem and possessing true vascular tissue. Using this precise definition as we take a closer look at true leaves in the tracheophytes, we see that there are two different types of leaves, very likely of different evolutionary origins.

The first type, the *simple leaf*, is usually small and only rarely has more than a single vascular strand, at least in plants alive today. Plants in the phylum Lycophyta (club mosses), of which only a few genera survive, have such leaves. The evolutionary origin of simple leaves is thought by some biologists to be sterile sporangia (Figure 28.12*a*). The principal characteristic of this type of leaf is that its vascular strand departs from the vascular system of the stem in such a way that the structure of the stem's vascular system is scarcely disturbed. This was true even in the fossil lycopod trees of the Carboniferous period, many of which had leaves many centimeters long.

28.12 The Evolution of Leaves
(*a*) Simple leaves are thought to have evolved from sterile sporangia. (*b*) The complex leaves of ferns and seed plants may have arisen as photosynthetic tissue developed between complex branching patterns.

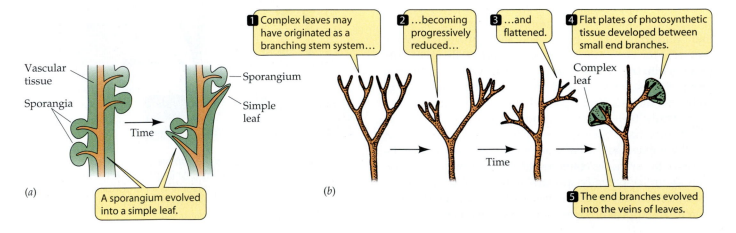

Vascular tissue

Sporangia

Sporangium

Simple leaf

Time

A sporangium evolved into a simple leaf.

(*a*)

1 Complex leaves may have originated as a branching stem system...

2 ...becoming progressively reduced...

3 ...and flattened.

4 Flat plates of photosynthetic tissue developed between small end branches.

Complex leaf

Time

(*b*)

5 The end branches evolved into the veins of leaves.

The other type of leaf is encountered in ferns and seed plants. This larger, more *complex leaf* is thought to have arisen from the flattening of a dichotomously branching stem system, with the development of extensive photosynthetic tissue between the branch members (Figure 28.12b). The complex leaf may have evolved several times, in different phyla of tracheophytes.

HOMOSPORY AND HETEROSPORY.   In the most ancient of the present-day tracheophytes, both the gametophyte and the sporophyte are independent and usually photosynthetic. Spores produced by the sporophytes are of a single type, and they develop into a single type of gametophyte that bears both female and male reproductive organs. Such plants, which bear a single type of spore, are said to be **homosporous** (Figure 28.13a). The sex organs on the gametophytes of homosporous plants are of two types. The female organ is a multicellular archegonium, typically containing a single egg. The male organ is an antheridium, containing many sperm.

A different system, with two distinct types of spores, evolved somewhat later. Plants of this type are said to be **heterosporous** (Figure 28.13b). One type of spore, the **megaspore**, develops into a larger, specifically female gametophyte (megagametophyte) that produces only eggs. The other type, the **microspore**, develops into a smaller, male gametophyte (microgametophyte) that produces only sperm. The sporophyte produces megaspores in small numbers in megasporangia on the sporophyte, and microspores in large numbers in microsporangia.

The most ancient tracheophytes were all homosporous. Heterospory evidently evolved independently several times in the early evolution of the tracheophytes descended from the rhyniophytes. The fact that heterospory evolved repeatedly suggests that it affords selective advantages. Subsequent evolution in the plant kingdom featured ever greater specialization of the heterosporous condition.

The Surviving Nonseed Tracheophytes

Today ferns are the most abundant and diverse phylum of nonseed tracheophytes, but club mosses and horsetails were once dominant elements of Earth's vegetation. A fourth phylum, the whisk ferns, contains only two genera. In this section we'll look at the characteristics of these four phyla and at some of the evolutionary advances that appeared in them.

The club mosses are sister to the other tracheophytes

The club mosses (lycopods, phylum Lycophyta) diverged earlier than all other living tracheophytes—that is, the remaining tracheophytes share an ancestor that was not ancestral to the Lycophyta. There are relatively few surviving species of club mosses. They have roots that branch dichoto-

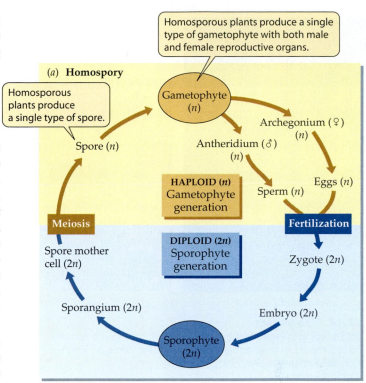

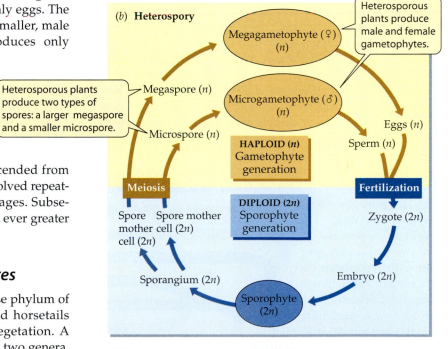

28.13 Homospory and Heterospory
(a) Homosporous plants bear a single type of spore. Each gametophyte has two types of sex organs, antheridia (male) and archegonia (female). (b) Heterospory, with two types of spores that develop into distinctly male and female gametophytes, evolved later.

mously. They bear only simple leaves, and the leaves are arranged spirally on the stem. Growth in club mosses comes entirely from groups of dividing cells at the tips of the stems and thus is apical, as it is in many flowering plants.

(a) *Lycopodium obscurum*   (b)

28.14 Club Mosses
(a) Strobili are visible at the tips of this club moss. Club mosses have simple leaves arranged spirally on their stems. (b) Thin section through a strobilus of another club moss.

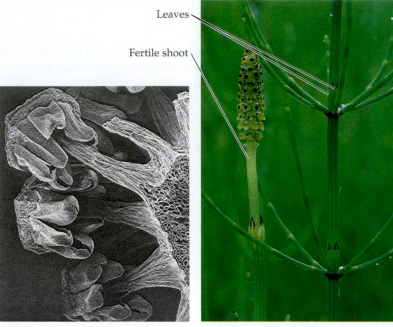

(a) *Equisetum arvense*   (b) *Equisetum palustre*

28.15 Horsetails
(a) Sporangia and sporangiophores of a horsetail. (b) Vegetative and fertile shoots of the marsh horsetail. Leaves form in whorls at nodes on the stems of the vegetative shoot on the right; the fertile shoot on the left is ready to disperse its spores.

The sporangia in most club mosses are contained within conelike structures called **strobili** (singular strobilus; Figure 28.14) and are tucked in the upper angle between a specialized leaf and the stem. This placement contrasts with the terminal sporangia of the rhyniophytes (see Figure 28.10). There are both homosporous species and heterosporous species of club mosses. Like all the nonseed tracheophytes, they have a large, independent sporophyte and a small, independent gametophyte.

Although only a minor element of present-day vegetation, the Lycophyta are one of two phyla that appear to have been the dominant vegetation during the Carboniferous period. One abundant type of coal (Cannel coal) is formed almost entirely from fossilized spores of a tree lycopod named *Lepidodendron*—which gives us an idea of the abundance of this genus in the forests of that time. The other major element of the Carboniferous vegetation was the phylum Sphenophyta, the horsetails.

Liverworts
Hornworts
Mosses
Club mosses
Horsetails
Whisk ferns
Ferns

Horsetails grow at the bases of their segments

Like the club mosses, the horsetails (phylum Sphenophyta) are represented by only a few present-day species. They are sometimes called scouring rushes because silica deposits found in the cell walls made them useful for cleaning. They have true roots that branch irregularly, as do the roots of all tracheophytes except the club mosses. Their sporangia curve back toward the stem on the ends of short stalks (sporangiophores) (Figure 28.15a). Horsetails have a large sporophyte and a small gametophyte, both independent.

The leaves of horsetails are simple and form distinct whorls (circles) around the stem (Figure 28.15b). Growth in horsetails originates to a large extent from discs of dividing cells just above each whorl of leaves, so each segment of the stem grows from its base. Such basal growth is uncommon in plants, although it is found in the grasses, a major group of flowering plants.

Present-day whisk ferns resemble the most ancient tracheophytes

There once was some disagreement about whether rhyniophytes are entirely extinct. The confusion arose because of the existence today of two genera of rootless, spore-bearing plants, *Psilotum* and *Tmesipteris*. *Psilotum nudum* (Figure 28.16) has only minute scales instead of true leaves, but plants of the genus *Tmesipteris* have flattened photosynthetic organs with well-developed vascular tissue. Are these two genera the living relics of the rhyniophytes, or do they have more recent origins?

Psilotum and *Tmesipteris* once were thought to be evolutionarily ancient descendants of anatomically simple ancestors. That hypothesis was weakened by an enormous hole in the geologic record between the rhyniophytes, which apparently became extinct more than 300 million years ago, and *Psilotum* and *Tmesipteris*, which are modern plants. DNA sequence data finally settled the question in favor

Psilotum nudum

28.16 A Whisk Fern
Aerial branches of a whisk fern, a plant once considered by some to be a surviving rhyniophyte and by others to be a fern. It is now included in the phylum Psilophyta, and is widespread in the tropics and subtropics.

of a more modern origin from fernlike ancestors. Most botanists now treat these two genera as their own phylum, the Psilophyta (whisk ferns) rather than as relatives of the rhyniophtes.

We now consider the whisk ferns to be highly specialized plants that evolved fairly recently from anatomically more complex ancestors. Whisk fern gametophytes live below the surface of the ground and lack chlorophyll. They depend upon fungal partners for their nutrition.

Ferns evolved large, complex leaves

The sporophytes of the ferns and seed plants have roots, stems, and leaves. Their leaves are typically large and have branching vascular strands. Some species have small leaves as a result of evolutionary reduction, but even the small leaves have more than one vascular strand.

The true ferns constitute the phylum Pterophyta, which first appeared during the Devonian period and today consists of about 12,000 species. The Pterophyta are probably not a monophyletic group. Ferns are characterized by

fronds (large leaves with complex vasculature; Figure 28.17a) and by a requirement for water for the transport of the male gametes to the female gametes. Most ferns inhabit shaded, moist woodlands and swamps. Tree ferns can reach heights of 20 meters. Tree ferns are not as rigid as woody plants, and they have poor root systems. Thus they do not grow in sites exposed directly to strong winds but rather in ravines or beneath trees in forests.

Liverworts
Hornworts
Mosses
Club mosses
Horsetails
Whisk ferns
Ferns

During its development, the fern frond unfurls from a tightly coiled "fiddlehead" (Figure 28.17b). Some fern leaves become climbing organs and may grow to be as much as 30 meters long. The sporangia are found on the undersurfaces of the leaves, sometimes covering the whole undersurface and sometimes only at the edges; in most species the sporangia are clustered in groups called **sori** (singular, sorus) (Figure 28.18).

The sporophyte generation dominates the fern life cycle

Inside the sporangia, fern cells undergo meiosis to form haploid spores. Once shed, spores travel great distances and eventually germinate to form independent gametophytes. Old World climbing fern, *Lygodium microphyllum,* is currently spreading disastrously through the Florida Everglades, choking off the growth of other plants. This rapid spread is testimony to the effectiveness of wind-borne spores.

28.17 Fern Fronds Take Many Forms
(a) Fronds of maidenhair fern form a pattern in this photograph. (b) The "fiddlehead" (developing frond) of a common forest fern; this structure will unfurl and expand to give rise to a complex adult frond such as those in (a). (c) The tiny fronds of a water fern.

(a) *Adiantum* sp.

(b)

(c) *Marsilea mutica*

Dryopteris intermedia

28.18 Fern Sori Contain Sporangia
Sori, each with many spore-producing sporangia, form on the underside of a frond of the Midwestern fancy fern.

Fern gametophytes produce antheridia and archegonia, although not necessarily at the same time or on the same gametophyte. Sperm swim through water to archegonia, often on other gametophytes, where they unite with an egg.

The resulting zygote develops into a new sporophyte embryo. The young sporophyte sprouts a root and can thus grow independently of the gametophyte. In the alternating generations of a fern, the gametophyte is small, delicate, and short-lived, but the sporophytes can be very large and can sometimes survive for hundreds of years (Figure 28.19).

Most ferns are homosporous. However, two groups of aquatic ferns, the Marsileales and Salviniales, are derived from a common ancestor that evolved heterospory. Megaspores and microspores of these plants (which germinate to produce female and male gametophytes, respectively) are produced in different sporangia, and the microspores are always much smaller and greater in number than the megaspores.

A few genera of ferns produce a tuberous, fleshy gametophyte instead of the characteristic flattened, photosynthetic structure described earlier. Like the gametophytes of whisk ferns, these tuberous gametophytes depend on a mutualistic fungus for nutrition;* in some genera, even the sporophyte embryo must become associated with the fun-

*In a mutualistic association, both partners—here, the gametophyte and the fungus—profit.

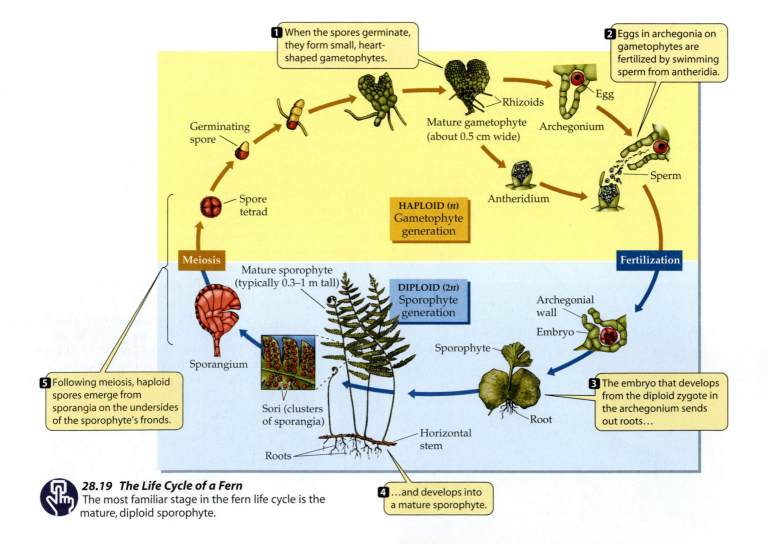

1 When the spores germinate, they form small, heart-shaped gametophytes.

2 Eggs in archegonia on gametophytes are fertilized by swimming sperm from antheridia.

Germinating spore

Rhizoids

Mature gametophyte (about 0.5 cm wide)

Egg

Archegonium

Spore tetrad

Antheridium

Sperm

HAPLOID (*n*)
Gametophyte generation

Meiosis

Fertilization

Mature sporophyte (typically 0.3–1 m tall)

DIPLOID (2*n*)
Sporophyte generation

Archegonial wall

Embryo

Sporophyte

Sporangium

Sori (clusters of sporangia)

Root

5 Following meiosis, haploid spores emerge from sporangia on the undersides of the sporophyte's fronds.

3 The embryo that develops from the diploid zygote in the archegonium sends out roots…

Horizontal stem

Roots

28.19 The Life Cycle of a Fern
The most familiar stage in the fern life cycle is the mature, diploid sporophyte.

4 …and develops into a mature sporophyte.

gus before extensive development can proceed. In Chapter 30 we will see that there are many important plant–fungus mutualisms.

All the tracheophytes we have discussed thus far disperse themselves by spores. In the next chapter we discuss the plants that dominate most of Earth's vegetation today, the seed plants, whose seeds afford new sporophytes protection unavailable to the nonseed tracheophytes.

Chapter Summary

▶ Plants are photosynthetic eukaryotes that use chlorophylls *a* and *b*, store carbohydrates as starch, and develop from embryos protected by parental tissue.

▶ Plant life cycles feature alternation of gametophyte (haploid) and sporophyte (diploid) generations.

▶ There are twelve surviving phyla of plants grouped into two main categories, nontracheophytes and tracheophytes. **Review Table 28.1**

▶ Plants arose from a common green algal ancestor, either of the stoneworts or of *Coleochaete*. Descendants of this ancestral charophyte colonized the land.

The Conquest of the Land

▶ Moving toward today's plants, early steps in plant evolution included the acquisition of a cuticle, gametangia, a protected embryo, protective pigments, and thick spore walls.

▶ Tracheophytes are characterized by possession of a vascular system, consisting of water- and mineral-conducting xylem and nutrient-conducting phloem. Nontracheophytes lack a vascular system. **Review Figure 28.2**

Nontracheophytes: Liverworts, Hornworts, and Mosses

▶ The nontracheophytes include the liverworts (phylum Hepatophyta), hornworts (phylum Anthocerophyta), and mosses (phylum Bryophyta). **Review Table 28.1**

▶ Nontracheophytes either lack vascular tissues completely or, in the case of certain mosses, have only a rudimentary system of water- and food-conducting cells.

▶ The nontracheophyte sporophyte generation is smaller than the gametophyte generation and depends on the gametophyte for water and nutrition. **Review Figures 28.3, 28.4**

▶ Liverwort sporophytes have no specific growing zone. Hornwort sporophytes grow at their basal end, and moss sporophytes grow at their apical end. **Review Figure 28.6**

▶ Beginning with hornworts, all plants have surface pores (stomata) that allow gas exchange and minimize water loss.

▶ Beginning with mosses, the sporophytes of all plants grow by apical cell division.

▶ The hydroids of mosses, through which water may travel, may be have arisen from cells also ancestral to the water-conducting cells of the tracheophytes.

Introducing the Tracheophytes

▶ The tracheophytes have vascular tissue with tracheids and other specialized cells designed to conduct water, minerals, and foods.

▶ Present-day tracheophytes are grouped into nine phyla that form two major groups: nonseed tracheophytes and seed plants. **Review Figure 28.8**

▶ In tracheophytes the sporophyte generation is larger than the gametophyte and independent of the gametophyte generation.

▶ The earliest tracheophytes, known to us only in fossil form, lacked roots and leaves. Roots may have evolved from branches that penetrated the ground. Simple leaves are thought to have evolved from sporangia, and complex leaves may have resulted from the flattening and reduction of a branching stem system. **Review Figures 28.10, 28.11, 28.12**

▶ Heterospory, the production of distinct female megaspores and male microspores, evolved on several occasions from homosporous ancestors. **Review Figure 28.13**

The Surviving Nonseed Tracheophytes

▶ Club mosses (phylum Lycophyta) have simple leaves arranged spirally. Horsetails (phylum Sphenophyta) have simple leaves in whorls. Whisk ferns (phylum Psilophyta) lack roots; one genus has minute scales rather than leaves, and the other has leaves with vascular tissue. Leaves with more complex vasculature are characteristic of all other phyla of tracheophytes. **Review Table 28.1**

▶ Ferns (phylum Pterophyta) are probably not a monophyletic group. They have complex leaves with branching vascular strands. **Review Figure 28.19**

For Discussion

1. Mosses and ferns share a common trait that makes water droplets a necessity for sexual reproduction. What is this trait?

2. Are the mosses well adapted to terrestrial life? Justify your answer.

3. Ferns display a dominant sporophyte stage (with large fronds). Describe the major advance in anatomy that enables most ferns to grow much larger than mosses.

4. What features distinguish club mosses from horsetails? What features distinguish these groups from rhyniophytes and psilophytes? From ferns?

5. Why did some botanists once believe that psilophytes should be classified together with the rhyniophytes?

6. Contrast simple leaves with complex leaves in terms of structure, evolutionary origin, and occurrence among plants.

29

The Evolution of Seed Plants

A VIOLENT THUNDERSTORM MOVES through forested hills and valleys where summer rain has been scarce. A jagged fork of lightning strikes a tree and it bursts into flame. Soon the flames reach dead and dry underbrush and fire spreads to surrounding trees. The fire rages rapidly through the forest, leaving a blackened and smoking landscape behind.

Though devastating, such fires are a natural part of the forest ecosystem. Life returns quickly following a fire in a natural grassland or forest, in part because some plants have adaptations that enable them to live with fire. One example, obvious from its common name, is fireweed. The seeds of fireweed not only survive fires, but are encouraged by high temperatures to break their dormancy and sprout. Another example is the lodgepole pine tree, which covers vast fire-prone areas in the Rocky Mountains and elsewhere. Its cones will not release their seeds unless the heat of a fire causes them to open.

Seeds are remarkable structures. They protect the plant embryo within them from environmental extremes through what may be a very long resting period. This and other properties contribute to making seed plants the predominant plants on Earth. All of today's forests are dominated by seed plants.

In this chapter we describe the defining characteristics of the seed plants as a group. We survey the diversity of seed plants and describe the flowers and fruits that are characteristic of the flowering plants. Finally, we consider some of the unsolved problems in seed plant evolution.

General Characteristics of the Seed Plants

The most recent group to appear in the evolution of the tracheophytes is the seed plants: the **gymnosperms** (such as pines and cycads) and the **angiosperms** (flowering plants). There are four living phyla of gymnosperms and one of angiosperms (Figure 29.1). The phylogenetic relationships among these five lineages have not yet been resolved.

In seed plants, the gametophyte generation is reduced even further than it is in the ferns (Figure 29.2). The haploid gametophyte develops partly or entirely while attached to and nutritionally dependent on the diploid sporophyte. Among the seed plants, only the earliest types of gymnosperms and their few survivors had swimming sperm. All other seed plants have evolved other means of bringing female and male gametes together. The culmination of this striking evolutionary trend in plants was independence from the liquid water that earlier plants needed for sexual reproduction.

Seed plants are heterosporous, forming separate megasporangia and microsporangia on structures that are grouped on short axes, such as the cones of conifers and the flowers of angiosperms.

As in other plants, the spores of seed plants are produced by meiosis within the sporangia, but in seed plants, the megaspores are not shed. Instead, the female gametophytes develop within the megasporangia and depend on them for food and water. In most species only one of the meiotic products in a megasporangium survives. The surviving haploid nucleus divides mitotically, and the resulting cells divide again to produce a multicellular female gametophyte. In the angiosperms, female gametophytes normally contain eight nuclei. The female gametophyte is retained within the megasporangium, where it matures and

A Forest Ablaze
Fires like this one in a northern Arizona forest can pose dangers to human life and property. But they play an essential role in the life cycles of many fire-adapted seed plants.

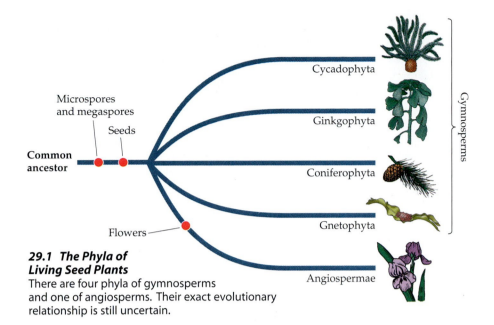

29.1 The Phyla of Living Seed Plants
There are four phyla of gymnosperms and one of angiosperms. Their exact evolutionary relationship is still uncertain.

When the tip of the pollen tube reaches the female gametophyte, two sperm are released from the tube and fertilization occurs. The resulting diploid zygote divides repeatedly, forming a young sporophyte that develops to an embryonic stage at which growth becomes temporarily suspended (often referred to as a dormant stage). The end product at this stage is a **seed**.

A seed may contain tissues from three generations. The seed coat and megasporangium develop from tissues of the diploid sporophyte parent (the integument). Within the megasporangium is the haploid female gametophytic tissue from the next generation. (This tissue is fairly extensive in most gymnosperm seeds. In angiosperm seeds its place is taken by a tissue called endosperm, which we will discuss shortly.) In the center of the seed package is the third generation, in the form of the embryo of the new diploid sporophyte.

The multicellular seed of a gymnosperm or an angiosperm is a well-protected resting stage. The seeds of some species may remain viable (capable of growth and development) for many years, germinating when conditions are favorable for the growth of the sporophyte. In contrast, the embryos of nonseed plants develop directly into sporophytes, which either survive or die, depending on environmental conditions; there is no resting stage in the life cycle.

houses the early development of the next sporophyte generation following fertilization of the egg. The megasporangium itself is surrounded by sterile sporophyte structures that form a protective **integument**.

Within the microsporangium, the meiotic products are microspores, which divide within the microspore wall one or a few times to form male gametophytes called **pollen grains** (Figure 29.3). Distributed by wind, an insect, a bird, or a plant breeder, a pollen grain that reaches the appropriate surface of a sporophyte develops further. It produces a slender **pollen tube** that elongates and digests its way through the sporophytic tissue toward the female gametophyte.

29.2 The Relationship between Sporophyte and Gametophyte Has Evolved
In seed plants, the gametophyte (shown in yellow) is nutritionally dependent on the sporophyte (shown in blue).

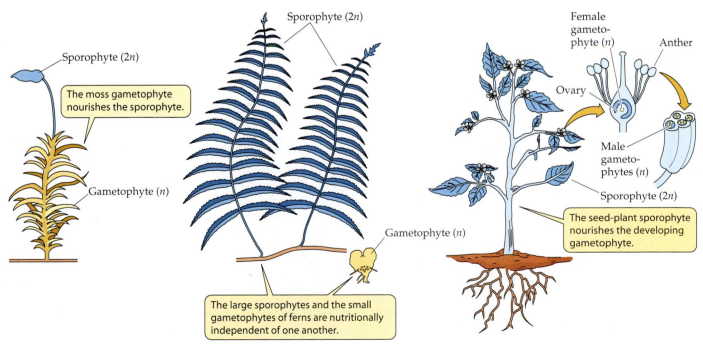

The moss gametophyte nourishes the sporophyte.

The large sporophytes and the small gametophytes of ferns are nutritionally independent of one another.

The seed-plant sporophyte nourishes the developing gametophyte.

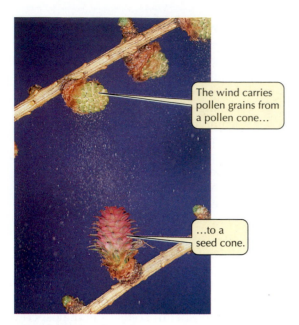

The wind carries pollen grains from a pollen cone...

...to a seed cone.

29.3 Pollen Grains
Pollen grains are the male gametophytes of seed plants. Conifers have separate seed cones (which contain the female gametophyte) and pollen cones; their pollen is dispersed by the wind.

During the dormant stage, the seed coat protects the embryo from excessive drying and may also protect against potential predators that would otherwise eat the embryo and its food reserves. Many seeds have structural adaptations that promote dispersal by wind or, more often, by animals. When the young sporophyte resumes growth, it draws on food reserves in the seed. The possession of seeds is a major reason for the enormous evolutionary success of seed plants, which are the dominant life forms of Earth's modern land flora in most areas.

The Gymnosperms: Naked Seeds

The gymnosperms are a group of seed plants that do not form flowers. Although there are probably fewer than 750 species of living gymnosperms, these plants are second only to the angiosperms (flowering plants) in their dominance of the terrestrial environment.

There are four phyla of living gymnosperms today. The **cycads** (phylum **Cycadophyta**) are palmlike plants of the tropics, growing as tall as 20 meters (Figure 29.4a). **Ginkgos** (phylum **Ginkgophyta**), which were common during the Mesozoic era, are represented today by a single genus and species, *Ginkgo biloba*, the maidenhair tree (Figure 29.4b). There are both microsporangiate and megasporangiate maidenhair trees. The difference is determined by X and Y sex chromo-

Cycadophyta
Ginkgophyta
Coniferophyta
Gnetophyta
Angiospermae

somes, as in humans; few other plants have sex chromosomes. The phylum **Gnetophyta** consists of three very different genera that share certain characteristics with the angiosperms. One of the gnetophytes is *Welwitschia* (Figure 29.4c), a long-lived desert plant with just two straplike leaves that sprawl on the sand and can become as long as 3 meters. Far and away the most abundant of the gymnosperms are the **conifers** (phylum **Coniferophyta**), cone-bearing plants such as pines and redwoods (Figure 29.4d).

All living gymnosperms have stems and roots that grow larger in diameter (called *secondary growth*), and all but the Gnetophyta have only tracheids as water-conducting and support cells in their xylem. Although the gymnosperm water transport and support system may seem less effective than that of the angiosperms, it serves some of the tallest trees known. The coastal redwoods of California are the tallest gymnosperms; the largest are well over 100 m tall. Secondary xylem—wood—produced by gymnosperms is the principal resource of the timber industry.

Before examining the conifer life cycle, we'll take a brief look at the fossil history of gymnosperms.

We know the early gymnosperms only as fossils

The earliest fossil evidence of gymnosperms is found in Devonian rocks. The early gymnosperms combined characteristics of rhyniophytes and heterosporous ferns, but they had tracheids of the same type found in modern gymnosperms. They also differed from the plants around them by their extensively thickened woody stems, which resulted from proliferation of xylem.

By the Carboniferous period, several new lines of gymnosperms had evolved, including various seed ferns that possessed fernlike foliage but had characteristic gymnosperm seeds attached to their leaves. The first true conifers appeared somewhat later. Either they were not dominant trees or they did not grow where conditions were right for fossilization, so we have few preserved examples. During the Permian period, however, the conifers and cycads flourished. Gymnosperm forests changed with time as the gymnosperm groups evolved, and they dominated the Mesozoic era, in which the continents drifted apart and dinosaurs strode the Earth. Gymnosperms dominated all forests until less than 100 million years ago, and they still dominate some present-day forests.

Conifers have cones but no motile cells

The great Douglas fir and cedar forests of the northwestern United States and the massive boreal forests of pine, fir, and spruce that clothe the northern continental regions and upper slopes of mountain ranges rank among the great vegetation formations of the world. All these trees belong to one phylum of gymnosperms, Coniferophyta—the conifers, or cone-bearers. A **cone** is an axis bearing a tight cluster of scales or leaves specialized for reproduction. Megaspores and microspores are produced in separate seed and pollen cones. Seed cones are much larger than pollen cones (see Figure 29.3).

(a) *Cycas* sp.

(d) *Sequoiadendron giganteum*

(b) *Ginkgo biloba*

(c) *Welwitschia mirabilis*

29.4 Diversity among the Gymnosperms

(a) This palm belongs to the cycads, the least changed group of present-day gymnosperms. Many cycads have growth forms that resemble both ferns and palms. (b) The characteristic fleshy seed coat and broad leaves of the maidenhair tree. (c) A gnetophyte growing in the Namib Desert of Africa. Two huge, straplike leaves grow throughout the life of the plant, breaking and splitting as they grow. (d) A dramatic conifer, this giant sequoia grows in Yosemite National Park, California.

We will use the life cycle of a pine to illustrate reproduction in gymnosperms (Figure 29.5). The production of male gametophytes in the form of pollen grains frees the plant completely from its dependence on liquid water for fertilization. Instead of water, wind assists conifer pollen grains in their first stage of travel to the female gametophyte inside the seed cone. The pollen tube provides the means for the last stage of travel by elongating and digesting its way through maternal sporophytic tissue. When it reaches the female gametophyte, it releases two sperm, one of which degenerates after the other unites with the egg.

The megasporangium, which will form the female gametophyte containing eggs within archegonia, is enclosed in a layer of sporophytic tissue—the integument—that will eventually develop into the seed coat. The integument, the megasporangium inside it, and the tissue attaching it to the maternal sporophyte constitute the **ovule**. The pollen grain enters through a small opening in the integument at the tip of the ovule, the **micropyle**.

Gymnosperms derive their name (which means "naked-seeded") from the fact that their ovules and seeds are not protected by flower or fruit tissue. Most conifer ovules (which upon fertilization develop into seeds) are borne ex-

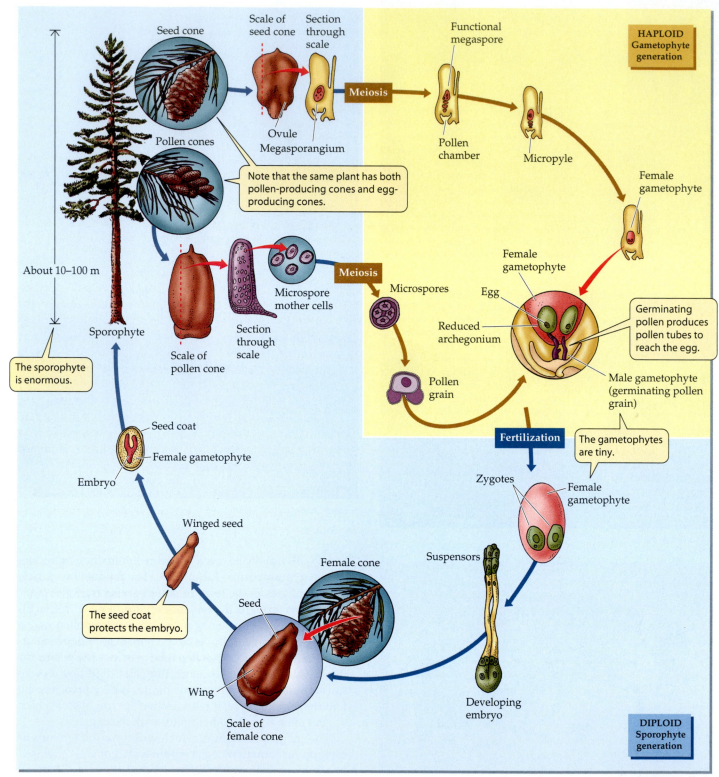

29.5 The Life Cycle of a Pine Tree
The gametophytes are microscopically small and nutritionally dependent on the sporophyte generation.

posed on the upper surfaces of modified branches called *cone scales*. Their only protection from the environment lies in the fact that the scales are tightly pressed against each other within the cone. As we have seen, some pines, such as

the lodgepole pine, have such tightly closed seed cones that only fire suffices to split them open and release the seeds.

About half of the conifer species have soft, fleshy fruitlike tissues associated with their seeds; examples are the "berries" of juniper and yew. Animals may eat these tissues and then disperse the seeds in their feces, often carrying them considerable distances from the parent plant. These tissues, however, are not true fruits, which are characteristic of the plant phylum that is dominant today: the angiosperms.

The Angiosperms: Flowering Plants

The phylum **Angiospermae** consists of the flowering plants, also commonly known as the **angiosperms**. This highly diverse phylum includes more than 230,000 species. The oldest evidence of angiosperms dates to the late Jurassic period, more than 140 mya. The angiosperms radiated explosively and, over a period of only about 60 million years, became the dominant plant life of the planet. In later chapters, when we mention "plants," we are generally referring to the angiosperms.

The angiosperms represent the current extreme of an evolutionary trend that runs throughout the tracheophytes: *The sporophyte generation becomes larger and more independent of the gametophyte, while the gametophyte generation becomes smaller and more dependent on the sporophyte.*

Angiosperms differ from other plants in several ways:

▶ They have double fertilization.
▶ They produce a triploid endosperm.
▶ Their ovules and seeds are enclosed in a carpel.
▶ They have flowers.
▶ They produce fruit.
▶ Their xylem contains vessel elements and fibers.
▶ Their phloem contains companion cells.

Double fertilization was long considered the single most reliable distinguishing characteristic of the angiosperms. Two male gametes, contained within a single microgametophyte (pollen grain), participate in fertilization events within the megagametophyte of an angiosperm. One sperm combines with the egg to produce a diploid zygote, the first cell of the sporophyte generation. In most angiosperms, the other sperm nucleus combines with two other haploid nuclei of the female gametophyte to form a triploid ($3n$) nucleus. This nucleus, in turn, divides to form a triploid tissue, the **endosperm**, that nourishes the embryonic sporophyte during its early development.

Double fertilization occurs in all present-day angiosperms. We are not sure when and how it evolved because there is no fossil evidence on this point. It probably first resulted in two embryos, as it does in the three existing genera of Gnetophyta: *Ephedra*, *Gnetum*, and *Welwitschia*. Both of the fertilizations in gnetophytes produce diploid products.

The formation of an extensive triploid endosperm is one of the most definitive angiosperm traits, although it is not universal.

The name angiosperm ("enclosed seed") is drawn from another diagnostic character: The ovules and seeds of these plants are enclosed in a modified leaf called a **carpel**. Besides protecting the ovules and seeds, the carpel often interacts with incoming pollen to prevent self-pollination, thus favoring cross-pollination and increasing genetic diversity. Of course, the most evident diagnostic feature of angiosperms is that they have flowers. Production of a fruit is another unique characteristic of the angiosperms.

Angiosperms are also distinguished by the possession of specialized water-transporting cells called **vessel elements** in their xylem, but these cells are also found, in anatomically different form, in gnetophytes and a few ferns. A second distinctive cell type in angiosperm xylem is the **fiber**, which plays an important role in supporting the plant body. Angiosperm phloem possesses another unique cell type, called a **companion cell**.

In the following sections we'll examine the structure and function of flowers, evolutionary trends in flower structure, the functions of pollen and fruits, the angiosperm life cycle, the two major groups of angiosperms, and the origin and evolution of flowering plants.

The sexual structures of angiosperms are flowers

If you examine any familiar flower, you will notice that the outer parts look somewhat like leaves. In fact, all the parts of a flower *are* modified leaves.

A generalized flower (for which there is no exact counterpart in nature) is shown in Figure 29.6 for the purpose of identifying its parts. The structures bearing microsporangia are called **stamens**. Each stamen is composed of a **filament** bearing an **anther** that contains pollen-producing microsporangia. The structures bearing megasporangia are the **carpels**. A structure composed of one carpel or two or more fused carpels is called a **pistil**. The swollen base of the pistil, containing one or more ovules (each containing a

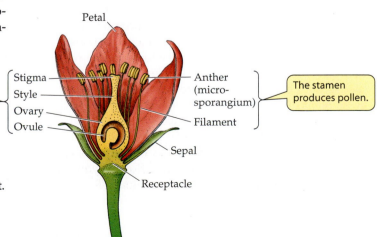

29.6 A Generalized Flower
Not all flowers possess all the structures shown here, but they must possess a stamen (male), pistil (female), or both in order to play their role in reproduction. Flowers that have both, as this one does, are referred to as perfect.

[Phylogenetic tree labels:]
Cycadophyta
Ginkgophyta
Coniferophyta
Gnetophyta
Angiospermae

[Flower diagram labels:]
Petal
The pistil receives pollen.
Stigma
Style
Ovary
Ovule
Anther (microsporangium)
Filament
The stamen produces pollen.
Sepal
Receptacle

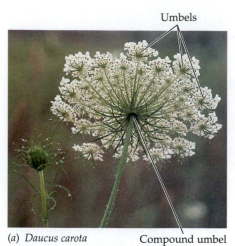

Umbels

Disk flowers (many)

Spikes

Ray flowers

(a) *Daucus carota*   Compound umbel   (b) *Echinacea purpurea*

(c) *Penniselum setaceum*

29.7 Inflorescences
(a) The inflorescence of Queen Anne's lace is an umbel. Each umbel bears flowers on stalks that arise from a common center. (b) Cornflowers are members of the aster family; their inflorescence is a head. In a head, each of the long, petal-like structures is a ray flower; the central portion of the head consists of dozens to hundreds of disc flowers. (c) Grasses such as this fountain grass have inflorescences called spikes.

megasporangium), is called the **ovary**. The apical stalk of the pistil is the **style**; and the terminal surface that receives pollen grains is called the **stigma**.

In addition, a flower often has several specialized sterile (non-spore-bearing) leaves: The inner ones are called **petals** (collectively, the **corolla**), and the outer ones **sepals** (collectively, the **calyx**). The corolla and calyx, which can be quite showy, often play roles in attracting animal pollinators to the flower. The calyx more commonly protects the immature flower in bud. From base to apex, the sepals, petals, stamens, and carpels (which are referred to as the floral organs; see Figure 15.11) are usually in circular arrangements called whorls and attached to a central stalk called the **receptacle**.

The generalized flower shown in Figure 29.6 has both megasporangia and microsporangia; such flowers are referred to as **perfect**. Many angiosperms produce two types of flowers, one with only megasporangia and the other with only microsporangia. Consequently, either the stamens or the carpels are nonfunctional or absent in a given flower, and the flower is referred to as **imperfect**.

Species such as corn or birch, in which both megasporangiate and microsporangiate flowers occur on the same plant, are said to be **monoecious** (meaning "one-housed"—but, it must be added, one house with separate rooms). Complete separation is the rule in some other angiosperm species, such as willows and date palms; in these species, a given plant produces either flowers with stamens or flowers with pistils, but never both. Such species are said to be **dioecious** ("two-housed").

Flowers come in an astonishing variety of forms, as you will realize if you think of some of the flowers you recognize. The generalized flower shown in Figure 29.6 has distinct petals and sepals arranged in distinct whorls. In nature, however, petals and sepals sometimes are indistinguishable. Such appendages are called **tepals**. In other flowers, petals, sepals, or tepals are completely absent.

Flowers may be single, or grouped together to form an **inflorescence**. Different families of flowering plants have their own, characteristic types of inflorescences, such as the umbels of the carrot family, the heads of the aster family, and the spikes of many grasses (Figure 29.7).

Flower structure has evolved over time

The flowers that are evolutionarily the most ancient have a large and variable number of tepals (or sepals and petals), carpels, and stamens (Figure 29.8a). Evolutionary change within the angiosperms has included some striking modifications from this early condition: reduction in the number of each type of organ to a fixed number; differentiation of petals from sepals; and change in symmetry from radial (as in a lily or magnolia) to bilateral (as in a sweet pea or orchid), often accompanied by an extensive fusion of parts (Figure 29.8b).

According to one theory, the first carpels to evolve were modified simple leaves, folded but incompletely closed, and thus differing from the scales of the gymnosperms. In the groups of angiosperms that evolved later, the carpels fused and became progressively more buried in receptacle tissue (Figure 29.9a). In the flowers of the latest groups to evolve, the other flower parts are attached at the very top of the ovary, rather than at the bottom as in Figure 29.6. The stamens of the most ancient flowers may have appeared leaflike (Figure 29.9b), little resembling those of the generalized flower in Figure 29.6.

Why do so many flowers have pistils with long styles and anthers with long filaments? Natural selection has favored length in both of these structures, probably because length increases the likelihood of successful pollination. Long filaments may bring the anthers into contact with insect bodies, or they may place the anthers in a better position to catch the wind. Similar arguments apply to long styles.

(b)

29.8 Flower Form and Evolution

(a) A magnolia flower shows the major features of early flowers: It is radially symmetrical, and the individual tepals, carpels, and stamens are separate, numerous, and attached at their bases. (b) Orchids have a bilaterally symmetrical structure that evolved much later than the form of the magnolia flower in (a). One of the three petals evolved into the complex lower "lip." Inside, the stamen and pistil are fused, and there is a single anther in this species.

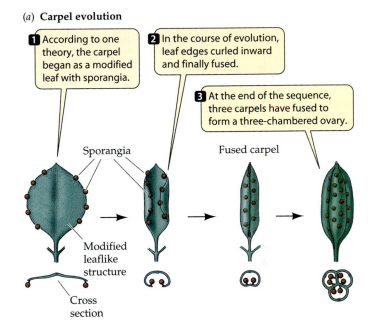

(a) **Carpel evolution**

1 According to one theory, the carpel began as a modified leaf with sporangia.

2 In the course of evolution, leaf edges curled inward and finally fused.

3 At the end of the sequence, three carpels have fused to form a three-chambered ovary.

Sporangia

Fused carpel

Modified leaflike structure

Cross section

(b) **Stamen evolution**

1 The leaflike portion of the structure was progressively reduced...

2 ...until only the microsporangia remained.

A long style may serve another purpose as well. If several pollen grains land on one stigma, a pollen tube will start growing from each grain toward the ovary. If there are more pollen grains than ovules, there is a "race" to fertilize the ovules. The race down the style can be viewed as "mate selection" by the plant bearing that style.

Angiosperms have coevolved with animals

Pollen has played another crucial role in the evolution of the angiosperms. Whereas many gymnosperms are wind-pollinated, most angiosperms are animal-pollinated. Animals visit flowers to obtain nectar or pollen, and in the process often carry pollen from one flower to another, or from one plant to another. Thus, in its quest for food, the animal contributes to the genetic diversity of the plant population. Insects, especially bees, are among the most important pollinators; birds and some species of bats also play major roles.

For more than 130 million years, angiosperms and their animal pollinators have coevolved in the terrestrial environment. The animals have affected the evolution of the plants, and the plants have affected the evolution of the animals. Flower structure has become incredibly diverse under these selection pressures.

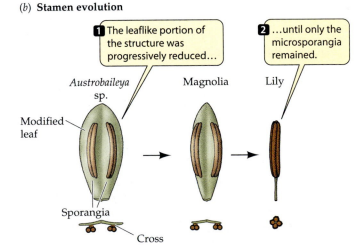

Austrobaileya sp.

Magnolia

Lily

Modified leaf

Sporangia

Cross section

29.9 Carpels and Stamens Evolved from Leaflike Structures

(a) Possible stages in the evolution of a carpel from a more leaflike structure. (b) The stamens of three modern plants show the various stages in the evolution of that organ.

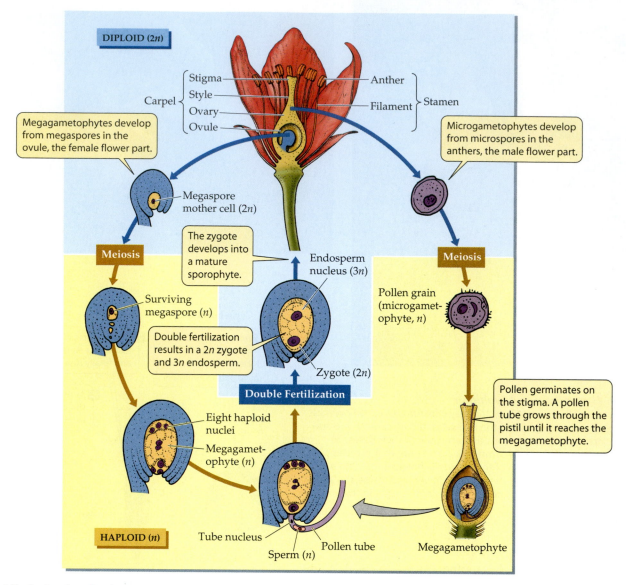

29.10 The Life Cycle of an Angiosperm
The formation of a triploid endosperm distinguishes the angiosperms from the gymnosperms.

Some of the products of coevolution are highly specific; for example, some yucca species are pollinated by only one species of moth. Pollination by just one or a very few animal species provides a plant species with a reliable mechanism for transferring pollen from one to another of its members.

Most plant–pollinator interactions are much less specific; that is, many different animal species pollinate the same plant species, and the same animal species pollinate many plant species. However, even these less specific interactions have developed some specialization. Bird-pollinated flowers are often red and odorless. Insect-pollinated flowers often have characteristic odors, and bee-pollinated flowers may have conspicuous markings, or *nectar guides*, that are evident only in the ultraviolet region of the spectrum, where bees have better vision than in the red region. Co-

evolution and other aspects of plant–animal interactions are covered in more detail in Chapter 55.

The angiosperm life cycle features double fertilization

The life cycle of the angiosperms is summarized in Figure 29.10. The angiosperm life cycle will be considered in detail in Chapter 38, but let's look at it briefly here and compare it with the conifer life cycle in Figure 29.5.

Like all seed plants, angiosperms are heterosporous. The female gametophyte is even more reduced than that of the gymnosperms. The ovules are contained within carpels, rather than being exposed on the surfaces of scales, as in most gymnosperms. The male gametophytes are, again, pollen grains.

The ovule develops into a seed containing the products of the double fertilization that characterizes angiosperms. The triploid endosperm serves as storage tissue for starch or lipids, proteins, and other substances that will be needed by the developing embryo.

(a)

(b)

29.11 Fleshy Fruits Come in Many Forms and Flavors
(a) A simple fruit (sour cherries). (b) An aggregate fruit (raspberries). (c) A multiple fruit (pineapple). (d) An accessory fruit (pear).

(c)

(d)

The diploid zygote develops into an embryo, consisting of an embryonic axis and one or two **cotyledons**. Also called seed leaves, the cotyledons have different fates in different plants. In many, they serve as absorptive organs that take up and digest the endosperm. In others, they enlarge and become photosynthetic when the seed germinates. Often they play both roles (see Chapter 37).

Angiosperms produce fruits

The ovary of a flowering plant (together with the seeds it contains) develops into a fruit after fertilization. A **fruit** may consist only of the mature ovary and its seeds, or it may include other parts of the flower or structures associated with it. A *simple fruit*, such as a cherry (Figure 29.11a), is one that develops from a single carpel or several united carpels. A raspberry is an example of an *aggregate fruit* (Figure 29.11b)—one that develops from several separate carpels of a single flower. Pineapples and figs are examples of *multiple fruits* (Figure 29.11c), formed from a cluster of flowers (an inflorescence). Fruits derived from parts in addition to the carpel and seeds are called *accessory fruits* (Figure 29.11d); examples are apples, pears, and strawberries. The development, ripening, and dispersal of fruits will be considered in Chapters 37 and 38.

Determining the oldest living angiosperm lineage

Which angiosperms were the first flowering plants was long a matter of great controversy. Two leading candidates were the magnolia family (see Figure 29.8a) and another family, the Chloranthaceae, whose flowers are much simpler than those of the magnolias. At the close of the twentieth century, an impressive convergence of evidence led to the conclusion that the base of the angiosperm phylogenetic tree belongs to neither of those families, but rather to a lineage that today consists of just a single species of the genus *Amborella* (Figure 29.12). This woody shrub, with cream-colored flowers, lives only on New Caledonia, an island in the South Pacific. Its 5 to 8 carpels are in a single whorl, and it has 30 to 100 stamens. The xylem of *Amborella* lacks vessel elements, which appeared later in angiosperm evolution. The characteristics of *Amborella* give us a good sense of what the first angiosperms might have been like.

(a)

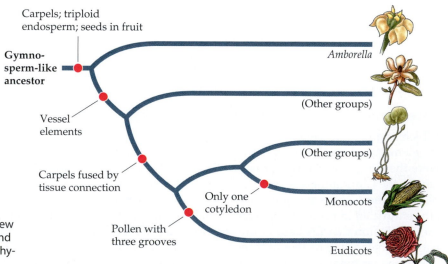

29.12 The First Angiosperm
(a) *Amborella*, a shrub, is the closest living relative of the first angiosperms. (b) A flower of *Amborella*.

(b)

There are two large monophyletic groups of angiosperms

There are two large lineages that include the great majority of angiosperm species: the **monocots** and the **eudicots**. Both are monophyletic groups (Figure 29.13). The monocots are so called because they have a single embryonic cotyledon; the eudicots have two. The cotyledons of some, but not all, eudicots store the reserves originally present in the endosperm. There are several other differences between the two lineages, which we will describe in Chapter 34. Some familiar plants, including magnolias and water lilies, belong to lineages more ancient than either the monocots or the eudicots.

The monocots (Figure 29.14) include grasses, cattails, lilies, orchids, and palm trees. The eudicots (Figure 29.15) include the vast number of familiar seed plants, including most of the herbs, vines, trees,

and shrubs. Among them are oaks, willows, violets, snapdragons, and sunflowers.

The origin of the angiosperms remains a mystery

We have learned a lot about evolution within the angiosperm lineage. The most important unanswered question about the evolution of seed plants is this: How did the angiosperms first arise—to which gymnosperm phylum are they sister? You might think that, given the advances in techniques of molecular genetics and computer technology, as well as new fossil finds, we would have opened this century with the answer to this question well in hand. A very few years ago, it seemed that we were on the verge of answering it. Although an answer may be agreed upon before the present decade ends, the puzzle is as vexing today as it was before.

Why should this be? Different phylogenetic methods, applied by different investigators, have produced apparently contradictory results. It might seem a simple matter to rectify this situation, but several questions complicate such efforts: What morphological characters should be selected as important, or should they all be treated as equally important? What algorithms should be applied to computerized analysis of data? Are all molecular differences and similarities significant, or are some of them incidental? Which fossils should be chosen for comparisons? What is the likelihood that we can find evidence of double fertilization in ancient fossils?

We are left with the question: Where did the first angiosperm come from? The angiosperms may be most closely related to the gnetophytes, or they may be more closely related to the conifers, or to another gymnosperm phylum. Current progress in methodology gives us reason to hope that our picture of seed plant evolution will be much more complete before this decade ends. We will see in Chapters 31–33 whether our understanding of animal evolution is more complete.

29.13 Evolutionary Relationships among the Angiosperms
The monocots and the eudicots are the largest monophyletic groups among the angiosperms. A few lineages that differ from the monocots, eudicots, and *Amborella* remain to be placed accurately on the phylogenetic tree.

Carpels; triploid endosperm; seeds in fruit

Gymno-sperm-like ancestor

Vessel elements

Carpels fused by tissue connection

Pollen with three grooves

Only one cotyledon

Amborella

(Other groups)

(Other groups)

Monocots

Eudicots

(a) *Phoenix dactylifera*

29.14 Monocots
(a) Palms are among the few monocot trees. Date palms are a major food source in some areas of the world. (b) Grasses such as this cultivated wheat and the fountain grass in Figure 29.7c are monocots. (c) Monocots include popular garden flowers such as these daylilies. Orchids (Figure 29.8b) are another highly prized monocot flower.

(b) *Triticum* sp.

(c) *Hemerocallis* sp.

(a)

(c) *Rosa rugosa*

(b) *Cornus florida*

29.15 Eudicots
(a) The cactus family is a large group of eudicots, with about 1,500 species in the Americas. This cactus bears scarlet flowers for a brief period of the year. (b) The flowering dogwood is a small eudicot tree. (c) Climbing Cape Cod roses are members of the eudicot family Rosaceae, as are the familiar roses from your local florist.

Chapter Summary

General Characteristics of the Seed Plants

▶ The seed plants (gymnosperms and angiosperms) are heterosporous and have greatly reduced gametophytes. **Review Figures 29.1, 29.2**

▶ Most modern seed plants have no swimming gametes and do not require liquid water for fertilization. The male gametophyte—the pollen grain—is dispersed by wind or by animals. **Review Figure 29.3**

▶ The seed is a well-protected resting stage that often contains food that supports the growth of the embryo.

The Gymnosperms: Naked Seeds

▶ The gymnosperms, once the dominant vegetation on Earth, still dominate forests in the northern parts of the Northern Hemisphere and at high elevations.

▶ The four surviving gymnosperm phyla are the Cycadophyta (the most ancient), Ginkgophyta (consisting of a single species, the maidenhair tree), Gnetophyta (which has some characters in common with the angiosperms), and Coniferophyta (the familiar cone-bearing trees).

▶ Modern gymnosperms all have abundant xylem and extensive secondary growth.

▶ Conifers have a life cycle in which naked seeds are produced on the scales of female cones. Pollen cones are smaller than seed cones. Pollen is transferred from pollen cones to seed cones by wind. **Review Figure 29.5**

The Angiosperms: Flowering Plants

▶ Angiosperms (phylum Angiospermae) are distinguished by double fertilization, which results in a triploid nutritive tissue, the endosperm. Double fertilization is also characteristic of the Gnetophyta. **Review Figure 29.10**

▶ The ovules and seeds of angiosperms are enclosed by a carpel. Angiosperms are also characterized by the production of flowers and fruits.

▶ The vascular tissues of angiosperms contain three characteristic cell types: vessel elements, fibers, and companion cells.

▶ Flowers are made up of various combinations of carpels, stamens, petals, and sepals. Perfect flowers have both carpels (female parts) and stamens (male parts). **Review Figure 29.6**

▶ Monoecious plant species have both female and male flowers on the same plant. Dioecious species have separate female and male plants.

▶ Carpels and stamens may have evolved from leaflike structures. **Review Figure 29.9**

▶ Angiosperms and the animals that pollinate them have coevolved.

▶ *Amborella*, a tropical shrub, is the sole living representative of the first angiosperm lineage.

▶ There are two major lineages of flowering plants: monocots and eudicots. **Review Figure 29.13**

▶ The evolutionary origin of the angiosperms remains a mystery.

For Discussion

1. In most seed plant species, only one of the products of meiosis in the megasporangium survives. How might this be advantageous?

2. Suggest an explanation for the great success of the angiosperms in occupying terrestrial habitats.

3. In many locales, large gymnosperms predominate over large angiosperms. Under what conditions might gymnosperms have the advantage, and why?

4. Not all flowers possess all of the following parts: sepals, petals, stamens, and carpels. What kind or kinds of flower parts do you think might be found in the flowers that have the smallest number of kinds? Discuss the possibilities, both for a single flower and for a species.

5. The problem of the origin of the angiosperms has long been "an abominable mystery," as Charles Darwin once put it. Scientists still do not know the nearest relatives of the angiosperms. It has often been suggested (correctly or incorrectly) that the gnetophytes are sister to the angiosperms. What pieces of evidence suggested this connection?

30 Fungi: Recyclers, Killers, and Plant Partners

 WHAT ARE THE LARGEST ORGANISMS you can think of? Whales? Trees? Some of the largest organisms on Earth are fungi. One such fungus, growing in Michigan, covers an area of 37 acres. Its effect on green plants is evident from the air, but from ground level, it is difficult to realize how large the fungus is. At the surface, you see only seemingly isolated clumps of mushrooms. But the vast body of the fungus *Armillariella*, which weighs approximately the same as a blue whale, grows underground and consists almost entirely of microscopic filaments.

Molecular studies indicate that this giant fungus is or was a single individual that arose from a single spore. It is possible that fragmentation over time may have broken it into a few separate—but still gigantic—individuals. Another, larger fungus of the same genus, growing in the state of Washington, occupies parts of three counties. But not all fungi are huge. Molds and mushrooms are fungi, as are the microscopic, unicellular yeasts.

Every breath we take contains large numbers of fungal spores. Some of those spores can be dangerous, and fungal diseases of humans, some of which are as yet uncurable, have become a major global threat. However, other fungi are of immense commercial importance to us. Fungi are essential to plants as well. Fungi interact with roots, greatly enhancing the roots' ability to take up water and mineral nutrients.

Earth would be a messy place without the fungi. They are at work in forests, fields, and garbage dumps, breaking down the remains of dead organisms (and even manufactured substances such as some plastics). For almost a billion years, the ability of fungi to decompose substances has been important for life on Earth, chiefly because by breaking down carbon compounds, they return carbon and other elements to the environment, where they can be used again by other organisms.

In this chapter we will examine the general biology of the kingdom Fungi, which differs in interesting ways from the other kingdoms. We will also explore the diversity of body forms, reproductive structures, and life cycles of the four phyla of fungi, as well as the mutually beneficial associations of certain fungi with other organisms. As we begin our study, recall that the fungi and the animals are descended from a common ancestor—we are more closely related to molds and mushrooms than we are to the flowers we admired in the last chapter.

General Biology of the Fungi

The fungi are superbly adapted for absorptive nutrition: They secrete digestive enzymes that break down large food molecules in the environment, then absorb the breakdown products. The kingdom Fungi encompasses *heterotrophic organisms with absorptive nutrition*. Many fungi are saprobes that absorb nutrients from dead matter, others are parasites that absorb nutrients from living hosts (Figure 30.1), and still others live in mutually beneficial symbioses with other organisms.

All fungi form spores, but only in one phylum (Chytridiomycota) do spores or gametes possess flagella. Fungi reproduce sexually in a variety of ways. Their cell walls contain at least some **chitin**, a polysaccharide that is also found in the skeletons of arthropods and in some protists. Most fungi have complex body forms.

These criteria enable us to distinguish between the fungi and some protists that resemble them. The slime molds consist of two protist groups whose members take up food

The Tip of the Iceberg
These fungal fruiting bodies of *Armillariella* are only a hint of the presence of a vast underground network of microscopic filaments extending over many acres.

30.1 Parasitic Fungi Attack Other Living Organisms

(a) The gray masses on this ear of corn are the parasitic fungus *Ustilago maydis*, commonly called corn smut. (b) The tropical fungus whose fruiting body is growing out of the carcass of this ant has developed from a spore ingested by the ant. The spores of this fungus must be ingested by insects before they will germinate and develop. The growing fungus absorbs organic and inorganic nutrients from the ant's body, eventually killing it, after which the fruiting body produces a new crop of spores. (c) An amoeba (below) being parasitized by a fungus (above) of the genus *Amoebophilus* ("amoeba lover").

(a)

Fungus

(b)

Fungal fruiting body

(c)

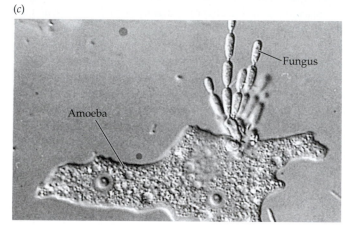

Fungus

Amoeba

by phagocytosis rather than by absorption, and a third protist group whose members have cells with flagella. Other funguslike protists (Oomycota) also have flagellated cells; they have cellulose rather than chitin in their cell walls.

The kingdom Fungi consists of four phyla: Chytridiomycota, Zygomycota, Ascomycota, and Basidiomycota (Table 30.1). We distinguish the phyla on the basis of their methods and structures for sexual reproduction and, to a lesser extent, on criteria such as the presence or absence of crosswalls separating their cell-like compartments. This morphologically based phylogeny has proved largely consistent with phylogenies based on DNA sequencing.

Some fungi, called **imperfect fungi** or **deuteromycetes**, do not form sexual structures by which they might be easily identified as members of one of the four phyla. However, techniques of molecular taxonomy, such as DNA sequencing, have allowed us to identify many imperfect fungi as asexual zygomycetes, ascomycetes, or basidiomycetes. The deuteromycetes are not considered a phylum, but rather a "holding group" for species whose status is yet to be resolved.

The fungi are an ancient kingdom. Fossil evidence suggests that they have been present since at least 600 million years ago, and perhaps much longer.

In the sections that follow, we'll consider some aspects of the general biology of the fungi, including their body structure and its intimate relationship with their environment, their nutrition, and some special aspects of their unusual sexual reproductive cycles.

Some fungi are unicellular

Unicellular forms are found in all of the fungal phyla, as well as among the deuteromycete group. Unicellular members of the Zygomycota, Ascomycota, and Basidiomycota are called **yeasts**. Yeasts may reproduce by budding, by fission, or by sexual means, which help us to place them in their appropriate phyla (Figure 30.2).

30.1 Classification of Fungi

PHYLUM	COMMON NAME	FEATURES	EXAMPLES
Chytridiomycota	Chytrids	Aquatic; gametes have flagella	*Allomyces*
Zygomycota	Zygomycetes	Zygosporangium; no regularly occurring septa; usually no fleshy fruiting body	*Rhizopus*
Ascomycota	Ascomycetes	Ascus; perforated septa	*Neurospora*, baker's yeast
Basidiomycota	Basidiomycetes	Basidium; perforated septa	*Puccinia*, mushrooms

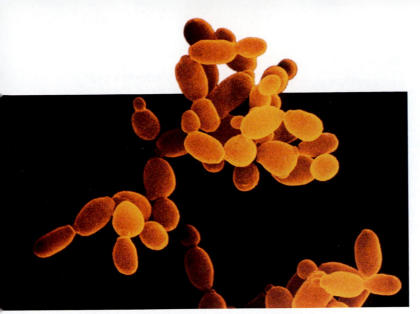

Saccharomyces sp.

30.2 Yeasts Are Microscopic, Unicellular Fungi
Unicellular members of the fungal phyla are known as yeasts. Many yeasts reproduce by budding, as those shown here are doing.

The body of a fungus is composed of hyphae

Most fungi are not unicellular, but whether they can truly be called multicellular is questionable. The vegetative (feeding) body of a fungus is called a **mycelium** (plural mycelia). It is composed of rapidly growing individual tubular filaments called **hyphae** (singular hypha). Within most hyphae, there is no division into separate cells, and organelles (even nuclei) can move around (Figure 30.3). Thus, it may be more appropriate to call these fungi *multinucleated* than multicellular. Some hyphae are subdivided into cell-like compartments by *incomplete* cross-walls called **septa** (singular septum). Other hyphae are **coenocytic** and have no septa.

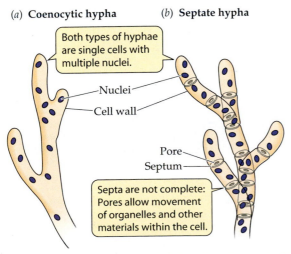

(a) **Coenocytic hypha**   *(b)* **Septate hypha**

Both types of hyphae are single cells with multiple nuclei.

Nuclei

Cell wall

Pore

Septum

Septa are not complete: Pores allow movement of organelles and other materials within the cell.

30.3 Most Hyphae Are Not Divided into Separate Cells
Even when septa are present, they do not block the movement of organelles within the hypha.

Certain modified hyphae, the **rhizoids**, anchor Chytridiomycota to their substrate (the dead organism or other matter upon which they feed). These rhizoids are not homologous to the rhizoids of plants. Parasitic fungi may have modified hyphae that take up nutrients from their host.

The total hyphal growth of a mycelium (not the growth of an individual hypha) may exceed 1 km per day. The hyphae may be highly dispersed or may clump together in a cottony mass. Sometimes, when sexual spores are produced, the mycelium becomes organized into elaborate fruiting bodies such as mushrooms.

The way in which a parasitic fungus attacks a plant illustrates the roles of some fungal structures (Figure 30.4). The hyphae of a fungus invade a leaf through the stomata, through wounds, or in some cases, by direct penetration of epidermal cells. Once inside the leaf, the hyphae form a mycelium. Some hyphae grow into the living plant cells, absorbing the nutrients within the cells. Fruiting bodies may form, either within the plant body or on its surface.

Fungi are in intimate contact with their environment

The tubular hyphae of a fungus give it a unique relationship with its physical environment. The fungal mycelium has an enormous surface area-to-volume ratio compared with that of most large multicellular organisms. This large ratio of surface area to volume is a marvelous adaptation for absorptive nutrition. Throughout the mycelium (except in fruiting bodies), all the hyphae are very close to their environmental food source.

Another characteristic of some fungi is their tolerance for highly hypertonic environments (those with a solute concentration higher than their own; see Chapter 5). Many fungi are more resistant than bacteria to damage in hypertonic surroundings. Jelly in the refrigerator, for example, will not become a growth medium for bacteria, because it is too hypertonic to the bacteria, but it may eventually harbor mold colonies. The refrigerator itself illustrates another trait of many fungi: tolerance of temperature extremes. Many fungi tolerate temperatures as low as 5–6°C below freezing, and some tolerate temperatures as high as 50°C or more.

Fungi are absorptive heterotrophs

All fungi are heterotrophs that obtain food by direct absorption from their immediate environment. The majority are **saprobes**, obtaining their energy, carbon, and nitrogen directly from dead organic matter through the action of enzymes they secrete. However, as we've learned already, some are parasites, and still others form mutualistic associations with other organisms.

Saprobic fungi, along with bacteria, are the major decomposers of the biosphere, contributing to decay and thus to the recycling of the elements used by living things. In the forest, for example, the invisible mycelia of fungi absorb nutrients from fallen trees, thus decomposing their wood. Fungi are the principal decomposers of cellulose and lignin,

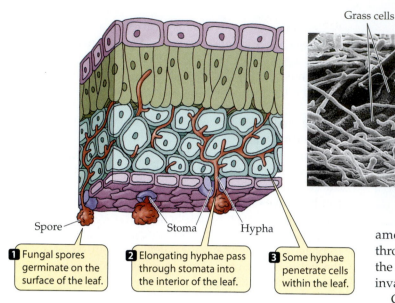

Grass cells

30.4 A Fungus Attacks a Leaf
The white structures in the micrograph are hyphae of the fungus *Blumeria graminis*, which is growing on the dark surface of the leaf of a grass.

Fungal hyphae

Spore   Stoma   Hypha

1 Fungal spores germinate on the surface of the leaf.

2 Elongating hyphae pass through stomata into the interior of the leaf.

3 Some hyphae penetrate cells within the leaf.

the main components of plant cell walls (most bacteria cannot break down these materials).

Because many saprobic fungi are able to grow on artificial media, we can perform experiments to determine their exact nutritional requirements. Sugars are their favored source of carbon. Most fungi obtain nitrogen from proteins or the products of protein breakdown. Many fungi can use nitrate (NO_3^-) or ammonium (NH_4^+) ions as their sole source of nitrogen. No known fungus can get its nitrogen directly from nitrogen gas, as can some bacteria and plant–bacteria associations (see Chapter 36). Nutritional studies also reveal that most fungi are unable to synthesize their own thiamin (vitamin B_1) or biotin (another B vitamin), and must absorb these vitamins from their environment. On the other hand, fungi can synthesize some vitamins that animals cannot. Like all organisms, fungi also require some mineral elements.

Nutrition in the parasitic fungi is particularly interesting to biologists. **Facultative** parasites can be grown by themselves on defined artificial media. **Obligate** parasites cannot be grown on any available medium; they can grow only on their specific living hosts, usually plants. Because their growth is limited to living hosts, they must have unusual nutritional requirements.

Some fungi have adaptations that enable them to function as active predators, trapping nearby microscopic protists or animals, from which they obtain nitrogen and energy. The most common strategy is to secrete sticky substances from the hyphae so that passing organisms stick tightly to them. The hyphae then quickly invade the prey, growing and branching within it, spreading through its body, absorbing nutrients, and eventually killing it.

A more dramatic adaptation for predation is the constricting ring formed by some species of *Arthrobotrys*, *Dactylaria*, and *Dactylella* (Figure 30.5). All of these fungi grow in soil. When nematodes (tiny roundworms) are present in the soil, these fungi form three-celled rings with a di-

ameter that just fits a nematode. A nematode crawling through one of these rings stimulates it, causing the cells of the ring to swell and trap the worm. Fungal hyphae quickly invade and digest the unlucky victim.

Certain highly specific associations between fungi and other organisms have nutritional consequences for the fungal partner. **Lichens** are associations of a fungus with a cyanobacterium, a unicellular photosynthetic eukaryote, or both. **Mycorrhizae** (singular mycorrhiza) are associations between specific fungi and the roots of plants. In such associations the fungus obtains organic compounds from its photosynthetic partner, but provides it with minerals and water so that the partner's nutrition is also promoted. We will discuss lichens and mycorrhizae more thoroughly later in this chapter.

Most fungi reproduce both asexually and sexually

Both asexual and sexual reproduction are common among the fungi. Asexual reproduction takes several forms:

▶ The production of (usually) haploid spores within structures called sporangia.

Roundworm   Fungal loop

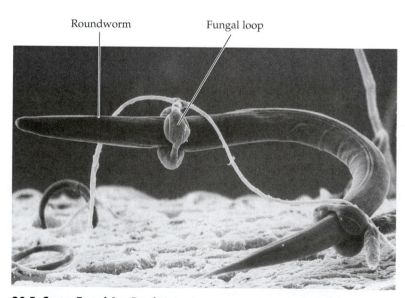

30.5 Some Fungi Are Predators
A nematode (roundworm) is trapped in sticky loops of the soil-dwelling fungus *Arthrobotrys anchonia*.

▶ The production of naked spores (not enclosed in sporangia) at the tips of hyphae; such spores are called **conidia** (from the Greek *konis*, "dust").

▶ Cell division by unicellular fungi—either a relatively equal division or an asymmetrical division in which a tiny bud is produced.

▶ Simple breakage of the mycelium.

Sexual reproduction in many fungi features an interesting twist. There is often no morphological distinction between female and male structures, or between female and male individuals. Rather, there is a genetically determined distinction between two or more **mating types**. Individuals of the same mating type cannot mate with one another, but they can mate with individuals of another mating type. This distinction prevents self-fertilization. Individuals of different mating types differ genetically from one another, but are often visually and behaviorally indistinguishable. Many protists also have mating type systems.

In many fungi, the zygote nuclei formed by sexual reproduction are the only diploid nuclei in the life cycle. These nuclei undergo meiosis, producing haploid nuclei that become incorporated into spores. Haploid fungal spores, whether produced sexually in this manner or asexually, germinate, and their nuclei divide mitotically to produce hyphae.

Many fungal life cycles include a dikaryon stage

The hyphae of some Zygomycota, Ascomycota, and Basidiomycota have a nuclear configuration other than the familiar haploid or diploid. In these fungi, sexual reproduction begins in an unusual way: The cytoplasms of two individuals of opposite mating types fuse (*plasmogamy*) long before their nuclei fuse (*karyogamy*), so that *two genetically different haploid nuclei exist within the same hypha*. This hypha is called a **dikaryon** (having *two* nuclei). Because the two nuclei differ genetically, the hypha is also called a **heterokaryon** (having *different* nuclei).

Eventually, specialized fruiting structures form, within which the pairs of dissimilar nuclei—one from each parent—fuse, giving rise to zygotes long after the original "mating." The zygote nucleus undergoes meiosis, producing four haploid nuclei. The mitotic descendants of those nuclei become the nuclei of the next generation of hyphae.

The reproduction of such fungi displays several unusual features. First, there are no gamete *cells*, only gamete *nuclei*. Second, there is never any true diploid tissue, although for a long period the genes of both parents are present in the dikaryon and can be expressed. In effect, these hyphae are neither diploid (2*n*) nor haploid (*n*); rather, they are *dikaryotic* (*n* + *n*). A harmful recessive mutation in one nucleus may be compensated for by a normal allele on the same chromosome in the other nucleus. Dikaryosis is perhaps the most significant of the genetic peculiarities of the fungi.

Finally, although Zygomycota, Ascomycota, and Basidiomycota grow in moist places, their gamete nuclei are not motile and are not released into the environment. Therefore, liquid water is not required for fertilization.

Some fungi are pathogens

Fungal pathogens are a major cause of death among people with compromised immune systems. Most patients with AIDS die of fungal diseases, such as the pneumonia caused by *Pneumocystis carinii* or the incurable diarrhea caused by some other fungi. *Candida albicans* and certain other yeasts also cause severe diseases in individuals with AIDS and in individuals taking immunosuppressive drugs. Such fungal diseases are a growing international health problem. Our limited understanding of the basic biology of these fungi still hampers our ability to treat the diseases they cause.

Various fungi cause other, less threatening human diseases, such as ringworm and athlete's foot. Still others are responsible for plant diseases that affect human food supplies. These diseases include black stem rust of wheat and other diseases of wheat, corn, and oats. Fungal diseases of plants have cost billions of dollars in crop losses.

Diversity in the Kingdom Fungi

Each of the four phyla of the kingdom Fungi appears to be monophyletic (Figure 30.6). Because the imperfect fungi (deuteromycetes) are polyphyletic, we will not give them phylum status. In this section on fungal diversity, we'll consider the four phyla—Chytridiomycota, Zygomycota, Ascomycota, and Basidiomycota—and we'll discuss the status of the deuteromycetes.

Chytrids probably resemble the ancestral fungi

The earliest-diverging fungal lineage is the **chytrids** (phylum **Chytridiomycota**). These aquatic microorganisms have sometimes been classified as protists. We place chytrids among the fungi because their cell walls consist primarily of chitin and because molecular evidence indicates that they and the other fungi form a monophyletic group.

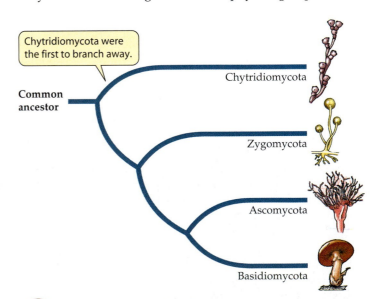

Chytridiomycota were the first to branch away.

Common ancestor

Chytridiomycota

Zygomycota

Ascomycota

Basidiomycota

30.6 Phylogeny of the Fungi
Four phyla are recognized among the fungi. In addition, the imperfect fungi, or Deuteromycetes, functions as a "holding group" for fungal species whose status is yet to be determined.

Chytrids are either parasitic (on organisms such as algae, mosquito larvae, and nematodes) or saprobic, obtaining nutrients by breaking down dead organic matter. (Chytrids in the compound stomachs of foregutfermenting animals may be an exception, living in a mutualistic association with their hosts.) Most chytrids live in freshwater habitats or in moist soil, but some are marine. Some chytrids are unicellular; others have mycelia made up of branching chains of cells. Chytrids reproduce both sexually and asexually, but they do not have a dikaryon stage.

Allomyces, a well-studied genus of chytrids, displays alternation generations. A haploid **zoospore** (spore with flagella) comes to rest on dead plant or animal material in water and germinates to form a small haploid organism. That produces female and male **gametangia** (gamete cases) (Figure 30.7). The male gametangia are smaller than the female gametangia and possess a light orange pigment. Mitosis in the gametangia results in the formation of haploid gametes, each with a single nucleus.

Both female and male gametes have flagella. The motile female gamete produces a *pheromone*, a chemical that attracts the swimming male gamete. The gametes fuse in pairs, and then their nuclei fuse to form a diploid zygote. Mitosis and cytokinesis in the zygote gives rise to a small diploid organism, which produces numerous diploid flagellate zoospores. These diploid zoospores disperse and germinate to form more diploid organisms. Eventually, the diploid organism produces thick-walled resting sporangia that can survive unfavorable conditions such as dry weather or freezing. Nuclei in the resting sporangia eventually undergo meiosis, giving rise to haploid zoospores that are released into the water and begin the cycle anew.

Chytrids are the only fungi that have flagella at any life cycle stage. We speculate that the protist ancestor of the fungi possessed flagella, because the phylum Chytridiomycota was the first fungal group to diverge from the others (see Figure 30.6). The same protist ancestor gave rise to the protist group Choanoflagellida and to the animal kingdom. A key event in the evolution of the fungi after the chytrids diverged from the others was the loss of flagella.

Zygomycetes reproduce sexually by fusion of two gametangia

Most **zygomycetes** (phylum **Zygomycota**) have coenocytic hyphae (hyphae without regularly occurring septa). They produce no motile cells, and only one diploid cell—the zygote—appears in the entire life cycle. The mycelium of a zygomycete spreads over its substrate, growing forward by means of specialized hyphae. Most zygomycetes do not form a fleshy fruiting body; rather, the hyphae spread in an apparently random fashion, with occasional stalked **sporangiophores** reaching up into the air (Figure 30.8).

Almost 900 species of zygomycetes have been described. A very important group of zygomycetes serves as the fungal partners in the most common type of mycorrhizal association with plant roots. A zygomycete that you may be more familiar with is *Rhizopus stolonifer*, the black bread mold. *Rhizopus* reproduces asexually by producing many stalked sporangiophores, each bearing a single sporangium containing hundreds of minute spores (Figure 30.9a). Other zygomycetes have sporangiophores with many sporangia. As in other filamentous fungi, the spore-forming structure is separated from the rest of the hypha by a wall.

Zygomycetes reproduce sexually when adjacent hyphae of two different mating types release pheromones, which cause them to grow together. These hyphae produce ga-

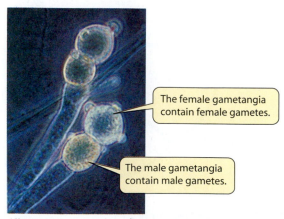

Allomyces sp.

30.7 Reproductive Structures of a Chytrid
The haploid gametes produced in these gametangia will fuse with another gamete to form a diploid organism.

Phycomyces sp.

30.8 A Zygomycete
This small forest of filamentous structures is made up of sporangiophores. The stalks end in tiny, rounded sporangia.

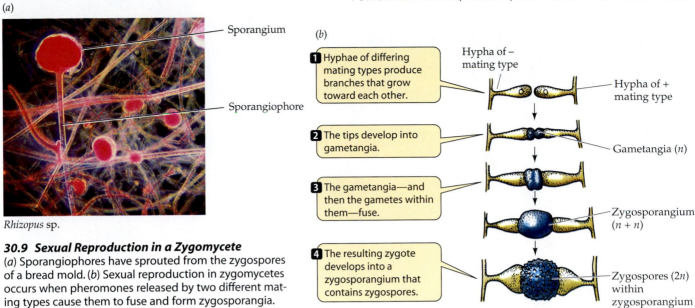

(a)

Sporangium

Sporangiophore

Rhizopus sp.

30.9 Sexual Reproduction in a Zygomycete
(*a*) Sporangiophores have sprouted from the zygospores of a bread mold. (*b*) Sexual reproduction in zygomycetes occurs when pheromones released by two different mating types cause them to fuse and form zygosporangia.

(b)

1 Hyphae of differing mating types produce branches that grow toward each other.

Hypha of – mating type

Hypha of + mating type

2 The tips develop into gametangia.

Gametangia (*n*)

3 The gametangia—and then the gametes within them—fuse.

Zygosporangium (*n* + *n*)

4 The resulting zygote develops into a zygosporangium that contains zygospores.

Zygospores (2*n*) within zygosporangium

metangia that fuse to form zygosporangia containing zygospores (Figure 30.9*b*). The zygosporangia develop thick, multilayered walls that protect the zygospores. The highly resistant zygospores may remain dormant for months before their nuclei undergo meiosis and a sporangium sprouts. The sporangium contains the products of meiosis: haploid nuclei that are incorporated into spores. These spores disperse and germinate to form a new generation of haploid hyphae.

The sexual reproductive structure of ascomycetes is an ascus

The **ascomycetes** (phylum **Ascomycota**) are a large and diverse group of fungi distinguished by the production of sacs called **asci** (singular ascus) (Figure 30.10). The ascus is the characteristic sexual reproductive structure of the ascomycetes. Ascomycete hyphae are segmented by more or less regularly spaced septa. A pore in each septum permits extensive movement of cytoplasm and organelles (including the nuclei) from one segment to the next.

The approximately 30,000 known species of ascomycetes can be divided into two broad groups, depending on whether the asci are contained within a specialized fruiting structure. Species that have this fruiting structure, the **ascocarp**, are collectively called **euascomycetes** ("true ascomycetes"); those without ascocarps are called **hemiascomycetes** ("half ascomycetes").

Chytridiomycota

Zygomycota

Ascomycota

Basidiomycota

HEMIASCOMYCETES. Most hemiascomycetes are microscopic, and many species are unicellular. Perhaps the best known are the ascomycete yeasts, especially baker's or brewer's yeast (*Saccharomyces cerevisiae*; see Figure 30.2). These yeasts are among the most important domesticated fungi. *S. cerevisiae* metabolizes glucose obtained from its environment to ethanol and carbon dioxide. It forms carbon dioxide bubbles in bread dough and gives baked bread its light texture. Although baked away in bread making, the ethanol and carbon dioxide are both retained in beer. Other yeasts live on fruits such as figs and grapes and play an important role in the making of wine.

Hemiascomycete yeasts reproduce asexually either by fission or by **budding** (the outgrowth of a new cell from the surface of an old one; see Figure 30.2). Sexual reproduction takes place when two adjacent haploid cells of opposite mating types fuse. (We discussed the genetics of yeast mating types in Chapter 14.) In some species, the resulting zygote buds to form a diploid cell population; in others, the zygote nucleus undergoes meiosis immediately. When these diploid nuclei undergo meiosis, the entire cell becomes an ascus. Depending on whether the products of meiosis then undergo mitosis, a yeast ascus usually has either eight or four **ascospores** (see Figure

Ascus

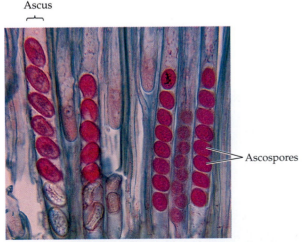

Ascospores

30.10 Asci and Ascospores
The ascomycetes are characterized by the production of ascospores within sacs called asci. Ascospores are the products of meiosis followed by a single mitotic division. Ascospores and asci do not mature all at once, and they may abort, so not every ascus in this micrograph contains eight mature ascospores.

(a) *Morchella esculenta*

(b) *Sarcoscypha coccinea*

30.11 Two Ascomycetes
(a) Morels, which have spongelike caps and a subtle flavor, are considered a delicacy by humans. The brilliant red cups in (b) are cup fungi, as are the three yellow morels in (a).

30.10). The ascospores germinate to become haploid cells. Hemiascomycetes have no dikaryon stage.

Yeasts, especially *Saccharomyces cerevisiae*, are frequently used in molecular biological research. Just as *E. coli* is the best-studied prokaryote, *S. cerevisiae* is the most completely studied eukaryote.

EUASCOMYCETES. The euascomycetes include the filamentous fungi known as molds. Among them are several common molds, including *Neurospora*, the pink molds, one of which Beadle and Tatum used in their pioneering work on biochemical genetics (see Figure 12.1). Many euascomycetes are parasites on higher plants. Chestnut blight and Dutch elm disease are caused by euascomycetes. The powdery mildews are euascomycetes that infect cereal grains, lilacs, and roses, among many other plants. They can be a serious problem to grape growers, and a great deal of research has focused on ways to control these agricultural pests.

The euascomycetes also include the cup fungi (Figure 30.11*a* and *b*). In most of these organisms the fruiting structures are cup-shaped and can be as large as several centimeters across. The inner surfaces of the cups are covered with a mixture of both sterile filaments and asci, and they produce huge numbers of spores. Although these fleshy structures appear to be composed of distinct tissue layers, microscopic examination shows that their basic organization is still filamentous—a tightly woven mycelium.

Two particularly delicious cup fungus fruiting structures are morels (Figure 30.11*a*) and truffles. Truffles grow underground, in a mutualistic association with the roots of some species of oaks. Europeans traditionally used pigs to find truffles because some truffles secrete a substance that has

an odor similar to a pig's sex attractant. Unfortunately, pigs also eat truffles, so dogs are now the usual truffle hunters.

Penicillium is a genus of green molds, of which some species produce the antibiotic penicillin, presumably for defense against competing bacteria. Two species, *P. camembertii* and *P. roquefortii*, are the organisms responsible for the characteristic flavors of Camembert and Roquefort cheeses, respectively.

Brown molds of the genus *Aspergillus* are important in some human diets. *A. tamarii* acts on soybeans in the production of soy sauce, and *A. oryzae* is used in brewing the Japanese alcoholic beverage sake. Some species of *Aspergillus* that grow on nuts such as peanuts and pecans produce extremely carcinogenic (cancer-inducing) compounds called aflatoxins.

The euascomycetes reproduce asexually by means of mating structures called conidia that form at the tips of specialized hyphae (Figure 30.12). Small chains of conidia are

30.12 Conidia
Chains of conidia are developing on stalks called hyphae arising from this powdery mildew growing on a leaf.

Erysiphe sp.

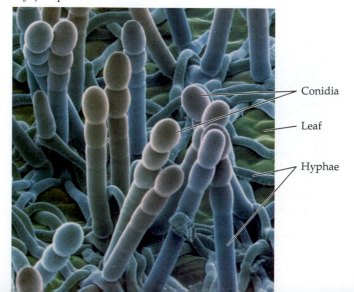

Conidia

Leaf

Hyphae

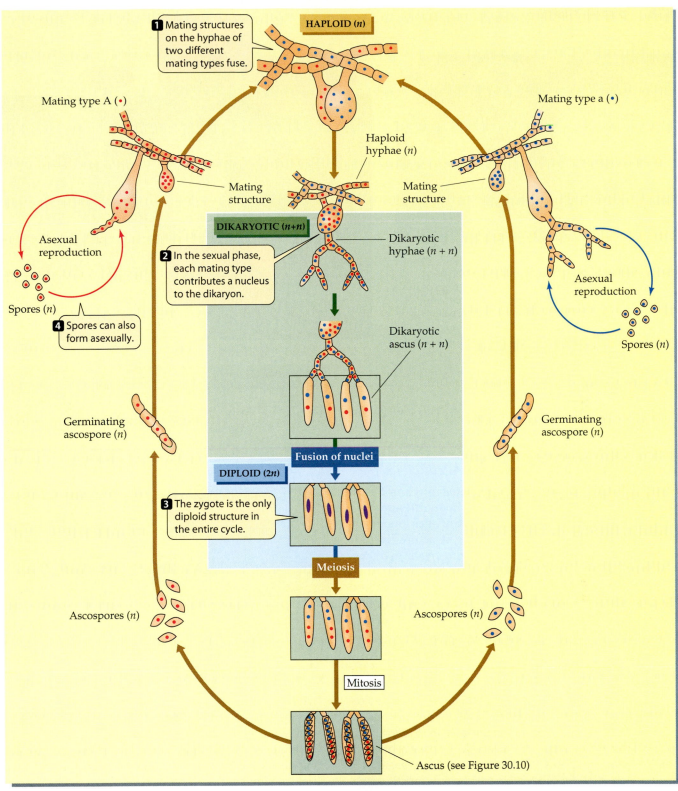

1 Mating structures on the hyphae of two different mating types fuse.

HAPLOID (*n*)

Mating type A (•)

Haploid hyphae (*n*)

Mating structure

Mating type a (•)

Mating structure

Asexual reproduction

DIKARYOTIC (*n+n*)

Dikaryotic hyphae (*n + n*)

2 In the sexual phase, each mating type contributes a nucleus to the dikaryon.

Spores (*n*)

4 Spores can also form asexually.

Asexual reproduction

Spores (*n*)

Dikaryotic ascus (*n + n*)

Germinating ascospore (*n*)

Germinating ascospore (*n*)

Fusion of nuclei

DIPLOID (*2n*)

3 The zygote is the only diploid structure in the entire cycle.

Meiosis

Ascospores (*n*)

Ascospores (*n*)

Mitosis

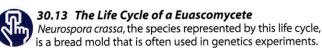

Ascus (see Figure 30.10)

30.13 The Life Cycle of a Euascomycete
Neurospora crassa, the species represented by this life cycle, is a bread mold that is often used in genetics experiments.

produced by the millions and can survive for weeks in nature. The conidia are what give molds their characteristic colors.

The sexual reproductive cycle of euascomycetes includes the formation of a dikaryon (Figure 30.13). Most euascomycetes form mating structures, some "female" and some "male."

Nuclei from a male structure on one hypha enter a female mating structure on a hypha of a compatible mating type. Dikaryotic *ascogenous* (ascus-forming) hyphae de-

velop from the now dikaryotic female mating structure. The introduced nuclei divide simultaneously with the host nuclei. Eventually asci form at the tips of the ascogenous hyphae. Only with the formation of asci do the nuclei finally fuse. Both nuclear fusion and the subsequent meiosis of the resulting diploid nucleus take place within individual asci. The meiotic products are incorporated into as-

(a) *Lycoperdon perlatum*

30.14 Basidiomycete Fruiting Structures
The fruiting structures of the basidiomycetes are probably the most familiar structures produced by fungi.
(a) When raindrops hit them, these puffballs will release
clouds of spores for dispersal. (b) A member of a highly
poisonous mushroom genus, *Amanita*. (c) This edible
bracket fungus is parasitizing a tree.

(b) *Amanita muscaria*

(c) *Laetiporus sulphureus*

cospores that are ultimately shed by the ascus to begin the
new haploid generation.

The sexual reproductive structure of basidiomycetes is a basidium

About 25,000 species of **basidiomycetes** (phylum **Basidiomycota**) have been described. Basidiomycetes produce
some of the most spectacular fruiting structures found anywhere among the fungi. These fruiting structures include
puffballs (which may be more than half a meter in diameter), mushrooms of all kinds, and the giant bracket fungi
often encountered on trees and fallen logs in a damp forest
(Figure 30.14). There are more than 3,250 species of mushrooms, including the familiar *Agaricus bisporus* you may
enjoy on your pizza, as well as poisonous species, such as
members of the genus *Amanita*. Bracket fungi do great
damage to cut lumber and stands of timber. Some of the
most damaging plant pathogens are basidiomycetes, including the smut fungi (see Figure 30.1a) that parasitize cereal grains. In contrast, other basidiomycetes contribute to
the well-being of plants as fungal partners in mycorrhizae.

Basidiomycete hyphae characteristically have septa with
small, distinctive pores. The **basidium** (plural basidia), a
swollen cell at the tip of a hypha, is the characteristic sexual
reproductive structure of
the basidiomycetes.
It is the site of nu-
clear fusion and meiosis.
Thus, the basidium plays
the same role in the basidiomycetes as the ascus does in the
ascomycetes and the zygosporangium
does in zygomycetes.

Chytridiomycota

Zygomycota

Ascomycota

Basidiomycota

The life cycle of the basidiomycetes is shown in Figure
30.15. After nuclei fuse in the basidium, the resulting

diploid nucleus undergoes meiosis, and the four resulting
haploid nuclei are incorporated into haploid **basidiospores**,
which form on tiny stalks. These basidiospores typically are
forcibly discharged from their basidia and then germinate,
giving rise to haploid hyphae. As these hyphae grow, haploid hyphae of different mating types meet and fuse, forming dikaryotic hyphae, each cell of which contains two nuclei, one from each parent hypha. The dikaryotic mycelium
grows and eventually produces fruiting structures. The
dikaryotic phase may persist for years—some basidiomycetes live for decades or even centuries.

The elaborate fruiting structure of some fleshy basidiomycetes, such as the gill mushroom in Figure 30.15, is
topped by a cap, or *pileus*, which has structures called *gills*
on its underside. Enormous numbers of basidia develop on
the surfaces of the gills. The basidia discharge their spores
into the air spaces between adjacent gills, and the spores
sift down into air currents for dispersal and germination as
new haploid mycelia.

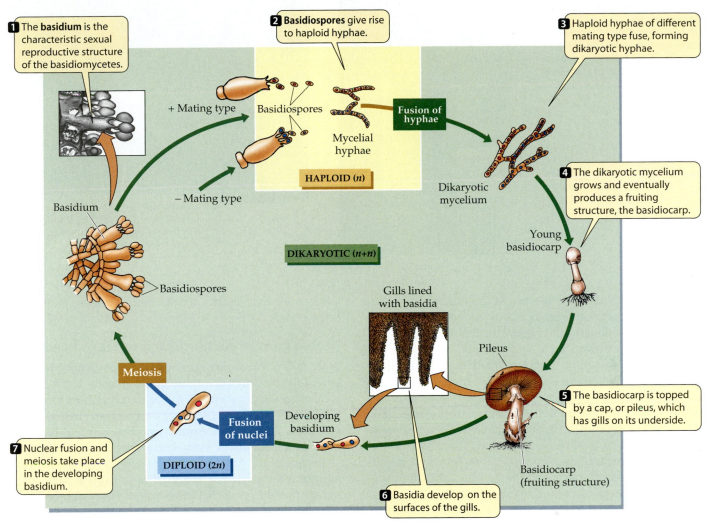

1 The **basidium** is the characteristic sexual reproductive structure of the basidiomycetes.

2 Basidiospores give rise to haploid hyphae.

3 Haploid hyphae of different mating type fuse, forming dikaryotic hyphae.

+ Mating type

Basidiospores

Fusion of hyphae

– Mating type

Mycelial hyphae

HAPLOID (n)

Basidium

Dikaryotic mycelium

4 The dikaryotic mycelium grows and eventually produces a fruiting structure, the basidiocarp.

Young basidiocarp

Basidiospores

DIKARYOTIC (n+n)

Gills lined with basidia

Pileus

Meiosis

5 The basidiocarp is topped by a cap, or pileus, which has gills on its underside.

Fusion of nuclei

Developing basidium

DIPLOID (2n)

Basidiocarp (fruiting structure)

7 Nuclear fusion and meiosis take place in the developing basidium.

6 Basidia develop on the surfaces of the gills.

30.15 The Basidiomycete Life Cycle
Basidiospores form on tiny stalks and are then forcibly dispersed to germinate into haploid hyphae, from which the familiar fruiting structure eventually grows.

Imperfect fungi lack a sexual stage

Mechanisms of sexual reproduction readily distinguish members of the four phyla of fungi from one another. But many fungi, including both saprobes and parasites, lack sexual stages entirely; presumably these stages have been lost during evolution or have not yet been found. Classifying these fungi as belonging to any of the four major phyla was at one time difficult, but biologists now can classify most such fungi on the basis of DNA sequences.

Fungi that have not yet been placed in any of the existing phyla are grouped together as the imperfect fungi, or deuteromycetes. Thus, the deuteromycete group is a holding area for species whose status is yet to be resolved. At present, about 25,000 species are classified as imperfect fungi. Some taxonomists, preferring to emphasize convenience of identification over strict phylogenetic considerations, treat the imperfect fungi as a paraphyletic phylum, Deuteromycota.

If sexual structures are found on a fungus classified as a deuteromycete, the fungus is reassigned to the appropriate phylum. That happened, for example, with a fungus that produces plant growth hormones called gibberellins. Originally classified as the deuteromycete *Fusarium moniliforme*, this fungus was later found to produce asci, whereupon it was renamed *Gibberella fujikuroi* and transferred to the phylum Ascomycota.

Fungal Associations

Earlier in this chapter we mentioned mycorrhizae and lichens, in which fungi live in intimate association with other organisms. Now that we have learned a bit about fungal diversity, let's consider mycorrhizae and lichens in greater detail.

Mycorrhizae are essential to many plants

Almost all tracheophytes enjoy a mutually beneficial symbiotic association with fungi. Unassisted, the root hairs of such plants do not absorb enough water or minerals to sustain maximum growth. However, the roots usually become infected with fungi, forming an association called a mycorrhiza.

30.16 Mycorrhizal Associations

(a) Ectomycorrhizal fungi wrap themselves around the plant root, increasing the area available for absorption of water and nutrients. (b) Endomycorrhizae infect the root internally.

(a)

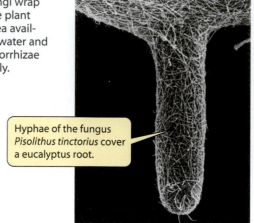

Hyphae of the fungus *Pisolithus tinctorius* cover a eucalyptus root.

(b)

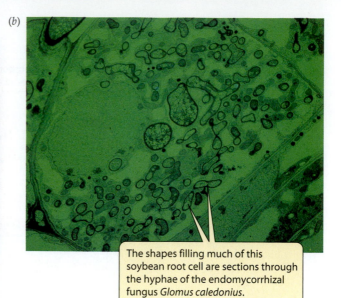

The shapes filling much of this soybean root cell are sections through the hyphae of the endomycorrhizal fungus *Glomus caledonius*.

In *ectomycorrhizae*, the fungus wraps around the root, and its mass is often as great as that of the root itself (Figure 30.16a). The hyphae of the fungi attached to the root increase the surface area for the absorption of water and minerals, and the mass of the mycorrhiza, like a sponge, holds water efficiently in the neighborhood of the root. Infected roots characteristically branch extensively and become swollen and club-shaped. In *endomycorrhizae*, the infection is internal to the root, with no hyphae visible on the root surface (Figure 30.16b).

The symbiotic fungus–plant association of a mycorrhiza is important to both partners. The fungus obtains important organic compounds, such as sugars and amino acids, from the plant. In return, the fungus greatly increases the absorption of water and minerals (especially phosphorus) by the plant. The fungus may also provide certain growth hormones, and may protect the plant against attack by microorganisms. Plants that have active mycorrhizae typically are a deeper green and may resist drought and temperature extremes better than plants of the same species that have little mycorrhizal development. Attempts to introduce some plant species to new areas have failed until a bit of soil from the native area (presumably containing the fungus necessary to establish mycorrhizae) was provided.

The partnership between plant and fungus results in a plant better adapted for life on land. It has been suggested that the evolution of this symbiotic association was the single most important step leading to the colonization of the terrestrial environment by living things. Fossils of mycorrhizal structures more than 300 million years old have been found, and some rocks dating back 460 million years contain structures that appear to be fossilized fungal spores.

Some liverworts, which are among the most ancient terrestrial plants (see Chapter 28), form mycorrhizae. Certain plants that live in nitrogen-poor habitats, such as cranberry bushes and orchids, invariably have mycorrhizae. Orchid seeds will not germinate in nature unless they are already infected by the fungus that will form their mycorrhizae. Plants that lack chlorophyll always have mycorrhizae, which they often share with the roots of green, photosynthetic plants.

Lichens grow where no eukaryote has succeeded

A lichen is not a single organism, but rather a meshwork of two radically different organisms: a fungus and a photosynthetic microorganism. Together the organisms constituting a lichen can survive some of the harshest environments on Earth. The flora of Antarctica, for example, features more than 100 times as many species of lichens as of plants.

In spite of this hardiness, lichens are very sensitive to air pollution because they are unable to excrete toxic substances that they absorb. Hence they are not common in industrialized cities. Because of their sensitivity, lichens are good biological indicators of air pollution.

The fungal components of most lichens are ascomycetes, but some are basidiomycetes or imperfect fungi (only one zygomycete serving as the fungal component of a lichen has been reported). The photosynthetic component may be either a cyanobacterium or a unicellular green alga. Relatively little experimental work has focused on lichens, perhaps because they grow so slowly—typically less than 1 centimeter per year.

There are about 13,500 "species" of lichens; their fungal components may constitute as many as 20 percent of all fungal species. Lichens are found in all sorts of exposed habitats: on tree bark, open soil, or bare rock. Reindeer "moss" (actually not a moss at all, but the lichen *Cladonia subtenuis*) covers vast areas in arctic, subarctic, and boreal regions, where it is an important part of the diets of reindeer and other large mammals. Lichens come in various forms and colors. *Crustose* (crustlike) lichens look like colored powder dusted over their substrate (Figure 30.17a); *foliose* (leafy) and *fruticose* (shrubby) lichens may have complex forms (Figure 30.17b).

The most widely held interpretation of the lichen relationship is that it is a type of mutually beneficial symbiosis. The hyphae of the fungal mycelium are tightly pressed against the photosynthetic cells of the alga or cyanobacterium and sometimes even invade them. The bacterial or

Foliose   Crustose

(a)

(b)

30.17 Lichen Body Forms
Lichens fall into three principal classes based on their body form. (a) Foliose and crustose lichens grow on otherwise bare rock. (b) A miniature jungle of fruticose lichens.

algal cells not only survive these indignities, but continue their growth and photosynthesis. In fact, algal cells in a lichen "leak" photosynthetic products at a greater rate than do similar cells growing on their own. On the other hand, photosynthetic cells from lichens grow more rapidly on their own than when combined with a fungus. On this basis, we could consider lichen fungi as parasitic on their photosynthetic partners.

Lichens can reproduce simply by fragmentation of the vegetative body, which is called the *thallus*, or by means of

specialized structures called **soredia** (singular soredium). Soredia consist of one or a few photosynthetic cells surrounded by fungal hyphae (Figure 30.18a). The soredia become detached, are dispersed by air currents, and upon arriving at a favorable location, develop into a new lichen. Alternatively, if the fungal partner is an ascomycete or a basidiomycete, it may go through its sexual cycle, producing either ascospores or basidiospores. When these spores are discharged, however, they disperse alone, unaccompanied by the photosynthetic partner, and thus may not be capable of reestablishing the lichen association. Nevertheless, many lichens produce characteristic fruiting structures containing asci or basidia.

Visible in a cross section of a typical foliose lichen are a tight upper region of fungal hyphae, a layer of cyanobacte-

30.18 Lichen Anatomy
(a) Soredia of a fruticose lichen. (b) Cross section showing the layers of a foliose lichen.

(a)   (b)

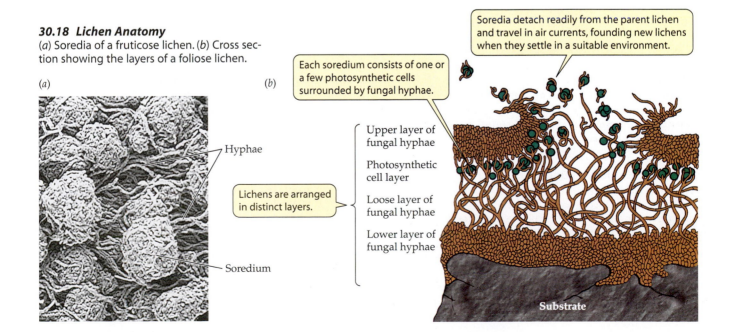

Hyphae

Soredium

Lichens are arranged in distinct layers.

Each soredium consists of one or a few photosynthetic cells surrounded by fungal hyphae.

Soredia detach readily from the parent lichen and travel in air currents, founding new lichens when they settle in a suitable environment.

Upper layer of fungal hyphae

Photosynthetic cell layer

Loose layer of fungal hyphae

Lower layer of fungal hyphae

Substrate

ria or algae, a looser hyphal layer, and finally hyphal rhizoids that attach the whole structure to its substrate (Figure 30.18b). The meshwork of fungal hyphae takes up some nutrients needed by the photosynthetic cells and provides a suitably moist environment for them by holding water tenaciously. The fungi derive fixed carbon from the photosynthesis of the algal or cyanobacterial cells.

Lichens are often the first colonists on new areas of bare rock. They satisfy most of their needs from the air and from rainwater, augmented by minerals absorbed from dust. A lichen begins to grow shortly after a rain, as it begins to dry. As it grows, the lichen acidifies its environment slightly, and this acid contributes to the slow breakdown of rocks, an early step in soil formation. After further drying, the lichen's photosynthesis ceases. The water content of the lichen may drop to less than 10 percent of its dry weight, at which point it becomes highly insensitive to extremes of temperature.

Whether living on their own or in symbiotic associations, fungi have spread successfully over much of Earth since their origin from a protist ancestor. That ancestor also gave rise to the choanoflagellates and the animal kingdom, the group we'll consider in the next three chapters.

Chapter Summary

▶ Fungi are the principal degraders of dead organic matter in the biosphere. Fungi are nutritional partners of almost all vascular plants. Some fungi are serious pathogens of plants and animals, including humans.

General Biology of the Fungi

▶ Fungi are heterotrophic eukaryotes with absorptive nutrition. They may be saprobes, parasites, or mutualists. **Review Figure 30.4**

▶ The yeasts are unicellular.

▶ The bodies of other fungi are composed of chitinous-walled, multinucleate hyphae, often massed to form a mycelium. The filamentous hyphae give fungi a large surface area-to-volume ratio, enhancing their ability to absorb nutrients. The hyphae usually have incomplete partitions (septa) that do not divide them into separate cells. **Review Figure 30.3**

▶ Fungi reproduce asexually by means of spores formed within sporangia, by conidia formed at the tips of hyphae, by budding, or by fragmentation.

▶ Fungi reproduce sexually when hyphae or motile cells of different mating types meet and fuse.

▶ In addition to the haploid and diploid states, many fungi demonstrate a third nuclear condition: the dikaryotic, or $n + n$, state. **Review Figure 30.13**

Diversity in the Kingdom Fungi

▶ The kingdom Fungi consists of four phyla: Chytridiomycota, Zygomycota, Ascomycota, and Basidiomycota. These phlya differ in their reproductive structures, mechanisms of spore formation, and less importantly, the presence and form of septa in their hyphae. **Review Figure 30.6, Table 30.1**

▶ Chytrids, with their flagellated zoospores and gametes, probably resemble the ancestral fungi.

▶ Zygomycetes reproduce sexually by fusion of gametangia. **Review Figure 30.9**

▶ The sexual reproductive structure of ascomycetes is an ascus containing ascospores. The ascomycetes are divided into two groups, euascomycetes and hemiascomycetes, on the basis of whether they have an ascocarp, or fruiting structure. **Review Figure 30.13**

▶ The sexual reproductive structure of basidiomycetes is a basidium, a swollen cell bearing basidiospores. **Review Figure 30.15**

▶ Imperfect fungi (deuteromycetes) lack sexual structures, but DNA sequencing can sometimes identify the phylum to which they belong.

Fungal Associations

▶ Mycorrhizae, associations of fungi with plant roots, enhance the ability of the roots to absorb water and nutrients.

▶ Lichens, mutualistic combinations of a fungus with a cyanobacterium or a green alga, are found in some of the most inhospitable environments on the planet. **Review Figure 30.18**

For Discussion

1. You are shown an object that looks superficially like a pale green mushroom. Describe at least three criteria (including anatomical and chemical traits) that would enable you to tell whether the object is a piece of a plant or a piece of a fungus.

2. Differentiate among the members of the following pairs of related terms:
   a. hypha/mycelium
   b. euascomycete/hemiascomycete
   c. ascus/basidium
   d. ectomycorrhiza/endomycorrhiza

3. For each type of organism listed below, give a single characteristic that may be used to differentiate it from the other, related organism(s) in parentheses.
   a. Zygomycota (Ascomycota)
   b. Basidiomycota (deuteromycetes)
   c. Ascomycota (Basidiomycota)
   d. baker's yeast (*Neurospora crassa*)

4. Many fungi are dikaryotic during part of their life cycle. Why are dikaryons described as $n + n$ instead of $2n$?

5. If all the fungi on Earth were suddenly to die, how would the surviving organisms be affected? Be thorough and specific in your answer.

6. How might the first mycorrhizae have arisen?

7. What might account for the ability of lichens to withstand the intensely cold environment of Antarctica? Be specific in your answer.

31 *Animal Origins and Lophotrochozoans*

IN 1822, NEARLY 40 YEARS BEFORE DARWIN wrote *The Origin of Species*, French naturalist E. Geoffroy Saint-Hilaire was examining a lobster. He noticed that when he viewed the lobster with its ventral surface up, its central nervous system was located above its digestive tract, which in turn was located above its heart—the same relative positions these systems have in mammals viewed *dorsally*. His observation led Saint-Hilaire to conclude that the differences between arthropods and vertebrates could be explained if the embryos of one of those groups had been inverted during development.

Saint-Hilaire's suggestion was regarded as totally preposterous at the time and was largely dismissed until recently. However, the discovery of two genes that influence a system of extracellular signals involved in development has lent new support to Saint-Hilaire's seemingly outrageous hypothesis.

A vertebrate gene called *chordin* helps establish cells on one side of the embryo as dorsal, the other as ventral. A probably homologous gene in fruit flies,* called *sog*, acts in a similar manner, but has the opposite effect. Fly cells where *sog* is active become ventral, whereas vertebrate cells expressing *chordin* become dorsal (see Figure 16.19). However, when *sog* mRNA is injected into the frog *Xenopus*, a vertebrate, it causes dorsal development. *Chordin* mRNA injected into flies promotes ventral development. In both cases, injection of the mRNA promotes the development of the portion of the embryo that contains the central nervous system!

Chordin and *sog* are among the many genes that appear to regulate similar functions in very different organisms. There

*Insects (such as fruit flies) are arthropods and belong to the same evolutionary lineage as crustaceans (such as the lobster), as we will discuss in Chapter 32.

Genes that Control Development
The human and the lobster carry similar genes that control the development of the body axis. A lobster's nervous system runs up its ventral (belly) surface, while its circulatory system is dorsal (down its back). In vertebrates such as humans, similar genes position these two systems inversely to those of the lobster.

are several almost universal animal genes that help transform a single-celled egg into a multicellular adult. Such genes are providing evolutionary biologists with information that can help them understand relationships among animal lineages that separated from one another in ancient times. As we saw in Chapter 23, new knowledge about gene functions and gene sequences provides some of the most powerful data being used in modern phylogenetic investigations to infer evolutionary relationships among organisms.

In this chapter we will first discuss how biologists infer evolutionary relationships among animals and review the defining characteristics of the animal way of life. Then we'll describe several lineages of simple animals. Finally, we'll describe the lophotrochozoans, one of the three great evolutionary lineages of animals. The next two chapters will discuss the other two great animal lineages, the ecdysozoans and the deuterostomes.

Descendants of a Common Ancestor

Biologists have long debated whether animals arose once or several times from protist ancestors, but enough molecular and morphological evidence has now been assembled to indicate that, with the possible exception of sponges (Porifera), the kingdom Animalia is a monophyletic group—that is, all animals are descendants of a single ancestral lineage.

This conclusion is supported by the fact that all animals share a set of derived traits:

▶ Similarities in their 5S and 18S ribosomal RNAs
▶ Special types of cell–cell junctions: tight junctions, desmosomes, and gap junctions (see Figure 5.6)
▶ A common set of extracellular matrix molecules, including collagen (see Figure 4.28)

Animals evolved from ancestral colonial flagellated protists as a result of division of labor among their aggregated cells. Within these ancestral colonies of cells—perhaps analogous to those still existing in the chlorophyte *Volvox* or some colonial choanoflagellates (see Figures 27.25a and 27.28)—some cells became specialized for movement, others for nutrition, and still others differentiated into gametes. Once the division of labor had begun, these units continued to differentiate while improving their coordination with other working groups of cells. Such coordinated groups of cells evolved into the larger and more complex organisms that we now call animals.

The Animal Way of Life

What traits characterize the organisms we call animals? Animals are multicellular organisms that must take in preformed organic molecules because they cannot synthesize them from inorganic chemicals. They acquire these organic molecules by ingesting other organisms, either living or dead, and digesting them inside their bodies. To acquire these organic molecules, animals must expend energy to move themselves through the environment to find food, to position themselves where food will pass by them, or to move the environment and the food it contains to them.

The foods animals eat include most other members of the animal kingdom, as well as members of all other evolutionary lineages. Much of the diversity of animal sizes and shapes evolved as animals acquired the ability to capture and eat many different kinds of foods, and to avoid becoming food for other animals.

The need to move in search for food has favored sensory structures that provide animals with detailed information about their environment, and nervous systems able to receive and coordinate this information. Consequently, most animals are behaviorally much more complex than plants. Because animals ingest chemically complex foods, they expend considerable energy to maintain relatively constant internal conditions while taking in foods that vary chemically.

A real appreciation of animal structure and functioning is best achieved through firsthand experience in the field and laboratory. The accounts in this chapter and the following two serve as an orientation to the major groups of animals, their similarities and differences, and the evolutionary pathways that resulted in the current richness of animal evolutionary lineages and species. But how do biologists infer evolutionary relationships among animals?

Clues to Evolutionary Relationships among Animals

Biologists use a variety of traits to infer animal phylogenies. As we discussed in Chapters 23 and 24, clues to these relationships are found in the fossil record, in patterns of embryonic development, in the comparative morphology and physiology of living and fossil animals, and in the structure of their molecules.

Patterns of early development evolved very slowly in some animal lineages. For this reason, biologists have traditionally based their classifications of the major lineages of animals on developmental patterns. More recently, comparative molecular data from small subunit rRNA and mitochondrial genes also have been used. These two types of evidence suggest similar animal phylogenies.

31.1 Animal Body Cavities
The three major types of animal body cavities. (a) Acoelomates do not have enclosed body cavities. (b) Pseudocoelomates have only one layer of muscle, lying outside the body cavity. (c) Coelomates have a peritoneum surrounding the internal organs; the body cavities of some, such as this earthworm, are segmented.

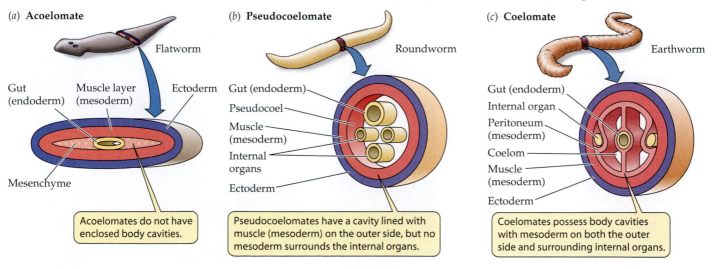

(a) **Acoelomate**   Flatworm

Gut (endoderm)   Muscle layer (mesoderm)   Ectoderm

Mesenchyme

Acoelomates do not have enclosed body cavities.

(b) **Pseudocoelomate**   Roundworm

Gut (endoderm)
Pseudocoel
Muscle (mesoderm)
Internal organs
Ectoderm

Pseudocoelomates have a cavity lined with muscle (mesoderm) on the outer side, but no mesoderm surrounds the internal organs.

(c) **Coelomate**   Earthworm

Gut (endoderm)
Internal organ
Peritoneum (mesoderm)
Coelom
Muscle (mesoderm)
Ectoderm

Coelomates possess body cavities with mesoderm on both the outer side and surrounding internal organs.

Using this wide variety of comparative data, zoologists have concluded that the sponges, cnidarians, and ctenophores separated from the remaining animal lineages early in evolutionary history. They have divided the remaining animals into two major lineages: the **protostomes** and the **deuterostomes**.

In the common ancestor of the protostomes and the deuterostomes, the pattern of early cell division in the fertilized egg—called *cleavage*—was radial. During *radial cleavage*, cells divide along a plane either parallel to or at right angles to the long axis of the fertilized egg. This pattern persisted during the evolution of deuterostomes and in many protostome lineages, but *spiral cleavage* evolved in one major protostome lineage. In spiral cleavage, the plane of cell division is oblique to the long axis of the egg, causing the cells to be arranged in a spiral pattern.

Other developmental patterns typically differ between protostomes and deuterostomes. Cleavage of the fertilized egg in protostomes is *determinate;* that is, if the egg is allowed to divide a few times and the cells are then separated, each cell develops into only a partial embryo. In contrast, cleavage in deuterostomes typically is *indeterminate;* cells separated after several cell divisions can still develop into complete embryos. We see this phenomenon in humans in identical twins. Among deuterostomes, the mouth of the embryo originates some distance away from the embryonic structure called the blastopore, which becomes the anus. Among protostomes, the mouth arises from or near the blastopore.

During development from a single-celled zygote to a multicellular adult, animals form layers of cells. The embryos of **diploblastic** animals have only two cell layers: an outer *ectoderm* and an inner *endoderm*. The embryos of **triploblastic** animals have a third layer, the *mesoderm*, which lies between the ectoderm and the endoderm.

Fluid-filled spaces, called **body cavities**, lie between the cell layers of the bodies of many kinds of animals. The type of body cavity an animal has strongly influences how it can move.

- ▶ Animals that lack an enclosed body cavity are called **acoelomates** In these animals, the space between the gut and the body wall is filled with masses of cells called *mesenchyme* (Figure 31.1*a*).
- ▶ Another group of animals, the **pseudocoelomates**, have a body cavity called the *pseudocoel*. The pseudocoel is a liquid-filled space in which many of the body organs

are suspended, but control over body shape is crude because a pseudocoel has muscles only on the outside (Figure 31.1*b*).

- ▶ **Coelomate** animals have a *coelom*, a body cavity that develops within the embryonic mesoderm. It is lined with a special structure called the *peritoneum*, and has muscles both inside and outside. The internal organs of coelomates are slung in pouches of peritoneum rather than being suspended within the body cavity (Figure 31.1*c*).

An animal with a coelom has better control over the movement of the fluids it contains, but control is limited if the animal has only a single, large body cavity. Control is improved if the coelom is separated into compartments or segments so that circular and longitudinal muscles in each individual segment can change its shape independently of the other segments. Segmentation of the coelom evolved several different times among both protostomes and deuterostomes.

The phylogeny of animals we adopt in this book is based on analyses of many developmental, structural (from both living and fossil animals), and molecular traits. Figure 31.2 shows the postulated order of splitting of the major lineages in animal evolution. New information continues to modify and refine our understanding of the details of phylogenetic relationships among animals. Nonetheless, the division of the animals into the lineages shown here is supported by many types of data.

Body Plans Are Basic Structural Designs

The entire structure of an animal, its organ systems, and the integrated functioning of its parts are known as its **body plan**. Animals in many (but not all) lineages have evolved greater body complexity over time.

A fundamental aspect of an animal's body plan is its overall shape, described as its **symmetry**. A symmetrical animal can be divided along at least one plane into similar halves. Animals that have no plane of symmetry are said to

31.2 A Probable Phylogeny of Animals
The evolutionary tree that we will use in this chapter and the following two postulates that animals are monophyletic. The traits highlighted by red circles on the tree will be explained as we discuss the different phyla.

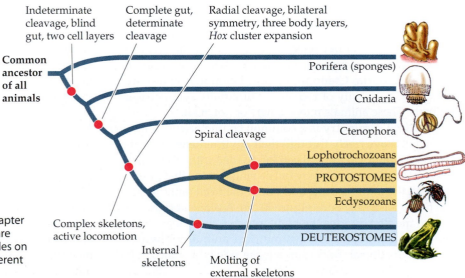

Indeterminate cleavage, blind gut, two cell layers

Complete gut, determinate cleavage

Radial cleavage, bilateral symmetry, three body layers, *Hox* cluster expansion

Common ancestor of all animals

Spiral cleavage

Porifera (sponges)

Cnidaria

Ctenophora

Lophotrochozoans

PROTOSTOMES

Ecdysozoans

DEUTEROSTOMES

Complex skeletons, active locomotion

Internal skeletons

Molting of external skeletons

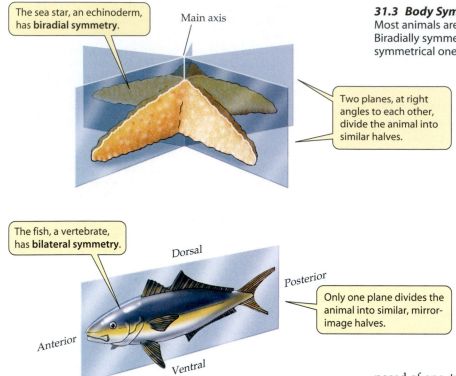

The sea star, an echinoderm, has **biradial symmetry.**

Main axis

Two planes, at right angles to each other, divide the animal into similar halves.

The fish, a vertebrate, has **bilateral symmetry.**

Dorsal

Posterior

Only one plane divides the animal into similar, mirror-image halves.

Anterior

Ventral

31.3 Body Symmetry
Most animals are either biradially or bilaterally symmetrical. Biradially symmetrical animals appear similar to radially symmetrical ones.

be **asymmetrical**. Many sponges are asymmetrical, but most animals have some kind of symmetry.

The simplest form of symmetry is **spherical symmetry**, in which body parts radiate out from a central point. An infinite number of planes passing through the central point can divide a spherically symmetrical organism into similar halves. Spherical symmetry is widespread among protists, but most animals possess other forms of symmetry.

An organism with **radial symmetry** has one main axis around which its body parts are arranged. A perfectly radially symmetrical animal can be divided into similar halves by any plane that contains the main axis. Some simple sponges and a few other animals, such as some sea anemones, have true radial symmetry.

Most radially symmetrical animals are modified such that only two planes, at right angles to each other, can divide them into similar halves. These animals are said to have **biradial symmetry** (Figure 31.3a). Three animal phyla—Cnidaria, Ctenophora, and Echinodermata—are composed primarily of radially or biradially symmetrical animals. These animals move slowly or not at all.

Bilateral symmetry is a common characteristic of animals that move freely through their environments. A bilaterally symmetrical animal can be divided into mirror images (left and right sides) by only a single plane that passes through the midline of its body from the front (anterior) to the back (posterior) end (Figure 31.3b). A plane at right angles to the first divides the body into two dissimilar sides; the side of a bilaterally symmetrical animal without a mouth is its dorsal (back) surface; the side with a mouth is its ventral (belly) surface.

Bilateral symmetry is strongly correlated with **cephalization**: the presence of a head, bearing sensory organs and central nervous tissues, at the anterior end of the animal. Cephalization may have been evolutionarily advantageous because the anterior end of a freely moving animal typically encounters new environments first.

Speed is often advantageous for both prey and the predators that pursue them. Fast-moving prey and predators had evolved by the early Cambrian period. To move rapidly, an animal needs some type of skeleton that supports its body and allows body parts to be moved relative to one another. A skeleton may be internal or external, rigid or flexible, and be composed of one, two, or more elements.

The fluid-filled body cavities of early animals functioned as **hydrostatic skeletons**. Because fluids are relatively incompressible, they move to another part of the cavity when muscles surrounding them contract. If the body tissues around the cavity are flexible, fluids moving from one region cause some other region to expand. Moving fluids can thus move specific body parts, or even the whole animal, provided that temporary attachments can be made to the substrate.

Other forms of skeletons developed in many animal lineages, either as substitutes for, or in combination with, hydrostatic skeletons. Some of these skeletons consist of a single element (snail shells); some have two elements (clam shells); others have many elements (centipedes). Some are internal (vertebrate bones); others are external (crab shells).

The form of the body cavity also changed in many animal lineages. Many became divided into compartments. The form of its skeleton and body cavities strongly influences the degree to which an animal can control and change its shape, and thus the complexity of the movements it can perform. What type of body plan and symmetry did the common ancestors of all animals possess? We are not certain, but because evidence suggests that animals evolved from colonies of flagellated cells, they may have been similar in structure to living flagellates.

Sponges: Loosely Organized Animals

The difference between protist colonies and simple multicellular animals is that the cells of animals are differentiated and their activities are coordinated. The lineage leading to modern sponges separated from the lineage leading to all other animals early during animal evolution. Some living

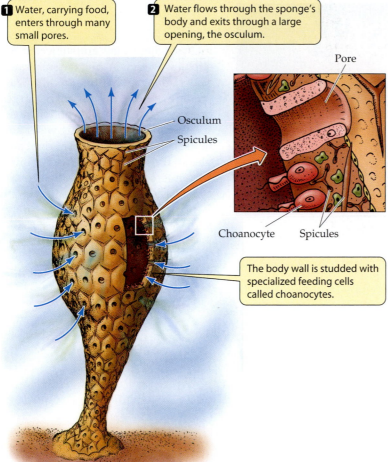

1 Water, carrying food, enters through many small pores.

2 Water flows through the sponge's body and exits through a large opening, the osculum.

Osculum

Spicules

Pore

Choanocyte   Spicules

The body wall is studded with specialized feeding cells called choanocytes.

31.4 The Sponge Body Plan
The flow of water through the sponge is shown by blue arrows. The body wall is studded with choanocytes, a type of specialized feeding cell that may be a link between animals and protists (see Figure 27.28).

The **sponges** (phylum **Porifera**, Latin for "pore bearers") are **sessile**. They live attached to the substrate and do not move about. The body plan of all sponges—even large ones, which may reach more than a meter in length—is an aggregation of cells built around a water canal system. A sponge feeds by drawing water into itself and filtering out the small organisms and nutrient particles that flow past the walls of its inner cavity.

Porifera
Cnidaria
Ctenophora
Lophotrochozoans
PROTOSTOMES
Ecdysozoans
DEUTEROSTOMES

Feeding cells with a collar and a flagellum, called *choanocytes*, line the inside of the water canals. By beating their flagella, the choanocytes cause water to flow into the animal, either by way of small pores that perforate special epidermal cells (in simple sponges) or through intercellular pores (in complex sponges). Water passes into small chambers within the body where food particles are captured by the choanocytes. Water then exits through one or more larger openings called *oscula* (Figure 31.4).

Between the thin epidermis and the choanocytes is a layer of cells, some of which are similar to amoebas and move about within the body. A supporting skeleton is also present, either in the form of simple or branching spines called *spicules* or as an elastic, often complex, network of fibers.

Most of the 10,000 species of sponges are marine animals; only about 50 species live in fresh water. Sponges come in a wide variety of sizes and shapes that are adapted to different patterns of water movement (Figure 31.5). Sponges living in intertidal or shallow subtidal environ-

sponges are still very similar to the probable ancestral colonial protists. Sponges are loosely organized. Even if a sponge is completely disassociated by being strained through a filter, its cells can reassociate into a new sponge.

(a) *Euplectella aspergillum*

(b) *Clathrina coriacea*

31.5 Sponges Differ in Size and Shape
(a) Glass sponges are named after their glasslike spicules, which are formed of silicon. (b) The spicules of this marine sponge are made of calcium carbonate. (c) The brown volcano sponge is typical of many simple marine sponges.

(c) *Aplysina fistularia*

(c) *Urticina lofotensis*

(a) *Gonothyraea loveni*

ments, where they are subjected to strong wave action, hug the substrate. Many sponges that live in calm waters are simple, with a single large opening on top of the body. Most sponges that live in flowing water are flattened and are oriented at right angles to the direction of current flow; they intercept water and the prey it contains as it flows past them.

Sponges reproduce both sexually and asexually. In most species, a single individual produces both eggs

(b) *Chrysaora fuscescens*

31.6 Diversity among Cnidarians
(a) The structure of the polyps on a North Atlantic coastal hydrozoan is visible here. (b) This sea nettle jellyfish illustrates the complexity of some scyphozoan medusae. (c) The nematocyst-studded tentacles of this white-spotted anemone from British Columbia are poised to capture large prey carried to the animal by water movement.

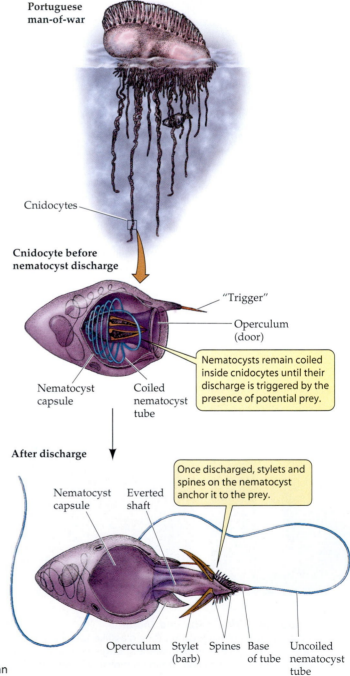

Portuguese man-of-war

Cnidocytes

Cnidocyte before nematocyst discharge

"Trigger"

Operculum (door)

Nematocyst capsule

Coiled nematocyst tube

Nematocysts remain coiled inside cnidocytes until their discharge is triggered by the presence of potential prey.

After discharge

Once discharged, stylets and spines on the nematocyst anchor it to the prey.

Nematocyst capsule

Everted shaft

Operculum   Stylet (barb)   Spines   Base of tube   Uncoiled nematocyst tube

31.7 Nematocysts Are Potent Weapons
Possessing a large number of nematocysts, cnidarians such as jellyfish can subdue and eat very large prey.

and sperm. Water currents carry sperm from one individual to another. Asexual reproduction is by budding and fragmentation.

Cnidarians: Cell Layers and Blind Guts

Animals in all phyla other than the Porifera have distinct cell layers and symmetrical bodies. The next lineage to split off from the main line of animal evolution after the sponges resulted in a phylum of animals—the **cnidarians** (phylum **Cnidaria**)—having only two cell layers (diploblastic) and a blind gut (with only one entrance). Within the constraints of this simple body organization, cnidarians evolved a wide variety of ways of making a living.

Cnidarians are simple but specialized carnivores

Cnidarians (phylum **Cnidaria**) appeared early in evolutionary history and radiated in the late Precambrian. About 10,000 cnidarian species—jellyfishes, sea anemones, corals, and hydrozoans—are living today (Figure 31.6). All but a few are marine. The smallest cnidarians can hardly be seen without a microscope; the largest known jellyfish is 2.5 meters in diameter. These animals are simple but specialized carnivores. The cnidarian body plan combines a low metabolic rate with the ability to capture large prey. These traits allow cnidarians to survive in environments where prey are scarce.

Porifera
Cnidaria
Ctenophora
Lophotrochozoans
PROTOSTOMES
Ecdysozoans
DEUTEROSTOMES

A key feature of cnidarians is tentacles that bear *cnidocytes*, specialized cells that contain stinging structures called *nematocysts* that can discharge toxins into their prey (Figure 31.7). Cnidocytes allow cnidarians to capture prey larger and more complex than themselves. Nematocysts are responsible for the sting that some jellyfishes and other cnidarians can inflict on human swimmers.

The mouth of a cnidarian is connected to a blind sac called the *gastrovascular cavity*, which functions in digestion, circulation, and gas exchange. The single opening serves as both mouth and anus. Cnidarians also have epithelial cells with muscle fibers whose contractions enable the animals to move, as well as nerve nets that integrate their body activities.

 ## Cnidarian life cycles

The generalized cnidarian life cycle has two distinct stages, the polyp and the medusa (Figure 31.8), although many species lack one of the stages.

▶ The sessile **polyp** stage has a cylindrical stalk attached to the substrate, with tentacles surrounding a mouth located at the opposite end from the site of attachment. This stage is usually asexual, but individual polyps may reproduce by budding, thereby forming a colony.

▶ The **medusa** is a free-swimming, sexual stage shaped like a bell or an umbrella. It typically floats with its mouth and tentacles facing downward. Medusae produce eggs and sperm and release them into the water. When an egg is fertilized, it develops into a free-swimming, ciliated larva called a **planula** that eventually settles to the bottom and transforms into a polyp.

Although the polyp and medusa stages appear very different, they share a similar body plan. A medusa is essentially a polyp without a stalk. Most of the outward differences between polyps and medusae are due to the *mesoglea*, a mass of jellylike material that lies between the two cell layers. The mesoglea contains few cells and has a low metabolic rate. In polyps, the mesoglea is usually thin; in medusae it is very thick, constituting the bulk of the animal.

HYDROZOANS. Life cycles are diverse among the **hydrozoans** (class **Hydrozoa**), a group containing the only freshwater cnidarians. The polyp commonly dominates the life cycle, but some species have only medusae and others only polyps. A few species have solitary polyps, but most hydrozoans are colonial. A single planula eventually gives rise to a colony of many polyps, all interconnected and sharing a continuous gastrovascular cavity (Figure 31.9). Within such a colony, some polyps have tentacles with many nematocysts; they capture prey for the colony. Others lack ten-

During the life cycle of many cnidarians, the usually sessile, asexual polyp alternates with the free-swimming, sexual medusa.

The mesoglea is a jellylike layer with few cells.

Medusa

Polyp

Tentacles
Mouth

Ectoderm
Mesoglea
Endoderm
Gastrovascular cavity

Mouth
Tentacles

As the positions of the mouth and tentacles indicate, the medusa is "upside down" from the polyp—or vice versa.

31.8 A Generalized Cnidarian Life Cycle
Cnidarians typically have two body forms, one asexual (the polyp) and the other sexual (the medusa).

31.9 Hydrozoans Often Have Colonial Polyps

The polyps with a hydrozoan colony may differentiate to perform specialized tasks.

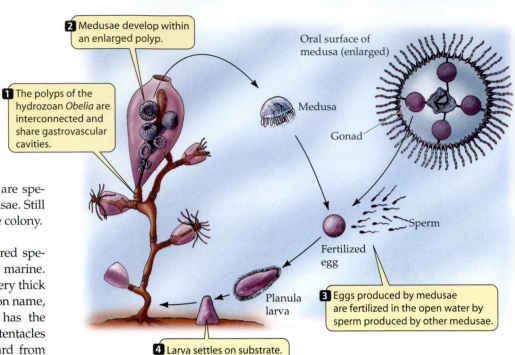

2 Medusae develop within an enlarged polyp.

1 The polyps of the hydrozoan *Obelia* are interconnected and share gastrovascular cavities.

Oral surface of medusa (enlarged)

Medusa

Gonad

Sperm

Fertilized egg

3 Eggs produced by medusae are fertilized in the open water by sperm produced by other medusae.

Planula larva

4 Larva settles on substrate.

tacles and are unable to feed, but are specialized for the production of medusae. Still others are fingerlike and defend the colony.

SCYPHOZOANS.   The several hundred species of the class **Scyphozoa** are all marine. The mesoglea of their medusae is very thick and firm, giving rise to their common name, jellyfishes. The medusa typically has the form of an inverted cup, and the tentacles with nematocysts extend downward from the margin of the cup.

The medusa, rather than the polyp, dominates the life cycle of scyphozoans. An individual medusa is male or female, releasing eggs or sperm into the open sea. The fertilized egg develops into a small planula that quickly settles on a substrate and changes into a small polyp. This polyp feeds and grows and may produce additional polyps by budding. After a period of growth, the polyp begins to bud off small medusae (Figure 31.10). These small medusae feed, grow, and transform themselves into adult medusae, which are commonly seen during summer in harbors and bays. Thus a polyp that grows from a single fertilized egg is capable of producing many genetically identical medusae that will eventually reproduce sexually.

ANTHOZOANS.   The roughly 6,000 species of sea anemones and corals that constitute the **anthozoans** (class **Anthozoa**) are all marine. Unlike other cnidarians, anthozoans entirely lack the medusa stage of the life cycle. The polyp produces eggs and sperm, and the fertilized egg develops into a planula that develops directly into another polyp. Many species can also reproduce asexually by budding or fission.

Sea anemones (see Figure 31.6c) are solitary. They are widespread in both warm and cold ocean waters. Many sea anemones are able to crawl slowly on the discs with which they attach themselves to the substrate. A few species can swim; some can burrow.

Corals, by contrast, are usually sessile and colonial. The polyps of most corals secrete a matrix of organic molecules upon which calcium carbonate—the eventual skeleton of the coral colony—is deposited. The forms of

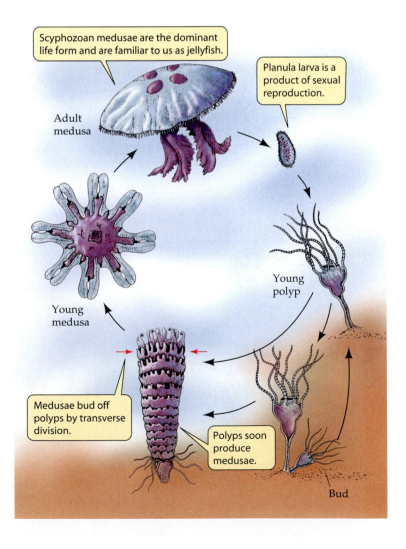

Scyphozoan medusae are the dominant life form and are familiar to us as jellyfish.

Planula larva is a product of sexual reproduction.

Adult medusa

Young polyp

Young medusa

Medusae bud off polyps by transverse division.

Polyps soon produce medusae.

Bud

31.10 Medusae Dominate Scyphozoan Life Cycles

Scyphozoan medusae are the familiar jellyfish of coastal waters. The small, sessile polyps quickly produce medusae.

(a)

(b)

31.11 Corals

(a) Many different species of corals and sponges grow together on this reef in the Bahama Islands. (b) The green plates of cabbage coral (*Turbinaria* sp.) and the branching staghorn coral (*Acrophora* sp.) are oriented to intercept sunlight in this Papua New Guinea reef.

coral skeletons are species-specific and highly diverse (Figure 31.11a). The common names of coral groups—horn corals, brain corals, staghorn corals, organ pipe corals, sea fans, and sea whips, among others—describe their appearance.

As a coral colony grows, old polyps die, but their calcareous skeletons remain. The living members form a layer on top of a growing reef of skeletal remains, eventually forming chains of islands and reefs. Corals are especially abundant in the Indo-Pacific region. The Great Barrier Reef along the northeastern coast of Australia is a system of coral formations more than 2,000 km long and as wide as 150 km. A continuous coral reef hundreds of kilometers long in the Red Sea has been calculated to contain more material than all the buildings in the major cities of North America combined.

Corals flourish in nutrient-poor, clear, tropical waters. For a long time scientists wondered how corals obtain enough nutrients to grow rapidly. The answer is that photosynthetic dinoflagellates live symbiotically within a coral's cells. They provide the corals with products of photosynthesis and contribute to calcium deposition. In turn, the corals protect the dinoflagellates from predators. This symbiotic relationship explains why reef-forming corals are restricted to clear surface waters, where light levels are high enough to allow photosynthesis (Figure 31.11b).

Coral reefs throughout the world are being threatened by both global warming, which is raising the temperatures of tropical shallow ocean waters, and nutrient runoff from developments on adjacent shorelines. An overabundance of

nitrogen gives an advantage to algae, which overgrow the corals and smother them.

Ctenophores: Complete Guts and Tentacles

Ctenophores (phylum **Ctenophora**) were the next lineage to separate from the lineage leading to all other animals. Ctenophores, also known as comb jellies, have body plans that are superficially similar to those of cnidarians. Both have two cell layers separated by a thick, gelatinous mesoglea, and both have radial symmetry and feeding tentacles. Like cnidarians, ctenophores have low metabolic rates because they are composed primarily of inert mesoglea. Unlike cnidarians, however, ctenophores have a complete gut. Food enters through a mouth and wastes are voided through two anal pores.

Ctenophores have eight comblike rows of fused plates of cilia, called *ctenes*. They move by beating these cilia rather than by use of muscular contractions. Ctenophoran tentacles do not have nematocysts; rather, they are covered with sticky filaments

Porifera

Cnidaria

Ctenophora

Lophotrochozoans
PROTOSTOMES
Ecdysozoans

DEUTEROSTOMES

to which prey adhere (Figure 31.12). After capturing its prey, a ctenophore retracts its tentacles to bring the food to its mouth. In some species, the entire surface of the body is coated with a sticky mucus that captures prey. All of the 100 known species of ctenophores are marine carnivores. They are common in open seas, where prey are often scarce. Most ctenophores cannot capture large prey.

Ctenophore life cycles are simple. Gametes from gonads located on the walls of the gastrovascular cavity are released into the cavity and then discharged through the

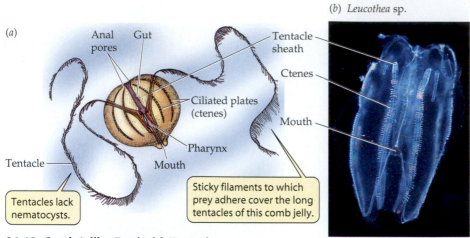

(b) Leucothea sp.

(a)

Anal pores
Gut

Ciliated plates (ctenes)

Pharynx

Tentacle

Mouth

Tentacles lack nematocysts.

Tentacle sheath

Ctenes

Mouth

Sticky filaments to which prey adhere cover the long tentacles of this comb jelly.

31.12 Comb Jellies Feed with Tentacles
(a) The body plan of a typical ctenophore. (b) This comb jelly has much shorter tentacles than many other ctenophores.

mouth or through pores. Fertilization takes place in the open seawater. In nearly all species, the fertilized egg develops directly into a miniature ctenophore that gradually grows into an adult.

The Evolution of Bilaterally Symmetrical Animals

The phylogenetic tree in Figure 31.2 postulates a common ancestor of all bilaterally symmetrical animals, but it does not tell us what that common ancestor looked like. Evolutionary biologists have attempted to infer the nature of those ancestral animals, which they call **urbilateria**, using evidence from the genes, development, and structure of existing animals.

One clue is provided by the fact that all living bilaterally symmetrical animals have an array of intercellular signaling systems and many homeobox gene families (see Chapter 16). The simplest bilaterally symmetrical animals have only a few homeobox genes, but some of them are shared with more complex animals. The mechanisms that regulate embryonic development in the protostomes and deuterostomes are governed by homologous homeobox genes. Such regulatory genes with similar functions are unlikely to have evolved independently in several different animal lineages.

Some evidence that urbilaterians may have been relatively complex is provided by fossilized traces of their movements. Fossilized trails from late Precambrian times exhibit complex search patterns, transverse furrows, and longitudinal ridges (Figure 31.13). They were made by organisms that were at least several centimeters long. The complexity of the movements recorded by the tracks suggests that urbilaterians had circulatory systems, systems of antagonistic muscles, and a tissue or fluid-filled body cavity. Some of their descendants subsequently lost some of those traits, but they retain signatures of their past in their genes.

Protostomes and Deuterostomes: An Early Lineage Split

The next major lineage split in the evolution of animals separated two groups that have been evolving separately ever since the Cambrian period. These two major lineages—the protostomes and deuterostomes—dominate today's biota. Members of both lineages are bilaterally symmetrical and have definite heads (cephalization). Because their skeletons and body cavities are more complex than those of the animals we have discussed so far, they are capable of more elaborate movements.

The most important shared, derived traits that unite the protostomes are a central nervous system consisting of an anterior brain that surrounds the entrance to the digestive tract; a ventral nervous system consisting of paired or fused longitudinal nerve cords; and a free-floating larva with a food-collecting system consisting of compound cilia on multiciliate cells. The major shared, derived traits that unite the deuterostomes are a dorsal nervous system and larvae with a food-collecting system consisting of cells with a single cilium.

Porifera
Cnidaria
Ctenophora
Lophotrochozoans
PROTOSTOMES
Ecdysozoans
DEUTEROSTOMES

Hiemalora

1 cm

31.13 Fossilized Trail of an Urbilaterian
These tracks indicate that their maker was able to crawl.

31.14 A Phylogeny of Lophotrochozoans
Three major lineages, including the lopho-phorate and spiralian phyla, dominate the tree. Some small phyla are not included in this diagram.

Three cell layers

Lopho-trochozoan ancestor

Lophophore

Spiralian body plan; active locomotion

Platyhelminthes

Rotifera

Bryozoa

Brachiopoda

Phoronida

Pterobranchia

Nemertea

Annelida

Mollusca

Lophophorates

Spiralians

Most of the world's living animal species are protostomes. The diversity of protostome body plans and lifestyles has posed many challenges to zoologists attempting to infer the evolutionary re-lationships among these animals. Developmental, structural, and molecular data all suggest that protostomes split into two major lineages that have been evolving independently since ancient times: the lophotrochozoans and the ecdysozoans.

Lophotrochozoans, the animals we will discuss in the remainder of this chapter, grow by adding to the size of their skeletal elements. They use cilia for loco-motion, and many lineages have a type of free-living larva known as a **trochophore**. The phylogeny of lophotro-chozoans we will use in this chapter is shown in Figure 31.14. In contrast, **ecdysozoans**, the animals we will discuss in the next chapter, increase in size by molting their external skeletons. They move by mechanisms other than ciliary ac-tion, and they share a common set of homeobox genes.

Simple Lophotrochozoans

Flatworms move by beating cilia

Members of the phylum **Platyhelminthes**, or **flatworms**, are the simplest lopho-trochozoans (Figure 31.15). They are bilat-erally symmetri-cal animals that have no enclosed body cavity. They lack organs for transporting oxygen to internal tissues, and they have only simple organs for excreting metabolic wastes. This body plan dictates that each cell must be near a body surface, a re-quirement met by the flattened body form.

Platyhelminthes

Rotifera

Bryozoa

Brachiopoda

Phoronida

Pterobranchia

Nemertea

Annelida

Mollusca

The digestive tract of a flatworm consists of a mouth opening into a blind sac. However, the sac is often highly branched, forming intricate patterns that increase the sur-face area available for absorption of nutrients. All living flatworms feed on animal tissues—living or dead. Motile flatworms glide over surfaces, powered by broad bands of cilia. This form of movement is very slow, but it is sufficient

31.15 Flatworms Live Freely and Parasitically
(a) Some flatworm species are free-living, like this marine flatworm of the South Pacific. (b) The flatworm diagrammed here lives para-sitically in the gut of sea urchins. It is representative of parasitic flukes. Because their hosts provide all the nutrition they need, intestinal parasites do not require elaborate feeding or digestive organs.

(b)

Anterior

Pharynx

Intestine

Egg capsule

Testis

Yolk gland

Seminal receptacle

Ovary

Vagina

The flatworm gut has a single exterior opening. The pharynx is both "mouth" and "anus."

As is typical of internal parasites, the flatworm's body is filled primarily with sex organs.

Posterior

(a)

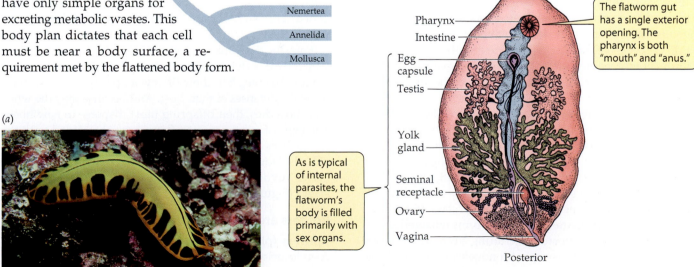

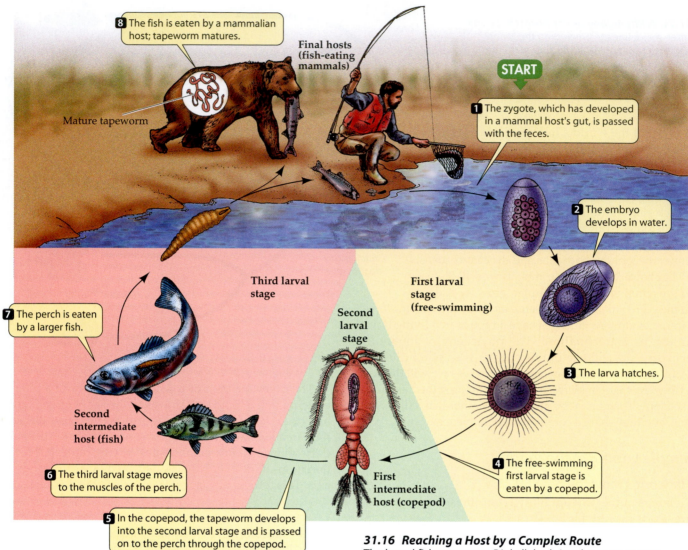

8 The fish is eaten by a mammalian host; tapeworm matures.

Final hosts (fish-eating mammals)

START

Mature tapeworm

1 The zygote, which has developed in a mammal host's gut, is passed with the feces.

2 The embryo develops in water.

Third larval stage

First larval stage (free-swimming)

Second larval stage

7 The perch is eaten by a larger fish.

3 The larva hatches.

Second intermediate host (fish)

6 The third larval stage moves to the muscles of the perch.

First intermediate host (copepod)

4 The free-swimming first larval stage is eaten by a copepod.

5 In the copepod, the tapeworm develops into the second larval stage and is passed on to the perch through the copepod.

31.16 Reaching a Host by a Complex Route
The broad fish tapeworm *Diphyllobothrium latum* must pass through the bodies of a copepod (a type of crustacean) and a fish before it can reinfect its primary host, a mammal. Such complex life cycles assist the flatworm's recolonization of hosts, but they also offer opportunities for humans to break the cycle with hygienic measures.

for small, scavenging animals. Parasitic species that absorb digested food from their hosts do not have a digestive tract.

The flatworms probably most similar to the ancestral forms are the turbellarians (class Turbellaria), which are small, free-living marine and freshwater animals (a few live in moist terrestrial habitats). Freshwater turbellarians of the genus *Dugesia*, better known as planarians, are the most familiar species of flatworms. At one end they have a head with chemoreceptor organs, two simple eyes, and a tiny brain composed of anterior thickenings of the longitudinal nerve cords.

Although the earliest flatworms were free-living (Figure 31.15*a*), many species evolved a parasitic existence. A likely evolutionary transition was from feeding on dead organisms, to feeding on the body surfaces of dying hosts, to invading and consuming parts of living, healthy hosts. Most of the 25,000 species of living flatworms—including the tapeworms (class Cestoda) and flukes (class Trematoda; Figure 31.15*b*)—are parasitic. These worms inhabit the bodies of many other species, including vertebrates; some cause serious human diseases. Monogeneans (class Mono-

genea) are external parasites of fishes and other aquatic vertebrates.

Parasitic flatworms live in nutrient-rich environments in which food is delivered to them, but they face other challenges. To complete their life cycle, parasites must overcome the defenses of their host. And because they die when their host dies, their offspring must disperse to new hosts. The eggs of some parasitic flatworms are voided with the host's feces and later ingested directly by other host individuals. However, most parasitic species have complex life cycles involving two or more hosts and several larval stages (Figure 31.16).

Rotifers are small but structurally complex

Rotifers (phylum **Rotifera**) are bilaterally symmetrical, pseudocoelomate, unsegmented animals that have three cell

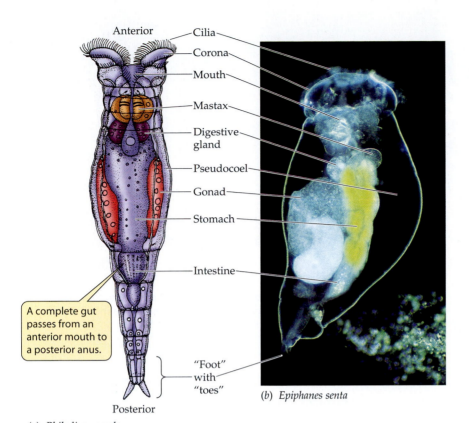

Anterior — Cilia
— Corona
— Mouth
— Mastax
— Digestive gland
— Pseudocoel
— Gonad
— Stomach
— Intestine
"Foot" with "toes"
Posterior

A complete gut passes from an anterior mouth to a posterior anus.

(a) *Philadina roseola*

(b) *Epiphanes senta*

31.17   Rotifers
(a) This rotifer reflects the general structure of many free-living species in this phylum. (b) The internal anatomy of a rotifer is clear in this micrograph.

Some rotifers are marine, but most of the 1,800 known species live in fresh water. Members of a few species rest on the surface of mosses and lichens in a desiccated, inactive state until it rains. When rain falls, they absorb water and become motile, feeding in the films of water that temporarily cover the plants. Most rotifers live no longer than 1 or 2 weeks.

Lophophorates: An Ancient Body Plan

After the platyhelminthes and rotifers diverged from it, the lophotrochozoan lineage divided into two branches. The descendants of these branches became the **lophophorates**—the subject of this section—and the **spiralians**, which we will describe in the following section.

Four phyla of lophophorate animals survive today: Phoronida, Brachiopoda, Ectoprocta, and Pterobranchia. Nearly all members of these phyla are marine; only a few species live in fresh water. About 4,500 living species are known, but many times that number existed during the Paleozoic and Mesozoic eras. These animals all have a body divided into three parts: *prosome* (anterior), *mesosome* (middle), and *metasome* (posterior). In most species, each region has a separate coelomic compartment: the *protocoel, mesocoel,* and *metacoel,* respectively.

Lophophorate animals obtain food by filtering it from ocean waters, a trait that they shared with many other protostomes. The most conspicuous feature of these animals, the **lophophore**, is a circular or U-shaped ridge around the mouth that bears one or two rows of ciliated, hollow tentacles (Figure 31.18). This large and complex structure is an organ for both food collection and gas exchange. All adult lophophorate animals are sessile; they use the tentacles and cilia of the lophophore to capture plankton. Lophophorates also have a U-shaped gut; the anus is located close to the mouth, but outside the tentacles.

The 20 known species of **phoronids** (phylum **Phoronida**) are sedentary worms that live in muddy or sandy sediments or attached to a rocky substrate. Phoronids are found in waters ranging from intertidal zones to about 400 meters deep. They range in size from 5 to 25 cm in length, and they secrete chitinous tubes in which they live. The lophophore is the most conspicuous external feature of

layers. Most rotifers are tiny (50–500 μm long)—smaller than some ciliate protists—but they have highly developed internal organs (Figure 31.17). A complete gut passes from an anterior mouth to a posterior anus; the pseudocoel functions as a hydrostatic skeleton. Most rotifers propel themselves through the water by means of rapidly beating cilia rather than by muscular contraction. This type of movement is effective because rotifers are so small.

The most distinctive organs of rotifers are those used to collect and process food. A conspicuous ciliated organ called the *corona* surmounts the head of many species. Coordinated beating of the cilia provides the force for locomotion and also sweeps particles of organic matter from the water into the mouth and down to a complex structure (the *mastax*) where the food is ground. By contracting the muscles that surround the pseudocoel, a few rotifer species that prey on protists and small animals can protrude the mastax through the mouth and seize small objects with it.

Platyhelminthes
Rotifera
Bryozoa
Brachiopoda
Phoronida
Pterobranchia
Nemertea
Annelida
Mollusca

Platyhelminthes
Rotifera
Bryozoa
Brachiopoda
Phoronida
Pterobranchia
Nemertea
Annelida
Mollusca

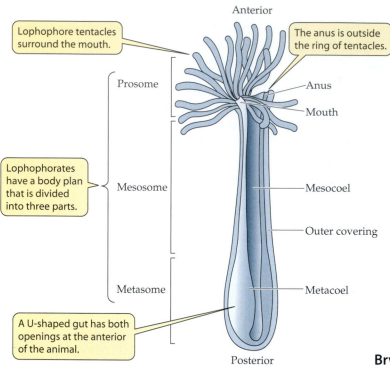

Anterior

Lophophore tentacles surround the mouth.

The anus is outside the ring of tentacles.

Prosome

Anus

Mouth

Lophophorates have a body plan that is divided into three parts.

Mesosome

Mesocoel

Outer covering

Metasome

Metacoel

A U-shaped gut has both openings at the anterior of the animal.

Posterior

31.18 Lophophore Artistry
The lophophore dominates the anatomy of a phoronid. The phoronid gut is U-shaped.

phoronids (see Figure 31.18). Cilia drive water into the top of the lophophore. Water exits through the narrow spaces between the tentacles. Suspended food particles are caught and transported by ciliary action to the food groove and into the mouth.

There are only 10 living species of **pterobranchs** (phylum **Pterobranchia**). Pterobranchs are sedentary animals up to 12 mm in length that live in tubes secreted by a proboscis, which is homologous to the prosome of phoronids. Some species are solitary; others form colonies of individuals joined together (Figure 31.19). Behind the proboscis is a collar with 1–9 pairs of arms bearing long tentacles that capture prey and permit gas exchange.

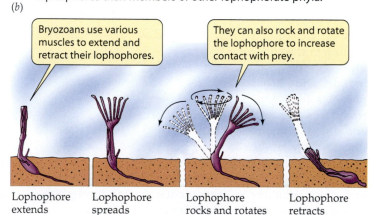

Arms

Tentacles

Tube

Proboscis

Collar

Stalk

Retracted animal

Tissue connecting colony members

31.19 Pterobranchs May Be Colonial or Solitary
This drawing of *Rhabdopleura* depicts two members of a colony.

Bryozoans Are Colonial Lophophorates

Bryozoans (phylum **Ectoprocta**) are colonial lophophorates that live in a "house" secreted by the body wall. A colony consists of many small individuals connected by strands of tissue along which materials can be moved (Figure 31.20a). Most bryozoans are marine, but a few live in fresh water. They are the only lophophorates able to completely retract the lophophore, which they can also rock and rotate to increase contact with prey (Figure 31.20b).

A colony of bryozoans is created by the asexual reproduction of its founding members. One colony may contain as many as 2 million individuals. In some species, individual colony members are specialized for feeding, reproduction, defense, or support. Bryozoans reproduce sexually by releasing sperm into the water, where they are collected by other individuals. Eggs are fertilized internally, and devel-

31.20 Bryozoans
(a) Branching colonies of bryozoans may appear plantlike.
(b) Bryozoans have greater control over the movement of their lophophores than members of other lophophorate phyla.

(a) *Iodyticium* sp.

(b)

Bryozoans use various muscles to extend and retract their lophophores.

They can also rock and rotate the lophophore to increase contact with prey.

Lophophore extends

Lophophore spreads

Lophophore rocks and rotates

Lophophore retracts

oping embryos are brooded before they exit as larvae to seek suitable sites for attachment.

Brachiopods Superficially Resemble Bivalve Mollusks

Brachiopods (phylum **Brachiopoda**) are solitary, marine lophophorate animals that superficially resemble bivalve mollusks (Figure 31.21). Most brachiopods are between 4 and 6 cm long, but some are as long as 9 cm. Brachiopods have a shell divided into two parts connected by a ligament. The two halves can be pulled shut to protect the soft body. The shell differs from that of mollusks in that the two halves are dorsal and ventral rather than lateral. The two-armed lophophore of a brachiopod is located within the shell. The beating of cilia on the lophophore draws water into the slightly opened shell. Food is trapped in the lophophore and directed to a ridge along which it is transferred to the mouth.

Brachiopods are either attached to a solid substrate or embedded in soft sediments. Most species are attached by means of a short, flexible stalk that holds the animal above the substrate. Gases are exchanged across body surfaces, especially the tentacles of the lophophore. Most brachiopods release their gametes into the water, where they are fertilized. The larvae remain in the plankton for only a few days before they settle and change into adults.

Brachiopods reached their peak abundance and diversity in Paleozoic and Mesozoic times. More than 26,000 fossil species have been described. Only about 350 species survive, but they are common in some marine environments.

Spiralians: Wormlike Body Plans

The spiralian lineage gave rise to many phyla. Members of more than a dozen of these phyla are wormlike; that is, they are bilaterally symmetrical, legless, soft-bodied, and at least several times longer than they are wide. This body form enables animals to move efficiently through muddy and sandy marine sediments. Most of these phyla have no more than several hundred species, even though the lineages have been evolving independently since early animal evolution.

The carnivorous **ribbon worms** (phylum **Nemertea**) are dorsoventrally flattened and have nervous and excretory systems similar to those of flatworms but, unlike flatworms, they have a complete digestive tract with a mouth at one end and an anus at the other. Food moves in one direction through the digestive tract of a ribbon worm and is acted on by a series of digestive enzymes. Small ribbon worms move by beating their cilia. Larger ones employ waves of contraction of body mus-

Platyhelminthes
Rotifera
Bryozoa
Brachiopoda
Phoronida
Pterobranchia
Nemertea
Annelida
Mollusca

Laqueus sp.

31.21 Brachiopods
You can see the lophophore of this North Pacific brachiopod between the valves of its shell.

cles to move on the surface of sediments or to burrow. Movement by both of these methods is slow.

Within the body of almost all 900 species of ribbon worms is a fluid-filled cavity called the *rhynchocoel*, within which floats a hollow, muscular *proboscis*. The proboscis, which is the feeding organ, may extend much of the length of the worm. Contraction of the muscles surrounding the rhynchocoel causes the proboscis to be everted explosively through an anterior opening (Figure 31.22) without moving

(a)

Floating in a cavity called the rhynchocoel, the proboscis can be moved rapidly. The worm, however, moves slowly.

Rhynchocoel   Proboscis   Proboscis retractor muscle
Proboscis pore   Mouth   Intestine   Anus

Retractor muscle
Everted proboscis   Mouth   Intestine

(b)

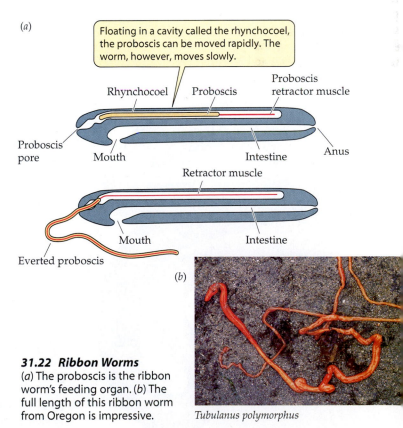

31.22 Ribbon Worms
(a) The proboscis is the ribbon worm's feeding organ. (b) The full length of this ribbon worm from Oregon is impressive.

Tubulanus polymorphus

the rest of the animal. The proboscis of most ribbon worms is armed with a sharp stylet that pierces the prey. Paralysis-causing toxins produced by the proboscis are discharged into the wound made by the stylet.

Segmented Bodies: Improved Locomotion

A body cavity divided into segments allows an animal to alter the shape of its body in complex ways and to control its movements precisely. Fossils of segmented worms are known from the middle Cambrian; the earliest forms are thought to have been burrowing marine animals. Segmentation evolved several times among spiralians.

Annelids have many-segmented bodies

The **annelids** (phylum **Annelida**) are a diverse group of segmented worms (Figure 31.23). The approximately 15,000 known annelid species live in marine, freshwater, and terrestrial environments. A separate nerve center called a *ganglion* controls each segment, but the ganglia are connected by nerve cords that coordinate their functioning. The coelom in each segment is isolated from those in other segments. Most annelids lack a rigid, external protective covering. The thin body wall serves as a surface for gas exchange in most species, but this thin, permeable body surface restricts annelids to moist environments; they lose body water rapidly in dry air.

POLYCHAETES. More than half of all annelid species are members of the class **Polychaeta**. Nearly all polychaetes are marine animals. Most have one or more pairs of eyes and one or more pairs of tentacles at the anterior end of the body. The body wall in most segments extends laterally as a series of thin outgrowths, called *parapodia*, that contain many blood vessels. The parapodia function in gas exchange, and some species use them to move. Stiff bristles called *setae* protrude from each parapodium, forming temporary attachments to the substrate and preventing the animal from slipping backward when its muscles contract.

Many polychaete species live in burrows in soft sediments and filter prey from the surrounding water with elaborate feathery tentacles (Figure 31.24*a*). Typically, males and females release gametes into the water, where the eggs are fertilized and develop into a trochophore larva. As the larva develops, it forms body segments at its posterior end, eventually changing into a small adult worm.

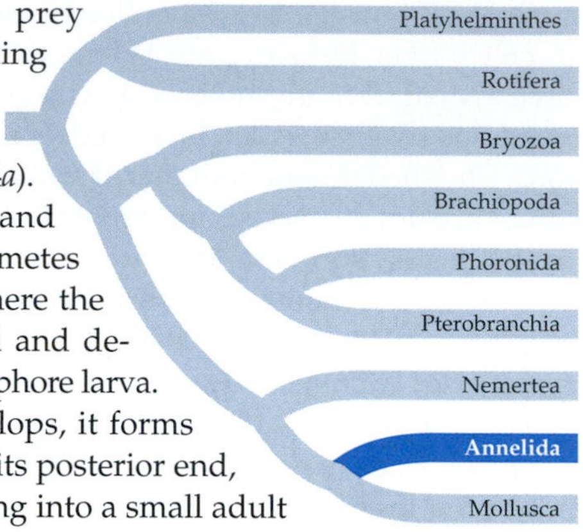

Platyhelminthes
Rotifera
Bryozoa
Brachiopoda
Phoronida
Pterobranchia
Nemertea
Annelida
Mollusca

OLIGOCHAETES. More than 90 percent of the approximately 3,000 described species of **oligochaetes** (class **Oligochaeta**) live in freshwater or terrestrial habitats. Oligochaetes have no parapodia, eyes, or anterior tentacles, and they have relatively few setae. Earthworms—the most familiar oligochaetes—are scavengers and ingesters of soil, from which they extract food particles.

Unlike polychaetes, all oligochaetes are *hermaphroditic*: Each individual is both male and female. Sperm are exchanged simultaneously between two copulating individuals (Figure 31.24*b*). Eggs are laid in a cocoon outside the adult's body. The cocoon is shed, and when development is complete, miniature worms emerge and begin independent life.

LEECHES. **Leeches** (class **Hirudinea**) probably evolved from oligochaete ancestors. Most species live in freshwater or terrestrial habitats and, like oligochaetes, lack parapodia and tentacles. Like oligochaetes, leeches are hermaphroditic. The coelom of leeches is not divided into compartments, and the coelomic space is largely filled with mesenchyme tissue. Groups of segments at each end of a leech are modified to form suckers, which serve as temporary anchors that aid in movement (Figure 31.24*c*). With its posterior sucker attached to a substrate, the leech extends its body by contracting its circular muscles. The anterior sucker is

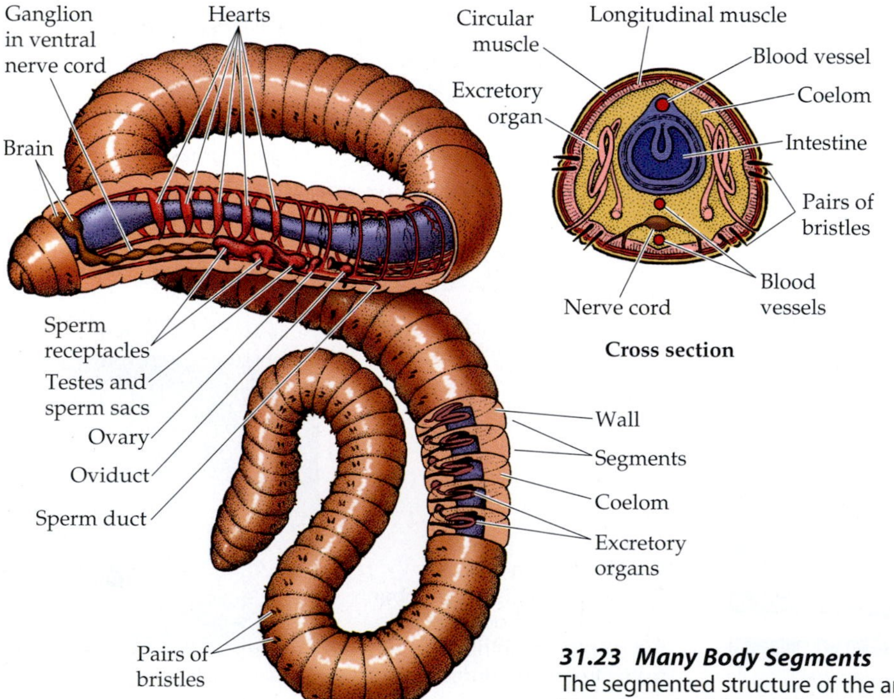

Ganglion in ventral nerve cord
Hearts
Brain
Sperm receptacles
Testes and sperm sacs
Ovary
Oviduct
Sperm duct
Pairs of bristles
Circular muscle
Longitudinal muscle
Blood vessel
Excretory organ
Coelom
Intestine
Pairs of bristles
Blood vessels
Nerve cord
Cross section
Wall
Segments
Coelom
Excretory organs

31.23 Many Body Segments
The segmented structure of the annelids is apparent both externally and internally. Most organs of this earthworm are repeated serially.

(a) *Spirobranchus* sp.

(b) *Lumbricus* sp.

(c) Australian tiger leech

31.24 Diversity among the Annelids
(a) The "feather duster" worm is a marine annelid with striking feeding tentacles. (b) Individual earthworms are hermaphroditic (simultaneously both male and female). When they copulate, each individual both donates and receives sperm. (c) This Australian tiger leech is attached to a leaf by its posterior sucker as it waits for a mammalian "victim." (d) Vestimentiferans live around thermal vents deep in the ocean. Their skin secretes chitin and other substances, forming tubes from which they extend feeding tentacles.

(d) *Riftia* sp.

om of a vestimentiferan consists of an anterior compartment, into which the tentacles can be withdrawn, and a long, subdivided cavity that extends much of the length of its body. Experiments using radioactively labeled molecules have shown that vestimentiferans take up dissolved organic matter at high rates from either the sediments in which they live or the surrounding water.

Vestimentiferans were not discovered until the twentieth century, when deep-ocean exploration revealed them living many thousands of meters below the surface. In these deep oceanic sediments they are abundant, reaching densities of many thousands per square meter. About 145 species have been described. The largest and most remarkable vestimentiferans, which grow to 2 meters in length, live near deep-ocean hydrothermal vents—openings in the seafloor through which hot, sulfide-rich water pours. The tissues of these species harbor endosymbiotic prokaryotes that fix carbon using energy obtained from the oxidation of hydrogen sulfide (H_2S).

then attached, the posterior one detached, and the leech shortens itself by contracting its longitudinal muscles.

Many leeches are external parasites of other animals, although some species eat snails and other invertebrates. A parasitic leech makes an incision in its host to expose its blood. It can ingest so much blood in a single feeding that its body may enlarge several times. A substance called hirudin secreted by the leech into the wound keeps the host's blood flowing (and gives this phylum its name; see the opening page of Chapter 6). For hundreds of years leeches were widely employed in medicine for bloodletting. Even today leeches are used to reduce fluid pressure and prevent blood clotting in damaged tissues and to eliminate pools of coagulated blood.

VESTIMENTIFERANS. Members of one lineage of annelids, the **vestimentiferans**, evolved into burrowing forms with a crown of tentacles through which gases are exchanged, and entirely lost their digestive systems (Figure 31.24d). The coel-

Mollusks lost segmentation but evolved shells

Mollusks (phylum **Mollusca**) range in size from snails only a millimeter high to giant squids more than 18 meters long—the largest known protostomes. Beginning with a segmented common ancestor, mollusks underwent one of the most

Platyhelminthes

Rotifera

Bryozoa

Brachiopoda

Phoronida

Pterobranchia

Nemertea

Annelida

Mollusca

Generalized molluscan body plan

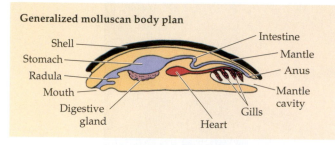

Chitons

In all mollusk lineages, a **mantle** covers the internal organs of the visceral mass.

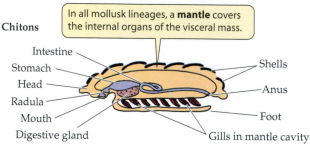

Gastropods

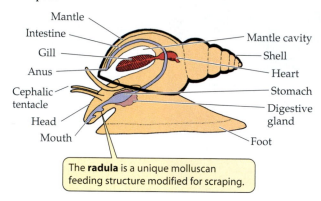

The **radula** is a unique molluscan feeding structure modified for scraping.

Bivalves

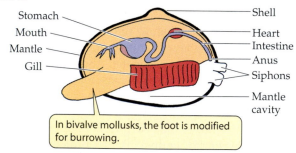

In bivalve mollusks, the foot is modified for burrowing.

Cephalopods

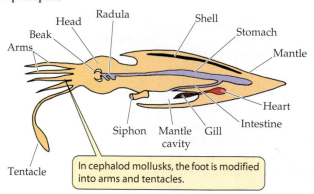

In cephalod mollusks, the foot is modified into arms and tentacles.

dramatic of animal evolutionary radiations, based on a body plan with three major structural components: a foot, a mantle, and a visceral mass. Animals that appear very different, such as snails, clams, and squids, are all built from these three components (Figure 31.25).

▶ The molluscan **foot** is a large, muscular structure that originally was both an organ of locomotion and a support for the internal organs.

In the lineage leading to squids and octopuses, the foot was modified to form arms and tentacles borne on a head with complex sensory organs. In other groups, such as clams, the foot was transformed into a burrowing organ. In some lineages the foot is greatly reduced.

▶ The **mantle** is a fold of tissue that covers the **visceral mass** of internal organs. In many mollusks, the mantle extends beyond the visceral mass to form a *mantle cavity*.

The gills, which are used for gas exchange and, in some species, for feeding, lie in the mantle cavity. When the cilia on the gills beat, they create a flow of oxygenated water over the gills.

The coelom of mollusks is much reduced, but the open circulatory system has large fluid-filled cavities that are major components of a hydraulic skeleton.

The mollusks also developed a rasping feeding structure known as the **radula**. The radula was originally an organ for scraping algae from rocks, a function it retains in many living mollusks. However, in some mollusks, it has been modified into a drill or a poison dart. In others, such as clams, it is absent.

Although individual components have been lost in some lineages, these three unique shared derived characteristics are why zoologists believe that all 100,000 species of mollusks share a common ancestor. A small sample of these species is shown in Figure 31.26.

MONOPLACOPHORANS. Monoplacophorans (class **Monoplacophora**) were the most abundant mollusks during the Cambrian period, but today there only a few surviving species. Unlike all other living mollusks, the surviving monoplacophorans have multiple gills, muscles, and excretory structures that are repeated over the length of the body. The gills are located in a large cavity under the shell, through which oxygen-bearing water circulates.

CHITONS. Chitons (class **Polyplacophora**) have multiple gills and segmented shells, but their other body parts are not segmented (Figure 31.26*a*). The chiton body is bilaterally symmetrical, and its internal organs, particularly the digestive and nervous systems, are relatively simple. The

31.25 Molluscan Body Plans
The diverse modern mollusks are all variations on a general body plan that includes a foot, a mantle, and a visceral mass of internal organs.

(a) *Tonicella lineata*

(c) *Hypsclodoris* sp.

(b) *Tridacna gigas*

(d) *Monadenia fidelis*

(f) *Nautilus belavensis*

(e) *Octopus cyanea*

31.26 Diversity among the Mollusks

(a) Chitons are common in the intertidal zones of the North American coast. (b) The giant clam of Indonesia is among the largest of the bivalve mollusks. (c) Slugs are terrestrial and marine gastropods that have lost their shells; this shell-less sea slug is very conspicuously colored. (d) Land snails are shelled, terrestrial gastropods. (e) Cephalopods such as the octopus are active predators. (f) The boundaries of its chambers are clearly visible on the outer surface of this shelled *Nautilus*, another cephalopod.

trochophore larvae of chitons are almost indistinguishable from those of annelids. Most chitons are marine herbivores that scrape algae from rocks with their sharp radulae. An adult chiton spends most of its life glued tightly to rock surfaces by its large, muscular, mucus-covered foot. It moves slowly by means of rippling waves of muscular contraction in the foot.

BIVALVES. One lineage of early mollusks developed a hinged, two-part shell that extended over the sides of the body as well as the top, giving rise to the **bivalves** (class **Bivalvia**), which include the familiar clams, oysters, scallops, and mussels (Figure 31.26*b*). Bivalves are largely sedentary and have greatly reduced heads. The foot is compressed and, in many clams, is used for burrowing into mud and sand. Bivalves feed by bringing water in through an opening called a *siphon* and extracting food from the water using their large gills, which are also the main sites of gas exchange. Water exits through another siphon.

GASTROPODS. Another lineage of early mollusks gave rise to the **gastropods** (class **Gastropoda**), which includes the snails. Most gastropods are motile, using the large foot to move slowly across the substrate or to burrow through it. Gastropods are the most species-rich and widely distributed of the molluscan classes (Figure 31.26*c,d*). Some species, such as snails, whelks, limpets, slugs, abalones, and the often brilliantly ornamented nudibranchs, can crawl. Others—the sea butterflies and heteropods—have a modified foot that functions as a swimming organ with which they move through open ocean waters. The only mollusks that live in terrestrial environments—land snails and slugs—are gastropods. In these terrestrial species the mantle cavity is modified into a highly vascularized lung.

CEPHALOPODS. In one lineage of mollusks, the **cephalopods** (class **Cephalopoda**), the exit siphon, which initially may have simply improved the flow of water over the gills, became modified to allow the early cephalopods to control the water content of the mantle cavity. The modification of the mantle into a device for forcibly ejecting water from the cavity enabled cephalopods to move rapidly through the water. Furthermore, as fluid moves out of a chamber, gases diffuse into it, changing the buoyancy of the animal. Thus, by pumping out water, the animals could also control their buoyancy. Together, these adaptations allowed cephalopods to live in open water.

With their greatly enhanced mobility, some cephalopods, such as squids and octopuses, became the major predators in open ocean waters (Figure 31.26*e*). They are still important marine predators today. Cephalopods capture and subdue their prey with their tentacles; octopuses use theirs to move over the substrate. As is typical of active predators, cephalopods have complex sensory organs, most notably eyes that are comparable to those of vertebrates in their ability to resolve images. The cephalopod head is closely associated with a large, branched foot that bears tentacles and a siphon. The large, muscular mantle is a solid external supporting structure. The gills hang within the mantle cavity.

Cephalopods appeared about 600 million years ago, near the beginning of the Cambrian period, and by the Ordovician period a wide variety of types were present. They were the first large, shelled animals able to move vertically in the ocean. The earliest cephalopod shells were divided by partitions penetrated by tubes through which liquids could be moved. Nautiloids (genus *Nautilus*) are the only cephalopods with external chambered shells that survive today (Figure 31.26*f*). Increases in size and reductions in external hard parts characterize the subsequent evolution of many lineages.

Chapter Summary

Descendants of a Common Ancestor
▶ All members of the kingdom Animalia are believed to have a common flagellated protist ancestor.
▶ The specialization of cells by function made possible the complex, multicellular body plan of animals.

The Animal Way of Life
▶ Animals obtain their food—complex organic molecules—by active expenditure of energy.

Clues to Evolutionary Relationships among Animals
▶ Morphological, developmental, and molecular data support similar animal phylogenies.
▶ The body cavity of an animal is strongly correlated with its ability to move. On the basis of their body cavities, animals are classified as acoelomates, pseudocoelomates, or coelomates. **Review Figure 31.1**
▶ The two major animal lineages—protostomes and deuterostomes—are believed to have separated early in animal evolution; they differ in several components of their early embryological development. **Review Figure 31.2**

Body Plans Are Basic Structural Designs
▶ Most animals have either radial or bilateral symmetry. Radially symmetrical animals move slowly or not at all. Bilateral symmetry is strongly correlated with more rapid movement and the development of sensory organs at the anterior end of the animal. **Review Figure 31.3**

Sponges: Loosely Organized Animals
▶ Sponges (phylum Porifera) are simple animals that lack cell layers and body symmetry, but have several different cell types.
▶ Sponges feed via choanocytes, feeding cells that draw water through the sponge body and filter out small organisms and nutrient particles. **Review Figure 31.4**

Cnidarians: Cell Layers and Blind Guts
▶ Cnidarians (phylum Cnidaria) are radially symmetrical and have only two cell layers, but with their nematocyst-studded tentacles they can capture prey larger and more complex than themselves. **Review Figure 31.7**

▶ Most cnidarian life cycles have a sessile polyp and a free-swimming, sexual medusa stage, but some species lack one of the stages. **Review Figures 31.8, 31.9, 31.10**

Ctenophores: Complete Guts and Tentacles

▶ Ctenophores (phylum Ctenophora), descendants of the first split in the lineage of bilaterally symmetrical animals, are marine carnivores that have simple life cycles. **Review Figure 31.12**

The Evolution of Bilaterally Symmetrical Animals

▶ The common ancestors of bilateral animals, called urbilaterians, were probably simple, bilaterally symmetrical animals composed of flattened masses of cells.

Protostomes and Deuterostomes: An Early Lineage Split

▶ Protostomes and deuterostomes are monophyletic lineages that have been evolving separately since the Cambrian period. Their members are structurally more complex than cnidarians and ctenophores. Protostomes have a ventral nervous system, paired nerve cords, and larvae with compound cilia. Deuterostomes have a dorsal nervous system and larvae with single cilia.

▶ Protostomes split into two major clades—lophotrochozoans and ecdysozoans. **Review Figure 31.14**

Simple Lophotrochozoans

▶ Flatworms (phylum Platyhelminthes) have no body cavity, lack organs for oxygen transport, have only one entrance to the gut, and move by beating their cilia. Many species are parasitic. **Review Figures 31.15, 31.16**

▶ Although no larger than many ciliated protists, rotifers (phylum Rotifera) have highly developed internal organs. **Review Figure 31.17**

Lophophorates: An Ancient Body Plan

▶ The lophotrochozoan lineage split into two branches whose descendants became the lophophorates and the spiralians.

▶ The lophophore dominates the anatomy of many lophophorate animals. **Review Figure 31.18**

▶ Bryozoans are colonial lophophorates that can move their lophophores. **Review Figure 31.20**

▶ Brachiopods, which superficially resemble bivalve mollusks, were much more abundant in the past than they are today.

Spiralians: Wormlike Body Plans

▶ The spiralian lineage gave rise to many phyla, most of whose members have wormlike body forms.

▶ Ribbon worms (phylum Nemertea) have a complete digestive tract and capture prey with an eversible proboscis. **Review Figure 31.22**

Segmented Bodies: Improved Locomotion

▶ Annelids (phylum Annelida) are a diverse group of segmented worms that live in marine, freshwater, and terrestrial environments. **Review Figure 31.23**

▶ Mollusks (phylum Mollusca) evolved from segmented ancestors but subsequently became unsegmented. The molluscan body plan has three basic components: foot, mantle, and visceral mass. **Review Figure 31.25**

▶ The molluscan body plan has been modified to yield a diverse array of animals that superficially appear very different from one another.

For Discussion

1. Differentiate among the members of each of the following sets of related terms:
   a. radial symmetry/bilateral symmetry
   b. protostome/deuterostome
   c. indeterminate cleavage/determinate cleavage
   d. spiral cleavage/radial cleavage
   e. coelomate/pseudocoelomate/acoelomate

2. For each of the types of organisms listed below, give a single trait that may be used to distinguish them from the organisms in parentheses:
   a. cnidarians (sponges)
   b. gastropods (all other mollusks)
   c. polychaetes (other annelids)

3. In this chapter we listed some of the traits shared by all animals that convince most biologists that all animals are descendants of a single common ancestral lineage. In your opinion, which of these traits provides the most compelling evidence that animals are monophyletic?

4. Describe some features that allow animals to capture prey that are larger and more complex than they themselves are.

5. Animals in many phyla have wormlike, or vermiform, shapes. Why has this body form met the needs of species in so many different lineages of animals? In what types of environments does the worm shape function well? Why?

6. Having a complete digestive tract in which materials enter at an anterior mouth and move in one direction until they exit from a posterior opening would appear to be a very efficient way to digest food and rid the body of the indigestible residues. Nonetheless, several successful phyla of animals with a blind gut must void their digestive wastes via the same opening through which food entered. Why has this type of digestive system persisted? What limitations does it impose on the types of food animals can eat and the way in which the food is treated?

32 *Ecdysozoans: The Molting Animals*

A FIRM, NONLIVING COVERING THAT IS DIFFICULT to penetrate—an **exoskeleton**—provides an animal with both protection and support. Its very attributes, however, pose a huge problem: An exoskeleton cannot grow as the animal body inside it grows. Ancestors of today's ecdysozoan animals evolved a solution. They shed, or **molt**, the outgrown exoskeleton and expand and harden a new, larger one.

The new exoskeleton is already in place, growing underneath the old one. Directly after molting, the animal is very vulnerable. With its soft, new armor it can move only very slowly for a while. Despite this constraint, the lineages of Ecdysozoa—the molting animals—have more species than all other animal lineages combined.

An increasingly rich array of molecular and genetic evidence, including a common set of homeobox genes, suggests that molting may have evolved only once during animal evolution. The exoskeletons of ecdysozoan animals range from thin and flexible to thick, hard, and rigid.

The presence of an exoskeleton presented new problems and opportunities in other areas of the body plan besides growth. Unlike the lophotrochozoans, ecdysozoans cannot use cilia for locomotion; new forms of locomotion evolved in these lineages. And because hard exoskeletons impede the passage of oxygen into the animal, new mechanisms for respiration evolved.

In this chapter we will review the characteristics of animals in the various ecdysozoan phyla and show how developing an exoskeleton has influenced the evolution of these animals. The phylogeny we follow here is presented in Figure 32.1. The latter half of the chapter details the characteristics of several ecdysozoan lineages that have traditionally been classified in the phylum Arthropoda— the arthropods.

Collectively, arthropods (which include the terrestrial insects and the marine crustaceans) are the dominant animals on Earth, both in number of species (some 1.5 million) and number of individuals (estimated at some 10^{18} individuals, or a billion billion). The highly successful arthropod body plan is based on three elements: the rigid exoskeleton that marks them as ecdysozoans; segmentation; and jointed appendages, which immensely enhance their powers of locomotion, and which we will encounter again in Chapter 33 when we cover another major lineage, the vertebrates.

We close the chapter with an overview of evolutionary themes found in the evolution of the protostomate phyla, including both the lophotrochozoan and ecdysozoan lineages.

Cuticles: Flexible, Unsegmented Exoskeletons

Some ecdysozoans have wormlike bodies covered by exoskeletons that are relatively thin and flexible. These exoskeletons, called **cuticles**, protect the animal, but do not provide support for the bodies. The action of circular and longitudinal muscles on fluid in the body cavity provides a hydrostatic skeleton for many of these animals, which can

Molting the Exoskeleton
This green darner dragonfly (genus *Anax*) has just emerged from its larval exoskeleton and is pumping fluids into its expanding wings. At this stage the insect can move only very slowly.

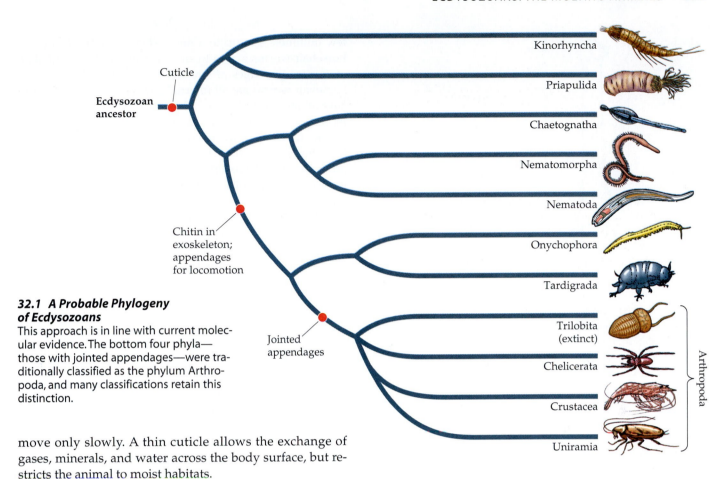

32.1 A Probable Phylogeny of Ecdysozoans
This approach is in line with current molecular evidence. The bottom four phyla—those with jointed appendages—were traditionally classified as the phylum Arthropoda, and many classifications retain this distinction.

Labels in figure:
Ecdysozoan ancestor
Cuticle
Chitin in exoskeleton; appendages for locomotion
Jointed appendages
Kinorhyncha
Priapulida
Chaetognatha
Nematomorpha
Nematoda
Onychophora
Tardigrada
Trilobita (extinct)
Chelicerata
Crustacea
Uniramia
Arthropoda

move only slowly. A thin cuticle allows the exchange of gases, minerals, and water across the body surface, but restricts the animal to moist habitats.

Some marine phyla have few species

Several phyla of marine wormlike animals (that is, they are long and slender, without appendages) branched off early within the ecdysozoan lineage. These phyla contain only a few species. They have relatively thin cuticles that are molted periodically as they grow to full size. Their bodies are supported primarily by their hydrostatic skeletons, not by the cuticle.

PRIAPULIDS AND KINORHYNCHS. The 16 species of **priapulids** (phylum **Priapulida**) are cylindrical, unsegmented, wormlike animals that range in size from half a millimeter to 20 centimeters in length. They burrow in fine marine sediments.

About 150 species of **kinorhynchs** (phylum **Kinorhyncha**) have been described. They are all less than 1 millimeter in length and live in marine sands or muds. Their bodies are divided into 13 segments by a series of cuticular plates that are periodically molted during growth (Figure 32.2). Kinorhynchs feed by ingesting the substratum and digesting the organic material found within it, which may include living algae as well as dead matter.

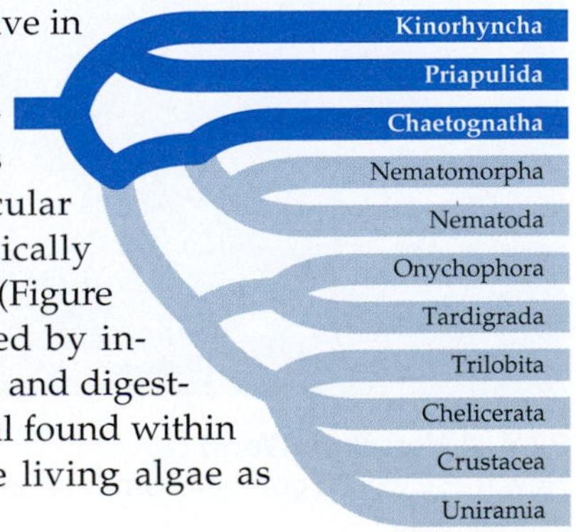

Inset phylogeny labels:
Kinorhyncha
Priapulida
Chaetognatha
Nematomorpha
Nematoda
Onychophora
Tardigrada
Trilobita
Chelicerata
Crustacea
Uniramia

ARROW WORMS. Arrow worms (phylum **Chaetognatha**) have three-part, streamlined bodies. Their body plan is based on a coelom that is divided into head, trunk, and tail compartments. Most of them swim in the open sea, but a few live on the seafloor. Their abundance as fossils indicates that they were already common more than 500 million years ago. The 100 or so living species of arrow worms are so small—less than 12 centimeters long—that their gas exchange and excretion requirements can be met by diffusion through the body surface. Arrow worms lack a circulatory system. Wastes and nutrients are moved around the body in the coelomic fluid, which is propelled by cilia that line the coelom. Arrow

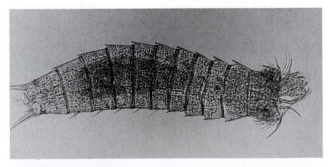

32.2 A Kinorhynch
Kinorhynchs are tiny (less than a millimeter long) marine worms. Their segmented bodies are covered with plates of cuticle that are periodically molted.

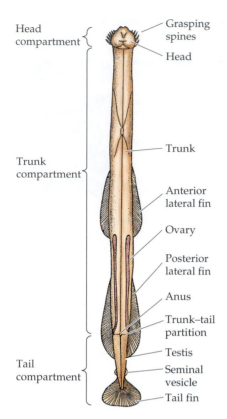

Head compartment — Grasping spines

— Head

Trunk compartment — Trunk

— Anterior lateral fin

— Ovary

— Posterior lateral fin

— Anus

— Trunk–tail partition

Tail compartment — Testis

— Seminal vesicle

— Tail fin

32.3 An Arrow Worm
Arrow worms have a three-part body plan. The fins and grasping spines are adaptations for a predatory life.

worms are stabilized in the water by means of one or two pairs of lateral fins and a "tail" fin (Figure 32.3). There is no distinct larval stage; miniature adults hatch directly from eggs that are released into the water.

Arrow worms are major predators of small organisms in the open oceans. Their prey range from small protists to young fish as large as an arrow worm. An arrow worm typically lies motionless in the water until movement of the water signals the approach of prey. The arrow worm then darts forward and grasps the prey with the stiff spines adjacent to its mouth.

Tough cuticles evolved in some unsegmented worms

Tough external cuticles evolved in some members of an ecdysozoan lineage whose descendants have colonized freshwater and terrestrial environments as well as marine ones. Two extant phyla represent this lineage.

Kinorhyncha
Priapulida
Chaetognatha
Nematomorpha
Nematoda
Onychophora
Tardigrada
Trilobita
Chelicerata
Crustacea
Uniramia

HORSEHAIR WORMS.   About 230 species of horsehair worms (phylum **Nematomorpha**) have been described. As their name implies, they are extremely thin and range in length from a few millimeters to up to a meter (Figure 32.4). Most adult horsehair worms live in fresh water among litter and algal mats near the edges of streams and ponds. The larvae of horsehair worms are all internal parasites of terrestrial and aquatic insects and crabs. The much reduced gut has no mouth opening and is probably nonfunctional. Horsehair worms may feed only as larvae, absorbing nutrients from their hosts across their body wall, but many continue to grow after they have left their hosts, suggesting that adults may also absorb nutrients from their environment.

ROUNDWORMS.   Roundworms (phylum **Nematoda**) have a thick, multilayered cuticle secreted by the underlying epidermis that gives their body its shape (Figure 32.5). As a roundworm grows, it sheds and re-secretes its cuticle four times. The largest known roundworm, which reaches a length of 9 meters, is a parasite in the placentas of female sperm whales. About 20,000 species of roundworms have been described, but the actual number of living species may be more than a million.

Roundworms exchange oxygen and nutrients with their environment through both the cuticle and the intestine, which is only one cell layer thick. Materials are moved through the gut by rhythmic contraction of a highly muscular organ, the *pharynx*, at the worm's anterior end. Roundworms move by contracting their longitudinal muscles.

Roundworms are one of the most abundant and universally distributed of all animal groups. Countless roundworms live as scavengers in the upper layers of the soil, on the bottoms of lakes and streams, and as parasites in the bodies of most kinds of plants and animals. The flesh of a single rotting apple found on the ground in an orchard contained 90,000 roundworms, and 1 square meter of mud off the coast of the Netherlands yielded 4,420,000 individuals. The topsoil of rich farmland has up to 3 billion nematodes per acre.

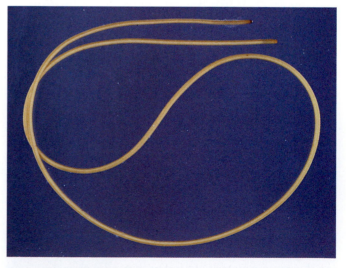

32.4 A Horsehair Worm
How these worms got their name is evident from this photograph.

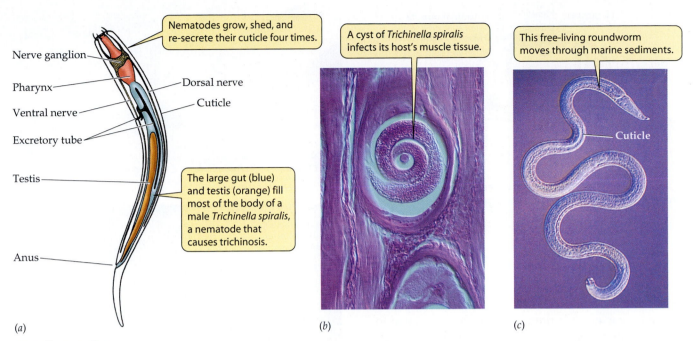

Nematodes grow, shed, and re-secrete their cuticle four times.

Nerve ganglion

Pharynx

Dorsal nerve

Ventral nerve

Cuticle

Excretory tube

Testis

The large gut (blue) and testis (orange) fill most of the body of a male *Trichinella spiralis*, a nematode that causes trichinosis.

Anus

(a)

A cyst of *Trichinella spiralis* infects its host's muscle tissue.

(b)

This free-living roundworm moves through marine sediments.

Cuticle

(c)

32.5 Nematodes
(a,b) *Trichinella* is an example of a parasitic roundworm that infects mammals, including humans. (c) Free-living roundworms have a body plan similar to the adult parasite's.

The diets of roundworms are as varied as their habitats. Many are predators, preying on protists and other small animals (including other roundworms). Many roundworms live parasitically within their hosts. The roundworms that are parasites of humans (causing diseases such as trichinosis, filariasis, and elephantiasis), domestic animals, and economically important plants have been studied intensively in an effort to find ways of controlling them. One soil-inhabiting nematode species, *Caenorhabditis elegans*, has been intensely studied in the laboratory by geneticists and developmental biologists.

The structure of parasitic roundworms is similar to that of free-living species, but the life cycles of many parasitic species have special stages that facilitate their transfer among hosts. *Trichinella spiralis*, the species that causes the human disease trichinosis, has a relatively simple life cycle. A person may become infected by eating the flesh of an animal (usually a pig) containing larvae of *Trichinella* encysted in its muscles.

The larvae are activated in the mammalian digestive tract, leave their cysts, and attach to the person's intestinal wall, where they feed. Later they bore through the intestinal wall and are carried in the bloodstream to the muscles, where they form cysts (Figure 32.5b). If present in great numbers, these cysts can cause severe pain or even death.

Trichinella can infect a number of mammal species; there is no special stage in the life cycle that lives in a particular alternate host. Other roundworm life cycles are more complex, involving one or more alternate hosts.

Arthropods and Their Relatives: Segmented External Skeletons

In Precambrian times, the body coverings of some wormlike ecdysozoan lineages became thickened by the incorporation of layers of protein and a strong, flexible, waterproof polysaccharide called **chitin**. After this change, which initially probably had a protective function, the rigid body covering acquired both support and locomotory functions.

A rigid body covering precludes wormlike movement. To move, these animals require **appendages** that can be manipulated by muscles. Such appendages evolved several times in late Precambrian times, leading to the phyla collectively called **arthropods**. The divisions among arthropod lineages are so ancient that we divide them into a number of phyla. However, as indicated at the opening of this chapter, many zoologists treat these groups as members of a single phylum, **Arthropoda**.

The bodies of arthropods are divided into segments. Their muscles attach to the inside of the skeleton, and each segment has muscles that operate that particular segment and the appendages attached to it (Figure 32.6). The appendages of most present-day arthropods have joints, although those of some lineages do not. Arthropod appendages serve many functions, including walking and swimming, food capture and manipulation, copulation, and sensory perception.

The sturdy exoskeleton had a profound influence on arthropod evolution. Encasement within armor provides support for walking on dry land, and, with special waterproofing, it keeps the animal from dehydrating in dry air. Aquatic arthropods were, in short, excellent candidates to invade the terrestrial environment, and as we will see, they did so several times.

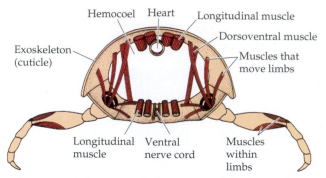

32.6 Arthropods Have Rigid, Segmented Exoskeletons
This cross section through a segment of a generalized arthropod shows the typical structure of an arthropod body, which is characterized by a rigid exoskeleton and jointed appendages.

Related lineages had unjointed legs

Although they were once thought to be closely related to segmented annelid worms, the molecular evidence links the 80 species of **onychophorans** (phylum **Onychophora**) to the arthropod lineages. Onychophorans have soft bodies that are covered by a thin, flexible cuticle that contains chitin. Onychophorans use their body cavities as hydrostatic skeletons. Their soft, fleshy, unjointed, claw-bearing legs are formed by outgrowths of the body (Figure 32.7a). They are probably similar in appearance to ancestral arthropods.

Like the onychophorans, **water bears** (phylum **Tardigrada**) have fleshy, unjointed legs and use their fluid-filled body cavities as hydrostatic skeletons (Figure 32.7b). Unlike onychophorans, water bears are all extremely small (0.1–0.5 mm in length), and they lack circulatory systems and gas exchange organs. The 600 extant species of water bears live in marine sands and on temporary water films on plants. When these films dry out, the water bears also lose water and shrink to small, barrel-shaped objects that can survive for at least a decade in a dehydrated resting state. They may occur at densities as high as 2,000,000 per square meter of moss.

Jointed legs appeared in the trilobites

Once the dominant line of arthropods, the **trilobites** (phylum **Trilobita**) flourished in Cambrian and Ordovician seas but were extinct by the close of the Paleozoic era. Trilobites were heavily armored, and their body segmentation and appendages followed a relatively simple, repetitive plan. But their appendages were jointed, giving them added flexibility, and the beginnings of specialization—using different appendages for different functions—can be discerned.

(a) *Peripatodes novazealaniae*

(b) *Echinisucus springer*   50 μm

32.7 Arthropod Relatives with Unjointed Appendages
(a) Onychophorans have unjointed legs and use the body cavity as a hydrostatic skeleton. (b) The appendages and general anatomy of a water bear (phylum Tardigrada) superficially resemble those of onychophorans.

Why trilobites declined in abundance and eventually became extinct is unknown. However, because their heavy external skeletons provided ideal material for fossilization, they left behind a vivid record of their presence (Figure 32.8).

Odontochile rugosa

32.8 A Trilobite
The relatively simple, repetitive segments of the now-extinct trilobites are illustrated here by a fossil trilobite from the shallow seas of the Devonian period.

Cladogram labels: Kinorhyncha, Priapulida, Chaetognatha, Nematomorpha, Nematoda, **Onychophora**, **Tardigrada**, **Trilobita**, Chelicerata, Crustacea, Uniramia

Figure 32.6 labels: Hemocoel, Heart, Longitudinal muscle, Dorsoventral muscle, Muscles that move limbs, Exoskeleton (cuticle), Longitudinal muscle, Ventral nerve cord, Muscles within limbs

(a) *Decalopoda* sp.

32.9 Minor Chelicerate Phyla
(a) Although they are not true spiders, it is easy to see why sea spiders were given their common name. (b) This spawning aggregation of horseshoe crabs was photographed on the New Jersey coast.

(b) *Limulus polyphemus*

Chelicerates Invaded the Land

The bodies of all **chelicerates** (phylum **Chelicerata**) are divided into two major regions. The anterior region bears two pairs of appendages, modified to form mouthparts, and four pairs of walking legs. The 63,000 described species are usually placed in three classes: Pycnogonida, Arachnida, and Merostomata. Only the class Arachnida contains many species.

The **pycnogonids** (class **Pycnogonida**), or sea spiders, are a small group of marine species that are seldom seen except by marine biologists (Figure 32.9a). The class **Merostomata** contains a single order, the Xiphosura, or horseshoe crabs. These marine animals, which have changed very little during their long fossil history, have a large horseshoe-shaped covering over most of the body. They are common in shallow waters along the eastern coasts of North America and Southeast Asia, where they scavenge and prey on bottom-dwelling invertebrates. Periodically they crawl into the intertidal zone to mate and lay eggs (Figure 32.9b).

Arachnids (class **Arachnida**) are abundant in terrestrial environments. Most arachnids have a simple life cycle in which miniature adults hatch from eggs and begin independent lives almost immediately. Some arachnids retain their eggs during development and give birth to live young. The most species-rich and abundant arachnids are the scorpions, harvestmen, spiders, mites, and ticks (Figure 32.10).

Spiders are important terrestrial predators. Some have excellent vision that enables them to chase and seize their prey. Others spin elaborate webs made of protein threads to snare prey. The webs of different groups of spiders are strikingly varied and enable spiders to position their snares in many different environments. Spiders also use protein threads to construct safety lines during climbing and as homes, mating structures, protection for developing young, and means of dispersal. The threads are produced by modified abdominal appendages that are connected to internal glands that secrete the proteins of which the threads are constructed.

Crustaceans: Diverse and Abundant

Crustaceans (phylum **Crustacea**) are the dominant marine arthropods. The most familiar crustaceans are decapods (shrimps, lobsters, crayfishes, and crabs; Figure 32.11a); isopods (sow bugs; Figure 32.11b); and amphipods (sand fleas; see Figure 1.11). Also included among the crustaceans are a wide variety of other small species, many of which superficially resemble shrimps (Figure 32.11c). The individuals of one group alone, the copepods (class Copepoda), are so numerous that they may be the most abundant of all animals.

Barnacles (class Cirripedia) are unusual crustaceans that are sessile as adults (Figure 32.11d). With their calcareous shells, they superficially resemble mollusks, but, as the zoologist Louis Agassiz remarked more than a century ago, a barnacle is "nothing more than a little shrimp-like animal, standing on its head in a limestone house and kicking food into its mouth."

Kinorhyncha
Priapulida
Chaetognatha
Nematomorpha
Nematoda
Onychophora
Tardigrada
Trilobita
Chelicerata
Crustacea
Uniramia

Kinorhyncha
Priapulida
Chaetognatha
Nematomorpha
Nematoda
Onychophora
Tardigrada
Trilobita
Chelicerata
Crustacea
Uniramia

(a) *Uroctonus mondax*

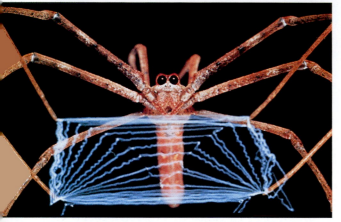

(b) *Deinopis* sp.

(d) *Ixodes ricinus*

(c) *Hadrobunus maculosus*

32.10 Diversity among the Arachnids

(a) Scorpions are nocturnal predators. (b) Net-casting spiders use their webs to snare and envelop their prey. (c) Harvestmen, often called daddy longlegs, are scavengers. (d) Ticks are blood-sucking, external parasites on vertebrates. This wood tick is piercing the skin of its human host.

32.11 Diversity among the Crustaceans

(a) This crayfish is a decapod crustacean. (b) This sow bug is a common isopod found in grasslands. (c) A typical planktonic copepod from the deep ocean. (d) The appendages of these gooseneck barnacles protrude from their shells to capture prey.

(a) *Orconectes palmeri*

(c) *Megacalanus princeps*

(b) *Armadillidium vulgare*

(d) *Lepas pectinata*

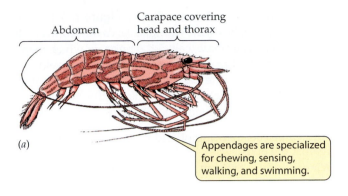

(a)

> Appendages are specialized for chewing, sensing, walking, and swimming.

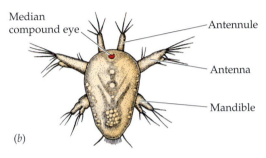

(b)

32.12 Crustacean Structure
(a) The bodies of most crustaceans are divided into three regions, each segment of which bears appendages. (b) A nauplius larva has one simple eye and three pairs of appendages.

Most of the 40,000 described species of crustaceans have a body that is divided into three regions: *head*, *thorax*, and *abdomen*. The segments of the head are fused together, and the head bears five pairs of appendages. Each of the multiple thoracic and abdominal segments usually bears one pair of appendages. In many species, a fold of the exoskeleton, the *carapace*, extends dorsally and laterally back from the head to cover and protect some of the other segments (Figure 32.12a).

The fertilized eggs of most crustacean species are attached to the outside of the female's body, where they remain during their early development. At hatching, the young of some species are released as larvae; those of other species are released as juveniles that are similar in form to the adults. Still other species release fertilized eggs into the water or attach them to an object in the environment. The typical crustacean larva, called a **nauplius**, has three pairs of appendages and one simple eye (Figure 32.12b). In many crustaceans, the nauplius larva develops within the egg before it hatches.

Uniramians are Primarily Terrestrial

The body of a **uniramian** (phylum **Uniramia**) is divided into either two or three regions (in myriapods and insects, respectively). The anterior regions have few segments, but the posterior region—the abdomen—has many segments. Uniramians are primarily terrestrial animals; most have elaborate systems of channels that bring oxygen to the cells of their internal organs.

Myriapods have many legs

Centipedes, millipedes, and the two other groups of animals in the subphylum **Myriapoda** have two body regions: a head and a trunk. Centipedes and millipedes have a well-formed head and a long, flexible, segmented trunk that bears many pairs of legs (Figure 32.13). Centipedes prey on insects and other small animals. Millipedes scavenge and eat plants. More than 3,000 species of centipedes and 10,000 species of millipedes have been described; many more species probably remain unknown. Although most myriapods are less than a few centimeters long, some tropical species are ten times that size.

Kinorhyncha
Priapulida
Chaetognatha
Nematomorpha
Nematoda
Onychophora
Tardigrada
Trilobita
Chelicerata
Crustacea
Uniramia

Insects are the dominant uniramians

The 1.5 million species of **insects** (subphylum **Insecta**) that have been described are believed to be only a small fraction of the total number living on Earth today. Insects are found in nearly all terrestrial and freshwater habitats, and they utilize as food nearly all species of plants and many species of animals. Some are internal parasites of plants and animals; others suck their host's blood or consume its surface

(a) *Scolopendra heros*

(b) *Harapaphe haydeniana*

32.13 Myriapods
(a) Centipedes have powerful jaws for capturing active prey. (b) Millipedes, which are scavengers and plant eaters, have smaller jaws and legs.

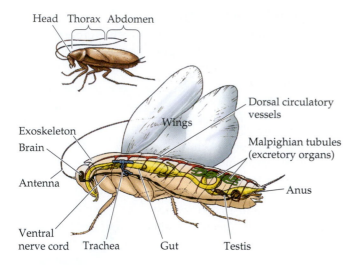

Head Thorax Abdomen

Wings

Dorsal circulatory vessels

Exoskeleton

Brain

Malpighian tubules (excretory organs)

Antenna

Anus

Ventral nerve cord   Trachea   Gut   Testis

32.14 Structure of an Insect
The body plan of an insect differs in many details from that of other arthropods, but the basic theme of a segmented body with modified jointed appendages is shared with most arthropod lineages.

body tissues. Insects transmit many viral, bacterial, and protist diseases among plants and animals. Very few insect species are oceanic. In freshwater environments, on the other hand, they are sometimes the dominant animals, burrowing through the muddy substrate, extracting suspended prey from the water, or actively pursuing other animals.

Insects, like the crustaceans, have three basic body parts: head, thorax, and abdomen. They have a single pair of antennae on the head, and three pairs of legs attached to the thorax (Figure 32.14). An insect exchanges gases by means of air sacs and tubular channels called *tracheae* (singular trachea) that extend from external openings inward to tissues throughout the body. The adults of most flying insects have two pairs of stiff, membranous wings attached to the thorax. The exceptions are flies, which have only one pair of wings, and beetles, in which the forewings form heavy, hardened wing covers.

Wingless insects (class **Apterygota**) include firebrats and collembolans (Figure 32.15a). Of the modern insects, they are probably the most similar in form to insect ancestors. Apterygote insects have a simple life cycle, hatching from their eggs looking like small adults.

Development in the winged insects (class **Pterygota**) is more complex. The hatchlings are not similar to adults, and they undergo substantial changes at each molt in the process of growing larger. The immature stages of insects between molts are called **instars**. A substantial change that occurs between one developmental stage and another is called **metamorphosis**. When the change from one instar to the next is gradual, an insect is said to undergo **incomplete metamorphosis**.

In some insect genera, the larvae and adult forms can appear to be completely different animals. The most familiar example of such **complete metamorphosis** occurs in members of the order Lepidoptera, when the larval caterpillar trans-

forms itself into the adult butterfly (see Figure 1.6). During complete metamorphosis, the wormlike larva transforms itself during a specialized phase, called the **pupa**, in which many larval tissues are broken down and the adult form develops.

Entomologists divide the winged insects into about 28 different orders. We can make sense out of this bewildering variety by recognizing three major lineages:

▶ Winged insects that cannot fold their wings back against the body.
▶ Winged insects that can fold their wings and that undergo incomplete metamorphosis.
▶ Winged insects that can fold their wings and that undergo complete metamorphosis.

Because they can fold their wings over their backs, flying insects belonging to the second and third lineages are able to tuck their wings out of the way upon landing and crawl into crevices and other tight places.

The only surviving groups of the first lineage are the orders Odonata (dragonflies and damselflies; Figure 32.15b) and Ephemeroptera (mayflies). All members of these two orders have aquatic larvae that metamorphose into flying adults after they crawl out of the water. Although many of these insects are excellent flyers, they require a great deal of open space in which to maneuver. Dragonflies and damselflies are active predators as adults, but adult mayflies lack functional digestive tracts and do not eat, living only long enough to mate and lay eggs.

The second lineage includes the orders Orthoptera (grasshoppers, crickets, roaches, mantids, and walking sticks; Figure 32.15c), Isoptera (termites), Plecoptera (stone flies), Dermaptera (earwigs), Thysanoptera (thrips), Hemiptera (true bugs; Figure 32.15d), and Homoptera (aphids, cicadas, and leafhoppers). Hatchlings are sufficiently similar in form to adults to be recognizable. They acquire adult organ systems, such as wings and compound eyes, gradually through several juvenile instars.

Insects belonging to the third lineage have different life stages specialized for living in different environments and using different food sources. In many species the larvae are adapted for feeding and growing, and the adults are specialized for reproduction and dispersal. The adults of some species do not feed at all, living only long enough to mate, disperse, and lay eggs. In many species whose adults do feed, adults and larvae use different food resources. About 85 percent of all species of winged insects belong to this lineage. Familiar examples are the orders Neuroptera (lacewings

32.15 Diversity among the Insects
(a) This silverfish is a typical member of the apterygote order Thysanura. (b) Unlike most insects, this adult dragonfly (order Odonata) cannot fold its wings over its back. Representatives of some of the largest insect orders are (c) a broad-winged katydid (Orthoptera), (d) harlequin bugs (Hemiptera), (e) a predaceous diving beetle (order Coleoptera), (f) a Great Mormon butterfly (Lepidoptera), (g) a hoverfly (Diptera), and (h) a honeybee (Hymenoptera).

(a) *Lepisma saccharina*

(e) *Dysticus marginalis*

(b) *Sympetrum vulgatum*

(f) *Papilio memnon*

(c) *Microcentrum rhombifolium*

(g) Family Syrphidae

(d) *Murgantia histrionica*

(h) *Apis mellifera*

32.1 General Characteristics of the Major Protostomate Phyla[a]

PHYLUM	BODY CAVITY	DIGESTIVE TRACT	CIRCULATORY SYSTEM
Lophotrochozoans			
Platyhelminthes	None	Dead-end sac	None
Rotifera	Pseudocoelom	Complete	None
Bryozoa	Coelom	Complete	None
Brachiopoda	Coelom	Complete	None
Phoronida	Coelom	Complete	None
Pterobranchia	Coelom	Complete	None
Nemertea	Coelom	Complete	Closed
Annelida	Coelom	Complete	Closed or open
Mollusca	Reduced coelom	Complete	Open except in cephalopods
Ecdysozoans			
Chaetognatha	Coelom	Complete	None
Nematomorpha	Pseudocoelom	Greatly reduced	None
Nematoda	Pseudoceolom	Greatly reduced	None
Chelicerata	Hemocoel	Complete	Closed or open
Crustacea	Hemocoel	Complete	Closed or open
Uniramia	Hemocoel	Complete	Closed or open

[a]All have bilateral symmetry.

and their relatives), Coleoptera (beetles; Figure 32.15*e*), Trichoptera (caddisflies), Lepidoptera (butterflies and moths; Figure 32.15*f*), Diptera (flies; Figure 32.15*g*), and Hymenoptera (sawflies, bees, wasps, and ants; Figure 32.15*h*).

There are also several orders of pterygote insects, including the Phthiraptera (lice) and Siphonaptera (fleas) that are parasitic. Although descended from flying ancestors, these insects have lost the ability to fly.

Why have the insects undergone such incredible evolutionary diversification? Insects may have originated from a centipede-like ancestor as far back as the Devonian period. The terrestrial environments penetrated by the arthropods were like a new planet, an ecological world with more complexity than the seas they came from, but one containing relatively few species other of animals. The evolution of the ability to fly allowed the insects to escape from potential predators and to traverse boundaries that might otherwise have been insurmountable—both very highly adaptive features. The numbers and diversity of insect species attest to the supreme success of this highly visible and dominant animal group.

Themes in Protostome Evolution

Most of protostome evolution took place in the oceans. As we have seen, early animals used fluids within their body cavities as the basis for support and movement. Subdivisions of the body cavity allowed better control of movement and permitted different parts of the body to be moved independently of one another. Thus some protostome lineages gradually evolved the ability to change their shape in complex ways and to move with greater speed on and through sediments or in the water.

During much of animal evolution, the only food available in the water consisted of dissolved organic matter and very small organisms. Consequently, many different lineages of animals evolved feeding structures designed to extract small prey from water, as well as structures for moving water through or over their prey-collecting devices. Animals that feed in this manner are abundant and widespread in marine waters today.

Because water flows readily, bringing food with it, sessile lifestyles evolved repeatedly during lophotrochozoan and ecdysozoan evolution. Most protostome phyla today have at least some sessile members. Sessile lifestyles have both advantages and disadvantages. A sessile animal gains access to local resources, but forfeits access to more distant resources. Sessile animals cannot come together to mate; instead, they must rely on the fertilization of gametes that they have ejected into the water. Some species eject both eggs and sperm into the water; others retain their eggs within their bodies and extrude only their sperm, which are carried by the water to other individuals. Species whose adults are sessile often have motile larvae, many of which have complicated mechanisms for locating suitable sites on which to settle. Many colonial sessile protostomes are able to grow in the direction of better resources or into sites offering better protection.

A frequent consequence of a sessile existence is competition for space. Such competition is intense among plants in most terrestrial environments. In the sea, especially in shallow waters, animals also compete directly for space. They have evolved mechanisms for overgrowing one another and for engaging in toxic warfare where they come into contact.

Individual members of sessile colonies, if they are directly connected, can share resources. The ability to share resources enables some individuals to specialize for particular functions, such as reproduction, defense, or feeding. The nonfeeding individuals derive their nutrition from their feeding associates.

Predation may have been the major selective pressure behind the development of external body coverings. Such coverings evolved independently in many lophotrochozoan and ecdysozoan lineages. In addition to providing protection, they became key elements in the development of new systems of locomotion. Locomotory abilities permitted prey to escape more readily from predators, but also allowed predators to pursue their prey more effectively. Thus, the evolution of animals has been, and continues to be, a complex arms race among predators and prey.

Although we have concentrated on the evolution of greater complexity in animal lineages, many lineages that remained simple have been very successful. Cnidarians are common in the oceans; roundworms abound in most aquatic and terrestrial environments. Parasites lost complex body plans but evolved complex life cycles.

The characteristics of the major existing phyla of protostomate animals are summarized in Table 32.1. All the phyla had evolved by the Cambrian period, but extinction and diversification within these lineages continue.

Many of the evolutionary trends demonstrated by protostomes also dominated the evolution of deuterostomes, the lineage that includes the chordates, the group to which humans belong. Hard external body coverings evolved and were later abandoned by many lineages. We will consider the evolution of the deuterostomes in the next chapter.

 Chapter Summary

▶ A major innovation during animal evolution was the development of a sturdy, nonliving external cover—an exoskeleton. An animal with an exoskeleton grows by periodically molting its exoskeleton and replacing it with a larger one.

▶ The presence of an exoskeleton opened avenues for the evolution of new body plans in the ecdysozoan lineage. **Review Figure 32.1**

Animals with Flexible Exoskeletons

▶ Tough cuticles are found in members of two phyla that live in freshwater, marine, and terrestrial environments.

▶ Roundworms (phylum Nematoda) are one of the most abundant and universally distributed of all animal groups. Many are parasites. **Review Figure 32.5**

Arthropods and Their Relatives: Segmented External Skeletons

▶ The body coverings of one ecdysozoan lineage, the arthropods, became thickened and made rigid by the incorporation of layers of protein and the polysaccharide chitin.

▶ Animals with rigid exoskeletons cannot move in a worm-like fashion. To move, they have appendages that can be manipulated by muscles. **Review Figure 32.6**

▶ Onychophorans have soft, fleshy, unjointed legs. They are probably similar to ancestral arthropods.

▶ The tiny and abundant water bears (phylum Tardigrada) also have unjointed legs.

▶ Jointed legs with specialized functions appeared among the trilobites (phylum Trilobita). Trilobites flourished in Cambrian and Ordovician seas, but became extinct by the close of the Paleozoic era.

Chelicerates: Invasion of the Land

▶ The bodies of all chelicerates (phylum Chelicerata) are divided into two major regions, the anterior of which bears four pairs of jointed legs.

▶ Arachnids—scorpions, harvestmen, spiders, mites, and ticks—are abundant in terrestrial environments.

Crustaceans: Diverse and Abundant

▶ Most of the 40,000 described species of crustaceans (phylum Crustacea) have a body that is divided into three regions: head, thorax, and abdomen. **Review Figure 32.12**

▶ The most familiar crustaceans are shrimps, lobsters, crayfishes, crabs, sow bugs, and sand fleas.

Uniramians are Primarily Terrestrial

▶ The body of a uniramian (phylum Uniramia) is divided into two or three regions; the posterior region has paired legs.

▶ Myriapods (centipedes and millipedes) have many segments and many pairs of legs.

▶ About 1.5 million species of insects (subphylum Insecta) have been described, but that is probably only a small fraction of the total number of species living on Earth today.

▶ Insects have three body regions (head, thorax, abdomen), a single pair of antennae on the head, and three pairs of legs attached to the thorax. **Review Figure 32.14**

▶ Wingless insects (class Apterygota) look like little adults when they hatch from their eggs. Hatchlings of many winged insects (class Pterygota) do not resemble adults and undergo substantial changes at each molt.

▶ Entomologists divide the winged insects into three major subgroups and about 28 different orders. Members of one subgroup cannot fold their wings back against the body; members of the other two groups can.

Themes in Protostome Evolution

▶ Most evolution of protostomes took place in the oceans.

▶ Early animals used fluid-filled spaces as hydrostatic skeletons. Subdivision of the body cavity allowed better control of movement and permitted different parts of the body to be moved independently of one another.

▶ Predation may have been the major selective pressure for the development of hard, external body coverings.

▶ During much of animal evolution, the only food in the water consisted of dissolved organic matter and very small organisms.

▶ Flowing water brings food with it, so many animals are sessile.

▶ All the phyla of protostomate animals had evolved by the Cambrian period.

For Discussion

1. Segmentation has arisen several times during animal evolution. What advantages does segmentation provide? Given these advantages, why do so many unsegmented animals survive?

2. Many animals extract food from the surrounding medium. What phyla contain animals that extract suspended food from the water column? What structures do these animals use to capture prey?

3. An animal that sheds its external skeleton in order to grow in size is virtually helpless during the time that its new, larger exoskeleton is hardening. Give some examples of how predators take advantage of this vulnerable stage of their prey. Include at least one example of predation by humans.

4. The British biologist J. B. S. Haldane is reputed to have quipped that "God was unusually fond of beetles." Beetles are, indeed, the most species-rich lineage of organisms. What features of beetles have contributed to the generation and survival of so many species?

5. In Part Three we pointed out that major structural novelties have arisen infrequently during the course of evolution. Which of the features of protostomes do you think are major evolutionary novelties? What criteria do you use to judge whether a feature is a major as opposed to a minor novelty?

6. A frequent consequence of sessile existence is competition for space. How do plants and animals differ in the ways in which they compete for space?

7. There are more described and named species of insects than of all other animals lineages combined. However, only a very few species of insects live in marine environments, and those species are restricted to the intertidal zone or the ocean surface. What factors may have contributed to the inability of insects to be successful in the oceans?

33 Deuterostomate Animals

THERE ARE ABOUT 25,000 SPECIES OF ray-finned fishes—more species than exist in all other vertebrate groups combined. Ray-finned fishes include almost all fish species with bony skeletons (as opposed to the sharks, whose skeletons are made of cartilage). Most ray-finned fishes have excellent color vision, and they use their brightly colored bodies to advertise their presence, species identity, and sex. Some go through dramatic color changes at different stages of their lives, and some are even able to change colors quickly when they are ready to mate, fight, or flee.

Part of the reason for the richness of ray-finned species may be that fishes are an ancient lineage that has had many millions of years in which to radiate in Earth's oceans and fresh waters. But part of the reason may be genetic. Most vertebrates have only four clusters of homeobox genes, but some ray-finned fishes have seven. The entire genome of these fishes was apparently duplicated about 300 million years ago, providing new opportunities for genetic variability that may have helped drive their explosive evolutionary radiation.

There are fewer major lineages and many fewer species among deuterostomes than among protostomes (Table 33.1), but we have a special interest in deuterostomes because we are members of that lineage. In this chapter, we first discuss some evolutionary themes shared by protostomes and deuterostomes, then describe and discuss the deuterostome phyla Echinodermata, Hemichordata, and Chordata, with special attention to the primate lineage of Chordata that gave rise to our own species.

Deuterostomes and Protostomes: Shared Evolutionary Themes

Deuterostome evolution paralleled protostome evolution in several important ways. Both lineages exploited the abundant food supplies buried in soft marine sediments, attached to rocks, or suspended in water. Because of the ease with which water can be moved, many groups in both lineages developed elaborate structures for moving water and extracting prey from it.

In lineages of both groups, the body became divided into compartments that allowed better control of shape and movement. Some members of both groups evolved mechanisms for controlling their buoyancy in water, using gas-filled internal spaces. Planktonic larval stages evolved in marine members of many protostome and deuterostome phyla; these all fed on tiny planktonic organisms while floating freely in the open water.

The ancestral traits shared by all members of the deuterostome lineage include indeterminate cleavage in the early embryo, a blastopore that becomes the anus, three body layers (they are triploblastic), formation of the mesoderm from an outpocketing of the embryonic gut, and a well-developed coelom (see Chapter 31). No fossils of ancestral deuterostomes that lived before the lineage split into two major lineages (echinoderms and chordates; Figure 33.1) have been found, so we can only deduce what they must have been like from these shared traits.

Both protostomes and deuterostomes colonized the land—the former via beaches, the latter via fresh water—but the consequences of these colonizations were very different. The jointed external skeletons of arthropods, although they provide excellent support and protection in air, cannot support large animals. The internal skeletons developed by deuterostomes are capable of supporting large bodies. The largest terrestrial deuterostomes to ever live were some of the dinosaurs; elephants are the largest living terrestrial animals.

Terrestrial deuterostomes recolonized aquatic environments a number of times. Suspension feeding re-evolved in several of these lineages. The largest living animals, the baleen (toothless) whales, feed on relatively small prey that they extract from the water with large straining structures in their mouths.

Two Colors, One Fish
The spotted puffer fish, *Arothron meleagris*, changes color during the course of its life cycle. The individuals shown here are in two different color phases of the cycle.

PHYLUM	NUMBER OF LIVING SPECIES DESCRIBED	SUBGROUPS
Porifera: Sponges	10,000	
Cnidaria: Cnidarians	10,000	Hydrozoa: Hydras and hydroids Scyphozoa: Jellyfishes Anthozoa: Corals, sea anemones
Ctenophora: Comb jellies	100	

PROTOSTOMES

PHYLUM	NUMBER OF LIVING SPECIES DESCRIBED	SUBGROUPS
Lophotrochozoans		
Platyhelminthes: Flatworms	20,000	Turbellaria: Free-living flatworms Trematoda: Flukes (all parasitic) Cestoda: Tapeworms (all parasitic) Monogenea (ectoparasites of fishes)
Rotifera: Rotifers	1,800	
Ectoprocta: Bryozoans	4,500	
Brachiopoda: Lamp shells	340	More than 26,000 fossil species described
Phoronida: Phoronids	20	
Pterobranchia: Pterobranchs	10	
Nemertea: Ribbon worms	900	
Annelida: Segmented worms	15,000	Polychaeta: Polychaetes (all marine) Oligochaeta: Earthworms, freshwater worms Hirudinea: Leeches
Mollusca: Mollusks	50,000	Monoplacophora: Monoplacophorans Polyplacophora: Chitons Bivalvia: Clams, oysters, mussels Gastropoda: Snails, slugs, limpets Cephalopoda: Squids, octopuses, nautiloids
Ecdysozoans		
Kinorhyncha: Kinorhynchs	150	
Chaetognatha: Arrow worms	100	
Nematoda: Roundworms	20,000	
Nematomorpha: Horsehair worms	230	
Onychophora: Onychophorans	80	
Tardigrada: Water bears	600	
Chelicerata: Chelicerates	70,000	Merostomata: Horseshoe crabs Arachnida: Scorpions, harvestmen, spiders, mites, ticks
Crustacea	50,000	Crabs, shrimps, lobsters, barnacles, copepods
Uniramia	1,500,000	Myriapoda: Millipedes, centipedes Insecta: Insects

DEUTEROSTOMES

PHYLUM	NUMBER OF LIVING SPECIES DESCRIBED	SUBGROUPS
Echinodermata: Echinoderms	7,000	Crinoidea: Sea lilies, feather stars Ophiuroidea: Brittle stars Asteroidea: Sea stars Concentricycloidea: Sea daisies Echinoidea: Sea urchins Holothuroidea: Sea cucumbers
Hemichordata: Hemichordates	85	Acorn worms
Chordata: Chordates	50,000	Urochordata: Sea squirts Cephalochordata: Lancelets Agnatha: Lampreys, hagfishes Chondrichthyes: Cartilaginous fishes Osteichthyes: Bony fishes Amphibia: Amphibians Reptilia: Reptiles Aves: Birds Mammalia: Mammals

[a]Some small phyla are not included.

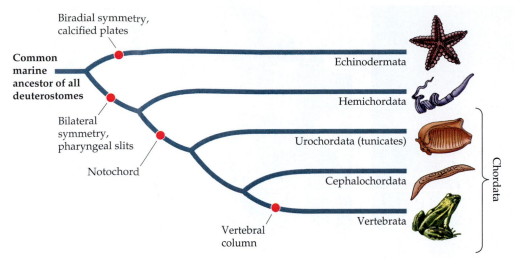

 33.1 A Probable Deuterostomate Phylogeny
There are fewer major lineages and many fewer species of deuterostomes than of protostomes.

Echinoderms: Complex Biradial Symmetry

The ancestors of one deuterostome lineage, the **echinoderms** (phylum **Echinodermata**), were probably sluggish animals. They evolved into more aggressive and active forms as a result of two major structural features. One is a system of calcified internal plates covered by thin layers of skin and some muscles. The calcified plates of early echinoderm ancestors became enlarged and thickened until they fused inside the entire body, giving rise to an internal skeleton.

The other major innovation was the evolution of a **water vascular system**, a network of calcified hydraulic canals leading to extensions called **tube feet**. The water vascular system functions in gas exchange, locomotion, and feeding (Figure 33.2a). Seawater enters the water vascular system through a perforated *sieve plate*. A calcified canal leads from the sieve plate to another canal that rings the esophagus. Other canals radiate from this *ring canal* extending through the arms (in species that have arms) and connecting with the tube feet. The development of these two structural innovations—calcified internal skeleton and water vascular system—resulted in one of the most striking of evolutionary radiations.

Echinoderms have an extensive fossil record. About 23 classes have been described, of which only 6 survive today. About 7,000 species of echinoderms exist today, but 13,000 species—probably only a small fraction of those that actually lived—have been described from their fossil remains. Nearly all living species have a bilaterally symmetrical, ciliated larva that feeds for some time as a planktonic organism before settling and transforming into a biradially symmetrical adult (Figure 33.2b).

The living echinoderms are divided into two lineages: **Pelmatozoa** and **Eleutherozoa**. The two lineages differ in the number of arms they have and the form of their water vascular systems. Pelmatozoa consists only of the crinoids, whereas several groups are included in the Eleutherozoa.

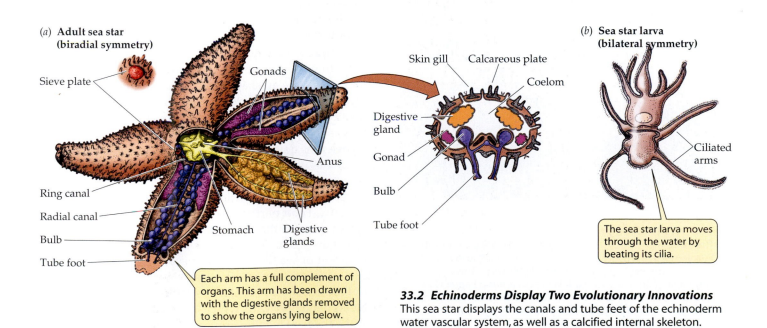

(a) **Adult sea star (biradial symmetry)**

Sieve plate
Gonads
Ring canal
Radial canal
Bulb
Tube foot
Stomach
Anus
Digestive glands

Each arm has a full complement of organs. This arm has been drawn with the digestive glands removed to show the organs lying below.

Skin gill
Calcareous plate
Coelom
Digestive gland
Gonad
Bulb
Tube foot

(b) **Sea star larva (bilateral symmetry)**

Ciliated arms

The sea star larva moves through the water by beating its cilia.

33.2 Echinoderms Display Two Evolutionary Innovations
This sea star displays the canals and tube feet of the echinoderm water vascular system, as well as a calcified internal skeleton.

(a) *Oxycomanthus bennetti*

(b) *Strongylocentrotus purpuratus*

(c) *Bohadschia argus*

(d) *Henricia leviuscula*

(e) *Opiothrix suemsonii*

33.3 Diversity among the Echinoderms

(a) The flexible arms of the golden feather star are clearly visible. (b) Purple sea urchins are important grazers of algae in the intertidal zone of the Pacific Coast of North America. (c) This sea cucumber lives on rocky substrates in seas around Papua New Guinea. (d) The blood sea star is typical of many sea stars; some species, however, have more than five arms. (e) This brittle star is resting on a sponge.

Pelmatozoans have jointed arms

Sea lilies and feather stars (class **Crinoidea**) are the only surviving pelmatozoans. Sea lilies were abundant 300–500 mya, but only about 80 species survive today. Most sea lilies attach to a substratum by means of a flexible stalk consisting of a stack of calcareous discs. The main body of the animal is a cup-shaped structure that contains a tubular digestive system. Five to several hundred arms, usually in multiples of five, extend outward from the cup. The jointed calcareous plates of the arms enable them to bend. A groove runs down the center of each arm to the mouth. On both sides of the groove are tube feet covered with mucus-secreting glands.

A sea lily feeds by orienting its arms in passing water currents. Food particles strike and stick to the tube feet, which transfer the particles to the grooves in the arms, where the action of cilia carries the food to the mouth. The tube feet of sea lilies are also used for gas exchange and elimination of nitrogenous wastes.

Feather stars are similar to sea lilies, but they have flexible appendages with which they grasp the substratum while they are feeding and resting (Figure 33.3a). Feather stars feed in much the same manner as sea lilies. They can walk on the tips of their arms or swim by rhythmically beating their arms. About 600 living species of feather stars have been described.

Eleutherozoans are the dominant echinoderms

Most surviving echinoderms are members of the eleutherozoan lineage. Biochemical data suggest that the ancestors of sea urchins and sand dollars (class **Echinoidea**) were the first to split off from the lineage leading to the other eleutherozoans. Sea urchins and sand dollars lack arms, but they share a five-part body plan with all other echinoderms. Sea urchins are hemispherical animals that are covered with spines attached to the underlying skeleton via ball-and-socket joints (Figure 33.3*b*). The spines of sea urchins come in varied sizes and shapes; a few produce highly toxic substances. Many sea urchins consume algae, which they scrape from the rocks with a complex rasping structure. Others feed on small organic debris that they collect with their tube feet or spines. Sand dollars, which are flattened and disc-shaped, feed on algae and fragments of organic matter on the seafloor.

The tube feet of sea cucumbers (class **Holothuroidea**; Figure 33.3*c*) are used primarily for attaching to the substratum rather than for moving. The anterior tube feet are modified into large, feathery, sticky tentacles that can be protruded around the mouth. Periodically, a sea cucumber withdraws the tentacles into its mouth, wipes off the material that has adhered to them, and digests it.

Sea daisies (class **Concentricycloidea**) were not discovered until 1986. Little is known about them. They have tiny disc-shaped bodies with a ring of marginal spines, and two ring canals, but no arms. Sea daisies are found on rotting wood in ocean waters. They apparently feed on prokaryotes, which they digest outside their bodies and absorb either through a membrane that covers the oral surface or via a shallow saclike stomach.

The most familiar echinoderms are the sea stars (class **Asteroidea**; Figure 33.3*d*; see also Figure 33.2). Their tube feet serve as organs of locomotion and, because their walls are thin, they are important sites for gas exchange. Each tube foot of a sea star is also an adhesive organ, consisting of an internal bulb connected by a muscular tube to an external sucker. A tube foot is moved by expansion and contraction of the circular and longitudinal muscles of the tube. It can adhere to a surface by secreting a sticky substance around the sucker.

Many sea stars prey on polychaetes, gastropods, bivalves, and fishes. They are important predators in many marine environments, such as coral reefs and rocky intertidal zones. With hundreds of tube feet acting simultaneously, a sea star can exert an enormous and continuous force. It can grasp a clam in its arms, anchor the arms with its tube feet, and, by steady contraction of the muscles in the arms, gradually exhaust the muscles with which the clam keeps its shell closed. Sea stars that feed on bivalves are able to push the stomach out through the mouth and then through the narrow space between the two halves of the shell. The stomach secretes digestive enzymes into the soft parts of the bivalve, digesting it.

Brittle stars (class **Ophiuroidea**) are similar in structure to sea stars, but their flexible arms are composed of jointed hard plates (Figure 33.3*e*). Brittle stars generally have five arms, but each arm may divide a number of times. Most of the 2,000 species of brittle stars ingest particles from the surfaces of sediments and assimilate the organic material from them, but some species remove suspended food particles from the water; others capture small animals. They eject the indigestible particles through their mouths because, unlike most other echinoderms, brittle stars have only one opening to their digestive tract.

Chordates: New Ways of Feeding

The second major lineage of deuterostomes, the phylum Chordata, evolved several different modifications of the coelomic cavity that provided new ways of capturing and handling food. Some living representatives of one early lineage—acorn worms—live buried in marine sand or mud, under rocks, or attached to algae. They may be similar to the ancestors of the chordate lineage, but are currently classed in their own lineage as hemichordates ("half-chordates"). Animals in the chordate lineage evolved a strikingly different body plan from the acorn worms, characterized by an internal dorsal supporting structure, which in the vertebrates evolved into the spinal column.

Acorn worms capture prey with a proboscis

The **acorn worms** (phylum **Hemichordata**) have a three-part body consisting of a proboscis, collar, and trunk (Figure 33.4). The 70 species of acorn worms live in burrows in muddy and sandy sediments.

Echinodermata
Hemichordata
Urochordata (tunicates)
Cephalochordata
Vertebrata

The large proboscis of acorn worms is a digging organ. It is coated with a sticky mucus that traps prey items in the sediment. The mucus and its attached prey are conveyed by cilia to the mouth. In the esophagus, the food-laden mucus is compacted into a ropelike mass that is moved through the digestive tract by ciliary action. Behind the mouth is a **pharynx** that opens to the outside through a number of **pharyngeal slits** through which water can exit. Highly vascularized tissue surrounding the pharyngeal slits serves as a gas exchange apparatus. An acorn worm breathes by pumping water into its mouth and out through its pharyngeal slits.

The pharynx becomes a feeding device

The same property required for effective gas exchange—a large surface area—also serves well for capturing prey. The pharyngeal slits, which originally functioned as sites for

Saccoglossus kowaleskii

33.4 A Hemichordate
The proboscis (right) of this acorn worm is modified for digging. This individual has been extracted from its burrow.

33.5 Tunicates
Pharyngeal baskets occupy most of the body cavities of these transparent sea squirts. The blue color is a reflection of the environment in this photograph.

gas exchange and eliminating water, as they do in modern acorn worms, were enlarged in a sister lineage. This enlargement of the pharyngeal slits eventually led to the remarkable evolutionary developments that gave rise to the chordates.

Chordates (phylum **Chordata**) are bilaterally symmetrical animals whose body plans are characterized by several shared features:

▶ Pharyngeal slits (at some stage of their development).

▶ A dorsal, hollow *nerve cord.*

▶ A ventral heart.

▶ A tail that extends beyond the anus.

▶ A dorsal supporting rod, the **notochord.**

The notochord is the most important derived trait of the Chordata and is unique to that phylum. In some species, such as tunicates, the notochord is lost during metamorphosis to the adult stage. In the vertebrates, it is replaced by skeletal structures that provide support for the body.

A notochord appears in tunicates and lancelets

The **tunicates** (subphylum **Urochordata**) may be similar to the ancestors of all chordates. All 2,500 species of tunicates are marine animals, most of which are sessile as adults. It is their swimming, tadpole-like larvae that reveal the close evolutionary relationships between tunicates and other chordates.

A tunicate larva has pharyngeal slits, a dorsal, hollow nerve cord, and a notochord. Muscles are attached to the notochord, providing the body with relatively rigid support. After a short time floating in the water, the larva settles on the seafloor and becomes a sessile adult. The nerve cord and notochord disappear in the adult animal, which

feeds by extracting plankton from the water. An adult's pharynx is enlarged into a *pharyngeal basket* lined with cilia, whose beating moves water through the animal.

More than 90 percent of known species of tunicates are sea squirts (class Ascidiacea). Some sea squirts are solitary,

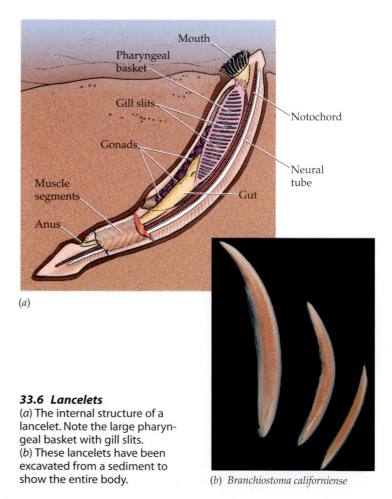

(a)

33.6 Lancelets
(a) The internal structure of a lancelet. Note the large pharyngeal basket with gill slits.
(b) These lancelets have been excavated from a sediment to show the entire body.

(b) *Branchiostoma californiense*

but others produce colonies by asexual budding from a single founder. Individual sea squirts range in size from less than 1 mm to 60 cm in length, but colonies may measure several meters across. The baglike body of an adult is surrounded by a tough tunic, composed of protein and a complex polysaccharide, which is secreted by the epidermal cells. Much of the body is occupied by the large pharyngeal basket (Figure 33.5).

The 25 species of **lancelets** (subphylum **Cephalochordata**) are small animals that rarely exceed 5 cm in length. Their notochord extends the entire length of the body throughout their lives, and they resemble small fishes. Lancelets live partly buried in soft marine sediments. They extract small prey from the water with their pharyngeal baskets (Figure 33.6).

Origin of the Vertebrates

In one chordate lineage, the pharyngeal basket became enlarged. With its many exit openings, an enlarged basket was effective in extracting prey from mud, which contains many inedible particles along with food. This lineage gave rise to the **vertebrates** (subphylum **Vertebrata**). In the late Cambrian period, these early vertebrates evolved improved structures for extracting food from mud and sand and for moving over the surface of the substratum.

Vertebrates take their name from a jointed, dorsal **vertebral column**, which replaced the notochord as their primary support. The vertebrate body plan (Figure 33.7) can be characterized as follows:

▶ With the vertebral column as its anchor, a rigid *internal skeleton* provides support and mobility.

▶ Two pairs of *appendages* are attached to the vertebral column.

▶ The faster locomotion made possible by appendages favored the evolution of an *anterior skull with a large brain* and highly developed sensory receptors.

▶ The internal organs are suspended in a large coelom.

▶ A well-developed *circulatory system*, driven by contractions of a ventral heart, delivers oxygen to internal organs.

The filter-feeding ancestral vertebrates lacked jaws. They probably swam over the sediments, sucking up mud and extracting microscopic food from it. These animals gave rise to the **fishes**. The lineage leading to modern hagfishes probably separated first from

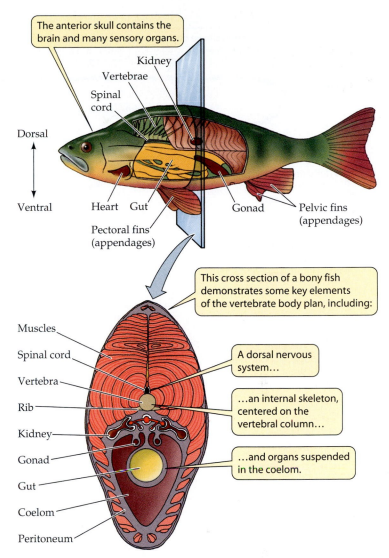

The anterior skull contains the brain and many sensory organs.

Dorsal

Ventral

Kidney
Vertebrae
Spinal cord

Heart   Gut   Gonad   Pelvic fins (appendages)
Pectoral fins (appendages)

This cross section of a bony fish demonstrates some key elements of the vertebrate body plan, including:

Muscles
Spinal cord
Vertebra
Rib
Kidney
Gonad
Gut
Coelom
Peritoneum

A dorsal nervous system…

…an internal skeleton, centered on the vertebral column…

…and organs suspended in the coelom.

33.7 The Vertebrate Body Plan
A bony fish is used here to illustrate the structural elements common to all vertebrates.

the other groups (Figure 33.8). One early group of jawless fishes, called ostracoderms, meaning "shell-skinned," evolved a bony external armor that protected them from predators. With their heavy armor, these small fishes could swim only slowly, but they could safely swim above the substratum, which was easier than having to burrow through it, as all previous sediment feeders had done.

The new mobility of jawless fishes enabled them to exploit their environments in new ways. They could attach to dead organisms and use the pharynx to create suction to pull fluids and partly decomposed tissues into the mouth. Hagfishes and lampreys, the only jawless fishes to survive beyond the Devonian period, feed on both dead and living organisms in this way (Figure 33.9). These fishes have tough, scaly skins instead of external armor. The round mouth is a sucking organ with which the animals attach to their prey and rasp at the flesh. Lampreys live in both fresh and salt water; many species move between the two environments, laying their eggs in rivers and maturing in the sea.

Echinodermata
Hemichordata
Urochordata (tunicates)
Cephalochordata
Vertebrata

Hagfishes
Lampreys
Cartilaginous fishes
Ray-finned fishes
Lobe-finned fishes
Lungfishes
Tetrapods

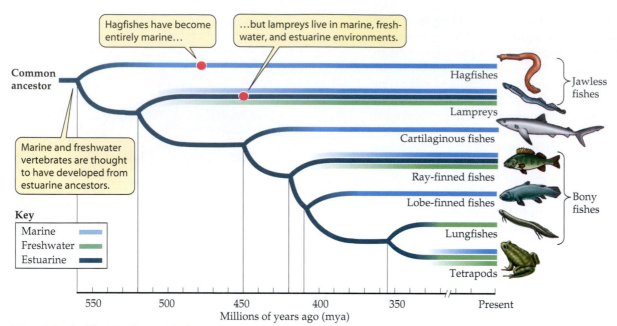

33.8 A Probable Vertebrate Phylogeny
This phylogeny incorporates the view that vertebrates evolved in estuaries, where their ability to handle varying salinities allowed them to exploit habitats not available to marine animals.

Jaws improve nutrition

During the Devonian period, many new kinds of fishes evolved in the seas, estuaries, and fresh waters. Although most of these were jawless, members of one lineage evolved jaws from some of the skeletal arches that supported the gill region (Figure 33.10). A jaw allows a fish to grasp and subdue relatively large, living prey. Further development of jaws and teeth among fishes led to the ability to chew both soft and hard body parts of prey. Chewing aided chemical digestion and improved the ability of fishes to obtain nutrients from prey.

Petromyzon marinus

33.9 A Modern Jawless Fish
This sea lamprey uses its large, jawless mouth to suck blood and flesh from other fishes.

The dominant early jawed fishes were the heavily armored **placoderms** (class **Placodermi**). Some of these fishes evolved elaborate appendages and relatively sleek body forms that improved their ability to maneuver in open water. A few became huge (10 meters long) and, together with squids (cephalopod mollusks), were probably the major predators in the Devonian oceans. Despite their early abundance, however, most placoderms disappeared by the end of the Devonian period; none survived to the end of the Paleozoic era.

Fins improve mobility

Two other groups of fishes—the bony fishes and the cartilaginous fishes, both of which survive today—became abundant during the Devonian period.

Cartilaginous fishes (class **Chondrichthyes**)—the sharks, skates and rays, and chimaeras (Figure 33.11)—have a skeleton composed entirely of a firm but pliable material called *cartilage*. Their skin is flexible and leathery, sometimes bearing bristly projections that give it the consistency of sandpaper. The loss of external armor increased their mobility and ability to escape from predators.

In the cartilaginous fishes and their descendants, swimming is controlled by pairs of unjointed appendages called *fins*: a pair of pectoral fins just behind the gill slits and a pair of pelvic fins just in front of the anal region. A dorsal median fin stabilizes the fish as it moves. Sharks move forward by means of their tail and pelvic fins. Skates and rays propel themselves by means of the undulating movements of their greatly enlarged pectoral fins.

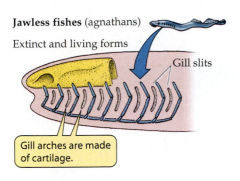

Jawless fishes (agnathans)
Extinct and living forms

Gill slits

Gill arches are made of cartilage.

Early jawed fishes (placoderms)

Extinct

Gill slits

Some anterior gill arches became modified to form jaws.

Modern jawed fishes (cartilaginous and bony fishes)

Living forms

Additional gill arches were incorporated to form heavier, more efficient jaws.

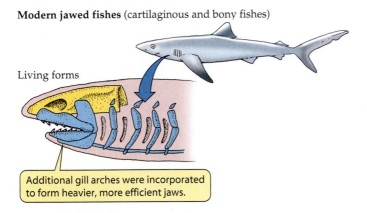

33.10 Jaws from Gill Arches
This illustrates one probable scenario for the evolution of jaws from the anterior gill arches of fishes.

Most sharks are predators, but some feed by filtering plankton from the water. The world's largest fish, the whale shark (*Rhincodon typhus*), is a filter feeder. It may grow to more than 15 meters in length and weigh more than 9,000 kilograms. Most skates and rays live on the ocean floor, where they feed on mollusks and other invertebrates buried in the sediments. Nearly all cartilaginous fishes live in the oceans.

Swim bladders allow control of buoyancy

The **bony fishes** (class **Osteichthyes**) have internal skeletons of bone rather than cartilage, giving them their common name. Their bony skeleton is lighter than that of the cartilaginous fishes. In most species, the outer surface is covered with flat, smooth, thin, lightweight scales that provide some protection. The gills of bony fishes open into a single chamber covered by a hard flap. Movement of the flap improves the flow of water over the gills, where gas exchange takes place.

Hagfishes
Lampreys
Cartilaginous fishes
Ray-finned fishes
Lobe-finned fishes
Lungfishes
Tetrapods

Early bony fishes also evolved gas-filled sacs that supplemented the action of the gills in respiration. These features enabled early bony fishes to live where oxygen was periodically in short supply, as it often is in estuarine and freshwater environments. They still serve this function in lungfishes and a few other modern fishes, but in the ray-finned fishes—a group that includes most of the many species of bony fish—these lunglike sacs evolved into **swim bladders**,

33.11 Cartilaginous Fishes
(*a*) Most sharks, such as this whitetip reef shark, are active marine predators. (*b*) Skates and rays, represented here by a stingray, feed on the ocean bottom. Their modified pectoral fins are used for propulsion.

(*a*) *Triaenodon obesus*

(*b*) *Trygon pastinaca*

(a) *Ocyurus chrysurus*

(b) *Plectorhinchus chaetodonoides*

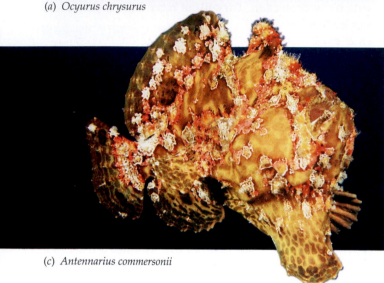

(c) *Antennarius commersonii*

(d) *Phyllopteryx taeniolatus*

33.12 Diversity among Bony Fishes
(a) The yellowtail snapper has a "typical" fish shape. (b) The coral grouper lives on tropical coral reefs. (c) Commerson's frogfish can change its color in a range from pale yellow to orange-brown to deep red, thus enhancing its camouflage abilities. (d) This weedy sea dragon is difficult to see when it hides in vegetation. It is a larger "cousin" of the more familiar seahorse.

which serve as organs of buoyancy. By adjusting the amount of gas in its swim bladder, a fish can control the depth at which it is suspended in the water without expending energy.

With their lighter skeletons and protective coverings and their swim bladders, ray-finned bony fishes evolved a remarkable diversity of sizes, shapes, and lifestyles (Figure 33.12). The smallest are less than 1 cm long as adults; the largest are ocean sunfishes that weigh up to 900 kilograms. Ray-finned fishes exploit nearly all types of aquatic food sources. In the oceans they filter plankton from the water, rasp algae from rocks, eat corals and other colonial invertebrates, dig invertebrates from soft sediments, and prey upon virtually all other vertebrates except large whales and dolphins. In fresh water they eat plankton, devour insects

of all aquatic orders, eat fruits that fall into the water in flooded forests, and prey on other aquatic vertebrates.

Some fishes live buried in soft sediments, capturing passing prey or emerging at night to feed. Many are solitary, but in open water others form large aggregations called *schools*. Many fishes perform complicated behaviors by means of which they maintain schools, build nests, court and choose mates, and care for their young.

With their fins and swim bladders, fishes can readily control their positions in open water, but their eggs tend to sink. Therefore, most fishes attach their eggs to the substratum, although a few species discharge their small eggs directly into surface waters where they are buoyant enough to complete their development before they sink very far. Most marine fishes move to food-rich shallow waters to lay their eggs, which is why coastal waters and estuaries are so important in the life cycles of many species. Some, such as salmon, abandon salt water when they breed, ascending rivers to spawn in freshwater streams and lakes.

Colonizing the Land: Obtaining Oxygen from the Air

Although the evolution of lunglike sacs was a response to the inadequacy of gills for respiration in oxygen-poor waters, it also set the stage for the invasion of land. Some early bony fishes probably used their lung sacs to supplement their gills when oxygen levels in the water were low. This ability would also have allowed them to breathe air, and to leave the water temporarily when pursued by predators unable to do so. But with their unjointed fins, bony fishes could only flop around on land, as most fishes do today if placed out of water. Changes in the structure of fins would help such fishes move about on land.

Two lineages of bony fishes evolved jointed fins: the **lobe-finned fishes** (subclass **Crossopterygii**) and the lungfishes. The lobe-fins flourished from the Devonian period until about 25 mya, when they were thought to have become extinct. However, in 1938, a lobe-fin was caught by commercial fishermen off the Comoro Islands in the Indian Ocean. Since that time, several dozen specimens of this extraordinary fish, *Latimeria chalumnae*, have been collected. *Latimeria*, a predator on other fishes, reaches a length of about 1.5 meters and weighs up to 82 kilograms (Figure 33.13). Although it belongs to the bony fish lineage, the skeleton of *Latimeria* is composed mostly of cartilage, not bone.

Some descendants of early fishes with jointed fins began to use terrestrial food sources and over time became more fully adapted to life on land. This lineage became the **tetrapods**: the four-legged amphibians, reptiles, birds, and mammals that are common today.

Amphibians invade the land

During the Devonian period, **amphibians** (class **Amphibia**) arose from ancestors they shared with the lungfishes. In this lineage, the stubby, jointed fins of their ancestors evolved into walking legs. The design of those legs has

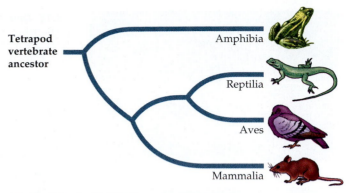

33.14 A Probable Phylogeny of Tetrapods
In the birds (Class Aves), the paired forelimbs evolved into wings.

remained largely unchanged throughout the evolution of terrestrial vertebrates (Figure 33.14).

The Devonian predecessors of amphibians were probably able to crawl from one pond or stream to another by pulling themselves along on their finlike legs, as do some modern species of catfishes. They gradually evolved to be able to live on swampy land and, eventually, on dry land. Living amphibians have relatively small lungs, and most species exchange gases through their skins. Most terrestrial species are confined to moist environments because they lose water rapidly through their skins when exposed to dry air.

About 4,500 species of amphibians live on Earth today, many fewer than the number known only from fossils. Living amphibians belong to three orders (Figure 33.15): the wormlike, tropical, burrowing caecilians (order Gymnophiona); frogs and toads (order Anura, which means "tailless"); and salamanders (order Urodela, which means "tailed"). Most species of frogs and toads live in tropical and warm temperate regions, although a few are found at very high latitudes and altitudes. Salamanders are more diverse in temperate regions, but many species are found in cool, moist environments in the mountains of Central America.

Most species of amphibians live in water at some time in their lives. In the typical amphibian life cycle, part or all of the adult stage is spent on land, usually in a moist habitat, but adults return to fresh water to lay their eggs (Figure 33.16). An amphibian egg can survive only in a moist environment because it is enclosed by a delicate envelope that cannot prevent water loss in dry conditions. The fertilized eggs of most species give rise to larvae that live in water until they change into terrestrial adults.

Amphibians are the focus of much attention today because populations of many species are declining rapidly. The golden toad, for example, has disappeared from the Monteverde Cloud Forest Reserve in Costa Rica, which was established primarily to protect this rare species. Several possible reasons for the declines, including drought, in-

33.13 A Modern Lobe-Fin
Latimeria chalumnae, found in deep waters of the Indian Ocean, is the sole surviving species of its lineage, which had been thought to be extinct.

(a) *Dermophus mexicanus*

(b) *Scaphiophryne gottlebei*

33.15 Diversity among the Amphibians
(a) Burrowing caecilians superficially look more like worms than amphibians.
(b) A rare frog species discovered in a national park on the island of Madagascar.
(c) A European fire salamander.

(c) *Salamandra salamandra*

Amniotes colonize dry environments

Most amphibians, as we have just seen, are limited to moist environments. Two morphological changes contributed to the ability of one lineage of vertebrates to control water loss and, therefore, to exploit a wider range of terrestrial habitats. One was the evolution of an egg with a shell that is relatively impermeable to water. The other was a combination of traits that reduced water loss, including a tough skin impermeable to water and kidneys that could excrete concentrated urine. The vertebrates that evolved both of these traits are called **amniotes**. They were the first vertebrates to become widely distributed over the terrestrial surface of Earth.

The amniote egg has a leathery or brittle calcium-impregnated shell that retards evaporation of the fluids inside, but permits O_2 and CO_2 to pass through. Such an egg does not require a moist environment, but can be laid anywhere. Within the shell and surrounding the embryo are membranes that protect the embryo from desiccation and assist its respiration and excretion of waste nitrogen. The egg also stores large quantities of food as yolk, permitting the embryo to attain a relatively advanced state of development before it hatches and must feed itself (Figure 33.17).

An early amniote lineage, the **reptiles** (class **Reptilia**) arose from the tetrapods during the Carboniferous period. As we discussed in Chapter 23, Rep-

creased ultraviolet radiation, and diseases, have been identified. Biologists are monitoring amphibian populations closely to learn more about the causes of their difficulties and to determine the implications of their declines for other organisms, including humans.

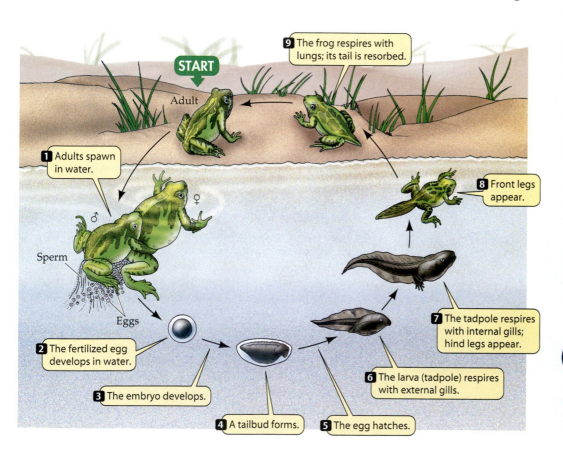

9 The frog respires with lungs; its tail is resorbed.

START

Adult

1 Adults spawn in water.

Sperm

♂

♀

Eggs

2 The fertilized egg develops in water.

3 The embryo develops.

4 A tailbud forms.

5 The egg hatches.

6 The larva (tadpole) respires with external gills.

7 The tadpole respires with internal gills; hind legs appear.

8 Front legs appear.

33.16 In and Out of the Water
Most stages in the life cycle of temperate-zone frogs take place in water. The aquatic tadpole is transformed into a terrestrial adult through metamorphosis.

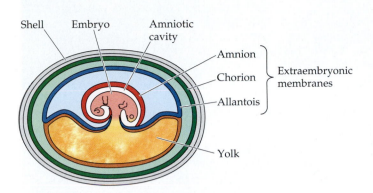

33.17 An Egg for Dry Places
The evolution of the amniote egg, with its shell, three extraembryonic membranes and embryo-nourishing yolk, was a major step in the colonization of the terrestrial environment.

The skin of a reptile is covered with horny scales that greatly reduce loss of water from the body surface. These scales, however, make the skin unavailable as an organ of gas exchange. In reptiles, gases are exchanged almost entirely by the lungs, which are proportionally much larger in surface area than those of amphibians. A reptile forces air into and out of its lungs by bellows-like movements of its ribs. Unlike the amphibian heart, the reptilian heart is divided into chambers that partially separate oxygenated from unoxygenated blood. With this type of heart, reptiles can generate higher blood pressures than amphibians and can sustain higher levels of muscular activity.

Reptilian lineages diverge

The lineages leading to modern reptiles began to diverge about 250 mya when the ancestors of the subclass **Squamata** (lizards, snakes, and amphisbaenians—a group of legless, wormlike, burrowing animals with greatly reduced eyes) diverged from the lineage leading to all other reptiles (Figure 33.19*a*, *b*). Most lizards are insectivores, but some are herbivores, and a few prey upon other vertebrates. The largest lizards, which may grow as long as 3 meters, are some species of monitors that live in the East Indies. Most lizards walk on four limbs, but some are limbless, as are all snakes, which are descendants of burrowing lizards.

All snakes are carnivores, and many can swallow objects much larger in diameter than themselves. The combination of poison glands and the ability to inject venom rapidly into their prey evolved several times. The largest snakes are pythons more than 10 meters long.

The tuataras (subclass **Sphenodontida**), which today are represented by only two species restricted to a few islands off the coast of New Zealand (Figure 33.19*c*), superficially resemble lizards, but differ from them in several internal anatomical features. Their phylogenetic relationships are uncertain.

Considerable uncertainty surrounds the next lineage split. Traditionally turtles were thought to have separated from other reptiles early in the history of the group, but new molecular analyses suggest that turtles are closely related to crocodilians, which diverged later. Both turtles and ancestral crocodiles have dorsal and ventral armored plates, and such plates have characterized those groups for many millions of years.

The dorsal and ventral bony plates of modern turtles and tortoises (subclass **Chelonia**) form a shell into which the head and limbs can be withdrawn (Fig-

tilia, as we use the term here, is a paraphyletic group because it does not include the birds, a major lineage that split off relatively recently during reptilian evolution. (Figure 33.18). However, because all reptiles are structurally similar, they serve as a convenient example for discussing the characteristics of amniotes. Therefore, we use this traditional classification as a basis for our discussion while recognizing that the birds should technically be included within it.

About 6,000 species of reptiles live today. Most reptiles do not care for their eggs after laying them. In some species the eggs do not develop shells, but are retained inside the female's body until they hatch. Still other species evolved structures called **placentas** that nourish the developing embryos.

33.18 The Reptiles Form a Paraphyletic Group
The traditional classification of the reptiles creates the paraphyletic group Reptilia. As used here, Reptilia does not include the birds (Aves), even though this major lineage split off from the crocodilian reptiles relatively recently (in evolutionary terms).

(a) *Trimeresurus sumatranus*

(d) *Chelonia mydas*

(b) *Ocyurus chrysurus*

(e) *Alligator mississippiensis*

Figure 33.19 Reptilian Diversity
(a) This Sumatran pit viper is prepared to strike. (b) This African chameleon, a lizard, has a long tail with which it can grasp branches and large eyes that move independently in their sockets. (c) The tuatara looks like a typical lizard, but it is one of only two survivors of a lineage that separated from lizards long ago. (d) The green sea turtle is widely distributed in tropical oceans. (e) Most crocodilians are tropical; alligators live in warm temperate environments in China and, like this one, in the southeastern United States.

(c) *Sphenodon punctatus*

ure 33.19d). Most turtles live in lakes and ponds, but tortoises are terrestrial; some live in deserts. Sea turtles spend their entire lives at sea except when they come ashore to lay eggs; all seven species are endangered. A few species of turtles and tortoises are carnivores, but most species are omnivores that eat a variety of aquatic and terrestrial plants and animals.

The crocodilians (subclass **Crocodylia**)—crocodiles, caimans, gharials, and alligators—are confined to tropical and warm temperate environments (Figure 33.19e). Crocodilians spend much of their time in water, but they build nests on land or on floating piles of vegetation. Their eggs are warmed by heat generated by the decay of organic matter that they place in the nest. Typically the eggs are guarded by the female until they hatch. All crocodilians are carnivorous; they prey on vertebrates of all classes, including large mammals.

Another lineage led to the *dinosaurs*, reptiles that rose to dominance about 215 mya and dominated terrestrial environments for about 150 million years. During this time, virtually all terrestrial animals more than 1 meter in length were dinosaurs. Some of the largest dinosaurs weighed up to 100 tons. Many were agile and could run rapidly. Some small predatory dinosaurs evolved feathers.

The ability to breathe and run simultaneously, which we take for granted, was a major innovation in the evolution of terrestrial vertebrates. Not until the evolution of the lineages leading to the mammals, dinosaurs, and birds did the legs assume vertical positions, which reduced the lateral forces on the body during locomotion. Special ventilatory muscles that enabled the lungs to be filled and emptied while the limbs moved also evolved. These muscles are visible in living birds and mammals; we can infer their existence in dinosaurs from the structure of the vertebral column in their fossils and the capacity of many dinosaurs for bounding bipedal (using two legs) locomotion.

Birds: Feathers and Flight

During the Mesozoic era, a dinosaur lineage gave rise to the birds (subclass **Aves**). The oldest known avian fossil, *Archaeopteryx*, which lived about 150 mya, was covered with feathers that are virtually identical to those of modern birds. It also had well-developed wings, a long tail (Figure 33.20*a*), and a wishbone, which in modern birds serves as an anchoring site for flight muscles. *Archaeopteryx* had typical perching bird claws, suggesting that it lived in trees and shrubs and used the clawed fingers on its forearms to assist it in clambering over branches.

Another early bird, *Confuciusornis sanctus*, which is known from hundreds of complete fossils from China, lived only slightly more recently than *Archaeopteryx*. Well-preserved fossils show that males had greatly elongated tail feathers (Figure 33.20*b*), which they probably used in communal courtship displays. Large numbers of these fossils have been found together, as would be expected if a number of males displayed together on communal display grounds.

Because the avian lineage separated from the other reptiles long before *Archaeopteryx* lived, existing data are insufficient to identify the ancestors of birds with certainty. Most paleontologists (scientists who study fossils) believe that birds evolved from terrestrial bipedal dinosaurs that used their forelimbs for capturing prey. According to this view, these small dinosaurs evolved feathers for insulation or display, and eventually were able to become airborne for short distances.

During the Cretaceous period, birds underwent an extensive evolutionary radiation. The dominant Cretaceous lineage was the "opposite birds," so named because the tarsal bones of their legs fused in the opposite direction from the way fusion happens in all modern birds. All lineages of opposite birds died out at the end of the Cretaceous,

(a)

(b)

33.20 Mesozoic Birds

(*a*) An artist's recreation of *Archaeopteryx* shows its modern feathers, arboreal habits, and flight. (*b*) The elongated tail feathers of a male *Confuciusornis sanctus* ("sacred bird of Confucius") fossil suggest that the males used them in courtship displays.

but scientists disagree over how many other avian lineages survived the mass extinction. Paleontologists believe that members of only one lineage, collectively known as the transitional shorebirds, survived the end of the Cretaceous, because no later fossils of other lineages have been found. Other scientists, who base their conclusions on molecular clocks, believe that at least some representatives of many Cretaceous avian lineages must have survived, because modern birds are too different to have come from a single lineage.

(a) *Pygoscelis papua*

33.21 Diversity among the Birds

(a) Penguins such as these gentoos are widespread in the cold waters of the southern hemisphere. They are expert swimmers, although they have lost the ability to fly. (b) Parrots are a diverse group of birds, especially in the tropics of Asia and the Pacific islands. This Australian king parrot is one member of Australia's rich parrot fauna. (c) Perching birds, represented here by a male northern cardinal, are the most species-rich of all the bird lineages.

(c) *Cardinalis cardinalis*

(b) *Alisterus scapularis*

As a group, birds eat almost all types of animal and plant material. A few aquatic species have bills modified for filtering small food particles from the water. Insects are the most important dietary item for terrestrial birds. In addition, they eat fruits and seeds, nectar and pollen, leaves and buds, carrion, and other vertebrates. Birds are major predators of flying insects during the day, and some species exploit that food source at night. By eating the fruits and seeds of plants, birds serve as major agents of seed dispersal.

The feathers developed by some dinosaurs may originally have had thermoregulatory or display functions. Birds use them for these purposes as well as for flying. The flying surface of the wings is created by large quills that arise from the forelimbs. Other strong feathers sprout like a fan from the shortened tail and serve as stabilizers during flight. The contour feathers and down feathers provide insulation to control loss of body heat.

The bones of birds are also modified for flight. They are hollow and have internal struts for strength. The sternum (breastbone) forms a large, vertical keel to which the breast muscles are attached. These muscles pull the wings downward during the main propulsive movement in flight.

Flight is metabolically expensive, and a flying bird consumes energy at a very high rate—about eight times the amount of energy per day as a lizard of the same weight! Because birds have such high metabolic rates, they generate large amounts of heat. They control the rate of heat loss using their feathers, which may be held close to the body or elevated to alter the amount of air trapped as insulation. The brain of a bird is larger in proportion to its body size than lizard or crocodile brains, primarily because the cerebellum, the center of sight and muscular coordination, is enlarged. The beaks of modern birds lack teeth.

Most birds lay their eggs in a nest, where they are warmed by the body heat of an adult that sits on them. Because birds have high body temperatures, the eggs of most species develop rapidly, hatching in less than 2 weeks. The offspring of many species hatch at a relatively helpless stage and are fed for some time by their parents. The young of other bird species, such as chickens, sandpipers, and ducks, can feed themselves shortly after hatching. Adults of all species attend their offspring for some time, warning them of and protecting them from predators, leading them to good foraging places, sheltering them from bad weather, and feeding them.

As adults, birds range in size from the 2-gram bee hummingbird of the West Indies to the 150-kg ostrich. Some flightless birds of Madagascar and New Zealand known from fossils were even larger, but they were exterminated by the humans that first colonized those islands. There are about 9,600 species of living birds, more than in any other major vertebrate group except fishes (Figure 33.21).

The Origin and Diversity of Mammals

Mammals (class **Mammalia**) appeared in the early part of the Mesozoic era, branching from a lineage of mammal-like reptiles. Small mammals coexisted with reptiles and dinosaurs for at least 150 million years. After the large reptiles and dinosaurs disappeared during the mass extinction at the close of the Mesozoic, mammals increased dramatically in numbers, diversity, and size.

Skeletal simplification accompanied the evolution of small mammals from their larger reptilian ancestors. During mammalian evolution, bones from the lower jaw were incorporated into the middle ear, leaving a single bone in the lower jaw, and the number of bones in the skull decreased. The bulk of both the limbs and the bony girdles from which they are suspended was reduced. Mammals have far fewer, but more highly differentiated teeth than reptiles. Differences in the number, type, and arrangement of teeth in mammals reflect their varied diets.

These skeletal features are readily preserved as fossils, but the soft parts of mammals are seldom fossilized. Therefore we do not know when mammalian features such as mammary glands, sweat glands, hair, and a four-chambered heart evolved. As is the case with birds, the mammalian fossil record suggests that most of the modern mammalian orders evolved rapidly after the end of the Cretaceous, whereas molecular data suggest that they had already evolved during the Cretaceous.

Mammals are unique among animals in providing their young with a nutritive fluid (milk) secreted by mammary glands. Mammalian eggs are fertilized within the female's body, and the embryos undergo some development within a uterus prior to being born. In addition, mammals have a protective and insulating covering of hair, which is luxuriant in some species but has been almost entirely lost in whales, dolphins, and humans. In whales and dolphins thick layers of insulating fat (blubber) replace hair. Clothing assumes the same role for humans.

Mammals range in size from tiny shrews weighing only about 2 grams to the endangered blue whale, which measures up to 31 meters long and weighs up to 160,000 kilograms—the largest animal ever to live on Earth. The approximately 4,000 species of living mammals are divided into two major subclasses: Prototheria and Theria.

The subclass **Prototheria** contains a single order, the Monotremata, with only three species, which are found only in Australia and New Guinea. These mammals, the duck-billed platypus and spiny anteaters, or echidnas, differ from other mammals in that they lack a placenta, lay eggs, and have legs that poke out to the side (Figure 33.22). Monotremes nurse their young on milk, but they have no nipples on their mammary glands; rather, the milk simply oozes out and is lapped off the fur by the offspring.

(a) *Tachyglossus aculeata*

(b) *Ornithorhyncus anatinus*

33.22 Monotremes
(a) The short-beaked echidna is one of two surviving species of echidnas. (b) The duck-billed platypus is the third surviving monotreme species.

Two major groups of mammals are members of the subclass **Theria**. Females of one group, the **Marsupialia**, have a ventral pouch in which they carry and feed their offspring (Figure 33.23a). Gestation (pregnancy) in marsupials is short; the young are born tiny but with well-developed forelimbs, with which they climb to the pouch. They attach to a nipple but cannot suck. The mother ejects milk into the tiny offspring until they grow large enough to suckle. Once her offspring have left the uterus, a female marsupial may become sexually receptive again. She can then carry fertilized eggs capable of initiating development and replacing the offspring in the pouch should something happen to them.

There are about 240 living species of marsupials. At one time marsupials were widely distributed on Earth, but today the majority of species are restricted to the Australian region, with a modest representation in South America

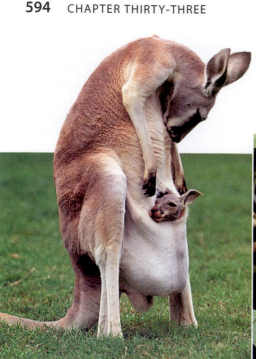

(a) *Macropus rufus*

(b) *Calumorys phicander*

(c) *Sarcophilus harrisii*

33.23 Marsupials
(a) Australia's kangaroos are thought of as the typical marsupial, but the marsupial radiation also produced (b) arboreal species such as this South American opossum and (c) carnivores such as the Tasmanian devil.

(Figure 33.23b). Only one species, the Virginia opossum, is widely distributed in the United States. Marsupials radiated to become terrestrial herbivores, insectivores, and carnivores, but no species live in the oceans or can fly, although some are gliders. The largest living marsupial is the red kangaroo of Australia, which weighs up to 90 kilo-

33.24 Diversity among the Eutherians
(a) The Arctic ground squirrel is one of many species of small, diurnal rodents of western North America. (b) Temperate-zone bats are all insectivores, but many tropical bats such as this leaf-nosed bat eat fruit. (c) Dolphins represent a eutherian lineage that returned to the marine environment. (d) Large hoofed mammals are important herbivores over much of Earth. This caribou bull is grazing by himself, although caribou are often seen in huge herds.

(a) *Citellus parryi*

(b) *Carollia perspicillapa*

(c) *Tursiops truncatus*

(d) *Rangifer tardanus*

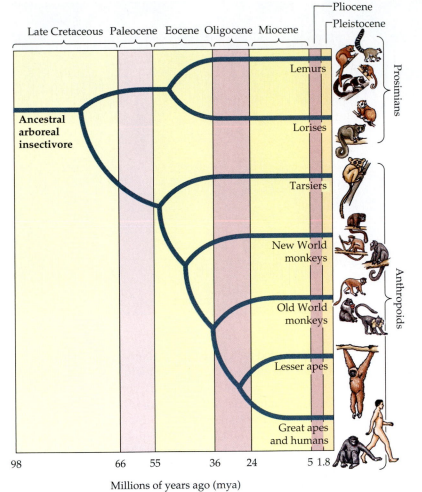

Late Cretaceous | Paleocene | Eocene | Oligocene | Miocene — Pliocene, Pleistocene

Ancestral arboreal insectivore

Lemurs — Prosimians
Lorises
Tarsiers
New World monkeys — Anthropoids
Old World monkeys
Lesser apes
Great apes and humans

98   66   55   36   24   5 1.8

Millions of years ago (mya)

33.25 A Probable Primate Phylogeny
Too few fossil primates have been discovered to reveal with certainty their evolutionary relationships, but this phylogenetic tree is consistent with existing evidence.

Primates and the Origins of Humans

Another eutherian lineage that has had dramatic effects on ecosystems worldwide is the **primate** lineage, to which humans belong. Primates have undergone extensive recent evolutionary radiation. They probably descended from small arboreal (tree-living) insectivores sometime during the Cretaceous period. The major traits that distinguish primates from other mammals are all adaptations to arboreal life. They include:

▶ Dexterous hands with opposable thumbs that can grasp branches and manipulate food.
▶ Nails rather than claws.
▶ Eyes on the front of the face that provide good depth perception.
▶ Very small litters of offspring (usually just one) that receive extended parental care.

Early in its evolutionary history, the primate lineage split into two main branches, prosimians and anthropoids (Figure 33.25). **Prosimians**—lemurs, bush babies, and lorises—once lived on all continents, but today they are restricted to Africa, Madagascar, and tropical Asia. All of the mainland species are arboreal and nocturnal (Figure 33.26). However, on Madagascar, the site of a remarkable prosimian radiation, there are also diurnal and terrestrial species.

The **anthropoids**—tarsiers, monkeys, apes, and humans—evolved from an early primate lineage about

grams, but much larger marsupials existed in Australia until they were exterminated by humans soon after they reached Australia about 50,000 years ago.

Most living mammals are **eutherians**. (Eutherians are sometimes called *placental mammals*, but this name is not accurate because some marsupials also have placentas.) Eutherians are more highly developed at birth than are marsupials, and no external pouch houses them after birth. The nearly 4,000 species of eutherians are divided into 16 major groups, the largest of which is the rodents, with about 1,700 species (Figure 33.24*a*). The next largest group, the bats, has about 850 species (Figure 33.24*b*), followed by the insectivores (moles and shrews) with slightly more than 400 species.

Eutherians are extremely varied in form and ecology. Several lineages of terrestrial mammals subsequently colonized marine environments, to become whales, dolphins, seals, and sea lions (Figure 33.24*c*). Eutherian mammals are—or were, until they were greatly reduced in numbers by humans—the most important grazers and browsers in most terrestrial ecosystems (Figure 33.24*d*). Grazing and browsing have been an evolutionary force intense enough to select for the spines, tough leaves, and difficult-to-eat growth forms found in many plants, a striking example of coevolution.

(*a*) *Propithecus verreauxi*   (*b*) *Galago senegalensis*

33.26 Prosimians
(*a*) The sifaka lemur is one of the many lemur species of Madagascar, where they are part of a unique asemblage of plants and animals. (*b*) The lesser bush baby is common in savannas over much of Africa. Its large eyes tell us that it is nocturnal.

(a) *Leontopithecus rosalia*

(b) *Macaca sylvanus*

33.27 Monkeys

(a) Golden lion tamarins, are endangered New World monkeys living in coastal Brazilian rainforests. (b) Many Old World species, such as these Barbary macaques, live in social groups. Here two members of a group groom each other.

55 million years ago in Africa or Asia. New World monkeys have been evolving separately from Old World monkeys long enough that they could have reached South America from Africa when those two continents were still close to each other. Perhaps because tropical America has been heavily forested for a long time, all New World monkeys are arboreal (Figure 33.27a). Many of them have long, prehensile tails with which they can grasp branches. Many Old World primates are arboreal as well, but a number of species are terrestrial. Some of these species, such as baboons and macaques, live and travel in large groups (Figure 33.27b). No Old World primates have prehensile tails.

About 20 mya, the lineage that leads to modern

(a) *Hylobates lar*

(b) *Pan paniscus*

(c) *Pongo pygmaeus*

(d) *Gorilla gorilla*

33.28 Apes

(a) Gibbons are the smallest of the apes. This common gibbon is found in Asia, from India to Borneo. (b) Chimpanzees, our closest relatives, are found in forested regions of Africa. (c) Orangutans live in the forests of Indonesia. (d) Gorillas, the largest apes, are restricted to humid African forests. This male is a lowland gorilla.

apes separated from the other Old World primates. The first apes were arboreal, but some species came to live in drier habitats with scattered trees, where they obtained most of their food on the ground. Apes are known to have lived in Africa, the Near East, and Asia 15–20 mya. Africa was especially rich in ape species, but the DNA sequences of living primates and some fossil evidence suggests that a European ape that dispersed into Africa about 10 mya may be the ancestor of modern apes. Four of the living genera of apes—gorillas (*Gorilla*), chimpanzees (*Pan*), orangutans (*Pongo*), and gibbons (*Hylobates*)—are restricted to tropical Africa and Asia (Figure 33.28). The fifth (*Homo*) has a worldwide distribution.

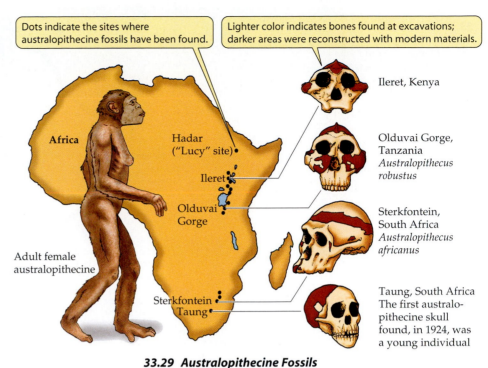

Dots indicate the sites where australopithecine fossils have been found.

Lighter color indicates bones found at excavations; darker areas were reconstructed with modern materials.

Africa

Adult female australopithecine

Hadar ("Lucy" site)
Ileret
Olduvai Gorge
Sterkfontein
Taung

Ileret, Kenya

Olduvai Gorge, Tanzania
Australopithecus robustus

Sterkfontein, South Africa
Australopithecus africanus

Taung, South Africa
The first australopithecine skull found, in 1924, was a young individual

33.29 Australopithecine Fossils
Few fossilized remains are complete, but skull shapes can be reconstructed accurately.

Human ancestors descended to the ground

The primate lineage that led to humans began with the **ardipithecines**. These apes had distinct morphological adaptations for **bipedalism**—locomotion in which the body is held erect and moved exclusively by movements of the hind legs. Bipedal locomotion frees the hands to manipulate objects and to carry them while walking. It also elevates the eyes, enabling the animal to see over tall vegetation to spot predators and prey. Both advantages were probably important for early ardipithecines and their descendants, the **australopithecines**.

The first australopithecine skull was found in South Africa in 1924; since then other fragments have been found in a number of sites in Africa (Figure 33.29). The most complete fossil skeleton of an australopithecine, approximately 3.5 million years old, was discovered in Ethiopia in 1974. That individual, a young female known to the world as Lucy, attracted a great deal of attention because her remains were so complete and well preserved. Lucy has been assigned to the species *Australopithecus afarensis*, the most likely ancestor of humans. All the evidence from different parts of her skeleton suggests that Lucy was only about 1 meter tall and walked upright.

From *Australopithecus afarensis* ancestors, a number of species of australopithecines evolved. Several million years ago, two distinct types of australopithecines lived together over much of eastern Africa. The larger type (about 40 kg) is represented by at least two species, both of which died out suddenly about 1.5 million years ago. The 25–30-kg *A. africanus* is much rarer as a fossil, suggesting that it was less common than the other species.

Humans arose from australopithecine ancestors

Many experts believe that the recently discovered *Australopithecus garhi* or a similar species gave rise to the genus *Homo*. *A. garhi*, a small-brained, big-toothed hominid with humanlike leg proportions, began using tools to obtain food about 2.5 mya. They butchered animals to get at bone marrow, which is rich in fat.

Early **hominids**—members of the genus *Homo*—lived contemporaneously with australopithecines for perhaps half a million years. Two major changes accompanied the evolution of *Homo* from *Australopithecus*: an increase in body size and a doubling of brain size.

The oldest fossil remains of a member of the genus *Homo*, named *H. habilis*, were discovered in the Olduvai Gorge, Tanzania, and are estimated to be 2 million years old. Other fossils of *H. habilis* have been found in Kenya and Ethiopia. Tools used by these early hominids were found with the fossils. *H. habilis* lived in relatively dry areas where, for much of the year, the main food reserves are subterranean roots, bulbs, and tubers. To exploit these food resources, an animal must dig into hard, dry soils, something that cannot be done with an unaided primate hand. However, roots can be dug in large quantities in a relatively short time by an individual with a simple digging tool. *H. habilis* females carrying infants could have done so, freeing males to hunt animal prey to provide the proteins that roots lack.

The only other known extinct species of our genus, *Homo erectus*, evolved in Africa about 1.6 million years ago. Soon thereafter it had spread as far as eastern Asia. As it expanded its range and increased in abundance, *H. erectus* may have exterminated *H. habilis*. Members of *H. erectus* were as large as modern people, but their bones were considerably heavier. *H. erectus* used fire for cooking and for hunting large animals, and made characteristic stone tools

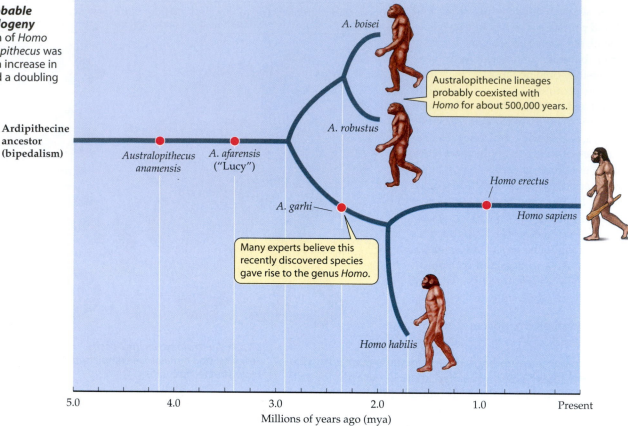

33.30 A Probable Human Phylogeny The evolution of *Homo* from *Australopithecus* was marked by an increase in body size and a doubling of brain size.

that have been found in many parts of the Old World. These tools were probably used for digging, capturing animals, cleaning and cutting meat, scraping hides, and cutting wood. Although *H. erectus* survived in Eurasia until about 250,000 years ago, it was replaced in tropical regions by our species, *Homo sapiens*, about 200,000 years ago.

Brains steadily became larger

The trends that accompanied the transition from *Australopithecus* to *H. erectus* continued during the evolution of our own species (Figure 33.30). The earliest members of *Homo sapiens* had larger brains than members of the earlier species of *Homo*, a change that was probably favored by an increasingly complex social life. The ability of group members to communicate with one another was valuable for cooperative hunting and gathering and for improving one's status in the complex social interactions that must have characterized those societies, just as they do ours today.

Several types of *H. sapiens* existed during the mid-Pleistocene epoch, from about 1.5 million to about 300,000 years ago. All were skilled hunters of large mammals, but plants continued to be important components of their diets. During this period another distinctly human trait emerged: rituals and a concept of life after death. Deceased individuals were buried with tools and clothing, presumably for their existence in the next world.

One type of *H. sapiens*, generally known as Neanderthals because they were first discovered in the Neander Valley in Germany, was widespread in Europe and Asia between about 75,000 and 30,000 years ago. Neanderthals were short, stocky, and powerfully built humans whose massive skulls housed brains somewhat larger than our own. They manufactured a variety of tools and hunted large mammals, which they probably ambushed and subdued in close combat. For a short time, their range overlapped that of a more modern form of *H. sapiens* known as Cro-Magnons, but then the Neanderthals abruptly disappeared. Many scientists believe that they were exterminated by the Cro-Magnons, just as *H. habilis* may have been exterminated by *H. erectus*.

Cro-Magnon people made and used a variety of sophisticated tools. They created the remarkable paintings of large mammals, many of them showing scenes of hunting, that have been discovered in caves in various parts of Europe (Figure 33.31*a*). The animals depicted were characteristic of the cold steppes and grasslands that occupied much of Europe during periods of glacial expansion. Cro-Magnon people spread across Asia, reaching North America perhaps as early as 20,000 years ago, although the date of their arrival in the New World is still uncertain. Within a few thousand years they had spread southward through North America to the southern tip of South America.

Humans evolved language and culture

As our ancestors evolved larger brains, their behavioral capabilities increased, especially the capacity for language. Most animal communication consists of a limited number of signals, which pertain mostly to immediate circumstances

(a)

(b)

33.31 Hunting, Pastoralism, and Agriculture

(a) Cro-Magnon cave drawings such as those found in Lascaux Cave in France typically depict the large mammals that they hunted. (b) The Masai are a pastoral people living on East African savannas, where they and their cattle typically coexist with native grazing and browsing mammals. (c) Intense agricultural development totally transforms the landscape. These rice terraces are on the island of Bali in Indonesia.

(c)

and are associated with changed emotional states induced by those circumstances. Human language is far richer in its symbolic character than any other animal vocalizations. Our words can refer to past and future times and to distant places. We are capable of learning thousands of words, many of them referring to abstract concepts. We can rearrange words to form sentences with complex meanings.

The expanded mental abilities of humans are largely responsible for the development of **culture**, the process by which knowledge and traditions are passed along from one generation to another by teaching and observation. Culture can change rapidly because genetic changes are not necessary for a cultural trait to spread through a population. The primary disadvantage of culture is that its norms must be taught to each generation. The tools and other implements associated with human fossils, as well as the cave paintings early humans created, reveal cultural traditions.

Cultural learning greatly facilitated the spread of domesticated plants and animals and the resultant conversion of most human societies from ones in which food was obtained by hunting and gathering to ones in which *pastoralism* (herding large animals) and *agriculture* dominated (Figure 33.31b,c). The development of agriculture led to an increasingly sedentary life, the growth of cities, greatly expanded food supplies, a rapid increase in the human population, and the appearance of occupational specializations, such as artisans and healers.

Agriculture developed in the Middle East approximately 11,000 years ago. From there it spread rapidly northwest-ward across Europe, finally reaching the British Isles about 4,000 years ago. The first plants and animals to be domesticated were cereal grains such as wheat and barley; legumes (beans, lentils, and peas); and woody plant crops such as grapes and olives. Others, such as rye, cabbage, celery, and carrots, were domesticated later. Cattle, sheep, goats, horses, dogs, and cats were the most important domesticated animals.

Agriculture developed independently in eastern Asia, contributing to our modern diet soybeans, rice, citrus fruits,

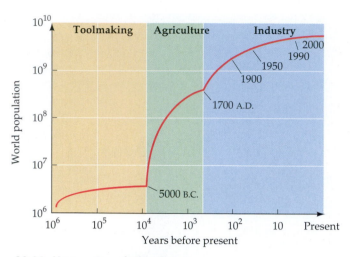

33.32 Human Population Surges
The human population surged following (1) the invention of tools, (2) the domestication of plants and animals, and (3) the Industrial Revolution.

mangoes, pigs, and chickens. There was some exchange, even at early times, among agricultural centers in the Old World, but when people spread across the cold and barren Bering land bridge into the New World, they apparently brought no domesticated plants with them. These people subsequently developed rich and varied agricultural systems based on corn, tomatoes, kidney and lima beans, peanuts, potatoes, chili peppers, and squashes. The largest animals domesticated by humans in the New World were llamas and alpacas in the Andes of South America—animals not large enough to carry a person. The Amerindians of Mexico and Central America had no domesticated animals larger than turkeys.

The human population has grown rapidly

The human population has experienced three major phases of increase (Figure 33.32). The first, stimulated by tool use, lasted about a million years. At the end of that period, the human population is estimated to have been approximately 5 million. During the second surge, which followed the domestication of plants and animals and the invention of agriculture, the human population may have increased to about 500 million people within 8,000 years.

We are currently in the middle of the third great population surge, triggered in the eighteenth century by the Industrial Revolution. In the industrialized countries of the world, death rates fell and life spans increased. By the end of the nineteenth century, human numbers had passed 1 billion. Despite the devastation of two World Wars and countless minor conflicts, the twentieth century saw the human population reach its current level of more than 6 billion. It is projected to increase to more than 11 billion by 2050.

The first two human population surges were followed by periods of relative stability. Whether the current surge will follow the same pattern, at what size it might level off, what hardships might ensue if it does not, and what the

consequences will be for other species are questions that are fiercely debated. We will discuss these issues further in Chapter 58.

 Chapter Summary

Deuterostomes and Protostomes

▶ The deuterostome lineage separated from the protostome lineage early in the history of animal life on Earth.

▶ There are fewer major lineages and fewer species of deuterostomes than protostomes, but as members of the deuterostome lineage we have a special interest in its members. **Review Figure 33.1**

Echinoderms: Complex Biradial Symmetry

▶ Echinoderms have a radially symmetrical body plan, a unique water vascular system, and a calcified internal skeleton. **Review Figure 33.2a**

▶ Nearly all living species of echinoderms have a bilaterally symmetrical, ciliated larva that feeds as a planktonic organism. **Review Figure 33.2b**

▶ Six major groups of echinoderms survive today, but 23 other lineages existed in the past. Some groups of echinoderms have arms, but others do not.

Chordates: New Ways of Feeding

▶ Evolution among the hemichordates and chordates led to new ways of capturing and handling food.

▶ The large proboscis of acorn worms is both a digging and a food-capturing organ.

▶ Members of the chordates evolved enlarged pharyngeal slits as feeding devices and a dorsal supporting rod, the notochord.

▶ Tunicates are sessile as adults and filter prey from seawater with large pharyngeal baskets. Their larvae have notochords and dorsal, hollow nerve cords.

Origin of the Vertebrates

▶ Vertebrates evolved jointed internal skeletons centered around a vertebral column, a body plan that enabled them to swim rapidly. Early vertebrates fed by filtering small animals from mud. **Review Figure 33.7**

▶ Jaws evolved from anterior gill arches and enabled their possessors to grasp and chew their prey, expanding food sources and improving nutrition. Jawed fishes rapidly became the dominant animals in both marine and fresh waters. **Review Figures 33.8, 33.10**

▶ Fishes evolved unjointed fins with which they could control their swimming movements and stabilize themselves in the water, and lunglike sacs that helped them stay suspended in open water.

▶ Bony fishes come in a wide variety of sizes and shapes, and many species have complex behaviors.

Colonizing the Land: Obtaining Oxygen from the Air

▶ Two fish lineages—lobe-fins and lungfishes—evolved jointed fins. Amphibians, the first terrestrial vertebrates, arose from one of these lineages. **Review Figure 33.14**

▶ Most amphibians live in water at some time in their lives, and their eggs must remain moist. **Review Figure 33.16**

▶ About 4,500 species of amphibians live today. They belong to three orders: caecilians, frogs and toads, and salamanders.

▶ Amniotes evolved eggs with shells impermeable to water and thus became the first vertebrates to be independent of water for breeding. **Review Figure 33.17**

▶ Modern reptiles are members of four lineages—turtles and tortoises; tuataras; snakes and lizards; and crocodilians. **Review Figure 33.18**

▶ Dinosaurs rose to dominance about 215 mya and dominated terrestrial environments for about 150 million years until their extinction at the end of the Mesozoic era.

Birds: Feathers and Flight

▶ Birds arose about 175 mya, but much controversy surrounds their origins.

▶ The 9,600 species of living birds are characterized by feathers, high metabolic rates, and parental care.

The Origin and Diversity of Mammals

▶ Mammals evolved during the Mesozoic era, about 225 mya.

▶ Eggs of mammals are fertilized within the bodies of females, and embryos develop there for some time before being born. Mammals are unique in suckling their young with milk secreted by mammary glands.

▶ The three species of mammals in subclass Prototheria lay eggs, but all other mammals give birth to developed young.

▶ Therian mammals are divided into two major groups, the marsupials: which give birth to tiny young that are raised in a pouch on the female's belly, and eutherians, which give birth to relatively well-developed offspring.

Primates and the Origins of Humans

▶ The primates split into two major lineages, one leading to the prosimians—lemurs, lorises, and pottos—and the other leading to the anthropoids—tarsiers, monkeys, apes, and humans. **Review Figure 33.25**

▶ Hominids evolved in Africa from terrestrial, bipedal ancestors. **Review Figures 33.29, 33.30**

▶ Early humans evolved large brains, language, and culture. They made and used tools, developed rituals, and domesticated plants and animals. In combination, these traits enabled humans to increase greatly in numbers.

▶ The human population has increased greatly three times. We are currently in the middle of the third population surge. When and how it will end is hotly debated. **Review Figure 33.32**

For Discussion

1. In what animal phyla has the ability to fly evolved? How do structures used for flying differ among these animals?

2. Extracting suspended food from the water column is a common mode of foraging among animals. Which groups contain species that extract prey from the air? Why is this mode of obtaining food so much less common than extracting prey from water?

3. Large size both confers benefits and poses certain risks. What are these risks and benefits?

4. Amphibians have survived and prospered for many millions of years, but today many species are disappearing and populations of others are declining seriously. What features of amphibian life histories might make them especially vulnerable to the kinds of environmental changes now happening on Earth?

5. The evolution of jaws allowed vertebrates to utilize a remarkably wide array of food types; yet jawless animals are able to eat most of those kinds of foods. Compare the ways that jawed and jawless animals would eat the following kinds of food:
   a. a snail
   b. an insect
   c. a fish
   d. a bird
   e. a plant leaf

6. The body plan of all vertebrates is based on four appendages. Describe the varied forms that these appendages take and how they are used. How do the vertebrates that have lost their four appendages move?

7. Compare the ways that different animal lineages colonized the land. How were those ways influenced by the body plans of animals in the different lineages?

Part Seven

ECOLOGY AND BIOGEOGRAPHY

53 *Behavioral Ecology*

WHEN A THOMSON'S GAZELLE SPOTS A PREDator such as a cheetah, it starts to run away but then it usually slows down a bit and performs a behavior called stotting. It jumps about half a meter off the ground with all four legs held straight and its white rump patch fully everted. Why would an animal that is fleeing from a predator slow down rather than speed up?

Several hypotheses have been proposed to explain why stotting may have evolved. First, the gazelle may be warning other nearby gazelles—particularly its relatives—that it has spotted a predator. Second, stotting individuals in a fleeing herd might distract and confuse a predator. Third, a stotting gazelle may be signaling to the predator that it has been seen and therefore pursuit will not be profitable.

Predictions based on these hypotheses have been tested in the field. The first and second hypotheses can be rejected because even solitary gazelles usually stot when they spot a predator. Only the third hypothesis is supported by existing data: Cheetahs, the most important predators on gazelles, are significantly more likely to abandon a pursuit if a gazelle stots than if it does not.

Individuals of all species, not only gazelles, interact in various ways with individuals of their own and other species and with their physical environments. The task of **ecology** is to understand the nature and consequences of these interactions. Ecologists may study these interactions by formulating and testing hypotheses, as they did for stotting in gazelles. Ecologists also study patterns of distribution and abundance of organisms to determine how these patterns are established and maintained, and how they change over short and long time periods. From its roots in descriptive natural history, ecology has developed into a complex field of inquiry dealing with levels of organization ranging from relationships of individual organisms with their physical and biological environments to the structure of communities and ecosystems.

As used by ecologists, the term *environment* includes *abiotic* (physical and chemical) factors such as water, nutrients, light, temperature, and wind. It also includes *biotic* factors: all other organisms that influence the lives of individuals. Because species are adapted for life in many different environments, their interactions with their biotic and abiotic environments are also varied. An environmental factor that exerts a strong influence on individuals of one species may have no influence on individuals of another species.

Interactions between organisms and their environments are two-way processes. Organisms both influence and are influenced by their environments. Indeed, managing environmental changes caused by our own species is one of the major problems of the modern world. For this reason, ecologists are often asked to help analyze causes of environmental problems and to assist in finding solutions for them. However, it is important not to confuse the science of ecology with the term "ecology" as it is often used in popular writing, referring to nature as a kind of superorganism.

In this first chapter on ecology, we will discuss **behavioral ecology**: how animals make decisions* that influence their survival and reproductive success, how ecologists study these decisions, and what they have learned from their studies.

Animals decide where to carry out their activities and how to select the resources they need—food, water, shelter, nest sites. Animals also respond to predators and competitors, and decide how to interact with other members of their own species (called *conspecifics*). Individual choices are the foundation of much of ecology, because changes in densities and distributions of populations are the cumulative results of the decisions of myriad individuals.

*The use of the words "decision" and "decide" here does not imply that the choices animals make are conscious, but rather that these choices influence their survival and reproductive success and thus are molded by natural selection.

A Stotting Gazelle
Although it is being pursued by a cheetah, this Thomson's gazelle (*Gazella thomsoni*) has slowed down to jump high off the ground.

Balancing Costs and Benefits of Behaviors

In their attempts to understand the evolution of behavior, ecologists often analyze their observations in terms of costs and benefits. A **cost–benefit analysis** is based on the principle that an animal has only a limited amount of time and energy to devote to its activities. Animals generally do not perform behaviors whose total costs are greater than the sum of their benefits—the improvements in survival and reproductive success that the animal achieves by performing the behavior.

Of course, animals do not consciously calculate costs and benefits, but over many generations, natural selection molds behavior in accordance with costs and benefits. A cost–benefit approach provides a framework that behavioral ecologists can use to make observations and design experiments that enable them to understand why behavior patterns evolve as they do.

The cost of behaving has three components. The *energetic cost* of a behavior is the difference between the energy the animal would have expended had it rested and the energy expended in performing the behavior. The *risk cost* of a behavior is the increased chance of being injured or killed as a result of performing it, compared with resting. The *opportunity cost* of a behavior is the sum of the benefits the animal forfeits by not being able to perform other behaviors during the same time interval. Opportunity costs measure the fitness trade-offs involved in performing different behaviors for different amounts of time. An animal that devotes all of its time to foraging, for example, does not achieve high reproductive success!

During the mating season in September and October, male Yarrow's spiny lizards maintain territories from which they exclude conspecific males. Behavioral ecologists assessed the costs of this behavior by stimulating territorial behavior in the lizards during June and July, a time of year when the lizards normally are only weakly territorial. They did this by inserting small capsules containing testosterone beneath the skin of some males. Control males were also captured and released, but received no testosterone implants.

The testosterone-implanted males patrolled their territories more, performed more threat displays, and expended about one-third more energy (energetic cost) than control males. As a result, they had less time to feed (opportunity cost), captured fewer insects, stored less energy, and died at a higher rate (risk cost) (Figure 53.1).

This experiment demonstrated that the costs of active territorial defense in June and July outweigh the benefits. Probably this is why the lizards normally reduce territorial behavior at that time of year. Presumably, during the mating season, when females are receptive and territorial behavior can result in increased reproductive success, its benefits are great enough to offset these costs.

Choosing Where to Live and Forage

Selecting a place in which to live is one of the most important decisions an individual makes. Where an individual chooses to live and how it uses that environment strongly

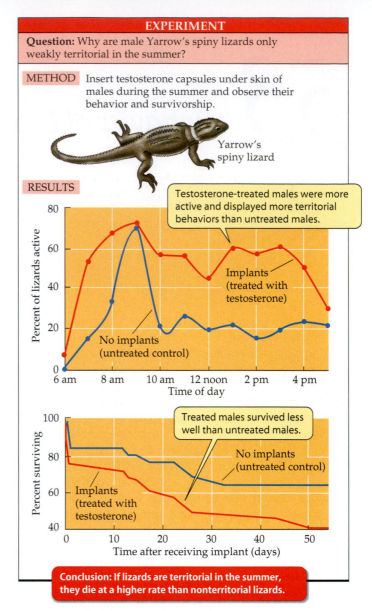

EXPERIMENT

Question: Why are male Yarrow's spiny lizards only weakly territorial in the summer?

METHOD   Insert testosterone capsules under skin of males during the summer and observe their behavior and survivorship.

Yarrow's spiny lizard

RESULTS

Testosterone-treated males were more active and displayed more territorial behaviors than untreated males.

Implants (treated with testosterone)

No implants (untreated control)

Treated males survived less well than untreated males.

No implants (untreated control)

Implants (treated with testosterone)

Conclusion: If lizards are territorial in the summer, they die at a higher rate than nonterritorial lizards.

53.1 The Costs of Defending a Territory
By using testosterone implants to increase territorial behavior, experimenters measured the costs to male Yarrow's spiny lizards of defending a territory during the summer months, when females are not receptive.

influence its survival and reproductive success. In this section, we will consider how animals choose environments that provide adequate food and shelter, and how they select their food.

Features of an environment may indicate its suitability

The environment in which an organism normally lives is called its **habitat**. Once a habitat is chosen, an animal seeks its food, resting places, nest sites, and escape routes from predators within that habitat.

The cues organisms use to select suitable habitats are as varied as the organisms themselves, but all habitat selection cues have a common feature: They are good predictors of general conditions suitable for future survival and reproduction. A simple example of habitat selection cues is that of the red abalone, a gastropod mollusk that begins its life as a fertilized egg in the open ocean. The egg hatches about 14 hours after fertilization, and a motile larva emerges with enough yolk to continue developing for another 7 days. During this time it swims in the open water without eating. At the end of 7 days, the larva stops developing, swims to the seafloor, chooses a place in which to settle, and metamorphoses into an adult.

Red abalone larvae settle only on coralline algae, upon which they feed. They recognize coralline algae by the presence of a specific chemical—a water-soluble peptide containing about 10 amino acids—that the algae produce. In the laboratory, abalone larvae will settle on any surface on which this chemical has been placed, but in nature *only* coralline algae produce it. By using this simple cue, these larvae always settle on a surface that is suitable for their future development.

Red abalone grow to adulthood feeding on a single patch of coralline algae; their habitat is their food. However, most animals must make many choices about where and how to seek and select food after they have settled in a habitat.

How do animals choose what to eat?

Because food is such an important resource, we consider it here in some detail. When an animal is looking for food, how much time should it spend searching in one area before moving to another site? When many different types of prey are available, which ones should a predator take, and which ones should it ignore? **Foraging theory** attempts to provide answers to these questions.

To construct a hypothesis about how a foraging animal should behave, a scientist first specifies the objective of the behavior and then attempts to determine the behavioral choices that would best achieve that objective. This approach is known as **optimality modeling**. Its underlying assumption is that natural selection has molded the behavior of animals so that they solve problems by making the best choices available to them. Many hypotheses are possible, because a forager may be attempting to maximize the rate at which it obtains energy, to get enough vitamins or minerals, or to minimize its risk while foraging.

As an example, consider how a predator should choose among the available prey. Let's assume that the predator chooses prey so as to maximize its rate of energy intake. This is a reasonable objective because the more efficiently a predator captures food, the more time and energy it will have for other activities, such as reproduction. Therefore, a more efficient predator should produce more offspring than a less efficient one, and animals should evolve to make prey choice decisions that maximize their rate of energy intake.

To test this hypothesis, we can characterize each type of prey available to the predator by two features: the amount of time it takes the predator to pursue, capture, and consume one of them, and the amount of energy an individual prey item contains. We then rank the prey according to the amount of energy the predator gets relative to the amount of time it spends pursuing, capturing, and handling the prey. The most valuable prey type is the one that yields the most energy per unit of time invested.

With this information, we can build an optimality model to determine the rate at which a predator would obtain energy given a particular prey selection strategy. We can then compare alternative foraging strategies and determine the one that yields the highest rate of energy intake. Such calculations show that, if the most valuable prey type is abundant enough, a predator gains the most energy per unit of time spent foraging by taking only the most valuable prey type and ignoring all others. However, as the abundance of the most valuable prey type decreases, an energy-maximizing predator adds less valuable prey types to its diet in order of the energy per unit of time that those prey yield.

Ecologists performed laboratory experiments with bluegill sunfish to measure the energy content of different prey types (water fleas of different sizes), the time needed to capture and eat different prey types, the energy spent pursuing and capturing prey, and actual encounter rates with prey under different prey densities.

Using these measurements, the investigators predicted the proportions of large, medium, and small water fleas that bluegills would capture in environments stocked with different densities and proportions of those prey types (Figure 53.2). Based on the optimality model, they predicted

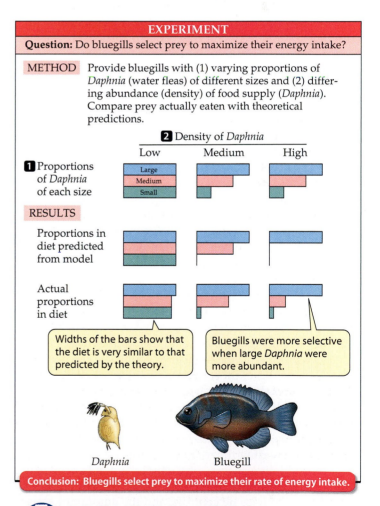

EXPERIMENT

Question: Do bluegills select prey to maximize their energy intake?

METHOD Provide bluegills with (1) varying proportions of *Daphnia* (water fleas) of different sizes and (2) differing abundance (density) of food supply (*Daphnia*). Compare prey actually eaten with theoretical predictions.

2 Density of *Daphnia*

Low | Medium | High

1 Proportions of *Daphnia* of each size
Large
Medium
Small

RESULTS

Proportions in diet predicted from model

Actual proportions in diet

Widths of the bars show that the diet is very similar to that predicted by the theory.

Bluegills were more selective when large *Daphnia* were more abundant.

Daphnia   Bluegill

Conclusion: Bluegills select prey to maximize their rate of energy intake.

53.2 Bluegills are Energy Maximizers
In an experiment, the prey choices of bluegill sunfish were very similar to those predicted by an optimality model based on the goal of maximizing the rate of energy intake.

that in an environment stocked with low densities of all three types of prey, the fish would take every water flea that they encountered, but that in an environment with abundant large water fleas, the bluegills would ignore smaller water fleas.

To test their predictions, the investigators put the bluegills in environments containing three different prey densities and observed the proportions of the water fleas of different sizes they actually captured. The proportions of large, medium, and small water fleas taken by the fish were very close to those predicted by the model. Such tests of foraging theory using many different kinds of animals have provided ecologists with a set of rules showing how animals find and choose their prey. They have also provided estimates of the energetic costs and benefits of foraging behavior.

Mating Tactics and Roles

Individual animals choose their associates, how to interact with them, and when to leave them. The most important choice of associates an animal makes is mate selection.

Mating behavior involves only a small set of choices. The most basic mating decision is choosing a partner of the correct species. Once that has been determined, additional decisions can be based on the qualities of a potential mate, on the resources it controls—food, nest sites, escape places—or on a combination of the two. Among those species in which individuals do not control any resources, the traits of the partner are the only criteria for mate selection. Here we will discuss how individuals choose their mating partners and show why males and females approach courtship so differently.

Abundant sperm and scarce eggs drive mating behavior

The reproductive behavior of males and females is often very different. Males usually initiate courtship, and they often fight for opportunities to mate with females. Females seldom fight over males, and they often reject courting males. Why do males and females approach courtship and copulation so differently?

The answer lies in the costs of producing sperm and eggs. Because sperm are small and cheap to produce, one male produces enough to sire a very large number of offspring—usually many more than the number of eggs a female can produce or the number of young she can nourish. Therefore, males of most species can increase their reproductive success by mating with many females

Eggs, on the other hand, are typically much larger than sperm and are expensive to produce. Consequently, a female is unlikely to increase her reproductive output very much by increasing the number of males she mates with. The reproductive success of a female depends primarily upon the the quality of the genes she receives from her mate, the resources he controls, and the amount of assistance he provides in the care of her offspring Thus, females choose among males based on these criteria. By their

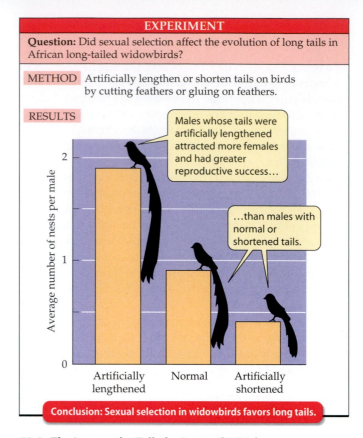

EXPERIMENT

Question: Did sexual selection affect the evolution of long tails in African long-tailed widowbirds?

METHOD   Artificially lengthen or shorten tails on birds by cutting feathers or gluing on feathers.

RESULTS

Males whose tails were artificially lengthened attracted more females and had greater reproductive success...

...than males with normal or shortened tails.

Average number of nests per male

Artificially lengthened   Normal   Artificially shortened

Conclusion: Sexual selection in widowbirds favors long tails.

53.3 The Longer the Tail, the Better the Male
Male widowbirds with shortened tails defended their display sites successfully but attracted fewer females than males with long tails.

choices, females may cause the evolution of exaggerated traits that signal male quality.

Sexual selection often leads to exaggerated traits

Traits may evolve among individuals of one sex as a result of **sexual selection**: the selection of traits that confer advantages to their bearers during courtship or when they compete for mates or resources. Successful competitors for resources may gain exclusive access to mates that are attracted to the resources they control. Traits that improve success in courtship may evolve as a result of mating preferences by individuals of the opposite sex.

Sexual selection is responsible for the evolution of the remarkable tails of African long-tailed widowbirds, which are longer than their heads and bodies combined. Male widowbirds compete for display sites, at which they perform courtship displays to attract females. To examine the role of the tail in sexual selection, an ecologist shortened the tails of some males by cutting them, and lengthened the tails of others by gluing on additional feathers. Both short-tailed and long-tailed males successfully defended their display sites, indicating that the long tail does not confer an advantage in male–male competition. However, males with artificially elongated tails attracted about four times more females than males with shortened tails (Figure 53.3).

Why do females prefer males with long tails? Probably because the ability to grow and maintain a long tail, which

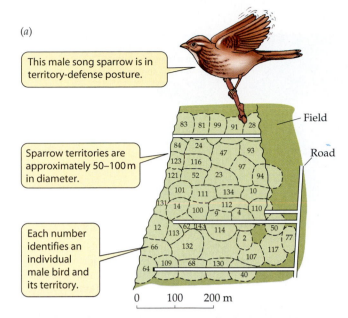

(a)

This male song sparrow is in territory-defense posture.

Sparrow territories are approximately 50–100 m in diameter.

Each number identifies an individual male bird and its territory.

Field

Road

0   100   200 m

(b) *Moranus bassanus*

53.4 Some Territories Provide Everything; Others Provide Only a Nest Site
(a) Male song sparrows defend large breeding territories that contain nesting sites, food resources, and protective cover. (b) The size of a breeding territory among these northern gannets is determined by how far an incubating bird can reach to peck its neighbors without leaving its eggs.

probably carries energetic costs, indicates that the male is vigorous and healthy. Why, then, don't male widowbirds have even longer tails than they do? A likely answer is that the costs of producing a longer tail would exceed the benefits, but the costs of long tails were not measured in these experiments.

Males attract mates in varied ways

Males employ a variety of tactics to induce females to copulate with them. Males of some species defend territories that contain food, nesting sites, or other resources. Some territories are all-purpose: They provide mating sites, nesting sites, and the food necessary to rear offspring (Figure 53.4a). Other territories include a breeding and nesting area, but do not supply all of the food necessary to rear young. The territories of many seabirds, such as gannets, penguins, and cormorants, are very small, consisting of only the area that individuals can defend while sitting on their nests (Figure 53.4b).

If a male controls no resources, he may use courtship behavior that signals in some way that he is in good health, that he is a good provider of parental care, or that he has a good genotype. For example, males of some species of hangingflies court females by offering them dead insects. By capturing an insect and defending it from other males, he demonstrates his ability as a forager and as a competitor. A female hangingfly will mate with a male only if he provides her with food. The bigger the food item, the longer she copulates with him, and the more of her eggs he fertilizes (Figure 53.5).

Whether a male fertilizes the eggs of a female with whom he has copulated depends on when they copulate and whether she copulates with other males. Males have evolved behavior patterns that increase the probability that it will be *their* sperm and no other male's that fertilize a female's eggs. The simplest method is to remain with the female for as long as she is fertile and prevent other males from copulating with her, but this method has high opportunity costs because a male cannot do anything else while he is guarding a female.

Males of many species have evolved behaviors that are more elaborate but take less time. A male black-winged damselfly grabs a female and, using his penis, scrubs out any sperm other males have deposited in her sperm storage chamber. The male removes 90–100 percent of competing sperm before he inserts his own sperm into the chamber. Males of some other insects deposit a plug that effectively seals the opening to the female's genital chamber and prevents other sperm from entering.

53.5 A Male Wins His Mate
The male hangingfly has just presented a moth to his mate, thus demonstrating his foraging skills. She feeds on the moth while they copulate.

Females are the choosier sex

As we have seen, females can improve their reproductive success if they can assess the genetic quality and health of potential mates, the quality of the resources they control, and the quantity of parental care they may provide. But how can females make such assessments when it is to the advantage of males to attempt to signal that they are good in all three of these traits?

By paying particular attention to those signals at which males cannot cheat, females have favored the evolution of "reliable" signals. Possession of a large dead insect indicates that a male hangingfly is a good forager and competitor. Likewise, a male widowbird with a very long tail is likely to be a high-quality mate.

Like tail length, the brightness of the plumage of birds may indicate their health and genetic quality. Male bluethroats (small Eurasian thrushes) have bright blue throat patches (Figure 53.6). Investigators in Norway tested the hypothesis that females use the throat patch as a sign of male quality by blackening the throats of some males and then measuring those males' success in fertilizing eggs. Although most birds form pair bonds, they sometimes engage in extra-pair copulations, so that some of the eggs in a nest may be fertilized by a male other than the one who attends it. A higher proportion of eggs in the nests of males whose throats had been blackened were fertilized by other males than of eggs in the nests of control males.

The investigators also tested the role of the ultraviolet reflectance of the blue feathers on mate choice by female bluethroats. They reduced the UV reflectance of some males by applying to their blue patches a mixture of fat from the glands the birds use to oil their own feathers and UV-absorbing sunblock. Control males received the fat coating with no sunblock. Although the two groups of males looked the same to human observers, female bluethroats could distinguish them. Females started laying eggs sooner on the territories of control males, and they preferred control males as partners for extra-pair copulations. These experiments show that females can use subtle clues to assess the quality of males.

Social and genetic partners may differ

Behavioral ecologists have known for many years that animals nearly always copulate with their mates—the individuals with which they have established pair bonds—but that they sometimes also copulate with other individuals. However, until the recent development of DNA fingerprinting methods, investigators had to assume that mated individuals were the parents of the offspring they raised.

By using the new molecular methods to compare the genomes of offspring with those of their supposed parents and other individuals, ecologists have found that nestling birds are nearly always the offspring of the female attending the nest—that is, females rarely lay eggs in other females' nests. However, the nestlings in a single nest often have different fathers. For example, 34 percent of nestlings

Luscinia svecica

53.6 Ultraviolet Reflectance Affects Mate Choice
Female bluethroats are attracted to males whose throat feathers have high reflectance, which signals a healthy, high-quality mate.

in nests of red-winged blackbirds in Washington state were fathered by a male other than the owner of the territory in which the nest was located. All these other fathers were males holding nearby territories; fertile females went to those territories and solicited copulations from the males.

Females that copulated with more than one male raised more offspring than females that remained faithful to their mates. Their reproductive success improved because neighboring males that had copulated with a female were more likely to defend her nest against predators than males that had not copulated with her. Males also let females with whom they had copulated look for food on their territories. Also, there were fewer infertile eggs in nests with multiple fathers than in nests with single fathers. Males try to prevent their mates from copulating with other males, but they must leave them unguarded at times, both to feed and to seek extra-pair copulations of their own.

Costs and Benefits of Social Behavior

Social behavior evolves when individuals that cooperate with others of the same species have, on average, higher rates of survival and reproductive success than those achieved by solitary individuals. Associations for mating may consist of little more than a coming together of eggs and sperm, but individuals of many species associate for longer times to provide care for their offspring. Associating with conspecifics may also improve survival for reasons unrelated to reproduction, such as by reducing the risk of being captured by a predator.

We describe only a few animal social systems, but these examples demonstrate two important concepts. First, social systems are best understood not by asking how they benefit the species as a whole, but by asking how the individuals that join together benefit. Second, social systems are dynamic: Individuals constantly communicate with one another and adjust their relationships.

53.7 Individuals Hunting Together Can Subdue Large Prey
By hunting as a group, lionesses can kill larger animals than a single female could subdue alone.

Group living confers benefits and imposses costs

Living in groups may confer many types of benefits on individuals. It may improve hunting success or expand the range of prey that can be captured. For example, by hunting together, social carnivores improve their efficiency in bringing down prey (Figure 53.7). Such cooperative hunting was a key component of the evolution of human social behavior. By hunting in groups, our ancestors were able to kill large mammals they could not have subdued alone. These social humans could also defend their prey and themselves from other carnivores.

Many small birds forage in flocks. To find out whether flocking provides protection against predators, an investigator released a trained goshawk near wood pigeons in England. The hawk was most successful in capturing a pigeon when it attacked solitary pigeons. Its success decreased as the number of pigeons in the flock increased (Figure 53.8). The larger the flock of pigeons, the sooner some individual in the flock spotted the hawk.

Living in a group typically imposes costs as well as benefits. Individuals in groups may compete for food, interfere with one another's foraging, injure one another's offspring, inhibit one another's reproduction, or transmit diseases to their associates.

The effects of group living on the survival and reproductive success of an individual also depend on its age, sex, size, and physical condition. Individuals may be larger or smaller than the average for their age and sex. Variation in skills, competitive abilities, and attractiveness to potential mates is often associated with these size differences.

An almost universal cost associated with group living is higher exposure to diseases and parasites. Long before the causes of disease were known, people knew that association with sick persons increased their chances of getting sick. Quarantine has been used to combat the spread of illness for as long as we have written records. The diseases of wild animals are not well known, but most of those that have been studied are spread by close contact.

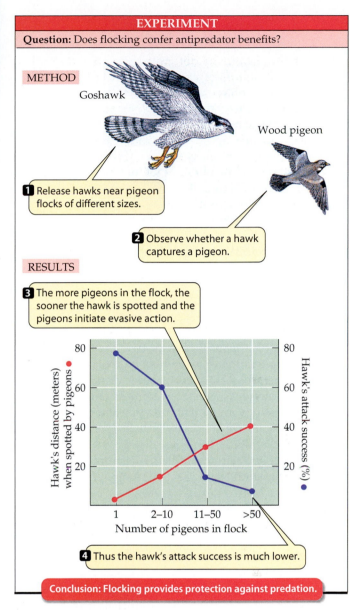

53.8 Flocking Provides Defense against Predators
The larger a flock of pigeons, the greater the distance at which they detect an approaching hawk, and the less likely the hawk is to succeed in capturing a pigeon.

Categories of Social Acts

Individuals living together perform many behaviors that are not performed by solitary animals. These acts can be grouped into four categories according to their effects on the interacting individuals:

▸ An **altruistic act** benefits another individual at a cost to the performer.
▸ A **selfish act** benefits the performer but inflicts a cost on some other individual.
▸ A **cooperative act** benefits both the performer and the recipient.
▸ A **spiteful act** inflicts costs on both the performer and the recipient.

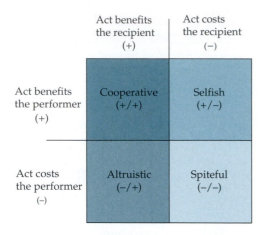

	Act benefits the recipient (+)	Act costs the recipient (−)
Act benefits the performer (+)	Cooperative (+/+)	Selfish (+/−)
Act costs the performer (−)	Altruistic (−/+)	Spiteful (−/−)

53.9 Types of Social Acts
Social acts can be divided into four categories, based on their effects on the performer and the recipient.

The types of social acts are summarized in Figure 53.9. The terms used are purely descriptive; they do not imply conscious motivation or awareness on the part of the animal. If a genetic basis for a cooperative or selfish act exists, and if performing it increases the fitness of the performer, then the genes governing that behavior will increase in frequency in the lineage. In other words, cooperative or selfish behavior can evolve.

How can behavior that inflicts a cost on the performer evolve? Behavioral ecologists believe spiteful behavior is rare in nature. Altruistic behavior, however, can evolve, both among close relatives and among unrelated individuals.

Altruism can evolve by means of natural selection

Altruistic behaviors evolve most easily when performers and recipients are genetically related. Genetic relatedness is important because an individual can influence its fitness in two different ways. First, it may produce its own offspring, contributing to its individual fitness. Second, it may help its relatives in ways that increase their fitness.

Because relatives are descended from a common ancestor, they are likely to bear some of the same alleles. Two offspring of the same parents, for example, are likely to share 50 percent of the same alleles; an individual is likely to share 25 percent of its alleles with its sibling's offspring. Therefore, by helping its relatives, an individual can increase the representation of some of its own alleles in the population. This process is called **kin selection**. Together, individual fitness and kin selection determine the **inclusive fitness** of an individual. Occasional altruistic acts may eventually evolve into altruistic behavior patterns if the benefits of increasing the reproductive success of relatives exceed the costs of decreasing the altruist's own reproductive success.

Many social groups consist of some individuals that are close relatives and others that are unrelated or distantly related. Individuals of some species recognize their relatives and adjust their behavior accordingly. White-fronted bee-

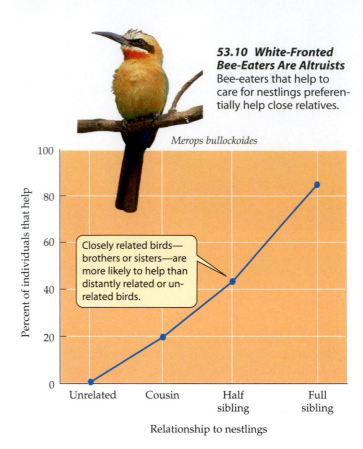

53.10 White-Fronted Bee-Eaters Are Altruists Bee-eaters that help to care for nestlings preferentially help close relatives.

Merops bullockoides

Closely related birds— brothers or sisters—are more likely to help than distantly related or unrelated birds.

y-axis: Percent of individuals that help
x-axis: Relationship to nestlings (Unrelated, Cousin, Half sibling, Full sibling)

eaters are African birds that nest colonially (Figure 53.10). Most breeding pairs are assisted by nonbreeding adults that help incubate their eggs and feed nestlings. Nearly all of these helpers assist close relatives. When helpers have a choice of two nests at which to help, about 95 percent of the time they choose the nest with the young most closely related to them.

Several other pieces of evidence suggest that the helping behavior of white-fronted bee-eaters evolved through kin selection. First, both males and females help to care for nestlings, but males help more often than females. Males remain in the social group in which they were born, but females join other social groups when they mature. Therefore, females typically live in social groups composed primarily of nonrelatives.

Second, individuals do not appear to gain anything other than inclusive fitness by helping—helpers are not, for example, more successful when they become breeders themselves than birds that do not help. Finally, nests with helpers produce more fledglings than do nests without helpers, showing that helpers do increase the number of fledglings produced by their close relatives. Notice that all these patterns are consistent with the principle that bee-eaters behave in ways that improve their fitness, not in ways that benefit the species.

Species whose social groups include sterile individuals are said to be **eusocial**. This extreme form of social behavior—the "ultimate altruism"— has evolved in termites and many hymenopterans (ants, bees, and wasps). In these species, worker females defend the group against predators

Eciton burchelli

53.11 Sterile Individuals are Extreme Altruists
Eusocial insect species contain classes of sterile individuals. These soldier ants from Panama protect their nests and nestmates with their large, powerful jaws.

or bring food to the colony, but do not reproduce. Some species have soldiers with large defensive weapons (Figure 53.11). These workers are at risk of being killed while defending the colony.

Both genetic and environmental factors facilitate the evolution of eusociality. The more closely related individuals are to one another, the greater the advantages they can receive by forgoing their own reproduction to help relatives reproduce. The British evolutionary biologist W. D. Hamilton first suggested that eusociality evolved among ants, bees, and wasps because members of the order Hymenoptera have an unusual sex determination system in which males are haploid but females are diploid.

Among the Hymenoptera, a fertilized egg hatches into a female; an unfertilized egg hatches into a male. If a female copulates with only one male, all the sperm she receives are identical because the haploid males have only one set of chromosomes, all of which are transmitted to each sperm cell. Therefore, a female's daughters share all of their father's genes. They also share, on average, half of the genes they receive from their mother. As a result, on average they share 75 percent of their alleles rather than 50 percent, as they would if both parents were diploid. Workers therefore can increase their fitness more by helping their sisters than by reproducing themselves, because they are genetically more similar to their sisters than they would be to their own offspring.

Mating between close relatives, known as *inbreeding*, can also generate close genetic relationships. Even if two mates are unrelated, but each is the product of generations of intense inbreeding, their offspring can be genetically nearly identical. These individuals would also benefit from helping to rear siblings. Genetic similarity generated by inbreeding could explain the evolution of eusociality among the many hymenopteran species in which queens mate with many males and among termites and naked mole-rats—the most extremely eusocial mammals—in which both sexes are diploid.

Eusociality may also be favored if establishment of new colonies is difficult and dangerous. Nearly all eusocial animals construct elaborate nests or burrow systems within which their offspring are reared (Figure 53.12). Naked mole-rats live in underground colonies containing 70 to 80 individuals. The tunnel systems are maintained by sterile workers. Breeding is restricted to a single queen and several kings that live in a nest chamber in the center of the colony. Individuals attempting to found new colonies are at high risk of being captured by predators, and most founding events fail. Thus, high predation rates, which favor cooperation among founding individuals, may facilitate the evolution of eusociality.

Unrelated individuals may behave altruistically toward one another

It is easy to understand how altruistic behavior can evolve among related individuals. It is more difficult to explain the existence of warnings of danger, sharing of food, and grooming among unrelated individuals of the same species, or even between members of different species. How can we explain the evolution of such behavior?

Such behavior among unrelated individuals could evolve through **reciprocal altruism**; that is, if helpers are in turn recipients of beneficial acts by the individuals they have helped. If there is a genetic basis for the acts, natural selection may increase the frequency of alleles governing this behavior. In order for reciprocal altruism to be a force, several social conditions must be present:

▶ Individuals in the group must know one another.
▶ They must associate for long periods.
▶ Individuals must be aware that their altruistic acts are being reciprocated.

53.12 Termite Mounds are Large and Complex
Immense termite mounds such as this one in Kenya are costly to construct and maintain. Elaborate nests or burrows are characteristic of nearly all eusocial animals.

Reciprocal altruism is especially highly developed among humans, in which these conditions prevail. It illustrates the subtle adjustments in behavior among individuals in social groups.

The Evolution of Animal Societies

The decisions animals make about where to live, with whom to mate, and whether and how to care for their offspring all help determine the type of social system they have. Today's social systems are the result of long periods of evolution, but there are few records of past social systems because behavior leaves few traces in the fossil record. Possible routes of the evolution of social systems must therefore be inferred primarily from current patterns of social organization. Fortunately, many stages of social system complexity exist among species, and the simpler systems provide clues about the stages through which the more complex ones may have passed.

Parents of many species care for their offspring

The origins of all animal societies lie in the association of parents with their offspring. Individuals of many species invest time and energy in caring for offspring. Parental care increases the chances of an offspring's survival, but it usually reduces the ability of the parent to produce additional offspring.

Parental care may also lower the chances of survival of the parent itself, because the parent could have used the time and energy to engage in other activities that would improve its own chances of survival. In other words, parents balance trade-offs between the success of their current offspring and their own future survival and reproduction.

Males and females often differ strikingly in the kinds and amounts of parental investment they can and do make. Birds, mammals, and fishes illustrate these differences and why they exist. Only female mammals have functional mammary glands; males cannot produce milk. Therefore parental care among mammals is usually given by females. On the other hand, among birds, all aspects of reproduction except production of eggs and sperm can be performed readily by both males and females. Not surprisingly, both males and females feed their offspring in about 90 percent of bird species.

Sex roles among fishes differ from those of birds and mammals because most fish species do not feed their young. Parental care consists primarily of guarding eggs and young from predators (Figure 53.13). In many fish species, males are the primary guardians. A male can guard a clutch of eggs while attracting additional females to lay eggs in his nest. A female, on the other hand, can produce another clutch of eggs sooner if she resumes foraging immediately after mating than if she spends time guarding her eggs.

The most widespread form of social system is the family, an association of one or more adults and their dependent offspring. If parental care lasts a long time, or if the breeding season is longer than the time it takes for offspring to

Abudeiduf saxatilis

53.13 A Sergeant Major Guards His Young
This male is defending the eggs a female has deposited. He can court other females while guarding the eggs.

The dark area is a large clutch of eggs.

mature, adults may still be caring for younger offspring when older offspring reach the age at which they could help their parents.

Many communal breeding systems, such as the white-fronted bee-eater families described above, most likely evolved via the extended family route. Most mammals evolved social behavior by this route. In simple mammalian social systems, solitary females or male–female pairs care for their young. In species whose young require a long period of parental care, older offspring are still present when the next generation is born, and they often help rear their younger siblings. In most social mammal species, female offspring remain in the group in which they were born, but males tend to leave—or are driven out—and must seek other social groups. Therefore, among mammals, most helpers are females.

53.14 Savanna Weaverbirds Nest Colonially
Many African weaverbirds nest in colonies in isolated trees. Although these nests are highly conspicuous, it is difficult for most predators to get to them at the tips of the small, thorny branches.

53.15 Cooperation among Florida Scrub Jays
These Florida scrub jay helpers, most of which are offspring from the previous breeding season, are helping to feed nestlings and defend the nest against predators, such as the approaching snake. By doing so, they are improving their inclusive fitness.

The environment influences the evolution of animal societies

The type of social organization a species evolves is strongly related to the environment in which it lives. Among the weaverbirds of Africa, species that live in forests eat insects, feed alone, and build well-hidden nests. Most of these species are monogamous, and males and females look alike. In marked contrast, weaverbirds that live in tree-studded grasslands called savannas eat primarily seeds, feed in large flocks, and nest in colonies, usually in isolated *Acacia* trees where their nests are large and conspicuous (Figure 53.14). In most colonial species, males have several mates and are more brightly colored than females.

These striking differences probably evolved because nesting sites and food in forests are common and widely dispersed. Solitary pairs can use these resources more efficiently than animals in groups can. In savannas, however, good nesting trees are scarce and highly clumped. Males compete for these limited nest sites; males that hold the best sites—near the tips of branches where they are safe from predators—attract the most females. Males spend their time attempting to attract additional mates rather than helping to rear the offspring they already have, which explains the evolution of brighter plumage among males.

Florida scrub jays live all year on territories, each of which is home to a breeding pair and up to six helpers (Figure 53.15). Nearly all the helpers are offspring from the previous breeding season that remain with their parents. This social system probably evolved because all suitable territories are occupied, and young individuals have little chance of establishing new territories on their own. By staying in their parents' territory and helping to raise their siblings, they both improve their inclusive fitness and have a chance of taking over the territory if one of their parents dies.

Among the herbivorous hoofed mammals of Africa, social organization and feeding ecology are correlated with the size of the animal. Smaller animals have higher metabolic demands per unit of body weight than do larger ones. Therefore, smaller hoofed mammals feed preferentially on high-protein foods such as buds, young leaves, and fruits. These foods are dispersed throughout forests, which also provide cover in which to hide from predators. Hiding is a tactic that is effective for solitary animals. In contrast, the largest hoofed mammal species are able to eat lower-quality food, but they must process great quantities of it each day. They feed in grasslands with abundant herbaceous vegetation, follow the rains to areas where grass growth is best, and live in large herds (Figure 53.16a).

Among primates, many diurnal species will eat insects and other animal food when they are available, but most of

(b) *Papio cynocephalus*

(a) *Connochaetes taurinus*

53.16 Many Mammals of Open Country Live in Large Groups
(a) East African wildebeest live in large herds that follow the rains to places with fresh grass. (b) Baboons are conspicuous as they forage but are seldom attacked because the formidable males cooperate in defending the group.

them eat fruits, seeds, and leaves. In Africa and Asia, primate group sizes are smallest among arboreal forest-dwelling species, whatever their diets, and largest among the ground-dwelling savanna species, such as baboons (Figure 53.16b). Troops of foraging baboons are conspicuous to predators, but their large males help protect the other troop members. Baboons have a complex social system. In troops with more than one male, strong dominance hierarchies exist among the males, and one or two of them father most of the offspring. Females may also have dominance relationships, and young females often assume the status of their mothers when they mature.

Chapter Summary

▶ Ecologists study the nature and consequences of interactions among organisms and their environments.

▶ Behavioral ecology is the study of how animals decide where to carry out different activities, select the resources they need, respond to predators and competitors, and interact with conspecifics.

Balancing Costs and Benefits of Behaviors

▶ Cost–benefit analyses of behavior are based on the principle that animals have only limited amounts of time and energy to devote to their activities.

▶ A cost of defending a territory may be increased risk of mortality. **Review Figure 53.1**

Choosing Where to Live and Forage

▶ Selecting a habitat in which to live is one of the most important decisions an individual makes.

▶ The cues animals use to select habitats are good predictors of conditions suitable for future survival and reproduction.

▶ Foraging theory was developed to understand how animals select prey. **Review Figure 53.2**

Mating Tactics and Roles

▶ Individuals choose their associates, how to interact with them, and when to leave them. The most important choice of associates is the choice of a mate.

▶ Because males produce enough sperm to fertilize many eggs, males typically increase their reproductive success by mating with many females. The reproductive success of females is typically limited by the cost of producing eggs. As a result, males usually initiate courtship and often fight for opportunities to mate with females. Females seldom fight over males and often reject courting males.

▶ Sexual selection often leads to exaggerated traits. **Review Figure 53.3**

▶ While courting, males signal their desirability as mating partners and may perform behaviors that increase the probability that their sperm will fertilize eggs.

▶ Males of some species defend territories that contain food, nesting sites, or other resources. **Review Figure 53.4**

▶ By paying particular attention to those signals at which males cannot cheat, females have favored the evolution of "reliable" signals of mate quality.

▶ DNA fingerprinting methods have shown that social fathers often are not genetic fathers.

Costs and Benefits of Social Behavior

▶ Benefits of social living include better opportunities to capture prey and to avoid predators. **Review Figure 53.8**

▶ Costs of social living include competition for food, interference by conspecifics, and transmission of diseases.

Categories of Social Acts

▶ Acts performed by individuals living together can be grouped into four descriptive categories: altruistic, selfish, cooperative, and spiteful. **Review Figure 53.9**

▶ Altruism among closely related individuals can evolve by means of kin selection because individuals that help close relatives can improve their inclusive fitness. **Review Figure 53.10**

▶ Eusocial systems with sterile individuals have evolved among termites, hymenopterans (ants, bees, and wasps), and in a mammal, the naked mole-rat.

The Evolution of Animal Societies

▶ The origin of most animal societies is the family, an association of one or more adults and their dependent offspring.

▶ The type of social organization a species evolves is strongly related to the environment in which it lives.

For Discussion

1. Most hawks are solitary hunters. Swallows often hunt in groups. What are some plausible explanations for this difference? How could you test your ideas?

2. Because costs and benefits of behaviors can seldom be measured directly, behavioral ecologists often use indirect measures such as correlations between behavior patterns and the presence of predators. What are the strengths and weaknesses of some of these indirect measures?

3. Polyandry is a mating system in which one female has a "harem" of several males. Why is polyandry much rarer among both birds and mammals than polygyny, the situation in which one male forms pair bonds with several females?

4. When frogs mate, a male clasps a gravid female behind her front legs and stays with her until she lays her eggs, at which time he fertilizes them. In most species of frogs, the male remains clasped to the female for a short time, usually no longer than a few hours. However, in some species, pairs may remain together for up to several weeks. In view of the fact that a male cannot court or mate with any other female while clasping one, and that a female lays only a single clutch of eggs, why is it advantageous for males to behave this way? What can you guess about the breeding ecology of frogs that remain clasped for long periods? Why should females permit males to clasp them for so long? (Females do not struggle!)

5. Among vertebrates, helpers are individuals capable of reproducing, and most of them later breed on their own. Among eusocial insects, sterile castes have evolved repeatedly. What differences between vertebrates and insects might explain the failure of sterile castes to evolve in the former?

6. The use of DNA fingerprinting technology has shown that in many species, social partners and genetic partners differ. Under what conditions do individuals benefit from copulating with individuals other than their social mates? Do males and females benefit equally from this behavior?

54 Population Ecology

LARGE SAGUARO CACTI ARE CONSPICUOUS features of the Sonoran Desert in southern Arizona. But finding a seedling cactus is difficult—at least, until you learn where to search. All the small cacti are found beneath trees or shrubs.

In the harsh environment of the Sonoran Desert, plants are exposed to intense daytime heat and wide temperature fluctuations. Their roots are in extremely dry soil much of the year. Small plants are most vulnerable to these conditions because they have small root systems, and daytime temperatures are extremely high at the soil surface. Therefore, although seeds of saguaro cacti are dispersed widely over the desert by birds, seedling cacti survive only in the shade of trees and shrubs, called *nurse plants*, where they are protected from the intense daytime heat. Thus, the density and distribution of saguaro cacti are strongly influenced by the number and location of trees and shrubs.

All the individuals of saguaro cacti—or of any other species—within a given area constitute a **population**. The sizes of populations continually change. To understand how and why these changes happen, population ecologists count individuals in different locations and try to determine the factors that influence birth, death, immigration, and emigration rates.

In addition to studying the dynamics of populations in a particular area, population ecologists also investigate changes over the entire ranges of species. They attempt to answer questions such as: What causes a species to be common or rare? Why is a species common in some parts of its geographic range and rare in others? What determines the limits of the ranges of species?

In this chapter we discuss how and why the sizes of populations vary over space and time, and show how this ecological knowledge is used to predict and manage the growth of populations. To set the stage for studies of populations, we present some background information on how the individuals of a population are distributed.

Population Structure: Patterns in Space and Time

At any given moment, an individual organism occupies only one spot and is of one particular age. The members of a population, however, are distributed over space and differ in age and size. These features are among the components of **population structure**. As we saw in Chapter 21, geneticists and evolutionary biologists also study population structure, but they are interested primarily in distributions of genotypes and their degree of isolation from one another, because that component of population structure influences how populations evolve.

Ecologists study population structure at different spatial scales, ranging from local subpopulations to entire species. They study the numbers and spatial distributions of individuals because these features influence the stability of populations and affect interactions among species.

Density is an important feature of populations

The number of individuals of a species per unit of area (or volume) is its **population density**. Ecologists are interested in population densities because dense populations often

A Cactus Needs Shade
This saguaro cactus (*Cereus giganteus*) grew in the shade of the palo verde tree, which protected it from the intense heat.

54.1 Modular Organisms

(a) Each member of this bryozoan animal colony is called a zooid and is genetically identical to an ancestor zooid from which it has budded. (b) A single modular organism can appear to be a population. This clump of quaking aspen trees looks like a grove of many individuals, but it is actually a single genetic individual.

(a) *Pectinatella magnifica*

(b) *Populus tremuloides*

from a cluster of genetically separate individuals (Figure 54.1b). The effects of modular organisms on their environment often depend primarily on the number and size of the modules. Therefore, ecologists studying modular organisms are often concerned primarily with the number, size, and shape of the modules rather than with the number of genetically distinct individuals.

Under some circumstances, the individuals in a population can be counted directly without missing any of them or counting any of them twice, but this process is usually impossible or too laborious. Ecologists commonly estimate population densities by sampling a population in a representative area and extending their findings to a larger area.

The size of a population can also be estimated by marking and recapturing individuals. For example, if we capture and mark 100 individuals in a population, we can take another sample later and count the individuals in that sample that are already marked. If, say, 10 percent of the individuals in our second sample are already marked, we would conclude that the population contains about 1,000 individuals. This estimate is based on the mathematical assumption that the number of individuals caught the first time is the same proportion of the total population as the propor-

exert strong influences on their own members as well as on populations of other species. Other scientists—such as those working in agriculture, conservation, or medicine—wish to manage species to raise (in the case of crop plants, aesthetically attractive species, or threatened or endangered species) or lower (in the case of agricultural pests and disease organisms) their densities. To manipulate population densities, we must know what factors make populations increase and decrease in size, and how those factors work.

Because species and their environments differ, population densities are measured in more than one way. Ecologists usually measure the densities of organisms in terrestrial environments by the number of individuals per unit of area, but number per unit of volume is generally a more useful measure for organisms living in water. For species whose members differ markedly in size, as do most plants and some animals (such as mollusks, fishes, and reptiles), the total mass of individuals—their *biomass*—may be a more useful measure of density than the number of individuals.

The fertilized egg of many organisms develops into a unit of construction called **module**, which produces additional modules much like itself. Many plants are modular, and there are many groups of modular protists, fungi, and animals (for example, sponges, corals, and bryozoans; Figure 54.1a). A modular organism may grow to a large size, and it is often difficult to distinguish a modular organism

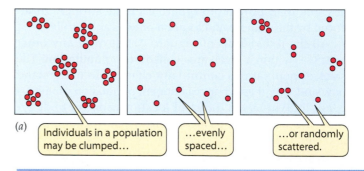

(a)   Individuals in a population may be clumped…   …evenly spaced…   …or randomly scattered.

(b)

54.2 Patterns of Spatial Distribution

(a) A diagrammatic representation of clumped, even, and random distribution patterns. (b) The relatively even spacing of these Australian desert plants results because each established plant removes so much water from the surrounding soil that no young plants can grow within its root zone.

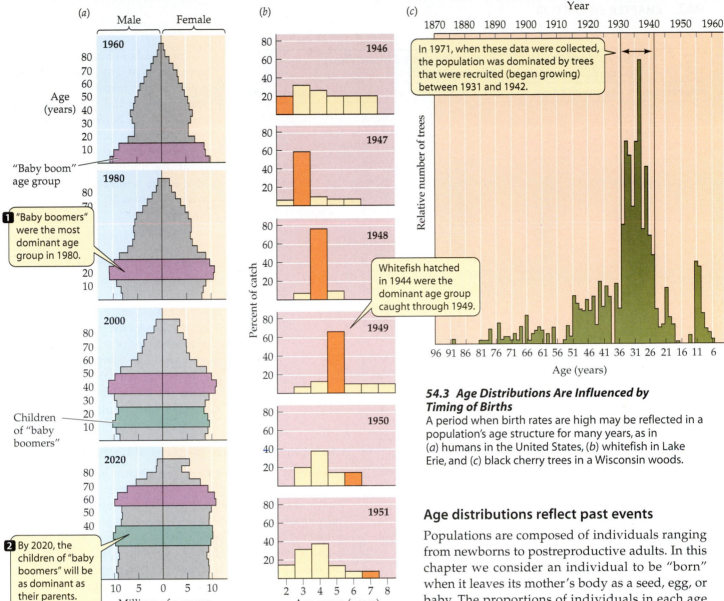

(a)

1960

"Baby boom" age group

1980

1 "Baby boomers" were the most dominant age group in 1980.

2000

Children of "baby boomers"

2020

2 By 2020, the children of "baby boomers" will be as dominant as their parents.

Millions of persons

(b)

1946

1947

1948

Whitefish hatched in 1944 were the dominant age group caught through 1949.

1949

1950

1951

Percent of catch

Age group (years)

(c)

In 1971, when these data were collected, the population was dominated by trees that were recruited (began growing) between 1931 and 1942.

Relative number of trees

Year

Age (years)

54.3 Age Distributions Are Influenced by Timing of Births

A period when birth rates are high may be reflected in a population's age structure for many years, as in (a) humans in the United States, (b) whitefish in Lake Erie, and (c) black cherry trees in a Wisconsin woods.

tion of marked individuals to the total number caught the second time

Spacing patterns reflect interactions among individuals

Ecologists studying population structure also look at the way in which the individuals in a population are spaced. Individuals in a population may be tightly clumped together, evenly spaced, or randomly scattered (Figure 54.2a). Distributions can become clumped when young individuals remain close to their birthplaces, when suitable habitat patches are "islands" separated by unsuitable areas, or by chance. The relatively even spacing of many plants is a result of competition for light, water, and soil nutrients (Figure 54.2b). Among animals, defense of space is the most common cause of even distributions (see Figure 53.4b). Random distributions may result when many factors interact to influence where individuals settle and survive. The saguaro cacti in the Sonoran Desert (discussed at the beginning of this chapter) are distributed randomly because that is how suitable nurse plants are distributed.

Age distributions reflect past events

Populations are composed of individuals ranging from newborns to postreproductive adults. In this chapter we consider an individual to be "born" when it leaves its mother's body as a seed, egg, or baby. The proportions of individuals in each age group in a population make up its **age distribution**. The density and spacing of individuals are spatial attributes of a population; age distribution is a *temporal* (time-oriented) attribute.

The timing and rates of births and deaths determine age distributions. If both birth rates and death rates are high, a population will be dominated by young individuals. If birth rates and death rates are low, a relatively even distribution of individuals of different ages results. The age distribution of a population thus reveals much about its past history of births and deaths.

The timing of births and deaths may influence age distributions for many years in populations of long-lived species. The human population of the United States is a good example. Between 1947 and 1964, the United States experienced what is called the post-World War II "baby boom." During these years, average family size grew from 2.5 to 3.8 children; an unprecedented 4.3 million babies were born in 1957. Birth rates declined during the 1960s, but Americans born during the baby boom will constitute the dominant age class into the twenty-first century (Figure 54.3a). "Baby boomers" became parents in the 1980s, producing another bulge in the age distribution—a baby boom

54.1	Life Table of the 1978 Cohort of Darwin's Ground Finch (Geospiza scandens) on Isla Daphne			
AGE IN YEARS (X)	NUMBER ALIVE	SURVIVORSHIP[a]	SURVIVAL RATE[b]	MORTALITY RATE[c]
0	210	1.000	0.434	0.566
1	91	0.434	0.855	0.143
2	78	0.371	0.898	0.102
3	70	0.333	0.928	0.072
4	65	0.309	0.955	0.045
5	62	0.295	0.678	0.322
6	42	0.200	0.545	0.455
7	23	0.109	0.651	0.349
8	15	0.071	0.944	0.056
9	14	0.067	0.776	0.224
10	11	0.052	0.923	0.077
11	10	0.048	0.396	0.604
12	4	0.019	0.737	0.263
13	3	0.014	0.714	0.004

[a]Survivorship = the proportion of newborns who survive to age x.

[b]Survival rate = the proportion of individuals of age x who survive to age $x + 1$.

[c]Mortality rate = the proportion of individuals of age x who die before the age of $x + 1$.

"echo"—but they had, on average, fewer children than their parents, so the bulge is not as large. Similarly, in Lake Erie, 1944 was such an excellent year for reproduction and survival of whitefish that individuals of that age group dominated whitefish catches in the lake for several years (Figure 54.3b). The population of black cherry trees in a Wisconsin woods is dominated by individuals that began growing between 1931 and 1941 (Figure 54.3c).

Population Dynamics: Changes over Time

At any moment in time, a population has a particular structure determined by the number and distribution of its members in space and their ages. However, as we have just seen, population structure is not static. Changes in the structure of a population influence whether it will increase or decrease; that is, they affect the *dynamics* of the population. We will now examine how ecologists measure birth and death rates and use that information to understand how population densities change. The study of changes in the size and structure of populations is known as **demography**.

Births, deaths, and movements drive population dynamics

Knowledge of when individuals are born and when they die provides a surprising amount of information about a population. Births, deaths, and movements of individuals are demographic events—that is, they determine the numbers of individuals in a population. Ecologists measure the *rates* at which these events take place—the number of such events per unit of time. These rates are influenced by environmental factors, by the life history traits of the species, and by population density.

The number of individuals in a population at any given time is equal to the number present at some time in the past, plus the number born between then and now, minus the number that died, plus the number that immigrated, minus the number that emigrated. That is, the number of individuals at a given time, N_1, is given by the equation

$$N_1 = N_0 + B - D + I - E$$

where N_1 is the number of individuals at time 1; N_0 is the number of individuals at time 0; B is the number of individuals born, D the number that died, I the number that immigrated, and E the number that emigrated between time 0 and time 1. If we measure these rates over many time intervals, we can determine how a population's density changes.

Life tables summarize patterns of births and deaths

Life tables summarize data about births and deaths that can be used to predict future growth rates of populations. We can construct a life table by determining for a group of individuals born at the same time—called a **cohort**—the number still alive at specific times and the number of offspring they produced during each time interval. An example, based on an intensive study of one population of Darwin's finches carried out on Isla Daphne in the Galápagos archipelago, is shown in Table 54.1.

The data in Table 54.1 are based on a cohort of 210 birds that hatched in 1978 and were followed until 1991, by which time all of them had died. This life table (which presents data only on survival, not on reproduction) shows that the mortality rate was high during the first year of life. It then dropped dramatically for several years, followed by a general increase in later years. Mortality rates fluctuated among years because survival of the birds depends on seed production, which is strongly correlated with rainfall. The Galápagos archipelago experiences both drought years, during which plants produce few seeds, birds do not nest, and adult survival is poor, and years of heavy rainfall, during which seed production is high, most birds breed several times, and adult survival is high. The life table reflects these fluctuations.

Ecologists often use graphs to highlight the most important changes in birth and death rates in populations. Graphs of **survivorship**—the mirror image of mortality—in relation to age show at what ages individuals survive well and at what ages they do not. To interpret survivorship data, ecologists have found it useful to compare real data with several hypothetical curves that illustrate a range of possible survivorship patterns (Figure 54.4a). At one extreme, nearly all individuals survive for their entire potential life span and die at about the same age (hypothetical curve I). At the other extreme, the survivorship of young individuals is very low, but survivorship is high for most of the remainder of the life span (hypothetical curve III). An intermediate possibility is that survivorship is the same throughout the life span (hypothetical curve II).

Survivorship data from real populations often resemble one of these hypothetical curves. For example, survivorship of humans in the United States remains high for many decades but then, as in hypothetical curve I, declines significantly in older individuals (Figure 54.4b). Many wild birds have survivorship curves similar to hypothetical curve II; the probability of their surviving is about the same over most of the life span once they are a few months old (Figure 54.4c).

A widespread survivorship pattern is found among organisms that produce many offspring, each of which receives few energy resources and no parental care. In these species, low survivorship of young individuals is followed by high survivorship during the middle part of the life span, and then low survivorship toward the end of the life span. *Spergula vernalis*, an annual plant that grows on sand dunes in Poland, illustrates this pattern (Figure 54.4d).

Patterns of Population Growth

If a single bacterium selected at random from the surface of this book, and all its descendants, were able to grow and reproduce in an unlimited environment, explosive population growth would result. In a month the bacterial colony would weigh more than the visible universe and would be expanding outward at the speed of light. Similarly, a single pair of Atlantic cod and their descendants, reproducing at the maximum rate of which they are capable, would fill the Atlantic Ocean in 6 years. But, as Darwin observed, this does not happen in nature.

All populations have the potential for explosive growth because as the number of individuals in the population increases, the number of new individuals added per unit of time accelerates, even if the rate per capita of population increase remains constant. This form of explosive increase is called **exponential growth**. If we ignore immigration and emigration and assume that births and deaths occur continuously and at constant rates, such a growth pattern forms a continuous curve (Figure 54.5a). This curve can be expressed mathematically in the following way:

Rate of increase in number of individuals

$$= \left(\begin{array}{c} \text{Average per capita birth rate} \\ - \text{ Average per capita death rate} \end{array} \right)$$

$\times$ Number of individuals

or, more concisely,

$$r = \frac{\Delta N}{\Delta t} = (b - d)N$$

where $\Delta N / \Delta t$ is the rate of change in the size of the population (ΔN = change in number of individuals; Δt = change in time). The difference between the average per capita birth rate (b) and the average per capita death rate (d) is the per capita rate of increase (r). In these equations, b includes

(a) **Hypothetical curves**

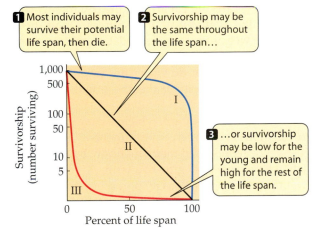

1 Most individuals may survive their potential life span, then die.

2 Survivorship may be the same throughout the life span…

3 …or survivorship may be low for the young and remain high for the rest of the life span.

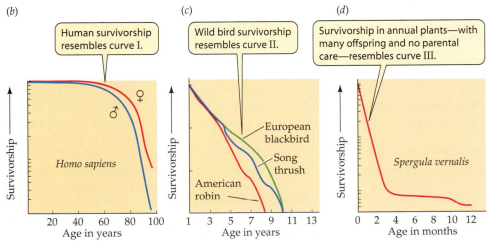

(b) Human survivorship resembles curve I.

(c) Wild bird survivorship resembles curve II.

(d) Survivorship in annual plants—with many offspring and no parental care—resembles curve III.

54.4 Survivorship Curves
Survivorship curves show the number of individuals in a cohort still alive at different times over the life span. (a) The range of possible survivorship patterns. Patterns for (b) humans in the United States, (c) some small wild birds, and (d) an annual plant, *Spergula vernalis*, on Polish sand dunes.

(a)

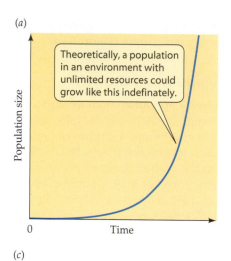

Theoretically, a population in an environment with unlimited resources could grow like this indefinately.

Population size

0   Time

(b)

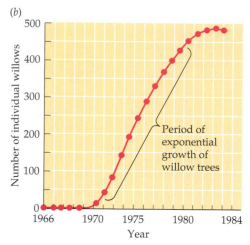

Number of individual willows

Period of exponential growth of willow trees

1966   1970   1975   1980   1984
Year

(c)

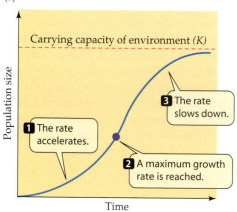

Carrying capacity of environment (K)

Population size

1 The rate accelerates.

2 A maximum growth rate is reached.

3 The rate slows down.

Time

 54.5 Exponential and Logistic Population Growth
(a) A theoretical exponential growth curve. (b) Growth curve of an actual population of willows at Newborough Warren, Wales. The trees experienced a surge of exponential growth when rabbits that fed on their leaves were decimated by disease. (c) A population in an environment with limited resources usually stops growing exponentially long before it reaches the environmental carrying capacity.

both births and immigration and d includes both deaths and emigration.

When there are no limits on population growth, r has its highest value, called r_{max}, the **intrinsic rate of increase**; r_{max} has a characteristic value for each species. Therefore, the rate of growth of a population under optimal conditions is

$$\frac{\Delta N}{\Delta t} = r_{max} N$$

But optimal conditions do not continue indefinitely, and growth rates eventually slow down, as we explore in the following section.

Population growth is influenced by the carrying capacity

Natural populations may experience exponential growth for short periods of time under favorable conditions. For example, one population of willows in Wales increased dramatically in the 1970s after the rabbit population, which had severely nibbled the willows, crashed due to an outbreak of the disease myxomatosis (Figure 54.5b). But no real population can maintain exponential growth for very long because environmental limitations cause birth rates to drop and death rates to rise. In fact, over long time periods, the sizes of most populations fluctuate around a relatively constant number.

The simplest way to picture the limits imposed by the environment is to assume that it can support no more than a certain number of individuals of a particular species. This number, called the **carrying capacity** of the environment, is determined by the availability of resources—food, nest sites, shelter—as well as by disease, predators, and, in some cases, social interactions. Rather than being exponential, population growth slows down as the population approaches the carrying capacity, so that the growth curve has an S shape (Figure 54.5c).

The S-shaped growth pattern, which is characteristic of many populations growing in environments with limited resources, can be represented mathematically by adding to the equation for exponential growth a term, $(K - N)/K$, that slows the population's growth as it approaches the carrying capacity (K). The simplest such equation is that for **logistic growth**:

$$\frac{\Delta N}{\Delta t} = r\left(\frac{K - N}{K}\right)N$$

The biological assumption in this equation is that each individual added to the population makes things slightly worse for the others because it competes with them for available resources, or for other reasons. Population growth stops when $N = K$ because then $(K - N) = 0$, so $(K - N)/K = 0$, and thus $\Delta N/\Delta t = 0$.

The logistic growth equation contains some important simplifications that are not true for most populations. Its most critical assumptions are that (1) each individual exerts its effects immediately at birth; (2) all individuals produce equal effects on the population; and (3) births and deaths are continuous. However, in nature, organisms grow during their lives, and their effects on others normally increase with age, so there may be a delay between the birth of an

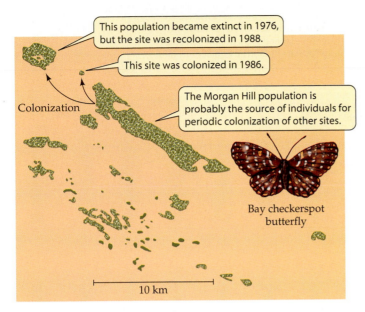

This population became extinct in 1976, but the site was recolonized in 1988.

This site was colonized in 1986.

Colonization

The Morgan Hill population is probably the source of individuals for periodic colonization of other sites.

Bay checkerspot butterfly

10 km

54.6 Subpopulation Dynamics
The population of the bay checkerspot butterfly *Euphydryas editha bayensis* is divided into a number of subpopulations confined to habitat patches that contain the plants its larvae feed on.

individual and the time at which it begins to affect the other members of its population. A seedling tree, for example, exerts a much smaller effect on its neighbors than a large adult tree does, and it does not begin to reproduce until it reaches a relatively large size.

In addition, the logistic equation models a population in a single habitat patch; it does not take immigration and emigration into account. Next we consider the dynamics of an assembly of local populations, which is more complex than the growth of a single population.

Many species are divided into discrete subpopulations

Many populations are divided into discrete *subpopulations* among which some exchange of individuals occurs. Such a pattern is often found where suitable habitat occurs in separated patches, or "habitat islands." Each subpopulation has a probability of "birth" (colonization) and "death" (extinction). Within each subpopulation, growth occurs in the ways we have just discussed, but because subpopulations are typically much smaller than the population as a whole, local disturbances and random fluctuations in numbers of individuals often cause the extinction of a subpopulation. However, if individuals frequently move between subpopulations, immigrants may prevent declining subpopulations from becoming extinct. This process is known as the **rescue effect**.

The bay checkerspot butterfly provides a good illustration of the dynamics of such divided populations. The larvae of this butterfly feed on only a few species of annual plants that are restricted to outcrops of a particular kind of rock on hills south of San Francisco. The bay checkerspot

has been studied for many years by Stanford University biologists. During drought years, most host plants die early in spring, before the butterfly larvae have pupated. At least three butterfly subpopulations became extinct during a severe drought in 1975–1977. The largest patch of suitable habitat, Morgan Hill, typically supported thousands of butterflies (Figure 54.6). It probably served as a source of individuals that dispersed to and colonized small habitat patches where the butterflies had become extinct.

Population Regulation

In a limited environment, population growth slows down as density increases because the members increasingly affect one another adversely. As a result, a population above the environmental carrying capacity is likely to decrease in density, and one that is below the carrying capacity is likely to increase. In this section we discuss how populations may be influenced by interactions between their density and the carrying capacity of their environment.

How does population density influence birth and death rates?

If per capita birth or death rates change in response to population density, they are said to be **density-dependent**. Death and birth rates may be density-dependent for several reasons:

▶ As a population increases in abundance, it may deplete its food supply, reducing the amount of food that each individual gets. Poor nutrition may increase death rates and decrease birth rates.

▶ Predators may be attracted to regions where densities of their prey have increased. If predators are able to capture a larger proportion of the prey than they did when prey were scarce, the per capita death rate of the prey rises.

▶ Diseases, which may increase death rates, spread more easily in dense populations than in sparse populations.

A population whose dynamics are influenced primarily by density-dependent factors is said to be *regulated*.

Factors that change per capita birth and death rates in a population independent of its density are said to be **density-independent**. A very cold spell in winter, for example, may kill a large proportion of the individuals in a population regardless of its density. However, even environmental factors whose frequency and severity are unrelated to population density may result indirectly in density-dependent mortality. Cold weather may not kill organisms directly, but may increase the amount of food individuals need to eat each day. Individuals pushed by population density into poorer foraging areas may be more likely to die than those in better foraging areas. Or the death rate may be related to the quality of sleeping places. If population density is high, a larger proportion of individuals may be forced to sleep in places that expose them to the cold.

Various combinations of density-dependent and density-independent factors can influence the density of a popula-

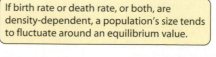

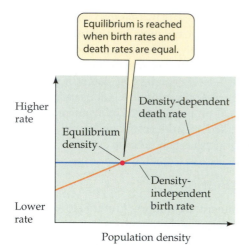

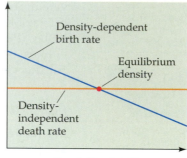

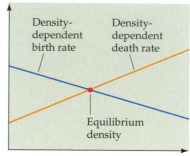

54.7 Density-Dependent Factors Regulate Population Size

The densities of all populations fluctuate, but they tend to return to equilibrium value if either birth rate and/or death rate are density-dependent.

tion. The hypothetical graphs in Figure 54.7 show how birth and death rates can change in relation to population density. When birth and death rates are equal (the point at which the two lines cross) the population neither grows nor shrinks. If birth or death rates, or both, are density-dependent, the population responds to increases or decreases in its density by returning toward the equilibrium density. If neither rate is density-dependent, there is no equilibrium and the population is not regulated.

Fluctuations in the density of a species' population are determined by all the factors and processes, density-dependent and density-independent, acting upon it. The combined action of density-independent and density-dependent factors is illustrated by the dynamics of a population of song sparrows on Mandarte Island, off the coast of British Columbia, Canada.

During recent years, in response to variable winter weather and other physical factors, the number of song sparrows on Mandarte has fluctuated between 4 and 72 breeding females and between 9 and 100 breeding males. Density-independent variation caused by weather is modulated by several density-dependent factors. The number of breeding males is limited by territorial behavior: The larger the number of males, the larger the number that fail to gain territories, so more live as "floaters" (Figure 54.8a). Also, the larger the number of breeding females, the fewer offspring each female fledges (Figure 54.8b). And, finally, the more offspring are fledged, the more poorly they survive over the winter (Figure 54.8c).

Disturbances affect population densities

Disturbances—short-term events that influence populations by changing their environment and, hence, its carrying capacity—regularly affect population densities. Common physical disturbances are fires, hurricanes, ice storms, wind storms, floods, landslides, and lava flows. Biological disturbances include tree falls, disease epidemics, and the burrowing and trampling activities of animals. Disturbances differ in their spatial distribution, frequency, predictability, and severity.

A disturbance typically decreases the environmental carrying capacity for some species while increasing it for others. A landslide, for example, may increase the carrying capacity of the environment for plants that require bare mineral soil for the germination of their seeds and survival of their seedlings. However, it will decrease the carrying capacity for species that require shade and rich organic litter for successful germination.

Populations themselves can influence the frequency of some disturbances. Immediately after a fire, for example, there is not enough combustible organic matter to carry another fire. However, as vegetation grows back, dead wood, branches, and leaves accumulate, gradually increasing the supply of fuel. Thus the frequency of fires may be proportional to the rate at which fuel accumulates through the growth of plant populations, or the rate at which herbivores consume plant materials that would otherwise accumulate. Similarly, as many trees age, their roots and trunks become weakened by fungal infections. Old, large trees are thus susceptible to being toppled by high winds. Therefore, the likelihood of a major blowdown increases as the forest ages.

Organisms cope with environmental changes by dispersing

A common response of animals to environmental changes is **dispersal**—movement to another habitat. If habitat quality declines greatly, individuals may be able to improve their survival and reproductive success by going elsewhere.

If repeated seasonal changes alter a habitat, organisms may evolve life cycles that appear to anticipate the changes. **Migration**—regular seasonal movement from one place to another—is most widespread among birds, but some insects, such as monarch butterflies, and some mammals also

(a)

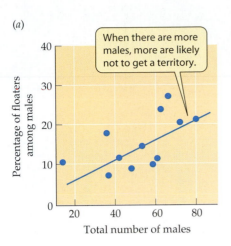

When there are more males, more are likely not to get a territory.

Percentage of floaters among males

Total number of males

(b)

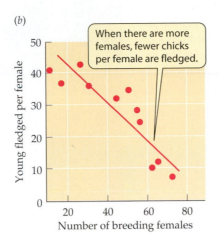

When there are more females, fewer chicks per female are fledged.

Young fledged per female

Number of breeding females

(c)

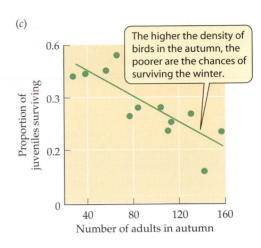

The higher the density of birds in the autumn, the poorer are the chances of surviving the winter.

Proportion of juveniles surviving

Number of adults in autumn

54.8 Regulation of an Island Population of Song Sparrows
The number of song sparrows on Mandarte Island is mainly determined by the severity of winter weather, but the weather's effect is modulated by several density-dependent factors, including (a) male territorial behavior, (b) the reproductive success of females, and (c) the survival of juveniles.

migrate (Figure 54.9). Most insectivorous birds leave high latitudes for more favorable wintering grounds in autumn, before conditions seriously deteriorate. In arctic regions, caribou migrate each year between winter and summer ranges.

Life Histories Influence Population Growth

The complete **life history** of an organism consists of its birth, growth to maturity, reproduction, and death. During its life, an individual organism ingests nutrients or food, grows, interacts with other individuals of the same and other species, reproduces, and usually moves or is moved so that it does not die exactly where it was born. Life histories describe how an organism divides its efforts among these activities. In previous discussions we have referred to various components of the life histories of organisms. Now we focus our attention specifically on life history patterns and why they have evolved to be as variable as they are.

The life history traits of organisms have been molded by natural selection acting over many generations. In each lineage, those traits that maximized reproductive success were favored, but natural selection has not produced a single, dominant pattern of reproduction. Some organisms, such as elephants and humans, usually give birth to a single offspring in each reproductive episode; others produce thou-

(a) *Danaus plexippus*

(b) *Rangifer tarandus*

54.9 Animals Migrate to Remain in Suitable Environments
(a) Most of North America's monarch butterflies migrate to central Mexico, where they aggregate on conifers in cool mountain valleys. They can survive the winter there without eating because their metabolic rates are low.
(b) These caribou in the American Arctic are migrating from open tundra to their winter foraging grounds at the edge of boreal forest. During the winter they feed on lichens on branches of trees.

sands or millions of eggs or seeds in one bout of reproduction. Some organisms begin to reproduce within days or weeks of being born; others live for many years before reproducing. All life histories are based on a certain set of traits, which includes:

▶ Size and energy supply of the individual at birth
▶ Its rate and pattern of growth and development
▶ How many times individuals disperse
▶ The number and timing of reproductive events
▶ The number, size, and sex ratio of offspring
▶ The ages at which individuals die

Life histories include stages for growth, change in form, and dispersal

For at least part of their lives, all organisms grow by gathering and assimilating energy and nutrients. Some organisms, such as birds and mammals, gather energy and nutrients throughout their lives, even after they reach adult size and stop growing. Energy gathered after growth stops maintains the organism and supports reproduction. In many species, however, energy gathering is confined to a particular life stage. Most moths, for example, feed only when they are larvae. The adults lack mouthparts and digestive tracts, live on the energy they gathered as larvae, and survive only long enough to disperse, mate, and lay eggs (Figure 54.10).

Individuals of many species also change form during their lives. Human babies are unmistakably human, but newborns of many species differ dramatically from adults. Some of the most striking changes are found among insects such as beetles, flies, moths, butterflies, and bees, which undergo metamorphosis from their larval to their adult forms. Many plants have resting stages, such as spores and seeds, that have low metabolic rates and are highly resistant to changes in the physical environment. Growth typically does not take place in these stages.

At some time in their lives, all organisms disperse. Some, such as plants and sessile animals, disperse as eggs, larvae, spores, or seeds. Others, such as insects and birds, disperse primarily as adults. Still others may disperse during several different life stages. Individuals of some species can change their location many times during their lives in response to environmental changes. Others remain in the first place they settle.

Life histories embody trade-offs

Life history evolution is influenced by inevitable trade-offs. These trade-offs exist because changes that improve fitness by means of one life history trait often reduce fitness by means of another. What are the major trade-offs in life history traits?

A universal trade-off exists between number and size of offspring. Every newborn individual begins to grow with energy and nutrients from its maternal parent, but how much energy and nutrients individuals receive from their mothers varies greatly. The larger the amount of energy

Polyphemus sp.

54.10 A Life Stage for Sex and Reproduction Only
This female moth has no mouth or digestive tract. She will live only a few days, just long enough to mate and lay her eggs

provided to each offspring, the larger it can grow before it must gather its own energy, but the fewer offspring a mother can produce for a given amount of energy—a major trade-off.

For animals, a trade-off also exists between the number of offspring produced and the amount of care parents provide to their offspring. The more parental care the parents provide, the fewer offspring they can produce for a given investment in reproduction. Birds and mammals produce fewer offspring at a time and provide extensive care for each one.

Some organisms invest so much in one reproductive effort that they have no energy left for another—or even for their own survival. If two individuals have the same amount of energy to invest in reproduction, and one reproduces only once while the other reproduces several times, the former can produce more offspring in a single episode than the latter because it reserves no energy for itself. Annual plants invest nearly all of the energy they gain during their single growing season in seed production; they do not survive long after reproducing. Some longer-lived organisms also reproduce once in their lifetimes and die soon afterward. Pacific salmon (genus *Oncorhynchus*) hatch in fresh water, migrate to the sea, spend a number of years at sea, return to fresh water, spawn, and die (Figure 54.11*a*). Most agaves (century plants) of the American Southwest likewise store up energy for many years before producing a large flowering stalk, forming many seeds, and dying (Figure 54.11*b*).

Trade-offs also exist between reproduction and growth. Members of many species do not begin to reproduce until they have reached full size, but others, such as most plants, mollusks, fishes, and reptiles, start to reproduce while they are still relatively small and continue to reproduce as they grow. Reproduction usually reduces growth because these two processes compete for the limited amount of energy an individual has at its disposal. Beech trees in Germany, for example, grew more slowly during years when they pro-

(a) *Oncorhyncus nerka*

54.11 A Single Reproductive Effort
(a) These sockeye salmon are ascending Hensen Creek, Alaska. They will lay their eggs in gravel beds in the stream and then die. (b) This century plant has mobilized the energy stored during its long life to produce a large flowering stalk with hundreds of flowers, literally reproducing itself to death.

Flowering stalk

Nonreproductive form of *Agave*

(b) *Agave* sp.

duced large crops of nuts than they did during years when their nut crops were small (Figure 54.12).

Offspring are like "money in the bank"

If reproduction compromises future growth and survival, why do some organisms start to reproduce when they are small or young? The potential contribution of an individual's offspring to future generations depends, in part, on when they are produced. A useful analogy compares the production of offspring to earning interest on money deposited in a bank. It pays to deposit money in the bank as soon as possible so that it can begin earning interest. Offspring produced early in an adult's life likewise "yield interest" quickly—that is, they can begin to reproduce sooner than offspring produced later.

Reproductive value is the average number of offspring that remain to be born to individuals of a particular age. A newborn individual does not have the highest reproductive value, even though it has its full reproductive potential ahead of it, because many newborn individuals will die before they have a chance to reproduce. Therefore we must discount the number of offspring an individual could produce if it survived by the chance that it will die before reaching reproductive age, or during reproduction. When we make the appropriate calculations, we find that the reproductive value of an individual steadily increases until it begins to reproduce. Once maturity is reached, reproductive value declines; in most species, it reaches zero when the individual has finished reproducing. However, individuals can still have positive reproductive value after they have stopped reproducing if they continue to assist the survival of their offspring and grandoffspring.

Because reproductive value declines after maturity, the power of natural selection grows increasingly weaker as an individual ages. Once reproductive value has dropped to zero, natural selection cannot act on alleles that first produce

54.12 Reproduction Slows Growth Rates in Beech Trees
The width of annual growth rings shows the growth rate of beech trees in different years. In general, the trees grew slowly in years when they produced large crops of nuts, but during unusually favorable years they grew rapidly and produced many nuts.

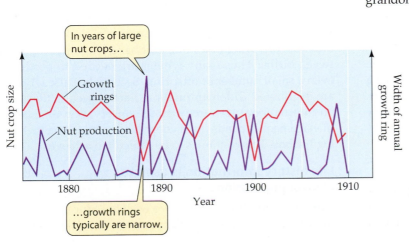

In years of large nut crops…

Growth rings

Nut production

Nut crop size

Width of annual growth ring

1880   1890   1900   1910
Year

…growth rings typically are narrow.

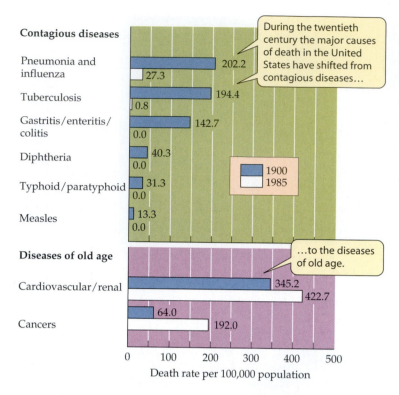

54.13 Causes of Human Death in the United States
Today most people die of diseases of old age because improved sanitation and public water supplies, as well as medical advances such as immunization, have greatly reduced the incidence of contagious diseases that formerly killed many young people.

their phenotypic effects at that age—even those that are highly detrimental to the individual's survival. As a result, increasing numbers of harmful alleles are expressed as individuals age, causing increased mortality rates, especially after reproduction has ceased. In this manner, **senescence**—an increased probability of dying per unit of time with increasing age—has evolved.

As a result of improved hygiene and nutrition, most people in modern industrial societies are now spared the contagious diseases that cause death rates to be high among people of all ages in nonindustrial societies. Most people live to the age when the so-called genetic diseases of old age begin to afflict them. Cancer and heart disease, the main killers in industrialized societies, are much more difficult to cure than the contagious diseases that formerly caused most deaths. For this reason, despite the expenditure of enormous resources to extend life, the average age of death in the United States has changed very little during the past 30 years. As one source of mortality is eliminated, another takes its place (Figure 54.13). Life history theory suggests that this situation is likely to continue indefinitely.

Can Humans Manage Populations?

For many centuries, people have tried to reduce populations of species they consider undesirable and increase populations of desirable species. Strategies for controlling and

managing populations of organisms are based on our understanding of how populations grow and are regulated.

A general principle of population dynamics is that the total number of births and the growth rates of individuals tend to be highest when a population is well below its carrying capacity. Therefore, if we want to maximize the number of individuals of a species that we wish to harvest, we should manage the population so that it is far enough below carrying capacity to have high birth and growth rates. Hunting seasons for game birds and mammals are established with this objective in mind.

Life history traits determine how heavily a population can be exploited

Populations with high reproductive capacities can sustain their growth despite a high rate of harvest. In such populations (many species of fish, for example), each female may lay thousands or millions of eggs. Another characteristic of these fast-reproducing populations is that individual growth is often density-dependent. If prereproductive individuals are harvested at a high rate, the remaining individuals may grow faster. Many fish populations can be harvested heavily for many years because only a modest number of females must survive to reproductive age to produce the eggs needed to maintain the population.

Fish can, of course, be overharvested. Many populations have been greatly reduced because so many individuals were harvested that too few reproductive adults survived to maintain the population. The Georges Bank off the coast of New England—a source of cod, halibut, and other prime food fishes—has been exploited so heavily that many fish stocks have been reduced to levels insufficient to support a commercial fishery.

The whaling industry has also engaged in excessive harvests. The blue whale, Earth's largest animal, was hunted nearly to extinction by the middle of the twentieth century. The industry then turned to smaller species of whales that were still numerous enough to support commercially viable whaling operations (Figure 54.14).

Management of whale populations is difficult for two reasons. First, unlike fish, whales reproduce at very low rates. They have long prereproductive periods, produce only one offspring at a time, and have long intervals between births. Thus many whales are needed to produce even a small number of offspring. Second, because whales are distributed widely throughout Earth's oceans, they are an international resource whose conservation and wise management depends upon cooperative action by all whaling nations. This continues to be difficult to achieve.

Life history information is used to control populations

The same principles apply if we wish to reduce the size of populations of undesirable species and keep them at low

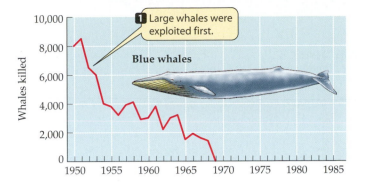

1 Large whales were exploited first.

Blue whales

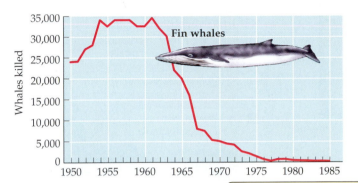

Fin whales

2 Progressively smaller whales were exploited more as populations of larger species decreased.

Sperm whales

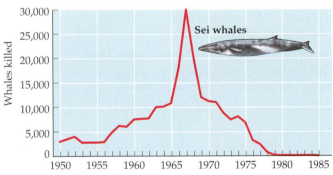

Year

Sei whales

54.14 Overexploitation of Whales

The graphs show the numbers of whales of four species killed each year from 1950 to 1985. As each species reached low population levels, the whaling industry turned to other species. All four species were driven to very low levels by sustained hunting.

Similarly, if we wish to preserve a rare species, the most important step usually is to provide it with suitable habitat. If habitat is available, the species will usually reproduce at rates sufficient to maintain its population. If the habitat is insufficient, preserving the species usually requires expensive and continuing intervention, such as providing extra food.

Humans have introduced many species to new habitats outside their native ranges. When these introduced species undergo population explosions, humans attempt to reduce their numbers by introducing new predators and parasites from the introduced species' original habitat. For example, the cactus *Opuntia*, introduced into Australia from South America, spread rapidly and became a common pest species over vast expanses of valuable sheep-grazing land. The cactus population was controlled by introducing a moth species (*Cactoblastis cactorum*) whose larvae feed on *Opuntia*. Once egg-laying females find a patch of cactus, their larvae completely destroy the patch (Figure 54.15).

But new patches of cactus arise in other places from seeds dispersed by birds. These new patches flourish until they are found and destroyed by *Cactoblastis*. Over a large region, the numbers of both *Opuntia* and *Cactoblastis* are today fairly constant and low, but in the local areas that make up the whole, there are extreme oscillations resulting

densities. At densities well below carrying capacity, populations have high birth rates and can therefore withstand higher death rates than they could closer to carrying capacity. Killing part of a population whose dynamics are influenced primarily by density-dependent factors only reduces it to the density at which it experiences the most rapid rate of growth. A far more effective approach to reducing the population of a species is to remove its resources, thereby lowering the carrying capacity of its environment. We can rid our dumps and cities of rats more easily by making garbage unavailable (reducing the carrying capacity of the rats' environment) than by poisoning rats.

54.15 Biological Control of an Introduced Pest
Cactoblastis caterpillars consume an *Opuntia* cactus in Australia.

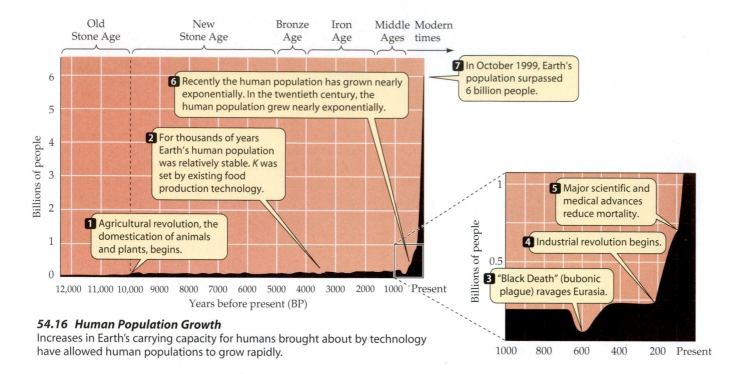

54.16 Human Population Growth
Increases in Earth's carrying capacity for humans brought about by technology have allowed human populations to grow rapidly.

from the extermination of first the plant and then the herbivore. This is another example of a series of subpopulations connected by occasional dispersing individuals.

Can we manage our own population?

Managing our own population has become a matter of great concern. For thousands of years, Earth's carrying capacity for human populations was set at a low level by food and water supplies and disease. Domestication of plants and animals and cultivation of the land enabled our ancestors to increase the resources at their disposal dramatically. These developments stimulated rapid population growth up to the next carrying capacity limit, which was determined by the agricultural productivity possible with only human- and animal-powered tools. Agricultural technology and artificial fertilizers, made possible by the tapping of fossil fuels, greatly increased agricultural productivity, further raising Earth's carrying capacity for humans. The development of modern medicine reduced the effectiveness of disease as a limiting factor on human populations, raising the global carrying capacity still further (Figure 54.16). Medicine and better hygiene have allowed people to live in large numbers in areas where diseases formerly kept numbers very low.

What is Earth's present carrying capacity for people? Today's carrying capacity is set by Earth's ability to absorb the by-products—especially CO_2—of our enormous consumption of fossil fuel energy and by whether we are willing to cause the extinction of millions of other species to accommodate our increasing use of environmental resources. We will explore some of the consequences of high human population densities for the survival of other species in Chapter 58.

Chapter Summary

Population Structure: Patterns in Space and Time

▶ A population consists of all the individuals of a species within a given area.

▶ The number of individuals of a species per unit of area (or volume) is its population density.

▶ Individuals in a population may have uniform, random, or clumped distributions. **Review Figure 54.2**

▶ The age distribution of individuals in a population reveals much about the recent history of births and deaths in the population. The timing of births and deaths may influence age distributions for many years. **Review Figure 54.3**

Population Dynamics: Changes over Time

▶ Births, deaths, immigration, and emigration drive changes in population density and distribution.

▶ Life tables help us visualize patterns of births and deaths in a population. **Review Table 54.1**

▶ Graphs of survivorship in relation to age show when individuals survive well and when they do not. **Review Figure 54.4**

Patterns of Population Growth

▶ All populations have the potential to grow exponentially. However, no population can maintain exponential growth for very long because environmental limitations cause birth rates to drop and death rates to rise.

▶ The number of individuals of a particular species that an environment can support—called the carrying capacity—is determined by the availability of resources and by disease and predators.

▶ A population in a limited environment at first grows rapidly, but growth rates decrease as the carrying capacity is approached. **Review Figure 54.5**

▶ The overall densities of many populations are determined by "births" (colonizations) and "deaths" (extinctions) of local subpopulations. Immigrants may prevent declining subpopulations from becoming extinct, a process known as the rescue effect. **Review Figure 54.6**

Population Regulation

▶ Regulation of a population by changes in per capita birth or death rates in response to density is said to be density-dependent.

▶ If per capita birth and death rates are unrelated to a population's density, the population is not regulated.

▶ The density of a population is determined by the combined effects of all density-dependent and density-independent factors affecting it. **Review Figures 54.7, 54.8**

Life Histories Influence Population Growth

▶ The life history of a species describes how it divides its efforts among growth, dispersal, and reproduction over time.

▶ Trade-offs inevitably exist between number and size of offspring, between number of offspring and parental care, between survival and reproduction, and between growth and reproduction. **Review Figure 54.12**

▶ Reproductive value is the average number of offspring that remain to be born to individuals of a particular age. Reproductive value rises to a peak when individuals first begin to reproduce and declines to zero after reproduction ceases.

▶ Senescence—an increased probability of dying with increasing age—evolves because natural selection cannot act on alleles that first produce their phenotypic effects after a reproductive value drops to zero. **Review Figure 54.13**

Can Humans Manage Populations?

▶ Humans use the principles of population dynamics to control and manage populations of species they consider desirable or undesirable. Nevertheless, many populations have been overexploited. **Review Figure 54.14**

▶ Earth's carrying capacity for humans has been increased several times by technological developments. **Review Figure 54.16**

For Discussion

1. Huntington's disease is a severe disorder of the human nervous system that generally results in death. It is caused by a dominant allele that does not usually express itself phenotypically until its bearer is 35 to 40 years old. How fast is the gene causing Huntington's disease likely to be eliminated from the human population? Would your answer change if the gene expressed itself when its bearer was 20 years old? 10 years old?

2. Many people have improperly formed wisdom teeth and must spend considerable sums of money to have them removed. Assuming, as is probably the case, that the presence or absence of wisdom teeth and their mode of development are partly under genetic control, will we gradually lose our wisdom teeth by evolutionary processes?

3. Some organisms, such as oysters, cod, and elm trees, produce vast quantities of offspring, nearly all of which die before they reach adulthood. What fraction of such deaths are likely to be selective—that is, dependent on the genotypes of the individuals dying? What does your answer imply for the rates of evolution of oysters, cod, and elms?

4. In this chapter we identified a number of trade-offs in life history evolution. Why are these trade-offs inevitable? Why is knowledge about trade-offs important when we attempt to manipulate the life histories of organisms?

5. Ecologists often use the concept of carrying capacity when studying the growth and regulation of populations in nature, even though carrying capacity often changes markedly over time. How can the concept be useful if its value changes so often?

6. Most organisms whose populations we wish to manage for higher densities are long-lived and have low reproductive rates, whereas most organisms whose populations we attempt to reduce are short-lived but have high reproductive rates. What is the significance of this difference for management strategies and the effectiveness of management practices?

7. In the mid-nineteenth century, the human population of Ireland was largely dependent upon a single food crop, the potato. When a disease caused the potato crop to fail, the Irish population declined drastically for three reasons: (1) a large percentage of the population emigrated to the United States and other countries; (2) the average age of a woman at marriage increased from about 20 to about 30 years; and (3) many families starved to death rather than accept food from Britain. None of these social changes was planned at the national level, yet all contributed to adjusting population size to the new carrying capacity. Discuss the ecological strategies involved, using examples from other species. What would you have done had you been in charge of the national population policy for Ireland?

8. From a purely ecological standpoint, can the problem of world hunger ever be overcome by improved agriculture alone? What components must a hunger-control policy include?

55

Community Ecology

SOME PEOPLE LIKE THEIR FOOD SPICY HOT; OTHERS don't. The spices that impart strong flavors to foods are actually antioxidant, antimicrobial, and antiviral chemicals. These chemicals evolved because they protected the plants that produced them from predators and diseases. When we add spices during cooking, we borrow the plants' survival "recipes" and use them to protect our own food. The protection that spices provide was especially important before refrigeration and freezing were widely available, but even today, spices help prevent contamination of foods in most parts of the world.

Organisms interact with one another in a variety of ways. Some of these interactions involve eating and being eaten, but organisms may also interact competitively, or they may benefit one another. All organisms are potentially or actually food for some other organism, and most of them have evolved defenses that make them more difficult to find and capture, or less palatable or nutritious if they are captured. Consumers of those organisms have, in turn, evolved ways of getting around the defenses of their prey.

The organisms that live together in a particular area constitute an **ecological community**. Each species interacts in unique ways with other species in its community and with its physical environment. Some of these interactions are strong and important; others are weak and affect the functioning of the community very little. The study of such interactions, and how they determine which and how many species live in a place, is the focus of **community ecology**.

For several decades, ecologists debated whether which species live together in communities is determined primarily by the interactions of individuals with the physical environment or by their interactions with other organisms. Some ecologists even suggested that a community is a superorganism in which each species plays a particular role, just as each organ plays a role within the body of an individual organism. That view has been abandoned because organisms, unlike organs, do not evolve under

the influence of natural selection to serve their community. Nevertheless, determining the roles of species interactions with the biotic and abiotic environment is a major challenge for community ecologists today.

In this chapter we do consider interactions between organisms and their physical environment, but we concentrate on the major biological interactions and show how they influence the structure and functioning of ecological communities.

Types of Ecological Interactions

Organisms interact with one another in five major ways:

▶ Two organisms may mutually harm one another. This type of interaction is common when two organisms use the same resources and those resources are insufficient to supply their combined needs. Such organisms are called *competitors*, and their interactions constitute **competition**.

▶ One organism, by its activities, may benefit itself while harming another, as when individuals of one species eat individuals of another. The eater is called a *predator* or *parasite*, and the eaten is its *prey* or *host*. These interactions are known as **predator–prey** or **parasite–host interactions**.

Some Like It Hot
The peppers on the right of this photo are the seeds of a tropical vine, *Piper nigrum*. The hot chiles (upper left) are the fruits of an unrelated pepper plant, *Capsicum anuum*. The chemicals in peppers and other "hot" spices are often antimicrobial agents.

55.1 Types of Ecological Interactions

		EFFECT ON ORGANISM 2		
		HARM	BENEFIT	NO EFFECT
EFFECT ON ORGANISM 1	HARM	Competition (–/–)	Predation or parasitism (–/+)	Amensalism (–/0)
	BENEFIT	Predation or parasitism (+/–)	Mutualism (+/+)	Commensalism (+/0)
	NO EFFECT	Amensalism (0/–)	Commensalism (0/+)	—

▶ If both participants benefit from an interaction, we call them *mutualists*, and their interaction is a **mutualism**.

▶ If one participant benefits but the other is unaffected, the interaction is a **commensalism**.

▶ If one participant is harmed but the other is unaffected, the interaction is an **amensalism.**

These categories of species interactions are summarized in Table 55.1. But they are not clear-cut, both because the strengths of interactions vary and because many cases do not fit the categories neatly. Nevertheless, most interactions fit these categories well enough for us to use them as a guide for exploring interactions among species in this chapter.

Resources and Consumers

Many interactions between organisms within communities center on resources and their consumers. A **resource** is anything directly used by an organism that can potentially lead to the growth of the population and whose availability is reduced when it is used. We usually think first of resources that can be consumed by being eaten, but space—including hiding places, nest sites, and establishment sites for sessile organisms—becomes unavailable if it is occupied, so it, too, is a resource. Factors such as temperature, humidity, salinity, and pH, even though they may strongly affect population size, are not resources because they can be neither consumed nor monopolized.

Some resources, such as nest sites, are not altered by being used and immediately become available for occupancy again when the user leaves. Other resources must regenerate before they are again available to consumers.

Biotic interactions influence the conditions under which species can persist

Each species can persist only under a certain set of environmental conditions, which define its **ecological niche**. If there were no competitors, predators, or disease organisms in its environment, a species would be able to persist under a broader array of physical conditions (its *fundamental niche*) than it can in the presence of other species that negatively affect it (its *realized niche*). On the other hand, the presence of beneficial species may increase the range of physical conditions in which a species can persist.

An experiment performed on two species of barnacles, *Balanus balanoides* and *Chthamalus stellatus*, demonstrated the importance of both abiotic and biotic factors in determining the fundamental and realized niches of these two species. These barnacles live between high tide and low tide levels on rocky North Atlantic shores. Adult *Chthamalus* generally live higher in the intertidal zone than do adult *Balanus*, but young *Chthamalus* settle in large numbers in the *Balanus* zone.

In the absence of *Balanus*, young *Chthamalus* survive and grow well in the *Balanus* zone, but if *Balanus* are present, the *Chthamalus* are smothered, crushed, or undercut by the larger and more rapidly growing *Balanus*. Young *Balanus* settle in the *Chthamalus* zone, but they grow poorly because they lose water rapidly when exposed to air. *Chthamalus* compete successfully with them there, but *Balanus* would persist slightly higher in the intertidal zone in the absence of *Chthamalus*. By experimentally removing one or the other species, researchers have shown that the vertical ranges of adults of both species are greater in the absence of the other species. The result of their interaction is intertidal zonation, with *Chthamalus* growing above *Balanus* (Figure 55.1).

Limiting resources determine the outcomes of interactions

Resources whose supply is less than the demand made upon them by organisms are called **limiting resources**. Resources that are not limiting may have little influence on a species' population dynamics. For example, most terrestrial animals have strict but similar requirements for a certain minimum level of oxygen. However, studying the use of oxygen reveals very little about the structure of terrestrial communities because the concentration of oxygen, which is 21 percent of the atmosphere, is nearly always above that minimum level.

The limiting resources that influence distributions and abundances of terrestrial species are those that are depletable and regenerate slowly, such as food. In freshwater aquatic environments, however, where the maximum concentration of dissolved oxygen is only 0.5 percent, organisms regularly deplete oxygen. Aquatic ecologists, unlike terrestrial ecologists, pay careful attention to oxygen levels.

Which resources are limiting differs among environments, but some kinds of resources, such as food supplies,

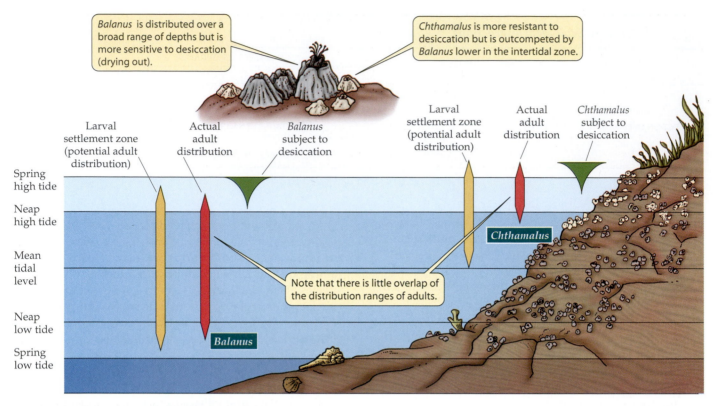

55.1 Potential and Actual Distributions of Two Barnacle Species

Interspecific competition makes the zone each species occupies smaller than the zone it could potentially occupy in the absence of the other species. The width of the red and gold bars is proportional to the density of the populations.

are often limiting. Because of the importance of resources in the lives of all species, we first examine competition that takes place among organisms needing scarce resources, and then consider predation.

Competition: Seeking and Using Scarce Resources

If two or more individuals use the same resources, and those resources are insufficient to meet their demands, the individuals are competitors, whether they are members of the same or a different species. **Intraspecific competition**—competition among individuals of the same species—may result in reduced growth and reproductive rates for some individuals, may exclude some individuals from better habitats, and may cause the deaths of others. **Interspecific competition**—competition among individuals of different species—affects individuals in the same way, but in addition, an entire species may be kept out of habitats where it cannot compete successfully, a phenomenon called **competitive exclusion**. In extreme cases, a competitor may cause the extinction of another species. In this section we will show how ecologists study competition and discuss how it influences species distributions and the composition of ecological communities.

Plants are good subjects for experiments to test the nature and results of competitive interactions because they compete for light, water, and nutrients, all of which can easily be manipulated. For example, the relative importance of root and shoot competition can be assessed by growing plants in shared or separate pots so that either their shoots or roots, or both, compete for resources: nutrients and water in the case of roots, light in the case of shoots. The results of an experiment measuring root and shoot competition between a clover (*Trifolium repens*) and a grass (*Lolium perenne*) are shown in Figure 55.2. The grass outcompeted the clover in both root and shoot competition because a rich supply of nutrients, which the grass was able to use more efficiently than the clover, was provided.

Competition can restrict species' ranges

The role of competition in restricting the ranges of species is illustrated by the interaction of the two barnacle species described in Figure 55.1. In some cases, competition can completely exclude a species from part or all of its range.

Parasitic wasps were introduced into southern California to control outbreaks of scale insects that were seriously damaging citrus orchards. The Mediterranean wasp *Aphytis chrysomphali* was established in southern California by 1900, but it did not effectively control scale insects. Therefore, a close relative from China, *A. lingnanensis*, which has a higher reproductive rate, was introduced in 1948. *A. lingnanensis* increased rapidly, and within a decade it had displaced *A. chrysomphali* from most of its California range (Figure 55.3).

EXPERIMENT

Question: What are the relative effects of root and shoot competition on plant growth?

METHOD

No interspecific competition (control)

Perennial ryegrass   White clover

Interspecific competition (experimental)

Root competition   Shoot competition   Root and shoot competition

RESULTS

□ White clover
■ Perennial ryegrass

Yield (percent of control)

No inter–specific competition | Root | Shoot | Root and shoot

Competition

Conclusion: In this experiment, both root and shoot competition were important but shoot competition was more important.

55.2 Plants Compete with their Roots and Shoots
By growing plants in separate pots or together, experimenters can distinguish the influences of root and shoot competition on plant growth. An experiment using white clover (*Trifolium repens*) and perennial ryegrass (*Lolium perenne*) showed that both roots and shoots of these plants are involved in competition.

Competition can reduce species' abundances

Often competition reduces the abundances of competing species rather than eliminating them from an area. Many species of seed-eating ants and rodents live together in the Sonoran Desert of Arizona. To determine whether competition between ants and rodents influences their abundances, ecologists removed ants from some sites, rodents from other sites, and both ants and rodents from a third set of sites. When they removed ants, the density of rodents increased slightly, but seed densities did not change. When they removed rodents, the density of ant colonies nearly doubled, but again, seed densities did not change. When they removed both ants and rodents, seed densities increased to five times their previous value. These results showed that ants and rodents were competing for and influencing their food supply, but that rodents had a much stronger effect on ants than vice versa.

To determine whether different species of rodents also compete with one another, the ecologists erected rodent-proof fences around 50 × 50-m desert plots. The fences around the experimental plots had holes that small rodents could pass through, but were too small to allow the passage of large kangaroo rats. The holes in the fences surrounding control plots were large enough for all rodents to pass through. Within 2 years of the exclusion of kangaroo

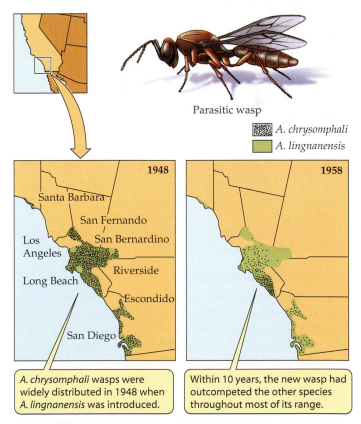

Parasitic wasp

▨ *A. chrysomphali*
▨ *A. lingnanensis*

1948   **1958**

Santa Barbara
San Fernando
Los Angeles   San Bernardino
Long Beach   Riverside
Escondido
San Diego

A. chrysomphali wasps were widely distributed in 1948 when *A. lingnanensis* was introduced.

Within 10 years, the new wasp had outcompeted the other species throughout most of its range.

55.3 A Species May Eliminate a Competitor from Parts of Its Range
Aphytis lingnanensis displaced *A. chrysomphali* over most of its range within a decade.

rats from the experimental plots, densities of small seed-eating rodents increased more than twofold, and the plots without kangaroo rats supported more rodent species than the control plots. These results showed that kangaroo rats reduce populations of some rodent species and eliminate others from places where they live. Kangaroo rats compete with other seed-eating rodents both by reducing their food supply—**exploitative competition**—and by aggressively defending space—**interference competition**.

In most natural communities, many species with similar ecological requirements live together, as seed-eating ants and rodents do in the Sonoran Desert. How can so many similar species share natural environments? Part of the answer is that natural environments are variable in space and time. That is why competing species often eliminate one another from some parts of the environment, but not from others. Also, natural environments typically provide many types of food, so that competing species do not overlap completely in their use of resources. In addition, other factors, such as predators, disease, and bad weather, may keep populations well below the environmental carrying capacity so that they rarely compete.

Predator–Prey and Parasite–Host Interactions

Competitive interactions in nature are often subtle, indirect, and difficult to detect. In contrast, predation is often direct, conspicuous, and easy to study. Consequently, our knowledge of predator–prey relationships is extensive. In this section we discuss how the dynamics of predator–prey and parasite–host interactions differ, and why interacting predator and prey populations typically fluctuate over time. Then we consider the evolutionary results of predator–prey interactions.

Predators are typically larger than and live outside the bodies of their prey. Generally they kill prey individuals when they eat them. Parasites are smaller than their hosts and may live inside or outside their bodies. Parasites often live in or on their hosts without killing them and may live for many generations within a single host. As a result, parasite–host interactions differ in interesting ways from predator–prey interactions.

Parasite–host interactions

Some parasites are only slightly smaller than their hosts, but others, such as viruses, bacteria, and protists, which are called **microparasites**, are much smaller than their hosts. Microparasites are able to reproduce within their hosts because their generation times are much shorter than those of their hosts. A host may harbor thousands or millions of them.

To understand the dynamics of microparasite–host interactions, it is useful to divide a host population into three distinct classes: susceptible; infected; or recovered and immune. Changes in the numbers of individuals in each class depend on births, deaths, infections, and development and

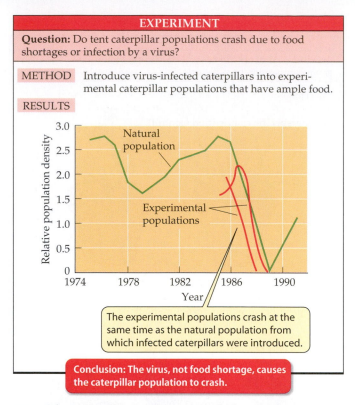

EXPERIMENT

Question: Do tent caterpillar populations crash due to food shortages or infection by a virus?

METHOD   Introduce virus-infected caterpillars into experimental caterpillar populations that have ample food.

RESULTS

The experimental populations crash at the same time as the natural population from which infected caterpillars were introduced.

Conclusion: The virus, not food shortage, causes the caterpillar population to crash.

55.4 Microparasites Can Cause Population Crashes
Transferring infected tent caterpillars from a natural population that was about to crash to uninfected populations caused those populations to decline even though food was abundant.

loss of immunity. For a microparasite population to survive in a host population, on average at least one new host individual must become infected before each infected host dies.

A microparasite can readily invade a host population dominated by susceptible individuals, but as the infection spreads, fewer and fewer susceptible individuals remain. Eventually a point is reached at which infected individuals do not, on average, transmit the infection to at least one other individual. Then the infection dies out in the host population. As a result, microparasite infections typically rise, then fall, and do not rise again until a sufficiently dense population of susceptible host individuals has reappeared.

Interactions between a microparasite—a virus—and populations of its host—the western tent caterpillar (*Malacosoma californicum*)—have been investigated experimentally (Figure 55.4). Larval western tent caterpillars eat the leaves of a variety of species of deciduous trees and shrubs in North America. Their populations fluctuate dramatically, with peak densities occurring about every 10 years. Food supply cannot, by itself, drive these cycles, because populations often collapse before they defoliate the trees. Rather, a virus appears to be responsible. This virus typically kills an infected caterpillar within a few weeks. When it dies, the caterpillar ruptures and releases millions of viruses onto the leaves and bark of trees, infecting many other caterpillars. After an outbreak of the virus, tent caterpillar populations remain low for many years until the remaining viruses are killed by exposure to sunlight.

To test whether viruses drove the population cycles,

ecologists introduced caterpillars from infected populations into experimental sites where caterpillar populations had abundant food and were starting to increase. Caterpillars in the experimental populations became infected and were killed by the introduced virus; consequently, these experimental populations declined at the same time as the populations from which the introduced caterpillars were drawn. In contrast, densities of caterpillars in control populations with abundant food continued to increase.

Predator–prey interactions

When a predator captures and eats a prey individual, it reduces the size of the prey population by one, but the effects of predators on prey population dynamics cannot be determined simply by counting the number of prey eaten. We also need to know how prey densities influence the ease with which prey are captured and how rapidly they reproduce. To understand the complex interactions between predators and their prey, it is useful to consider the process of predation from the perspective of an individual predator.

Consider, for simplicity, a predator species that eats only one kind of prey. An individual predator can find enough to eat if the rate at which it encounters prey is above a certain threshold value. Below that threshold, it will lose weight and eventually starve. Nevertheless, the predator may continue to eat prey while slowly losing weight, driv-

ing the prey population down further. Eventually the number of predators may be reduced by starvation or emigration, which may allow the prey population to increase its numbers. This increase may, in turn, permit the predator population to increase.

Because of this pattern, the recovery of a predator population often lags behind the recovery of its prey population. Predator–prey interactions often result in oscillations in the densities of both populations, just as microparasites often cause fluctuations in populations of their hosts.

Population density changes among small mammals and their predators living at high latitudes are the best-known examples of predator–prey oscillations. Populations of Canadian lynx and their principal prey—snowshoe hares—oscillate on a 9- to 11-year cycle (Figure 55.5). For many years, these oscillations were believed to be driven only by interactions between hares and lynxes. Recently, however, ecologists asked whether any part of the lynx–hare oscillation could be explained by fluctuations in the hares' food supply in addition to predation.

To answer this question, the ecologists selected nine 1-km² blocks of undisturbed coniferous forest in Yukon Territory, Canada. In two of the blocks, the hares were given supplemental food year-round. An electric fence with a mesh large enough to allow hares, but not lynxes, to pass through was erected around two other blocks. In one of these blocks, extra food was provided. In two other blocks, fertilizer was added to increase the quality of plant food for the hares. Three other blocks served as unmanipulated controls.

These experiments produced striking results. Excluding lynxes doubled, and adding food tripled, hare densities during the peak and decline phases of a cycle. Predator exclusion combined with food addition increased hare density 11-fold, but adding fertilizer had no effect on hare population density (Figure 55.6). Thus, the ecologists concluded that the cycle is driven both by predation by lynxes and by interactions between hares and their food supply.

Predators may eliminate prey from some environments but not others

In heterogeneous environments, predators may eliminate their prey in some places but not in others. In ponds on islands in Lake Superior,

55.5 Hare and Lynx Populations Cycle in Nature
The 9- to 11-year cycles of the snowshoe hare and its major predator, the Canadian lynx, were revealed by the number of pelts sold by fur trappers to the Hudson Bay Company.

Number of hares (prey) increases…

…followed by an increase in numbers of lynx (predator).

Each population cycle consists of an increase to a peak, a decline, and a low before another increase.

These predator–prey cycles follow a regular oscillating pattern.

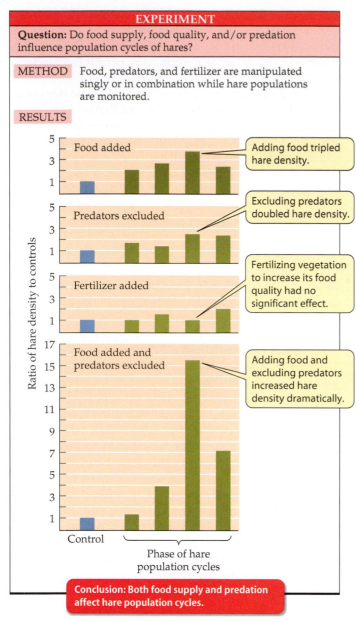

EXPERIMENT

Question: Do food supply, food quality, and/or predation influence population cycles of hares?

METHOD   Food, predators, and fertilizer are manipulated singly or in combination while hare populations are monitored.

RESULTS

Adding food tripled hare density.

Excluding predators doubled hare density.

Fertilizing vegetation to increase its food quality had no significant effect.

Adding food and excluding predators increased hare density dramatically.

Conclusion: Both food supply and predation affect hare population cycles.

55.6 Prey Population Cycles May Have Multiple Causes
Experiments showed that both food supply and predation (but not food quality) affect the population densities of snowshoe hares.

chorus frogs (*Pseudacris triseriata*) are found in only some of the habitats that seem suitable for them. Three major predators—the larvae of a salamander, nymphs of a large dragonfly, and dytiscid beetles—eat chorus frog tadpoles. An ecologist noticed that the tadpoles were common in ponds containing beetles, but were rare in ponds with salamander larvae and dragonfly nymphs. In laboratory experiments, he established that the salamander larvae could eat only small tadpoles, but that dragonfly nymphs could eat tadpoles of all sizes. Therefore, he hypothesized that dragonfly nymphs were responsible for eliminating chorus frogs from many ponds.

To test this hypothesis, he selected two large ponds that contained dragonfly nymphs but no tadpoles, and two other ponds that contained tadpoles but no nymphs. So that all the tadpoles would be handled equally, he removed the tadpoles from the two ponds that lacked dragonfly nymphs and then reintroduced them at the same densities. He introduced dragonfly nymphs into one of those ponds at typical densities. He also removed nearly all dragonfly nymphs from one of the ponds that had them and then introduced tadpoles into both of those ponds.

The dramatic results of the experiment supported his hypothesis (Figure 55.7). Tadpoles were eliminated from ponds with dragonfly nymphs, but survived well in ponds from which dragonfly nymphs were absent, or nearly so. This experiment shows that a particular predator may eliminate its prey in certain environments; however, it does not tell us why dragonfly nymphs were naturally absent from some of the ponds.

Predator–prey interactions change over evolutionary time

Because predators do not capture prey individuals randomly, they are agents of natural selection as well as agents of mortality. As a consequence, prey species have evolved a rich variety of adaptations that make them more difficult to capture, subdue, and eat. Among the evolutionary adaptations of prey are toxic hairs and bristles, tough spines, noxious chemicals and the means for ejecting them (Figure 55.8), camouflage, and mimicry of inedible objects or of larger or more dangerous organisms. Predators, in turn, may evolve to be more effective at overcoming prey defenses, leading to an evolutionary "arms race."

MIMICRY IS AN EVOLVED DEFENSE.   Among the best-studied adaptations of prey to predation is **mimicry**, an evolved resemblance to some inedible or unpalatable item. In **Batesian mimicry** a palatable species mimics an unpalatable or noxious one. Examples are the mimicry of ants by spiders and of bees and wasps by many different insects (Figure 55.9). Batesian mimicry works because a predator that captures an individual of an unpalatable or noxious species learns to avoid any prey of similar appearance.

However, if a predator captures a palatable mimic, it is rewarded with food, and it learns to associate palatability with the appearance of that prey. As a result, individuals of unpalatable species are attacked more often than they would be if they had no Batesian mimics. Because unpalatable individuals that differ from their mimics more than the average are less likely to be attacked by predators that have eaten a mimic, directional selection causes unpalatable species to evolve away from their mimics. Batesian mimicry systems are stable only if a mimic evolves toward the appearance of an unpalatable species faster than the unpalatable species evolves away from it, which usually requires that the mimic be less common than the unpalatable species.

Another type of mimicry is **Müllerian mimicry**, the convergence over evolutionary time in the appearance of two

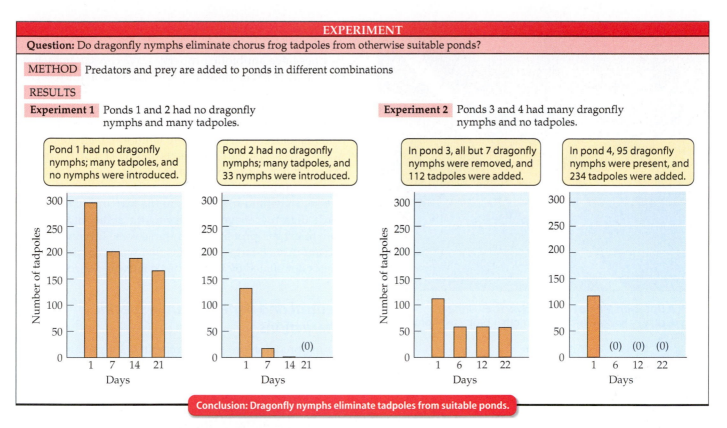

EXPERIMENT

Question: Do dragonfly nymphs eliminate chorus frog tadpoles from otherwise suitable ponds?

METHOD   Predators and prey are added to ponds in different combinations

RESULTS

Experiment 1 Ponds 1 and 2 had no dragonfly nymphs and many tadpoles.

Pond 1 had no dragonfly nymphs; many tadpoles, and no nymphs were introduced.

Pond 2 had no dragonfly nymphs; many tadpoles, and 33 nymphs were introduced.

Experiment 2 Ponds 3 and 4 had many dragonfly nymphs and no tadpoles.

In pond 3, all but 7 dragonfly nymphs were removed, and 112 tadpoles were added.

In pond 4, 95 dragonfly nymphs were present, and 234 tadpoles were added.

Conclusion: Dragonfly nymphs eliminate tadpoles from suitable ponds.

55.7 Predators Exclude Prey from Some Habitats
The speed with which dragonfly nymphs can eliminate tadpoles of the chorus frog from a pond is illustrated by the results of experiments in which dragonfly nymphs were added to ponds containing tadpoles.

(a) *Brachinus* sp.

(b) *Pterois volitans*

55.8 Defenses of Animal Prey
(a) A bombardier beetle ejects a noxious spray at the temperature of boiling water in the direction of a predator. The spray is ejected in high-speed pulses more than 20 times in succession. (b) The Indo-Pacific lionfish is among the most toxic of all reef fishes. Glands at the base of its spines can inject poison into an attacker. Its bright markings are thought to warn potential predators of this capability.

55.9 A Batesian Mimic Falsely Advertises Danger
By mimicking a wasp, this ctenucid moth is protected from predators.

plants against viruses and bacteria. As we saw at the beginning of this chapter, most of the spices used in human cuisines have antibiotic properties. We can safely include them in our food because they are toxic only to microorganisms.

Digestibility-reducing compounds are secondary compounds that make plant tissues difficult to digest. The most common of these substances are tannins, which are present in the leaves of some herbaceous and most woody species. When an herbivore chews on a leaf, tannins are released from the intracellular compartments in which they are stored. They bind to proteins in the leaf and to the herbivore's digestive enzymes, reducing the ability of the herbivore to extract proteins from the leaves. Tannins may be present in such large quantities that waters draining from forests dominated by tanniferous plants are tea-colored.

Neutral and Beneficial Interspecific Interactions

During predator–prey and competitive interactions, one or both participants in the interaction are harmed. Amensal-

or more unpalatable species. All species in a Müllerian mimicry system, including the predators, benefit when inexperienced predators eat individuals of any of the species because the predators learn rapidly that all species of similar appearance are unpalatable. Some of the most spectacular tropical butterflies are members of Müllerian mimicry systems (Figure 55.10), as are many kinds of bees and wasps.

EVOLVED CHEMICAL DEFENSES ARE WIDESPREAD AMONG PLANTS. In addition to physical defenses against herbivores, such as tough leaves, hairs, or spines, most plant tissues also contain defensive chemicals called **secondary compounds**. There are two types of defensive secondary compounds: acute toxins and digestibility-reducing compounds.

Acute toxins disrupt herbivore metabolism. Some of these toxins, such as nicotine, interfere with the transmission of nerve impulses to muscles. Others are hallucinogens, which cause individuals that ingest them to have a seriously distorted view of their environment. Some toxins imitate insect hormones and prevent insects from completing metamorphosis. Still others are unusual amino acids that become incorporated into herbivore proteins and interfere with their functioning. Other acute toxins defend

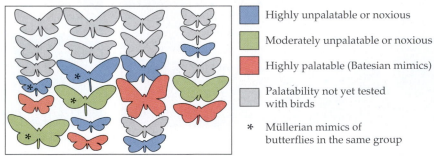

🟦 Highly unpalatable or noxious

🟩 Moderately unpalatable or noxious

🟥 Highly palatable (Batesian mimics)

⬜ Palatability not yet tested with birds

* Müllerian mimics of butterflies in the same group

55.10 Müllerian and Batesian Mimics
By converging in appearance, the unpalatable Müllerian mimics among these different species of Costa Rican butterflies and moths reinforce each other in deterring predators. The palatable Batesian mimics benefit because predators have learned to associate these color patterns with distasteful or noxious effects.

55.11 Danger Comes From Above
Shrubs and herbaceous plants are often damaged by branches falling from tall trees.

ism causes harm to one of the partners without affecting the other. In the other two types of interspecific interactions—commensalism and mutualism—neither partner is harmed, and one or both may benefit. We examine these interactions in the sections that follow.

In amensalism and commensalism, one participant is unaffected

Amensalisms, in which an individual harms another organism but is unaffected by the species it harms, are widespread and important in nature. Mammals, for example, create bare spaces around water holes. They benefit by drinking water, but not by trampling the plants they kill. Leaves and branches falling from trees damage smaller plants beneath them (Figure 55.11). The trees drop these old structures regardless of whether or not they damage other plants.

Commensalism benefits one partner but has no effect on the other. An example is the relationship between cattle egrets and grazing mammals. Cattle egrets are found throughout the tropics and subtropics. They typically forage on the ground around cattle or other large mammals, concentrating their attention near the heads and feet of the mammals, where they catch insects flushed by their hooves and mouths (Figure 55.12). Cattle egrets that forage close to grazing mammals capture more food for less effort than egrets that do not. The benefit to the egrets is clear; the mammals neither gain nor lose.

Mutualisms benefit both participants

Mutualisms are interactions that benefit both participants. Mutualistic interactions exist between plants and microorganisms, protists and fungi, plants and insects, and among plants. Animals also have mutualistic interactions with pro-

tists and with one another. As you learned in Chapter 27, the evolution of eukaryotic organisms is believed to be the result of mutualistic interactions between previously free-living prokaryotes and the cells they originally infected.

INTERGROUP MUTUALISMS. Some of the most complex and ecologically important mutualisms are between members of different kingdoms or domains.

Most plants have beneficial associations with soil-inhabiting fungi called mycorrhizae that enhance the plant's ability to extract minerals from the soil (see Figure 30.16). And in the critical mutualistic relationship of some plants with nitrogen-fixing bacteria of the genus *Rhizobium* (discussed at length in Chapter 36), the bacteria receive protection and nutrients from their host plant while providing the host with usable nitrogen.

Lichens are compound organisms consisting of highly modified fungi that harbor cyanobacteria or green algae among their hyphae (see Figure 30.18). The fungi absorb water and nutrients from the environment and provide these as well as a supporting structure for the microorganisms, which in turn provide the fungi with the products of photosynthesis.

Animals have important mutualistic interactions with protists. For example, corals, some anemones, and some tunicates gain most of their energy from photosynthetic protists that live within their tissues. In exchange, they provide the protists with nutrients from the small animals they capture (see Figure 4.16c).

Termites have nitrogen-fixing protists in their guts that help them digest the cellulose in the wood they eat. Young termites must acquire their protists by eating the feces of other termites. If prevented from doing so, they soon die. The protists are provided with a suitable environment in which to live and an abundant supply of cellulose.

55.12 Commensalism Benefits One Partner
Cattle egrets catch more insects with less effort when they forage around large grazing mammals, such as cape buffalos. The buffalos are neither harmed nor helped by the egrets.

55.13 A Plant–Animal Mutualism
Some *Acacia* species have large, swollen, hollow thorns in which ants build their nests.

thorns in which ants of the genus *Pseudomyrmex* construct their nests and raise their young (Figure 55.13). These ants live only on acacias. They feed on nectar that the trees produce at the bases of their leaf petioles and on special nutritive bodies on the leaves. The ants attack and drive off leaf-eating insects, eat the eggs and larvae of herbivorous insects, and even bite and sting browsing mammals. They also cut back the tips of other plants, particularly vines that grow over their host tree. The ants get room and board; the plants get protection against both predators and competitors. Experiments in which acacias are deprived of their ants demonstrate the amount of protection the ants provide (Figure 55.14).

Many angiosperms depend on animals to transport both their pollen and their seeds. The plants benefit from pollination mutualisms by having their pollen carried to other conspecific plants and by receiving pollen to fertilize their ovules. Animals benefit by obtaining food in the form of nectar and pollen (Figure 55.15*a*). Plants also benefit from having their seeds dispersed to sites where they are more likely to germinate and survive than directly under the parent plant (Figure 55.15*b*). Animal dispersers benefit by eating the nutritious fruits surrounding the seeds. The plants pay a price for the benefits they receive: The energy and materials a plant uses to produce nectar, fruits, and other rewards for animals cannot be used for growth or seed production.

ANIMAL–ANIMAL MUTUALISMS. Many species of ants have mutualistic relationships with aphids. Ants "milk" these small, plant-sucking insects by stroking them with their forelegs and antennae. The aphids respond by secreting droplets of partly digested plant sap that has passed through their guts. In return, the ants protect the aphids from predatory wasps, beetles, and other natural enemies. The aphids lose nothing, because plant sap is high in sugar but low in amino acids. Thus, aphids inevitably ingest more sugar than they can use.

PLANT–ANIMAL MUTUALISMS. Terrestrial plants have many mutualistic interactions with animals. A complex mutualism between trees and ants that live in Central America illustrates the benefits of such interactions. Trees of the species *Acacia cornigera* have large, hollow

55.14 An Experiment Demonstrates the Benefits of Housing Ants
Acacia trees that housed ant colonies grew back faster than acacias without ants.

EXPERIMENT

Question: Do ants provide effective protection for acacia trees?

METHOD   Acacia trees were severely pruned. Ants were allowed to recolonize some trees as they regrew, but not others.

RESULTS

With ants   Without ants

Trees grown without ants were heavily attacked by other insects and regained their leaves very slowly.

Conclusion: Ants provide very effective protection for acacia trees.

(a) *Glossophaea* sp.

(b) *Bombycilla garrulus*

55.15 Plants Incur Costs to Attract Mutualists
(a) Some animal pollinators such as this long-tongued bat are attracted by rewards of nectar or pollen. (b) The nutritious pulp of fruits are attractive to many birds, such as this cedar waxwing.

Interactions between plants and their pollinators and seed dispersers are clearly mutualistic, but they are not purely mutualistic. Many seed dispersers are also seed predators that destroy some of the seeds they remove from plants. Some organisms that collect these rewards are not mutualists at all. Many animals visit flowers without transferring any pollen. Some of them cut holes to get to the nectar-producing regions at the base of the flowers. On the other hand, some plants exploit their pollinators. The flowers of certain orchids, for example, mimic female insects, enticing male insects to copulate with them (Figure 55.16). The male insects neither sire any offspring nor obtain any reward, but they transfer pollen between flowers, benefiting the orchid.

Coevolution of Interacting Species

The relationships between plants and their pollinators and seed dispersers show how the evolution of species traits can be influenced by interactions with other species. Species that have mutually influenced one another's evolution are said to have **coevolved**. In **diffuse coevolution**, species traits are in-

fluenced by interactions with a wide variety of predators, parasites, prey, or mutualists. For example, most flowers are pollinated by a number of animal species, and most pollinators visit many species of flowers.

The traits of the fleshy fruits that surround many seeds are also the result of diffuse coevolution. Few fruits are adapted for dispersal by only a few species of animals. Most bird-dispersed fruits are red, or some combination of red and another color, and have no obvious odor to humans. Fruits dispersed by nonflying mammals, many of which lack color vision, are typically purple, are not highly visible to birds, and usually have a pleasant odor. Bat-dispersed fruits are typically green and have a fruity odor when ripe. They are inconspicuous during the day, but are easy for bats to detect at night. Many bird-dispersed fruits have unpleasant tastes to mammals.

Species-specific coevolution is much rarer than diffuse coevolution, but yucca plants and the moths of the genus *Tegeticula* that pollinate them have this kind of relationship. Female yucca moths lay their eggs only in the ovules of yucca flowers, and yucca flowers are pollinated only by *Tegeticula*. A female *Tegeticula* lays no more than five eggs in any one flower. After she has laid her eggs, she scrapes pollen from the flower's anthers, rolls it into a small ball, flies to another yucca plant, and places the pollen ball on the stigma of a flower before laying another batch of eggs. When the eggs hatch, the larvae burrow into the ovary and feed upon the developing seeds. Each yucca species has a specific moth species associated with it (Figure 55.17).

One feature of the coevolved relationship between *Tegeticula* and *Yucca* is surprising: Why do female moths lay so few eggs per flower? Wouldn't a female moth that laid more than five eggs per flower produce more surviving off-

55.16 Some Orchids Mimic Female Insects
The flowers of this orchid so closely resembles a female wasp that male wasps are tricked into attempting to copulate with them, as this male is doing. The orchid gets pollinated, but the wasp gets no reward.

(a) *Yucca brevifolia*

55.17 Yucca–Yucca Moth Coevolution
The Joshua tree (*a*) is pollinated only by the yucca moth (*b*), shown here on a flower. The moth's larvae feed on developing yucca seeds fertilized by the pollen she transports.

(b) *Tegeticula yuccasella*

spring than moths that lay only the usual number? The evolutionary reason for their restraint is that *Yucca* plants abort flowers in which more than five eggs are laid. As a result, no moth offspring are produced in such flowers. Thus, the mutualism is stabilized at a level that represents an "evolutionary compromise" between the fitness of the moths and the yuccas.

Some Species Have Major Influences on Community Composition

Organisms influence the communities in which they live through all of the types of interactions that we have just described. Through these interactions, they may influence the **species richness** of their communities—that is, the number of species that live there. They also influence their communities by altering microclimate, soil structure and chemistry, and water movement. These alterations affect the suitability of the physical environment for other organisms. Organisms also change the amounts and distributions of resources in the community. Species whose influences on ecological communities are greater than would be expected on the basis of their abundance are called **keystone species**. Keystone species may influence the species richness of communities, the flow of energy and materials in ecosystems, or both.

Some species have major influences on their communities simply through their abundance. For example, in terrestrial communities, plants form most of the structural environment, are the major modifiers of the physical environment, and are the pathway through which energy and nutrients enter communities. Anyone who has walked into the shade of a tree on a hot, sunny day knows that climate near the ground is strongly influenced by plants.

Animals may change vegetation structure and species richness

Animals that are able to change vegetation structure alter the environment for many other species. Beavers, for example, create meadows by cutting down trees and create ponds by building dams. Large grazing and browsing mammals may also dramatically change the structure and composition of vegetation.

To determine the influence of bison (*Bison bison*) on prairie vegetation, 30 individuals were introduced to the Konza Prairie Research Natural Area in northeastern Kansas, the largest tract of unplowed tallgrass prairie in North America. The herd, now about 200 animals in size, has unrestricted access to 10 watersheds that are regularly burned in spring. Bison prefer to eat grasses and eat few of the broad-leaved plants (**forbs**) that grow among the grasses. Bison also prefer to graze on recently burned areas.

Areas from which bison are excluded and which are burned annually are dominated by tall grasses and have few plant species. In contrast, regularly burned areas that are grazed by bison have more forbs—and many more species of forbs—because the bison, by preferentially grazing on the grasses, create space for them (Figure 55.18). Also, urea in bison urine is hydrolyzed to ammonium and its nitrogen is available to plants within a few days; decomposing plant litter releases nitrogen much more slowly. Therefore, plants in areas grazed by bison have higher leaf nitrogen levels and hence grow faster.

Predators may change marine community structure

Predation by the sea star *Pisaster ochraceous*, an abundant animal in rocky intertidal communities on the Pacific coast of North America, increases local species richness. In the

Bison bison

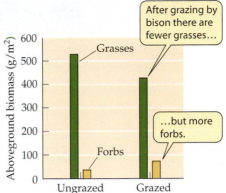

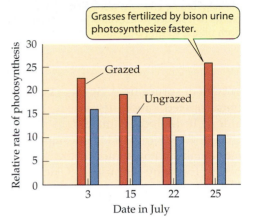

55.18 Bison Increase Plant Species Richness and Productivity
By grazing preferentially on grasses, bison increase the density of forbs and overall plant productivity.

After grazing by bison there are fewer grasses…

…but more forbs.

Grasses fertilized by bison urine photosynthesize faster.

absence of sea stars, their preferred prey, the mussel *Mytilus californianus*, crowds out other competitors in a broad belt of the intertidal zone. By consuming mussels, *Pisaster* creates bare spaces that are taken over by a variety of other species (Figure 55.19).

The influence of *Pisaster* on community composition was demonstrated by experimentally removing them from selected areas in the intertidal zone repeatedly over a 5-year period. These removals resulted in two major changes. First, the lower edge of the mussel bed extended farther down into the intertidal zone, showing that sea stars are able to eliminate mussels completely where they are covered with water most of the time. Second, and more dramatically, 28 species of animals and algae disappeared from the removal zones, until only *Mytilus*, the dominant competitor, occupied the entire substrate. By altering competitive relationships, predation by *Pisaster* largely determines which species live in these rocky intertidal communities.

55.19 Sea Stars are Keystone Predators
This sea star (*Pisaster ochraceus*) is resting on a bed of mussels (*Mytulis* sp.), its preferred prey.

Temporal Changes in Communities

Because organisms alter soil structure and chemistry and microclimates, the species composition of ecological communities changes constantly over time. The plants that first colonize a site after a disturbance, for example, differ from those that dominate the site later. Such a sequence of change in the species composition of a community is called **ecological succession**. Patterns and causes of ecological succession are varied, but the early colonists often alter the conditions under which later-arriving species grow.

Ecologists divide succession into two major types. **Primary succession** begins with the establishment of organisms on newly available sites that previously had no organisms. **Secondary succession** begins when organisms reestablish themselves on disturbed sites where some organisms survived the disturbance.

A good example of primary succession is the sequence of changes following the retreat of a glacier in Glacier Bay, Alaska, over the last 200 years. The retreating glacier left a series of *moraines*—gravel deposits formed where the glacial front was stationary for a number of years. No scientist was present to measure changes over the 200-year period, but ecologists have inferred the temporal pattern of succession by examining plant communities on moraines of different ages. The youngest moraines, close to the current glacial front, are populated with bacteria, fungi, and photosynthetic microorganisms. Slightly older moraines have lichens, mosses, and a few species of shallow-rooted herbs. Successively older moraines have shrubby willows, alders, and conifers.

By comparing moraines of different ages, ecologists deduced the pattern of plant succession and of changes in the soil at Glacier Bay. Succession was caused in part by changes in the soil brought about by the organisms themselves. An herbaceous plant, *Dryas*, and alder trees have nitrogen-fixing bacteria in nodules on their roots. Because nitrogen is virtually absent from glacial moraines, nitrogen fixation by *Dryas* and alders improved the soil for the

55.20 Primary Succession on a Glacial Moraine
As the plant community occupying an Alaskan glacial moraine changes from pioneering plants to a spruce forest, nitrogen accumulates both on the forest floor and in the mineral soil.

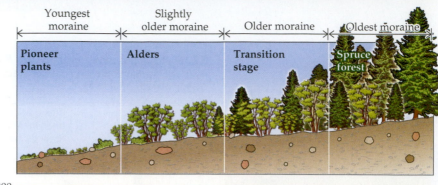

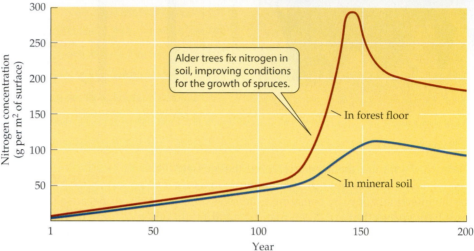

Alder trees fix nitrogen in soil, improving conditions for the growth of spruces.

In forest floor

In mineral soil

growth of spruce trees. Spruces then outcompeted and displaced the alders. If the local climate does not change dramatically, a forest community dominated by spruces is likely to persist for many centuries at Glacier Bay (Figure 55.20).

The changes that take place when all or part of the dead body of some plant or animal is decomposed are examples of secondary succession. The succession of fungal species that decompose pine needles in litter beneath Scots pines (*Pinus sylvestris*) is shown in Figure 55.21. New litter is continuously deposited under pines, so that the surface layer of litter is young and deeper layers are progressively older. Decomposition begins when the first group of fungi starts consuming the needles as soon as they fall. Each group of fungi decomposes certain compounds, converting them to other compounds that are decomposed by the next successional group. This process continues over about 7 years, by which time the last group of organisms—basidiomycetes—has decomposed the litter completely.

Indirect Effects of Interactions among Species

In the experiments we have described above, one member of a community was removed or added, and investigators measured the resulting changes. Single-species removals or additions can demonstrate the direct effects of species on one another, but to quantify indirect effects, observations and manipulations of several species are needed.

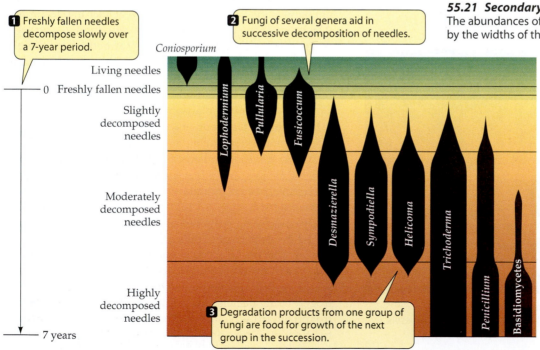

1 Freshly fallen needles decompose slowly over a 7-year period.

2 Fungi of several genera aid in successive decomposition of needles.

3 Degradation products from one group of fungi are food for growth of the next group in the succession.

55.21 Secondary Succession on Pine Needles
The abundances of ten types of fungi (indicated by the widths of the black bars) in pine needle litter change with time. The age of the needles increases with depth within the layer.

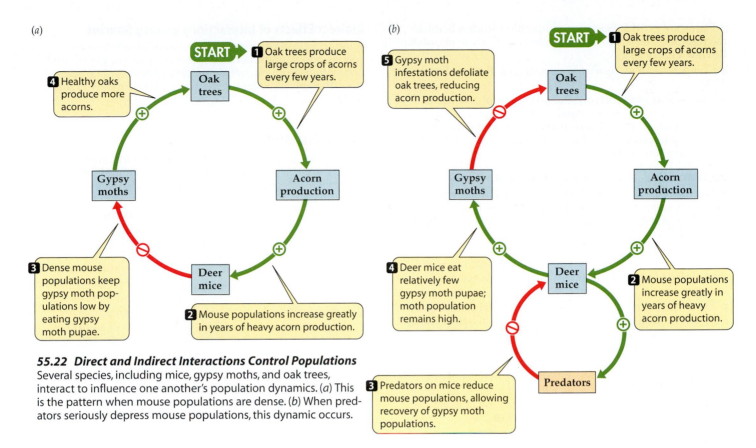

55.22 Direct and Indirect Interactions Control Populations
Several species, including mice, gypsy moths, and oak trees, interact to influence one another's population dynamics. (*a*) This is the pattern when mouse populations are dense. (*b*) When predators seriously depress mouse populations, this dynamic occurs.

Ecologists have assembled a variety of data to help them understand relationships between oak trees and the animals that eat their leaves and acorns. Most damage to leaves in the oak forests of eastern North America is caused by gypsy moths; most acorns are eaten by mice, chipmunks, deer, squirrels, and birds (Figure 55.22). The abundance of mice and chipmunks is controlled largely by acorn abundance, which varies greatly from year to year. Deer consume large quantities of acorns during years of high acorn production, but during poor acorn years they shift to other foods.

Gypsy moths eat oak leaves. During years when their populations are very large, they may defoliate large expanses of forest. Outbreaks of gypsy moths occur once every 6–10 years. Gypsy moth populations collapse after they defoliate a forest because most larvae die of starvation. In the year following defoliation, oak trees are full of leaves, but gypsy moth populations remain low for many years. Why don't they quickly rebound?

To determine whether predation by mice prevented gypsy moth populations from recovering after a crash, ecologists measured predation rates by attaching freeze-dried gypsy moth pupae to small squares of burlap. They placed the burlap panels on oak tree trunks at sites where gypsy moths typically pupate. During a year of moderately dense mouse populations, all of the pupae were eaten within 8 days. During a year of low mouse density, half of the pupae remained after 18 days, which is longer than it would take them to complete metamorphosis and emerge. These results suggest that in years when acorns are abundant, mice keep gypsy moth populations at low densities by eating most of their pupae. In so doing, they allow the oak trees to recover from defoliation and accumulate enough energy reserves to produce another large crop of acorns.

Predators typically reduce mouse populations within 1.5 years after a year of high acorn production, after which gypsy moth populations rebound and again defoliate the trees. If few mice were present, the gypsy moths might rebound so quickly that the oaks could never produce large crops of acorns. If the investigators had studied only the interactions between mice and acorns or between gypsy moths and oak trees, they would not have discovered the important influences of other species on these interactions.

Chapter Summary

Types of Ecological Interactions

▶ Species interact with one another in five major ways: competition, predator–prey or parasite–host interactions, mutualism, commensalism, and amensalism. **Review Table 55.1**

Resources and Consumers

▶ A resource is anything directly used by an organism that can potentially lead to the growth of the population and whose availability is reduced when it is used.
▶ Biotic interactions, as well as the physical environment, influence the conditions under which species can persist. **Review Figure 55.1**

Competition: Seeking and Using Scarce Resources

▶ If organisms use the same resources and those resources are in short supply, the individuals are competitors. Competition may be either intraspecific or interspecific.

▶ Plants are good subjects for competition studies because the resources for which they compete are easily manipulated. **Review Figure 55.2**

▶ Competition may restrict species ranges. **Review Figure 55.3**

▶ Species that use similar resources commonly coexist in nature because nature is spatially and temporally complex, many resources typically are available, and other factors often keep populations below carrying capacity so that they do not compete strongly.

Predator–Prey and Parasite–Host Interactions

▶ The relative sizes of predators and prey influence their interactions. Microparasites are typically much smaller than their prey and may live in or on their hosts without killing them.

▶ To understand the dynamics of host–microparasite interactions, it is useful to divide a host population into three distinct classes—susceptible, infected, and recovered and immune.

▶ Experimental manipulation of parasites and predators in nature reveals that they are often important in determining both numbers and distributions of their prey. **Review Figures 55.4, 55.5, 55.6, 55.7**

▶ Predators act as evolutionary agents by selecting for adaptations to protect against them. Prey have evolved many such adaptations, such as toxic hairs and bristles, tough spines, noxious chemicals, and mimicry of inedible objects or dangerous organisms. **Review Figure 55.10**

Neutral and Beneficial Interspecific Interactions

▶ Commensal interactions, in which one partner benefits while the other is unaffected, are common in nature.

▶ Mutualistic interactions, in which both participants benefit, are also common in nature. Mutualistic interactions occur between members of different groups of organisms (between plants and prokaryotes, between fungi and protists, and between animals and protists). Animals have mutualistic interactions with other animals and with plants, such as pollination and seed dispersal. **Review Figure 55.14**

Coevolution of Interacting Species

▶ Some mutualistic relationships, such as those between yuccas and yucca moths, are tightly coevolved, but diffuse coevolution between many species is much more common.

Some Species Have Major Influences on Community Composition

▶ Keystone species have influences on ecological communities that are greater than would be expected from their abundances, but abundant species also have major influences on community structure.

▶ Vascular plants, mammals that change vegetation structure, predators on dominant competitors, and microorganisms often have major influences on ecological communities. **Review Figure 55.18**

Temporal Changes in Communities

▶ Ecological succession involves changes in the species composition of a community over time. Early colonists often alter the conditions under which later-arriving species grow.

▶ Primary succession begins at sites that have never been modified by organisms. **Review Figure 55.20**

▶ Succession may take place when all or part of the dead body of some organism is decomposed. **Review Figure 55.21**

Indirect Effects of Interactions among Species

▶ Indirect effects of species interactions influence many species populations. For example, mice prevent gypsy moth populations from recovering quickly after they defoliate oak trees, thereby allowing the trees to recover. **Review Figure 55.22**

For Discussion

1. Environmental factors such as temperature, humidity, and salinity, even though they are important in the lives of many organisms, are not considered to be resources. Why not?

2. Kangaroo rats prevent smaller species of rodents from occupying some Sonoran desert habitats. They also reduce populations of seed-harvesting ants, but they do not cause elimination of ants from any habitats. Why can they competitively exclude other rodents, but not ants?

3. On the eastern side of the Sierra Nevada in California, four species of chipmunks occupy adjacent habitats from which they exclude one another by direct aggressive interference. In the San Jacinto Mountains of southern California, three other chipmunk species similarly occupy adjacent habitats, but no interspecific aggression is observed. Each species simply remains in its own habitat. Which of these two assemblages do you think is the older one? Why?

4. What features of predator–prey interactions tend to generate instabilities that lead to fluctuations in the densities of both species? Given that instabilities are expected, what keeps populations of either predator or prey from fluctuating to extinction?

5. Parasites usually have generation times much shorter than those of their hosts. Consequently, they should be able to evolve faster. What prevents them from evolving so fast that they completely overcome the resistance of their hosts and exterminate them?

6. Mimicry of inedible, toxic, or dangerous objects is widespread in nature, but most species are not mimics. Why don't more species evolve to mimic such objects?

7. Wind does not direct pollen toward conspecific stigmas. Given this inefficiency, why are there so many wind-pollinated plants? Similarly, if seeds that land close to the parent plant survive less well than those that are carried farther away, why do so many plants produce seeds lacking dispersal devices?

8. Some direct interactions between two species benefit only one of those species. Give examples of such "one-way" benefits in each of the following cases:
   a. between two species of plants (give one example of energetic and another example of physical support)
   b. between a plant and an eater of its leaves
   c. between a predator and its prey

9. Wood is an abundant food source that has been available for millions of years. Why have so few animals evolved to be able to eat wood?

10. A keystone species exerts a larger influence on the ecological community in which it lives than one would expect given its abundance. What traits are likely to result in a species having such major ecological effects?

56 Ecosystems

IN 1976 THE OUTLET OF SOUTHERN INDIAN Lake in northern Manitoba, Canada, was dammed, raising the lake level 3 meters. Engineers then diverted the Churchill River so that rather than flowing into the lake, it flowed southward across a drainage divide and through a series of hydroelectric generating stations.

Before the dam was built, ecologists studied the lake in detail to assess the likely consequences of raising its level and greatly reducing the flow of river water into it. They predicted that fewer nutrients would enter the lake, but that the reduction would be compensated for by nutrients derived from increased soil erosion along the newly submerged shoreline. Based on their predictions, they believed that the Southern Indian Lake whitefish fishery, the most important commercial fishery in northern Manitoba, would not be harmed.

The ecologists' predictions of the future nutrient status of the lake and amounts of photosynthesis by algae were correct. However, to everyone's surprise, the whitefish fishery was ruined. The greatly increased soil erosion on the new shoreline released large quantities of mercury into the lake. Mercury concentrations in fish in Southern Indian Lake now exceed Canadian safety standards and will probably remain above standard for many years. From 1977 to 1982, Manitoba Hydro, the builder of the dam, subsidized the commercial fishermen, and in 1982, it provided a one-time cash settlement of $2.5 million Canadian dollars for future losses to the fishermen.

Unexpected surprises commonly follow not only the damming of rivers and lakes, but most attempts to alter ecological systems. Surprises arise because the behavior of ecosystems is the result of interactions among many different processes, most of which are only incompletely understood. Ecologists now recognize that mercury pollution often results from the raising of lake levels, but they did not know that in the 1970s. As humans continue to alter Earth's ecological systems, new surprises confront us each year.

Dams: Some Surprising Aftereffects
Damming rivers and lakes in the expectation of creating benefits for the human population often results in unexpected and unpredictable detrimental effects on the ecosystem.

The organisms living in a particular area, such as Southern Indian Lake, together with the physical environment with which they interact, constitute an **ecosystem**. Ecosystems can be recognized and studied at many different spatial scales, ranging from local units—such as a lake—to the entire globe. At the global scale, Earth is a single ecosystem.

The dynamics of ecosystems are the result of the activities of myriad individual organisms, which are influenced by processes in the physical environment. Some of these processes are, in turn, altered by organisms, and some are not. Individuals of the many different species that interact do so by capturing energy and materials, transforming and retaining them, and transferring them to other organisms.

In this chapter we discuss climates on Earth and how they influence ecosystem processes. We then describe patterns of energy flow and the cycles of materials in ecosystems. We will see how knowledge about ecosystems can be used to understand how and why they respond as they do to human-caused disturbances, and how we can learn from our mistakes.

Climates on Earth

The energy of the sun drives the global circulation patterns of air and ocean waters. The warming and cooling of moving masses of air and water explain most of Earth's climatic patterns. Climates, in turn, exert a powerful influence on the distributions, abundances, and evolution of species.

Climates vary greatly from place to place on Earth, primarily because different places receive different amounts of solar energy. The amount of incident solar energy is nearly constant at the equator, but varies dramatically at high latitudes. In this section we will examine how differences in solar energy input determine atmospheric and oceanic circulation.

Solar energy inputs drive global climates

Every place on Earth receives the same total number of hours of sunlight each year—an average of 12 hours per day—but not the same amount of *heat*. The rate at which heat arrives on Earth per unit of ground area depends primarily on the angle of sunlight. If the sun is low in the sky, a given amount of solar energy is spread over a larger area (and is thus less intense) than if the sun is directly overhead. In addition, when the sun is low in the sky, sunlight must pass through more of Earth's atmosphere, with the result that more of its energy is absorbed and reflected before it reaches the ground. At high latitudes (closer to the poles), there is more variation in both day length and the angle of arriving solar energy over the course of a year than at latitudes closer to the equator. On average, mean annual air temperature decreases about 0.4°C for every degree of latitude (about 110 km) at sea level.

Air temperature also decreases with elevation. The effect of elevation on temperature is due to the properties of gases. As a parcel of air rises, it expands, its pressure drops, and energy is expended in pushing molecules apart. With that loss of energy, the temperature of the air drops. When the parcel of air descends, it is compressed, its pressure rises, the same amount of energy is recovered, and its temperature increases.

When wind patterns bring air into contact with a mountain range, the air rises to pass over the mountains, cooling as it does so. Because cool air cannot hold as much moisture as warm air, clouds frequently form, and moisture is released as rain or snow. On the leeward side of the range, the air, now containing little moisture, descends, warms, and picks up moisture. This pattern often results in a dry area, called a **rain shadow**, on the leeward side of a mountain range (Figure 56.1).

Global atmospheric circulation influences climates

Earth's climates are strongly influenced by global air circulation patterns. Air rises not only when it crosses mountains, but also when it is heated by the sun. Warm air rises in the tropics, which receive the greatest solar energy input. This air is replaced by air that flows toward the equator from the north and south. That air, in turn, is replaced by air from aloft that descends after having traveled away from the equator at great heights. At roughly 30° north and south latitudes, air that cooled and lost its moisture when it rose at the equator descends and warms. Many of Earth's deserts, such as the Sahara and the Australian deserts, are located at these latitudes.

At about 60° north and south latitudes, air rises again. Cold, dense air descends at the poles, where there is little

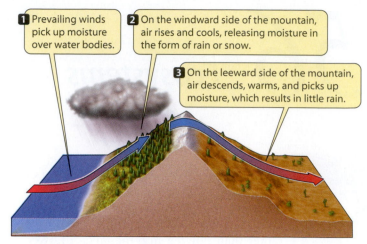

1 Prevailing winds pick up moisture over water bodies.

2 On the windward side of the mountain, air rises and cools, releasing moisture in the form of rain or snow.

3 On the leeward side of the mountain, air descends, warms, and picks up moisture, which results in little rain.

56.1 A Rain Shadow
Average annual rainfall tends to be lower on the leeward side of a mountain range than on the windward side.

input of solar energy. The black arrows around the edges of Figure 56.2 show these vertical patterns, which are one component of Earth's winds.

The spinning of Earth on its axis influences surface winds because Earth's spinning velocity is rapid at the equator, but relatively slow close to the poles. An air mass at a specific latitude has the same velocity as Earth has at that latitude. As an air mass moves toward the equator, it confronts a faster and faster spin, and its rotational movement is slower than that of Earth beneath it. Similarly, as an air mass moves poleward, it confronts a slower and slower spin, and it speeds up relative to Earth beneath it. Therefore, air masses moving latitudinally are deflected to the right in the Northern Hemisphere and to the left in the Southern Hemisphere. Winds blowing toward the equator from the north and south veer to become the northeast and southeast trade winds, respectively. Winds blowing away from the equator also veer and become the westerlies that prevail at mid-latitudes. The average directions of these surface winds are shown by the blue arrows in Figure 56.2.

Because Earth's axis is tilted, the amount of solar energy that reaches a given region varies seasonally as Earth orbits the sun. At any given location, the amount of solar energy input is at its maximum at the time of year when the sun is closest to being overhead at noon. The location of greatest solar energy input in the tropics, and the site where the trade winds converge and air rises, is called the **intertropical convergence zone** (see Figure 56.2). It shifts to the north during the northern summer (southern winter) and to the south during the southern summer (northern winter). Seasonal changes in climate (rainy and dry seasons) in the tropics are associated with the movement of the intertropical convergence zone because whenever an area is within the zone, air rises and heavy rains fall. When the zone is to the north or south of a tropical region, the prevailing winds are trade winds, which seldom yield rain unless forced to rise over mountains.

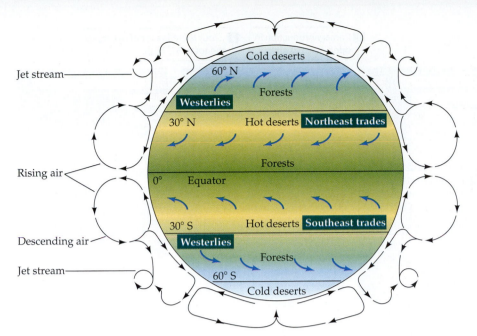

56.2 Circulation of Earth's Atmosphere
If we could stand outside Earth and observe its air movements, we would see vertical air movements similar to those indicated by the black arrows and surface winds similar to those shown by the blue arrows. The vertical and horizontal circulation patterns shift to the north during the northern summer and to the south during the northern winter. Thus the intertropical convergence zone is on the equator only twice during each year.

deflected laterally along its shores. In both hemispheres, water flows toward the equator along the west sides of continents, continuing to veer right or left until it meets at the equator and flows westward again.

The oceans play an important role in world climates, both because their waters move long distances and because water has a high specific heat. The **specific heat** of a substance is the amount of energy required to raise the temperature of 1 gram of the substance 1°C. For water, this value is 1 cal/g at 15°C. Similarly, 1 gram of water that cools 1°C gives off 1 cal/g. Air and land surfaces have a much lower specific heat. Consequently, in comparison with continents, oceans warm up more slowly in summer because it takes more heat to raise their temperature. Oceans cool off more slowly in winter because more heat must be released to cool them.

Global oceanic circulation is driven by winds

The global pattern of wind circulation drives the circulation of ocean water. Ocean water generally moves in the direction of the prevailing winds (Figure 56.3). Winds blowing toward the equator from the northeast and southeast cause water to converge at the equator and move westward until it encounters a continental land mass. At that point the water splits, some of it moving north and some of it moving south along continental shores. The poleward movement of ocean water that has been warmed in the tropics is a major mechanism of heat transfer to high latitudes. As it moves toward the poles, the water veers right in the Northern Hemisphere and left in the Southern Hemisphere. Thus water moves eastward until it encounters another continent and is

56.3 Global Oceanic Circulation
To see that ocean currents are driven primarily by winds, compare the surface currents shown here with the prevailing surface winds shown in Figure 56.2. Deep ocean currents differ strikingly from the surface ones shown here.

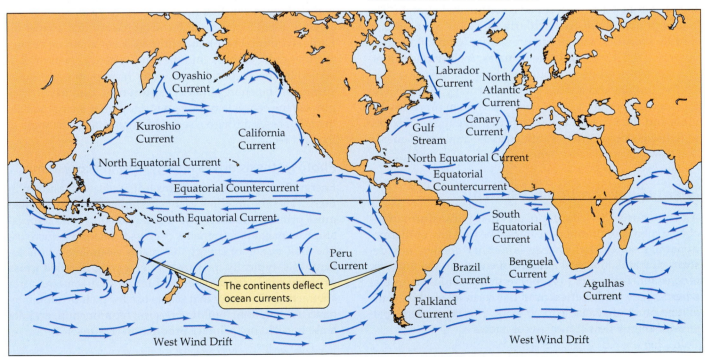

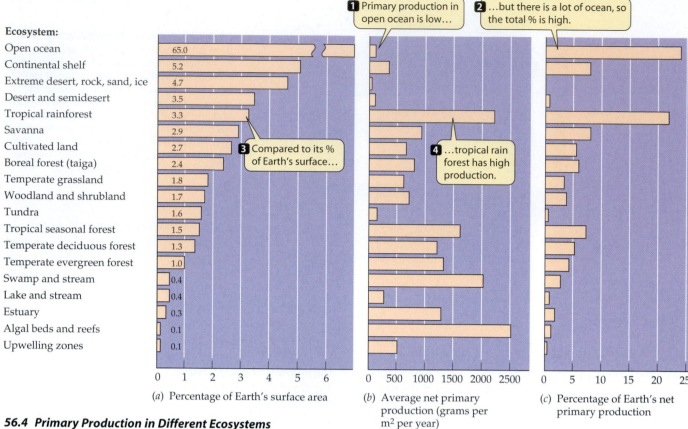

56.4 Primary Production in Different Ecosystems
The primary production of Earth's ecosystems is measured in several ways. (a) The geographic extent of the different ecosystems. (b) Net annual primary production and (c) the percentage of Earth's total primary production contributed by each ecosystem.

Both wind circulation patterns and the properties of ocean water affect terrestrial climates. At high latitudes, the temperatures of the interiors of large continents fluctuate greatly with the seasons, becoming very cold in winter and hot in summer, a pattern called a **continental climate**. The coasts of continents, particularly those on west sides at middle latitudes, where the prevailing winds blow from ocean to land, have **maritime climates**, with smaller differences between winter and summer temperatures.

The amount and annual pattern of energy input in a region determines the rates at which ecosystem processes operate and the kinds of organisms that live there. Next we will discuss how climates influence the amount of energy that flows through ecosystems.

Energy Flow through Ecosystems

As you learned in Chapter 6, all energy transformations obey the laws of thermodynamics. The first law of thermodynamics states that energy is neither created nor destroyed; that is, the total amount of energy in the universe is constant. The second law of thermodynamics states that when energy is converted from one form to another, some of it becomes unavailable to do work. This law governs patterns of energy flow through ecosystems.

Organisms depend on inputs of energy (in the form of sunlight or high-energy molecules), water, and minerals for metabolism and growth. Except for a few limited ecosystems (some caves, deep-sea hydrothermal systems) in which solar energy is not the main energy source, all energy utilized by organisms comes (or once came) from the sun. Even the fossil fuels—coal, oil, and natural gas—upon which the economy of modern human civilization is based are reserves of captured solar energy locked up in the remains of organisms that lived millions of years ago.

Only about 5 percent of the solar energy that arrives on Earth is captured by photosynthesis. The remaining energy is either radiated back into the atmosphere as heat or consumed by the evaporation of water from plants and other surfaces. The energy that *is* captured by photosynthesis powers the ecosystem processes. How that energy subsequently passes through a series of organisms is the topic of the next section.

Photosynthesis drives energy flow in ecosystems

Energy flow in most ecosystems originates with photosynthesis. The *rate* at which plants assimilate energy is called **gross primary productivity**. Water availability and temperature are major determinants of gross primary productivity. The total *amount* of energy that plants assimilate by photosynthesis, typically measured over a year, is called **gross primary production**. The production that remains after subtracting the energy that plants use for their own maintenance (respiration), building tissues, reproduction, and defense is called **net primary production** (Figure 56.4).

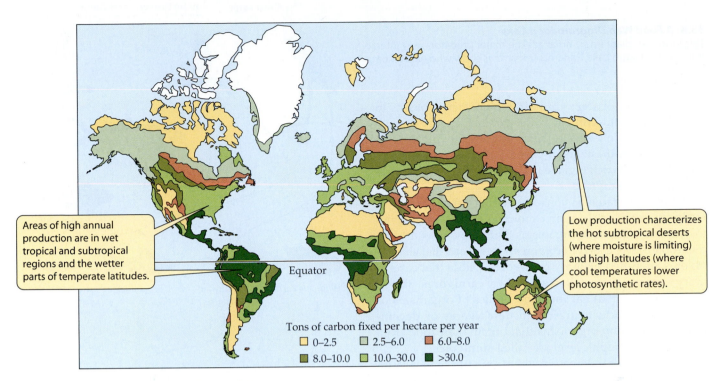

> Areas of high annual production are in wet tropical and subtropical regions and the wetter parts of temperate latitudes.

Equator

> Low production characterizes the hot subtropical deserts (where moisture is limiting) and high latitudes (where cool temperatures lower photosynthetic rates).

Tons of carbon fixed per hectare per year

☐ 0–2.5   ☐ 2.5–6.0   ☐ 6.0–8.0
■ 8.0–10.0   ☐ 10.0–30.0   ■ >30.0

56.5 Net Primary Production of Terrestrial Ecosystems
Variations in temperature and water availability over Earth's land surface affect the annual production of its ecosystems.

The distribution of primary production worldwide reflects the distribution of temperature and moisture on Earth. Close to the equator at sea level, temperatures are high throughout the year and typically the water supply is adequate much of the time. In these climates, highly productive forests thrive. In lower-latitude and mid-latitude deserts, where plant growth is limited by lack of moisture, primary production is low; plants of low stature dominate most landscapes. At higher latitudes, where there is more moisture and trees grow well, primary production is limited by low temperatures during much of the year. Production in aquatic systems is limited by light, which decreases rapidly with depth; by nutrients, which are scarce in open water; and by temperature.

Plants use most of the energy they capture to maintain themselves, to grow, and to reproduce. Some of this energy produces new tissues that can be eaten by herbivores or used by other organisms after the plants die. Because so much of the energy they capture goes to power their own metabolism, however, plants always contain much less energy than the total amount they have assimilated. Only the energy plants do not use to maintain themselves is available to be harvested by animals.

The global distribution of net primary production is shown in Figure 56.5.

Energy flows through a series of organisms

Because energy flows through ecosystems when organisms eat one another, it is useful to group organisms according to their source of energy. The organisms in an ecosystem that obtain their energy from a common source constitute a **trophic level** (Table 56.1). Organisms at a particular trophic level occupy a position in an ecosystem that is determined

56.1 *The Major Trophic Levels*

TROPHIC LEVEL	SOURCE OF ENERGY	EXAMPLES
Photosynthesizers (primary producers)	Solar energy	Green plants, photosynthetic bacteria and protists
Herbivores	Tissues of primary producers	Termites, grasshoppers, water fleas, anchovies, deer, geese
Primary carnivores	Herbivores	Spiders, warblers, wolves, copepods
Secondary carnivores	Primary carnivores	Tuna, falcons, killer whales
Omnivores	Several trophic levels	Humans, opossums, crabs, robins
Detritivores (decomposers)	Dead bodies and waste products of other organisms	Fungi, many bacteria, vultures, earthworms

56.6 A Food Web Diagram for a Lake

This food web diagram summarizes the major predator–prey interactions within Gatun Lake, Panama. The arrows show who eats whom.

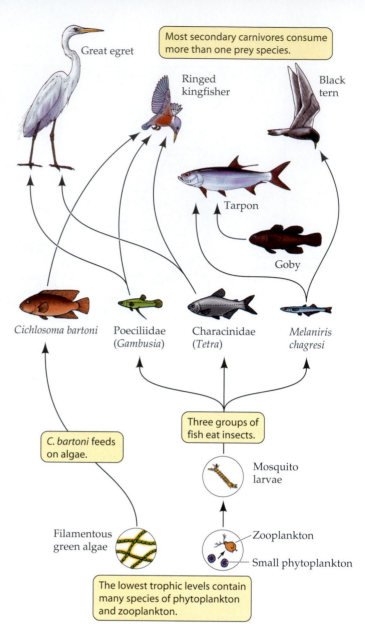

Great egret

Ringed kingfisher

Black tern

Most secondary carnivores consume more than one prey species.

Tarpon

Goby

Cichlosoma bartoni

Poeciliidae (*Gambusia*)

Characinidae (*Tetra*)

Melaniris chagresi

C. bartoni feeds on algae.

Three groups of fish eat insects.

Mosquito larvae

Filamentous green algae

Zooplankton

Small phytoplankton

The lowest trophic levels contain many species of phytoplankton and zooplankton.

by the number of steps through which energy passes to reach them. Photosynthetic plants get their energy directly from sunlight. Collectively, they constitute the trophic level called **photosynthesizers** or **primary producers**. They produce the energy-rich organic molecules upon which nearly all other organisms feed.

All other organisms consume, either directly or indirectly, the energy-rich organic molecules produced by photosynthetic organisms. Organisms that eat plants constitute the trophic level called **herbivores**. Organisms that eat herbivores are called **primary carnivores**. Those that eat primary carnivores are called **secondary carnivores**, and so on. Organisms that eat the dead bodies of organisms or their waste products are called **detritivores** or **decomposers**. The many organisms that obtain their food from more than one trophic level are called **omnivores**.

A sequence of linkages in which a plant is eaten by an herbivore, which is in turn eaten by a primary carnivore, and so on, is called a **food chain**. Food chains are usually interconnected in a **food web**, because most species in a community eat and are eaten by more than one other species.

A food web diagram is a useful summary of predator–prey interactions within a community. A simplified food web diagram (not including detritivores) for Gatun Lake, Panama, is shown in Figure 56.6. A complete food web diagram, showing the position of every species in a community, would be confusingly complex because most biological communities contain so many species. Therefore, similar species, especially those at lower trophic levels, are usually lumped together, as they are in the diagram of the Gatun Lake food web.

Much energy is lost between trophic levels

The energy that organisms use to maintain themselves is dissipated as heat, a form of energy that cannot be used by other organisms. For this reason, only a small portion of the energy captured at one trophic level is available to organisms at the next higher level. The energy content of an organism's **net production**—its growth plus reproduction—is available to organisms at the next trophic level (Figure 56.7).

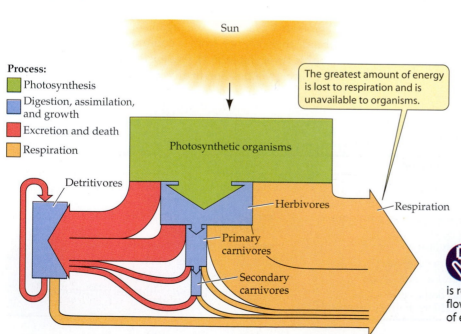

Sun

Process:

- Photosynthesis
- Digestion, assimilation, and growth
- Excretion and death
- Respiration

The greatest amount of energy is lost to respiration and is unavailable to organisms.

Photosynthetic organisms

Detritivores

Herbivores

Respiration

Primary carnivores

Secondary carnivores

56.7 Energy Flow through an Ecosystem

In this diagram, the width of each channel is roughly proportional to the amount of energy flowing through it. The arrows indicate directions of energy flow.

The efficiency of energy transfer through food webs depends on:

▶ The fraction of net production at one trophic level that is consumed by organisms at the next level

▶ How those organisms divide the ingested energy between production and maintenance

Birds and mammals have very low production efficiencies because they expend so much energy maintaining constant high body temperatures. Herbivores are less efficient than carnivores because plant tissues generally take more energy to digest than animal tissues do, but because of the low efficiency of energy transfer between trophic levels, a given amount of primary production can support many more herbivores than carnivores.

The amount of energy reaching a trophic level is determined by net primary production in the ecosystem and by the efficiencies with which food energy is converted to **biomass** (the total weight of organisms) at the trophic levels below it. To show how energy decreases in moving from lower to higher trophic levels, ecologists construct diagrams called **pyramids of energy**. A **pyramid of biomass**, which shows the mass of organisms existing at different trophic levels, illustrates the amount of biomass that is available at a given moment in time for organisms at the next trophic level (Figure 56.8).

Pyramids of energy and biomass for the same ecosystem usually have similar shapes, but sometimes they do not. The shapes depend on the dominant organisms and how they allocate their energies. Terrestrial ecosystems may differ strikingly in patterns of energy flow depending on the life forms of the dominant plants.

In grassland ecosystems, for example, where plants produce few hard-to-digest woody tissues, animals are able to consume most of the annual production of plant tissues each year. In grasslands, mammals—wild or domestic—may consume 30–40 percent of the annual aboveground net primary production. Insects may consume an additional 5–15 percent. Soil organisms, primarily nematodes, may consume 6–40 percent of the belowground biomass (Figure 56.8a).

In forest ecosystems, the dominant plants allocate a great deal of their energy to forming wood, which accumulates at high rates in growing forests. Wood, which is constructed of difficult-to-digest material (such as cellulose and lignin), is rarely eaten unless a plant is diseased or otherwise weakened. In most forests, leaves fall to the ground relatively undamaged at the end of the growing season. Although there are outbreaks of defoliating insects in forests, browsing rates are generally so low that forest ecologists often ignore losses to herbivores when calculating forest production (Figure 56.8b).

In most aquatic communities, on the other hand, the dominant photosynthesizers are bacteria and protists.

These organisms have such high rates of cell division that a small biomass of photosynthesizers can feed a much larger biomass of herbivores, which grow and reproduce much more slowly. This pattern can produce an inverted pyramid of biomass, even though the pyramid of energy for the same ecosystem has the typical shape (Figure 56.8c).

Much of the energy ingested by organisms is converted to biomass that is eventually consumed by detritivores (see Figure 56.7). Detritivores transform the remains and waste products of organisms (*detritus*) into carbon dioxide, water, and free mineral nutrients that can be taken up by plants again. If there were no detritivores, most nutrients would eventually be tied up in dead bodies, where they would be unavailable to plants. Therefore, continued ecosystem productivity depends on rapid decomposition of detritus.

Under the warm, wet conditions found in tropical forests, detritus is decomposed within a few weeks or months, and no litter accumulates on the soil surface. Rates of decomposition are slower under colder, drier, or highly acid conditions, such as those in many coniferous forests. At high altitudes and latitudes, decomposition of leaf litter

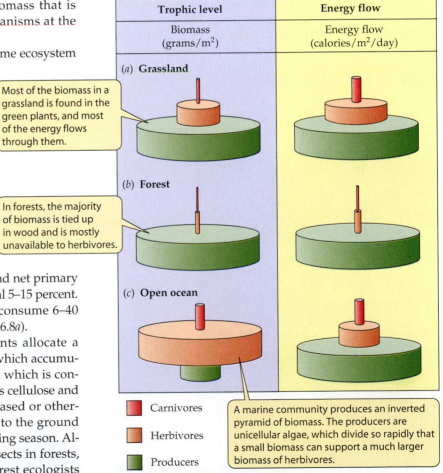

Most of the biomass in a grassland is found in the green plants, and most of the energy flows through them.

In forests, the majority of biomass is tied up in wood and is mostly unavailable to herbivores.

A marine community produces an inverted pyramid of biomass. The producers are unicellular algae, which divide so rapidly that a small biomass can support a much larger biomass of herbivores.

■ Carnivores

■ Herbivores

■ Producers

56.8 Pyramids of Biomass and Energy
Ecosystems can be compared in terms of the amount of material present in organisms at different trophic levels (left) and in terms of energy flow (right).

may take decades, and decomposition of tree trunks may take more than a century.

Some ecosystems are not powered by direct sunlight

Most ecosystems are powered by sunlight falling directly on them, but some depend upon sunlight that falls elsewhere. For example, marine ecosystems deeper than the level at which enough light penetrates for photosynthesis depend on biomass produced in the well-lit zone above them. The productivity of most deep-sea ecosystems is low because only small amounts of detritus descend through the water column to reach them.

Some deep-sea ecosystems are totally independent of sunlight. The most striking are those around hydrothermal vents associated with seafloor spreading zones. The energy base of these ecosystems is chemoautotrophy by sulfur-oxidizing bacteria. These bacteria obtain energy by oxidizing hydrogen sulfide in the hot water emitted from the vents. Most other organisms in these ecosystems, such as vestimentiferans, live directly or indirectly on these sulfur-oxidizing bacteria.

Ecologists recently discovered a cave ecosystem in southern Romania that is powered by bacteria that fix inorganic carbon by using hydrogen sulfide as an energy source. Chemoautotrophic production by these bacteria is the food base for 48 species of cave-adapted terrestrial and aquatic invertebrates. However, most cave ecosystems depend on imported photosynthetically produced organic matter.

Humans manipulate ecosystem productivity

Through agriculture, humans exploit ecosystems by replacing species of low economic value with species of high value. We do this by manipulating ecosystems so as to increase the yield of products useful to us. Agriculture has several intricately intertwined components:

▶ We eliminate competition between crops and unwanted plants by cultivating and by applying herbicides.
▶ We reduce competing herbivores and disease-causing organisms, usually by applying toxic chemicals.
▶ We augment photosynthesis by adding nutrients (fertilizing) and water (irrigating).
▶ We develop special high-yielding strains of plants that respond to additional fertilizer by increasing their growth or reproductive rates.

All these manipulations must work together, because "miracle" strains of crops typically do not yield more than other strains unless they are provided with fertilizers and protected from competitors, herbivores, and pathogens. Agriculture also depends on energy from outside the system for cultivation and harvesting. In modern agriculture, this energy comes from fossil fuels (Figure 56.9).

Modern agricultural ecosystems have spectacularly increased food production per hectare, but they have also created problems. Herbicides and insecticides have polluted lakes, rivers, and groundwater in most industrialized countries. Many agricultural pests have evolved resistances to pesticides. New methods of pest control, collectively known as **integrated pest management** (IPM), are becoming more common. IPM combines chemical applications with cultural practices—crop rotation, mixed plantings of crop plants, and mechanical tillage of the soil—and biological methods—development of pest-resistant strains of

(a)

56.9 Agriculture Requires Energy Input
(*a*) In traditional agriculture, people and domesticated animals supply most of the energy, as in this Indian rice paddy. (*b*) Modern agriculture is based on the use of toxic chemicals and high rates of consumption of fossil fuels during site preparation, growth of the crop, and at harvest time. The large machines harvesting this wheat field are typical of much modern agriculture.

(b)

crops, use of natural predators and parasites, and use of chemical attractants, such as pheromones—to control insect herbivores. The reduced use of toxic chemicals avoids most pollution problems and reduces the chance that pests will evolve resistance to pesticides.

Cycles of Materials through Ecosystem Compartments

As we have just seen, energy flows through ecosystems according to the second law of thermodynamics. At each transformation, much of it is dissipated as heat, a form that cannot be used by organisms to power their metabolism. Chemical elements, on the other hand, are not lost when they are transferred among organisms; they cycle repeatedly through organisms and the physical environment. The carbon and nitrogen atoms of which life is composed today are the same atoms that made up dinosaurs, insects, and trees in the Mesozoic era. The amounts of carbon, nitrogen, phosphorus, calcium, sodium, sulfur, hydrogen, oxygen, and other chemical elements on Earth do not change, but the quantities that are available to organisms are strongly influenced by how organisms get them, how long they hold onto them, and what they do with them while they have them.

To understand the cycling of elements, it is convenient to divide the global ecosystem into four **compartments**: oceans, fresh waters, atmosphere, and land. The physical environments in each compartment and the types of organisms living there are different. Therefore the amounts of elements found in the different compartments, what happens to those elements, and the rates at which they enter and leave the compartments differ strikingly. After we have described these compartments, we will consider them together to illustrate how elements cycle through the global ecosystem.

Oceans receive materials from the land and atmosphere

Oceans receive chemical elements from land as runoff from rivers. The immediate receivers of materials from land are often fresh waters and the atmosphere, but, on time scales of hundreds to thousands of years, oceans receive most of those materials, including those produced by human activity. Because of their huge size, and because oceans exchange materials with the atmosphere only at their surface, oceans respond very slowly to outside inputs.

Elements that enter the oceans from other compartments gradually sink to the seafloor and remain there, unless

something happens to bring them back to the surface. Most elements remain in the bottom sediments until they are elevated above sea level by movements of Earth's crust. This process may take many millions of years. For this reason, concentrations of mineral nutrients are very low throughout most of the oceans. Oxygen, however, is usually present at all depths, because even slow mixing suffices to replenish the oxygen consumed by the respiration and decomposition of the few organisms that live in the nutrient-poor ocean waters.

Except on the relatively shallow continental shelves, ocean waters mix very slowly and are strongly stratified by depth. Typically there is a well-defined depth zone—a **thermocline**—at which temperatures drop abruptly. Some elements are returned to the surface by cool bottom water that rises—**upwells**. Much upwelling occurs near the coasts of continents where offshore winds push surface waters away from shore (Figure 56.10). Waters in these *zones of upwelling* are rich in nutrients, and most of the world's great fisheries are concentrated there.

Lakes and rivers contain only a small fraction of Earth's water

Lakes and rivers contain much less water than oceans do, and because these bodies of water are relatively small, most mineral nutrients entering them are not buried in bottom sediments for long periods of time. Some mineral nutrients enter fresh waters in rainfall, but most are released by the weathering of rocks and are carried to lakes and rivers via

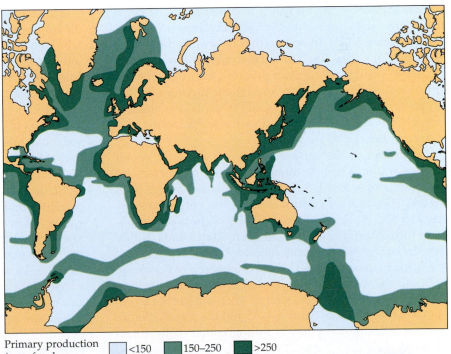

Primary production (mg of carbon per m² per day)   ☐ <150   ■ 150–250   ■ >250

56.10 Primary Production is High in Zones of Upwelling
Primary production in the oceans is highest near continents where surface waters, driven by prevailing winds, move offshore and are replaced by cool, nutrient-rich water upwelling from below.

56.11 Annual Temperature and Oxygen Cycles in a Temperate Lake
These vertical temperature and oxygen profiles are typical of temperate zone lakes that freeze in winter.

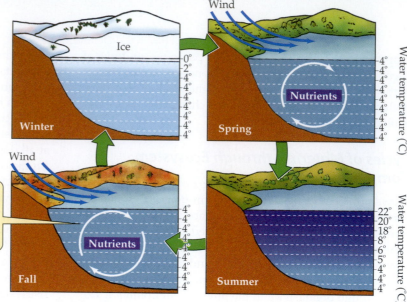

Turnovers occur in the spring and fall and allow nutrients and oxygen to become evenly distributed in the water column.

groundwater (the water that resides in soil and rocks) or by surface flow.

After entering rivers, mineral nutrients are usually carried rapidly to lakes or to the oceans. In lakes they are taken up by organisms and incorporated into their cells. These organisms eventually die and sink to the bottom, where decomposition of their tissues consumes the oxygen. Surface waters of lakes thus quickly become depleted of nutrients, while deeper waters become depleted of oxygen. However, this stratification process is countered by vertical movements of water—**turnover**—that bring nutrients and dissolved CO_2 to the surface and oxygen to deeper water. Wind is an important agent of turnover in shallow lakes, but in deeper lakes it usually mixes only surface waters.

Lakes in temperate regions turn over because water is most dense at 4°C; above and below that temperature it expands (Figure 56.11). In spring, the sun warms the surface layer of a lake. The depth of the warm layer gradually increases as spring and summer progress. However, there is still a well-defined thermocline where the temperature drops abruptly to about 4°C. Only if the lake is shallow enough to warm to the bottom does the temperature of the deepest water rise above 4°C.

In autumn, as the surface of the lake cools, the cooler surface water, which is denser than the warmer water below it, sinks, and is replaced by warmer water from below. This process continues until the entire lake has reached 4°C. At this point, the density of the water is uniform throughout the lake, and even modest winds readily mix the entire water column. As colder weather then cools the surface water below 4°C, that water becomes less dense than the 4°C water below it. Therefore, it floats at the top. Another turnover occurs in spring, when the surface layers above the thermocline warm to 4°C and the water column, again being of uniform density throughout, is easily mixed by wind.

Deep tropical and subtropical lakes may be permanently stratified because they never become cool enough to have uniformly dense water. Their bottom waters lack oxygen because decomposition quickly depletes any oxygen that reaches them. However, many tropical lakes are overturned at least periodically by strong winds so that their deeper waters are occasionally oxygenated. Arctic lakes turn over only once each year.

The atmosphere regulates temperatures close to Earth's surface

The *atmosphere* is a thin layer of gases surrounding Earth. About 80 percent of the mass of the atmosphere lies in its lowest layer, the **troposphere**, which extends upward from Earth's surface about 17 km in the tropics and subtropics, but only about 10 km at higher latitudes. Most global air circulation takes place within the troposphere, and virtually all atmospheric water vapor is located there.

The **stratosphere**, which extends upward from the top of the troposphere to about 50 km above Earth's surface, contains about 99 percent of the remaining atmospheric mass, but it is extremely dry. Materials enter the stratosphere from the troposphere near the equator, where air rises to high altitudes. These materials tend to remain in the stratosphere for a relatively long time because stratospheric air circulation is horizontal. Ozone (O_3) in the stratosphere absorbs most incoming short-wavelength ultraviolet radiation, shielding organisms from its damaging effects.

The atmosphere is composed of 78.08 percent nitrogen as N_2, 20.95 percent oxygen, 0.93 percent argon, and 0.03 percent carbon dioxide. It also contains traces of hydrogen gas, neon, helium, krypton, xenon, ozone, and methane. The atmosphere contains Earth's biggest pool of nitrogen and large supplies of oxygen. Although carbon dioxide constitutes a very small fraction of the atmosphere, it is the source of the carbon used by terrestrial photosynthetic organisms.

The atmosphere plays a decisive role in regulating temperatures at and close to Earth's surface. Without an atmosphere, the average surface temperature of Earth would be about –18°C, rather than the actual +17°C. The atmosphere is relatively transparent to visible light, but it traps a large part of the outgoing infrared radiation (heat) that is emitted by Earth. It traps heat at Earth's surface in a way analogous

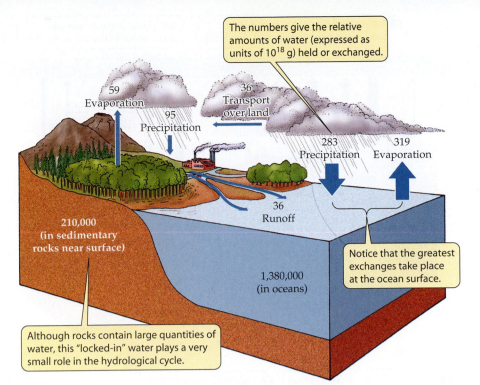

The numbers give the relative amounts of water (expressed as units of 10^{18} g) held or exchanged.

59 Evaporation

95 Precipitation

36 Transport over land

283 Precipitation

319 Evaporation

36 Runoff

210,000 (in sedimentary rocks near surface)

1,380,000 (in oceans)

Notice that the greatest exchanges take place at the ocean surface.

Although rocks contain large quantities of water, this "locked-in" water plays a very small role in the hydrological cycle.

56.12 The Global Hydrological Cycle
The numbers show the relative amounts of water (expressed as units of 10^{18} g) held in or exchanged annually by ecosystem compartments. The widths of the arrows are proportional to the sizes of the fluxes.

to the glass in a greenhouse, which is why this phenomenon is called the **greenhouse effect**. Water vapor, carbon dioxide, and ozone are especially important trappers of infrared radiation. That is why, as we will see below, elevated concentrations of atmospheric carbon dioxide caused by human activities may lead to major climatic changes.

Land covers about one-fourth of Earth's surface

About one-fourth of Earth's surface, most of it in the Northern Hemisphere, is currently above sea level. Most of this land is covered by a layer of soil. Even though the global supply of nutrients is constant, regional and local deficiencies strongly affect ecosystem processes on land. Such deficiencies occur because, unlike their behavior in air and water, elements on land move slowly, and they usually move only short distances.

The terrestrial compartment is connected to the atmospheric compartment by terrestrial organisms that take chemical elements from and release them to the air. Chemical elements in soils are carried in solution into the groundwater and eventually into rivers and oceans, where they are lost to organisms until geological processes raise marine sediments above sea level, and a new cycle of erosion and weathering begins. The type of soil that forms in an area and the elements it contains depend on the underlying rock, as well as on climate, topography, the organisms living there, and the length of time that soil-forming processes have been acting. Very old soils contain fewer nutrients than most young soils.

 Biogeochemical Cycles

The chemical elements organisms need in large quantities—carbon, hydrogen, oxygen, nitrogen, phosphorus, and sulfur—cycle through organisms to the physical envi-

ronment and back again. The pattern of movement of a chemical element through organisms and compartments of the physical environment is called its **biogeochemical cycle**. Some chemical elements circulate continually, but large quantities of other elements are temporarily lost from circulation through deposition in deep-sea sediments.

Each chemical element used by organisms has a distinctive biogeochemical cycle whose properties depend on the physical and chemical nature of the element and how organisms use it. All chemical elements cycle quickly through organisms because no individual, even of the longest-lived species, lives very long in geological terms. Chemical elements, such as carbon and nitrogen, that exist in the atmosphere as a gas cycle faster than nongaseous elements. After discussing the movements of water, we discuss the cycling of the most abundant chemical elements in organisms.

Water cycles through the oceans, fresh waters, atmosphere, and land

The cycling of water through the four ecosystem compartments is known as the **hydrological cycle** (Figure 56.12). Although water is a compound, not an element, we describe its cycle here together with those of individual elements because of its importance to life.

The hydrological cycle is driven by the evaporation of water, most of it from ocean surfaces. Some water returns to the oceans as precipitation, but much less falls back on the oceans than is evaporated from them. The remaining evaporated water is carried by winds over the land, where it falls as rain or snow.

Water also evaporates from soils, from freshwater lakes and rivers, and from the leaves of plants (transpiration), but the total amount evaporated is less than the amount that falls on land as precipitation. The excess water eventually returns to the oceans via rivers, coastal runoff, and subterranean flows.

Organisms profoundly influence the carbon cycle

Organisms are triumphs of carbon chemistry. To survive, they must have access to carbon atoms. Nearly all the carbon in organisms comes from carbon dioxide (CO_2) in the atmosphere or dissolved carbonate ions (HCO_3^-) in water. In the cells of some bacteria, photosynthetic protists, and the leaves of plants, carbon is incorporated into organic mol-

56.13 The Global Carbon Cycle
The numbers show the quantities of carbon (expressed as units of 10^{15} g) in organisms and in ecosystem compartments, and the amounts that move annually between them.

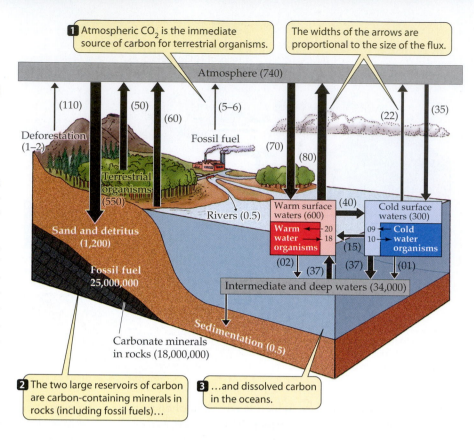

ecules by photosynthesis. All organisms in other groups get their carbon by consuming other organisms, their remains, or their waste products.

Although marine organisms contain very little of Earth's total carbon, they have a profound influence on the distribution of carbon in the oceans. They convert soluble carbonate ions in seawater into insoluble ocean sediments by depositing carbon in their shells and skeletons, which eventually sink to the bottom.

Biological processes on land move carbon between the atmospheric and terrestrial compartments, removing it from the atmosphere during photosynthesis and returning it to the atmosphere during respiration (Figure 56.13). Growing plants at mid- and high latitudes in the Northern Hemisphere incorporate so much carbon into their bodies that they reduce the concentration of atmospheric CO_2 from about 350 parts per million in winter to 335 parts per million in midsummer. This carbon is released back into the atmosphere by decomposition in autumn.

At times in the remote past, large quantities of carbon were removed from the cycling when organisms died in large numbers and were buried in sediments lacking oxygen. In such environments, detritivores do not reduce organic carbon to carbon dioxide. Instead, organic molecules accumulate and eventually are transformed into oil, natural gas, coal, or peat (decomposed remains of plants in high-latitude bogs). Humans have discovered and used these deposits, known as **fossil fuels**, at ever increasing rates during the past 150 years. As a result, carbon dioxide, the final product of burning these fuels, is being released into the atmosphere faster than it is being transferred to the oceans and incorporated into terrestrial biomass (Figure 56.14).

Based on a variety of calculations, atmospheric scientists believe that 150 years ago, before the Industrial Revolution, the concentration of atmospheric CO_2 was probably about 265 parts per million. Today it is 350 parts per million. This increase has been caused primarily by combustion of fossil fuels and secondarily by the burning of forests. If current trends in both these activities continue, atmospheric CO_2 is expected to reach 580 parts per million by the middle of the twenty-first century. This CO_2 will eventually be transferred to the oceans and deposited in sediments as calcium carbonate ($CaCO_3$), but the rate of transfer is much slower than the rate at which humans are introducing CO_2 into the atmosphere.

Enough carbon is being released by the burning of fossil fuels to alter the heat balance of Earth. As we saw above, carbon dioxide is one of the components of the atmosphere that traps infrared radiation at Earth's surface. Scientists have measured the concentration of CO_2 in air trapped in the Antarctic and Greenland ice caps during the last Ice Age—between 15,000 and 30,000 years ago—when the climate was much colder. The CO_2 concentration then was as low as 200 parts per million. During a warm interval that occurred some 5,000 years ago, it may have been slightly higher than it is today. This long-term record, which

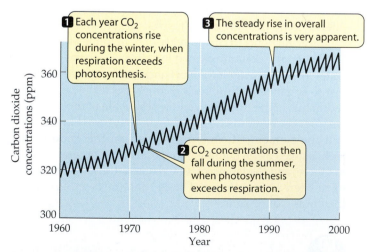

56.14 Atmospheric Carbon Dioxide Concentrations Are Increasing
These carbon dioxide concentrations, expressed as parts per million by volume of dry air, were recorded on top of Mauna Loa, Hawaii.

Question: What effect do CO_2 levels have on a community of soil organisms?

METHOD

1. Establish multiple plots containing the same communities of organisms. Maintain half of the units at experimentally high levels of atmospheric CO_2, the others at normal atmospheric levels.

2. Maintain communities for three generations (9 months).

3. Measure abundances of three species of collembolans (springtails), prominent members of the community.

RESULTS With elevated CO_2, species 1 became more prevalent, while species 2 remained the same and species 3 dramatically decreased.

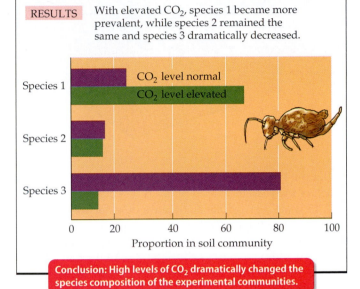

Conclusion: High levels of CO_2 dramatically changed the species composition of the experimental communities.

56.15 Increased Atmospheric CO_2 Concentrations Alter Soil Communities
After 9 months of exposure to elevated CO_2 concentrations, the relative abundances of three collembolan insect species (springtails) differed dramatically from those in communities exposed to normal atmospheric CO_2 concentrations.

demonstrates a relationship between climate and atmospheric CO_2 levels, is a major reason why scientists expect **global warming** as atmospheric CO_2 levels continue to increase.

A doubling of atmospheric CO_2 would shift climate zones toward the poles. Complex computer models predict there would probably be droughts in the central regions of continents while precipitation would increase in coastal areas.

Current evidence indicates such warming is already underway. The average global temperature has risen steadily over the past 20 years, and the sheets of sea ice surrounding the polar land masses have become measurably smaller and thinner. Sea levels have risen by an average of a few centimeters worldwide, and some islands, such as Bermuda, have seen even greater effects. If global warming were to result in the melting of the Greenland and Antarctic ice caps, rising sea levels could flood coastal cities and agricultural lands.

In addition to studying the potential effects of global warming on ecosystem processes, ecologists are investigat-

ing the direct effects of elevated CO_2 concentrations on ecosystems. Increases in CO_2 affect plants directly because CO_2 is the source of carbon for photosynthesis. Their effects on other organisms are indirect.

To determine how elevated concentrations of atmospheric CO_2 might influence soil biota, ecologists established and maintained sixteen 1-m^2 experimental plots, into which they introduced the same community of plants, herbivores, carnivores, and soil microorganisms. All sixteen plots were maintained under the same conditions, except that CO_2 levels were allowed to vary naturally in half of them but were kept high in the other half. After three plant generations, the abundance and species composition of soil microorganisms were about the same under the two conditions. However, populations of cellulose-decomposing fungi increased dramatically under the elevated CO_2 condition, and the species composition also changed. Populations of springtails—wingless insects that eat fungi—also increased, and different species dominated the biota (Figure 56.15). The sequence of these changes appeared to be:

▸ increased plant photosynthesis →
▸ increased transport of carbon belowground →
▸ increased dissolved organic carbon →
▸ changes in soil fungal assemblages →
▸ changes in abundances and species composition of springtails.

To ecologists, these changes are striking. However, these experiments inform us only about relatively short-term changes. Other experiments are being run to assess long-term changes in the functioning of ecological systems in response to elevated atmospheric concentrations of CO_2.

Few organisms can use elemental nitrogen

Although nitrogen gas (N_2) makes up about 78 percent of the atmosphere, it cannot be used by most organisms in this form. It can be converted into biologically useful forms only by a few species of bacteria and cyanobacteria, which convert N_2 into ammonia by a process called nitrogen fixation (see Chapter 36). So, despite the abundance of N_2, usable nitrogen is often in short supply in ecosystems.

Why don't nitrogen-fixing organisms increase in numbers so that nitrogen is no longer a limiting resource? Fixing nitrogen is energetically expensive; as a result, nitrogen-fixing organisms often lose out in competition with non-fixers when nitrogen becomes more readily available. Nitrogen tends to be lost rapidly from ecosystems by leaching, vaporizaton of ammonia, and denitrification (the return of nitrogen to the atmosphere as nitrogen gas, N_2, by some organisms), but is released only slowly by decomposition of detritus.

Organisms called nitrifiers neither take up nitrogen gas directly from the atmosphere nor respire nitrogen back to the atmosphere. Instead, they convert organic molecules containing nitrogen to inorganic molecules. This process is carried out in several stages by different organisms. Most of the resulting nitrogen-containing compounds, such as ni-

56.16 The Global Nitrogen Cycle

The numbers show the quantities of nitrogen (expressed as units of 10^9 kg) in organisms and in ecosystem compartments, and the amounts that move annually between compartments. The widths of the arrows are proportional to the sizes of the fluxes.

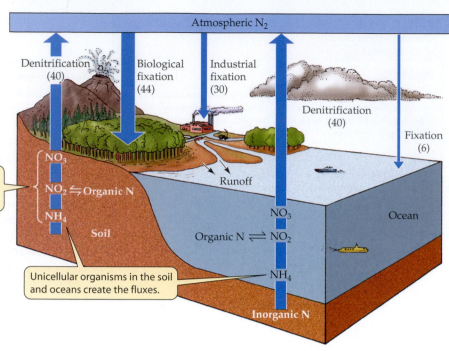

The several stages of inorganic nitrogen are nitrate (NO_3), nitrite (NO_2), and ammonium (NH_4).

Unicellular organisms in the soil and oceans create the fluxes.

trates or ammonia, are taken up by plants. This movement of nitrogen among organisms accounts for about 95 percent of all nitrogen fluxes on Earth (Figure 56.16).

Organisms drive the sulfur cycle

Emissions of volatile sulfur dioxide (SO_2) and hydrogen sulfide (H_2S) from volcanoes and fumaroles (vents for hot gases) are the only significant natural nonbiological fluxes of sulfur. These emissions release, on average, between 10 and 20 percent of the total natural flux of sulfur to the atmosphere, but they vary greatly in time and space. Large volcanic eruptions spread great quantities of sulfur over broad areas, but they are rare events.

Volatile sulfur compounds are also emitted by both terrestrial and marine organisms. Certain marine algae produce large amounts of dimethyl sulfide (CH_3SCH_3), which accounts for about half of the biotic component of the sulfur cycle. The other half is produced by terrestrial organisms.

Sulfur is apparently always abundant enough to meet the needs of living organisms. It also plays an important role in global climate patterns. Even if air is moist, clouds do not form readily unless there are small particles around which water can condense. Dimethyl sulfide is the major source of such particles. Therefore, increases or decreases in sulfur emissions can change cloud cover and hence climate.

Humans have altered the nitrogen and sulfur cycles, largely through the burning of fossil fuels. An important regional effect of this alteration is **acid precipitation**—rain or snow whose pH is lowered by the presence of sulfuric acid (H_2SO_4) and nitric acid (HNO_3). These acids enter the atmosphere, primarily from industrial smokestacks and automobile emissions, and may travel hundreds of kilometers before they settle to Earth in precipitation or as dry particles.

Acid precipitation now characterizes all major industrialized countries and is particularly widespread in eastern North America and Europe. The normal pH of precipitation in New England is about 5.6, but precipitation there now averages about pH 4.1, and there are occasional storms with a precipitation pH as low as 3.0. Precipitation with a pH of about 3.5 or lower causes direct damage to the leaves of plants and reduces photosynthetic rates. Fortunately, as a result of the establishment of a flexible regulatory system

under the 1990 Clean Air Act Amendments, precipitation in much of the eastern United States is less acid today than it was 15 years ago, primarily because of reductions in emissions of sulfur (Figure 56.17).

Ecologists in Canada studied the effects of acid precipitation by adding enough sulfuric acid to two small lakes to reduce their pH from about 6.6 to 5.2. In both lakes, nitrifying bacteria failed to adapt to these moderately acidic conditions. As a result, the nitrogen cycle was blocked and ammonium accumulated in the water. When the ecologists stopped adding acid to one of the lakes, its pH increased to 5.4, and nitrification resumed. After about 1 year, the pH of the lake returned to its original value. These experiments show that lakes are very sensitive to acidification, but can recover quickly when pH returns to normal values.

The phosphorus cycle has no gaseous phase

The phosphorus cycle differs from the other biogeochemical cycles discussed in this section in that it lacks a gaseous phase. Some phosphorus is transported on dust particles, but in general the atmosphere plays a very minor role in the phosphorus cycle. Phosphorus exists mostly as phosphate (PO_4^{3-}) or similar compounds. Most phosphate deposits are of marine origin. On land, phosphorus becomes available through the slow weathering and dissolution of rocks and minerals.

Organisms need phosphorus as a component of the energy-rich molecules involved in cellular metabolism. Phosphorus is often a limiting nutrient in soils and lakes. That is why phosphate is a component of most fertilizers, and why adding phosphate to lakes causes marked increases in their biological productivity. The extra phosphorus from fertilizers and detergents that enters lakes through runoff allows algae and bacteria to multiply, forming blooms that turn water green. The decomposition of the extra biomass produced by this increased biological activity consumes all the oxygen in the lake, and anaerobic organisms come to dominate the sediments. These anaerobic organisms do not

1955

>5.3

5.0

4.6

1 Because of prevailing winds, acid rains affect areas far from the pollution sources.

2 In 30 years, pH of precipitation fell substantially (became more acidic).

1985

4.6 4.2

5.0

>5.3

3 Regulation of emission sources has raised the pH of precipitation in some areas of the Northeast…

1998

>5.3

4.6

4.2

4.6

5.0

4 … but in many areas of the West, precipitation is becoming more acidic.

56.17 Acid Precipitation is Decreasing in the Eastern United States
Thanks to emission controls, precipitation in many parts of the eastern United States is less acid than it was two decades ago. The figure shows the regional average pH of precipitation.

As a result, the amount of phosphate added to Lake Erie has decreased more than 80 percent from the maximum level, and phosphorus concentrations in the lake have declined substantially. The deeper waters of Lake Erie still become poor in oxygen during the summer months, but the rate of oxygen depletion is declining. Algal blooms have decreased, as have populations of small fishes that feed on algae.

The rate at which a eutrophic lake recovers depends on the replacement rate of its waters. Because water flows slowly through Lake Erie, it will take many years for the lake to recover from the heavy pollutant loads it has received. By contrast, the water in Lake Washington, a smaller lake adjacent to the city of Seattle, is replaced every 3 years. When sewage was diverted away from Lake Washington, the lake returned to its former condition within a decade.

Humans influence other biogeochemical cycles

In addition to modifying the great natural biogeochemical cycles we have just described, human activities are increasing the cycling of elements such as lead and creating cycles of synthetic chemicals, such as DDT. These changes are large enough to cause serious environmental problems. However, the results of experiments show that ecosystems have the capacity to recover from many disturbances if the alterations have not been too great and the disturbing forces are reduced or eliminated. Controlling our manipulations of biogeochemical cycles so that ecosystems can continue to provide the goods and services upon which humanity depends is one of the major challenges facing modern societies.

 Chapter Summary

▶ The organisms living in a particular area, together with the physical environment with which they interact, constitute an ecosystem. At a global scale, Earth is a single ecosystem.

Climates on Earth

▶ Air and water movements on Earth are driven primarily by solar radiation.

▶ Climates determine the amount of heat, moisture, and sunlight available to living organisms in different places on Earth.

▶ Rising air expands and cools, releasing moisture. Descending air warms and dries and takes up moisture, creating rain shadows. **Review Figure 56.1**

▶ Global air circulation is driven by solar radiation and the spinning of Earth on its axis. **Review Figure 56.2**

break down organic compounds all the way to carbon dioxide. Many of the end products of their activities have unpleasant odors.

Lake Erie is an example of such a **eutrophic** (enriched) lake. Two hundred years ago it had only moderate levels of photosynthesis and clear, oxygenated water. Today more than 15 million people live in the Lake Erie basin. Nearby cities pour more than 250 billion liters of domestic and industrial wastes into the lake annually. The entire basin is intensely farmed and heavily fertilized.

In the early part of the twentieth century, nutrients in the lake increased greatly, and algae proliferated. At the water filtration plant in Cleveland, algae increased from 81 per milliliter in 1929 to 2,423 per milliliter in 1962. Algal blooms and populations of bacteria also increased. Numbers of the human fecal bacterium *Escherichia coli* increased enough to cause the closing of many of the lake's beaches as health hazards.

Since 1972, the United States and Canada have invested more than 8 billion dollars to improve municipal waste facilities and reduce discharges of phosphorus into Lake Erie.

▶ Oceanic currents are driven primarily by prevailing winds. Review Figure 56.3

Energy Flow through Ecosystems

▶ The capture of solar radiation by photosynthesis powers ecosystem productivity.

▶ The annual production of an area is determined primarily by temperature and moisture. **Review Figures 56.4, 56.5**

▶ Energy flows through ecosystems as organisms capture and store energy and transfer it to other organisms when they are eaten. Organisms are grouped into trophic levels according to the number of steps through which energy passes to get to them. **Review Table 56.1**

▶ Who eats whom in a ecosystem can be diagrammed as a food web. **Review Figure 56.6**

▶ The amount of energy flowing through an ecosystem depends on net primary production and on the efficiency of transfer of energy from one trophic level to another. **Review Figures 56.7, 56.8**

▶ A few deep-sea and cave ecosystems are powered by chemosynthesis rather than photosynthesis.

▶ Through agriculture, humans manipulate ecosystem productivity so as to increase the yield of products useful to us. In modern agriculture, the energy required to do this is provided by fossil fuels.

Cycles of Materials through Ecosystem Compartments

▶ The main compartments of the global ecosystem are the oceans, fresh waters, land, and atmosphere, among which materials are constantly being exchanged.

▶ Primary production in the oceans is highest adjacent to continents in zones of upwelling, where nutrient-rich waters rise to the surface. **Review Figure 56.10**

▶ Temperate-zone lakes turn over twice each year as water cools and warms. **Review Figure 56.11**

▶ The two lowest layers of Earth's atmosphere differ from each other in their circulation patterns, the amount of moisture they contain, and the amount of ultraviolet radiation they receive.

Biogeochemical Cycles

▶ The elements organisms need in large quantities cycle though organisms to the environment and back again.

▶ The hydrological cycle is driven by evaporation of water, most of it from ocean surfaces. **Review Figure 56.12**

▶ Atmospheric carbon dioxide is the immediate source of carbon for terrestrial organisms, but only a small part of Earth's carbon is found in the atmosphere. **Review Figure 56.13**

▶ Increasing concentrations of carbon dioxide in the atmosphere are changing climates and influencing ecological processes. **Review Figures 56.14, 56.15**

▶ Although nitrogen makes up 78 percent of Earth's atmosphere, nitrogen can be converted into biologically useful forms only by a few species of bacteria and cyanobacteria. **Review Figure 56.16**

▶ Acid precipitation, an important regional consequence of human modifications of the nitrogen and sulfur cycles, is being addressed in the United States by flexible regulations. **Review Figure 56.17**

▶ The phosphorus cycle differs from the cycles of carbon and nitrogen in that it lacks a gaseous phase.

▶ The most striking effect of altering the phosphorus cycle is lake eutrophication.

For Discussion

1. Continental climates are largely confined to the Northern Hemisphere. Why are there so few continental climates in the Southern Hemisphere?

2. The amount of energy flowing through a food chain declines more or less rapidly depending upon the nature of the organisms in the chain. Which of the following simplified food chains is likely to be more efficient? Why? What criteria of efficiency are you using?

   a. phytoplankton → zooplankton → herring
   b. shrubs → moose → wolf

3. What principles of ecosystem functioning underlie modern agricultural practices? Which components of those practices are most likely to be unsustainable in the long term?

4. How would you expect temperature and oxygen profiles to appear in a broad, shallow tropical lake? In a very deep tropical lake? A subarctic lake? Why?

5. The waters of Lake Washington, adjacent to the city of Seattle, rapidly returned to their preindustrial condition when sewage was diverted from the lake to Puget Sound, an arm of the Pacific Ocean. Would all lakes being polluted with sewage clean themselves up as rapidly as Lake Washington if pollutant input were stopped? What characteristics of a lake are most important to its rate of recovery following removal of pollutant inputs? What are the likely ecological effects of the diverted sewage in Puget Sound?

6. Tropical forests currently are being cut at a very rapid rate. Does this necessarily mean that deforestation is a major source of input of carbon dioxide to the atmosphere? If not, why not?

7. Why do elevated atmospheric concentrations of carbon dioxide have direct effects on plants but not on animals? What are the most important indirect effects on animals, and by what ecological pathways do they operate?

8. The two drawings below represent pyramids of biomass for (a) an old field in Georgia and (b) the English Channel. Explain the significance of the inversion of the second pyramid compared with the first.

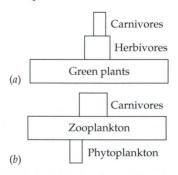

9. A government official authorizes the construction of a large power plant in a former wilderness area. Its smokestacks discharge great quantities of waste resulting from the combustion of coal. List and describe all likely ecological results at local, regional, and global levels. Now suppose the wastes were thoroughly scrubbed from the stack gases. Which of the ecological results you have just outlined would still happen?

57 Biogeography

WHEN THE FIRST EUROPEANS ARRIVED IN AUStralia, they saw plants and animals that differed in perplexing ways from the ones they knew at home. Among the oddities they found were flowers pollinated by brush-tongued parrots and mammals that hopped around on their hind legs, carrying their offspring in pouches. The first Europeans to visit North America felt more at home because the plants and animals of North America were more similar to those of Europe.

During their worldwide travels, European explorers found many vegetation types—tropical forests, mangrove forests, and deserts with tall cacti—that were unfamiliar to them, but they also found many areas where the vegetation forms and species were similar to what they knew back home. The study of the distribution of organisms over Earth's surface began when those eighteenth-century travelers first noted intercontinental differences in biotas and attempted to understand them.

Biogeography is the science that attempts to explain patterns of variation among individuals, populations, species, and communities across Earth. In this chapter, we will show how biogeographers study both events in the remote past and current ecological processes to discover how they influence the distribution patterns we see today.

Why Are Species Found Where They Are?

Explaining species' distributions might seem to be a simple matter because the question of why a species is or is not found in a certain location has only a few possible answers:

▶ If a species occupies a particular area, either it evolved there, or it evolved elsewhere and dispersed to that area.

▶ If a species is not found in a particular area, either it evolved elsewhere and never dispersed to that area, or it was once present in that area but no longer lives there.

Unfortunately, determining which of these answers is correct turns out to be far from simple.

To explain the distributions of organisms, biogeographers must draw upon and interpret a broad array of knowledge. Answering the questions listed above requires information about the evolutionary histories of species, which comes from fossils and from knowledge of their phylogenetic relationships. It also requires information about changes in Earth itself—continental drift, glacial advances and retreats, sea level changes, and mountain building—during the time the organisms were evolving. Such geological information can tell us whether organisms evolved where they are currently found or dispersed to colonize new areas from a distant area of origin. In this section we show how the acceptance of continental drift and the use of new methods of reconstructing phylogenies changed the science of biogeography.

Strange Organisms in New Places
Kangaroos that hopped on large hind legs and carried their offspring in pouches were novel sights to the Europeans who colonized Australia.

Ancient events influence current distributions

Early biogeographers believed in an unchanging Earth that was too young to account for the diversity and distribution of life by any means except divine creation. Linnaeus, for example, believed that all organisms had been created in one place—which he called Paradise—from which they later dispersed. Indeed, because most people believed that the continents were fixed in their positions, the only way to account for current distributions was to invoke massive dispersal.

The notion that the continents might have moved was not seriously considered until 1912. Alfred Wegener, the German meteorologist who proposed the idea of continental drift, based his theory on several observations:

▶ the shapes of continents (the outlines of Africa and South America seemed to fit together like pieces of a puzzle)

▶ the alignment of mountain chains, rock strata, coal beds, and glacial deposits on different continents

▶ the distributions of organisms (the distributions of species in Africa and South America were hard to explain if one assumed that the continents had never moved)

When Wegener proposed his ideas, few scientists took them seriously, primarily because there were no known mechanisms to move continents and because no convincing geological evidence of such movements existed. As we learned in Chapter 20, geological evidence and plausible mechanisms were eventually discovered, and the broad pattern of continental movement is now clear.

About 280 million years ago, the continents were united to form a single land mass, Pangaea (see Figure 20.13). By the early Mesozoic era (about 245 million years ago), when the continents were still very close to one another, many groups of nonmarine organisms, including insects, freshwater fishes, frogs, and vascular plants, had already evolved. The ancestors of some organisms that live on widely separated continents today were probably present on those land masses when they were part of Pangaea.

By 100 mya, Pangaea had separated into northern (Laurasia) and southern (Gondwana) land masses, and the southern continents were drifting away from each other (see Figure 20.15). Eventually, continental drift, which continues today, brought India from Africa to southern Asia, Australia closer to Southeast Asia, and South America, which had drifted as an island for 60 million years, into contact with North America. Continental drift has thus influenced the evolution and mixing of species throughout the history of life on Earth.

Modern biogeographic methods

As the great age of Earth and the fact of evolution began to be understood, two groups of investigators developed new methods for generating testable hypotheses about geographic distributions. **Ecological biogeographers** study how current distributions are influenced by interactions among species and by interactions between species and their physical environments. They examine species interactions of the types discussed in Chapter 55 to explain patterns of local and regional species diversity. We will see some examples of their work later in this chapter.

The other group of investigators consists of *historical biogeographers*, who concentrate on longer time frames and larger spatial scales. They ask questions such as

▶ Where and when did evolutionary lineages originate?

▶ How did they spread?

▶ What do the present-day distributions of organisms tell us about their past histories?

An important technique developed by historical biogeographers was the transformation of taxonomic phylogenies into "area phylogenies" by substituting the taxa's geographic distributions for their names. Distribution patterns identified in this manner may suggest routes of dispersal or point to the splitting of biotas due to the appearance of barriers to dispersal. For example, by combining the phylogenetic relationships and current distribution pattern of the horse family, we can better understand why its members are found where they are and where past barriers probably influenced speciation events among them (Figure 57.1).

If we compare the distribution patterns of many evolutionary lineages, we may detect similarities and differences. Similarities suggest common responses to physical events, such as continental drift, mountain building, and sea level changes. Differences suggest that organisms in different lineages responded in different ways, or at different times, to past events, or that they dispersed in different ways and at different times.

The Role of History in Biogeography

Past events influence today's patterns of distribution. We can never know past events with complete certainty, but by using a variety of types of evidence, historical biogeographers can develop and test hypotheses in which they eventually have a high degree of confidence. As we have just seen, biogeographers often base their interpretations on phylogenies, which show the evolutionary relationships among the organisms in a lineage (see Chapter 23). Phylogenies are most useful to biogeographers if the approximate times of evolutionary and geographic separations of lineages can be estimated.

Biogeographers use several approaches to infer the approximate times of separation of taxa within a lineage. First, if a "molecular clock" has been ticking at a relatively constant rate, the degree of difference in the molecules of species will be strongly correlated with the length of time their lineages have been evolving independently (see Chapter 24).

Second, fossils can help to show how long a taxon has been present in an area and whether its members formerly lived in areas where they are no longer found. The fossil record is helpful, but it is always incomplete. The first and

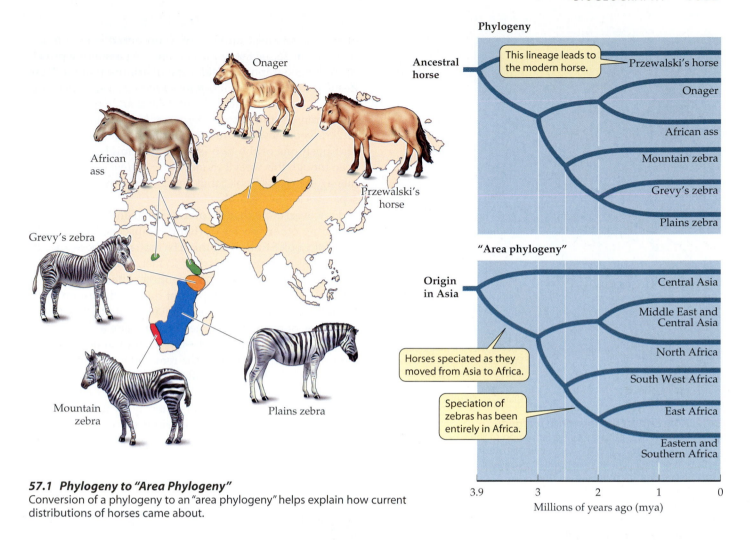

57.1 Phylogeny to "Area Phylogeny"
Conversion of a phylogeny to an "area phylogeny" helps explain how current distributions of horses came about.

last members of a taxon to live in an area are extremely unlikely to have become fossils that are discovered and described. Therefore, dates of colonization and extinction cannot be estimated accurately.

A third valuable source of information is the distributions of living species. Much more complete and extensive information can be gathered on current distributions than will ever be available from fossils. Much can be learned by examining the distribution patterns of many different groups of living organisms. Similarities in their distributions provide clues about past events that affected many of them.

Vicariance and dispersal can both explain distributions

As we have seen, a species may be found in an area either because it evolved there or because it dispersed to that area. But what if a species is found in two or more different places? What accounts for such a split distribution? There are two possibilities.

▶ A barrier may appear that splits a species' distribution. This is called a **vicariant event**, and no dispersal need be postulated to account for a split distribution.

▶ Members of a species may cross an already existing barrier and establish a new population. In this case, the species' split range must be attributed to dispersal.

By studying a single evolutionary lineage, a biogeographer may discover evidence suggesting that the distributions of its ancestors were influenced by a vicariant event, such as a change in sea level, mountain building, or continental movement. If that inference is correct, species in other lineages are likely to have been influenced by the same event and should therefore have similar distribution patterns.

Differences in distribution patterns among lineages indicate either that the lineages responded differently to the same vicariant events, that they separated at different times, or that they have very different dispersal abilities. By analyzing such similarities and differences among lineages, biogeographers can discover the relative roles of vicariant events and dispersal in determining today's distribution patterns.

Species, genera, and families found in only one place are said to be **endemic** to that location. As far as we know, all species are endemic to Earth. Some species are endemic to one continent. Others are restricted to very small areas, such as tiny islands or single mountaintops. Because a species may disperse widely and then die out where it originated,

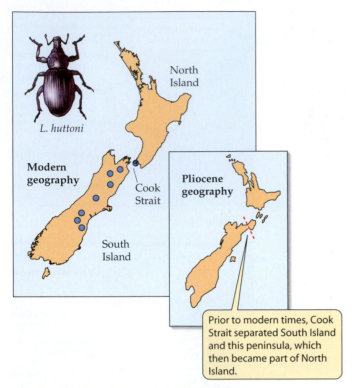

Prior to modern times, Cook Strait separated South Island and this peninsula, which then became part of North Island.

57.2 A Vicariant Distribution Explained
Blue circles indicate the current distribution of the weevil *Lyperobius huttoni*. Compare New Zealand's present geography with that of the Pliocene (inset), when the southern part of today's North Island was part of South Island.

quires the smallest number of unobserved events to account for it. To see the application of the parsimony principle to biogeography, consider the distribution of the New Zealand flightless weevil *Lyperobius huttoni*, a species that is found in the mountains of South Island and on sea cliffs at the extreme southwestern corner of North Island (Figure 57.2). If you knew only its current distribution and the current positions of the two islands, you might guess that, even though this weevil cannot fly, *L. huttoni* had somehow managed to cross Cook Strait, the 25-km body of water that separates the two islands.

However, more than 60 other animal and plant species, including other species of flightless insects, live on both sides of Cook Strait. It is unlikely that all of these species made the same ocean crossing. In fact, that assumption is unnecessary to explain the distribution patterns. Geological evidence indicates that the present-day southwestern tip of North Island was formerly united with South Island. Therefore, none of the 60 species need have made a water crossing. A single vicariant event—the separation of the northern tip of South Island from the remainder of the island by the newly formed Cook Strait—could have split all of the distributions. Although organisms do cross oceanic and terrestrial barriers, biogeographers often apply the parsimony principle when interpreting distribution patterns, just as evolutionists do when reconstructing phylogenies.

biogeographers cannot assume that a species now endemic to a location originated there. Endemic taxa can be very old ones that are in the process of becoming extinct, or very young taxa that have recently evolved in a restricted area.

The longer an area has been isolated from other areas by a vicariant event, the more endemic taxa it is likely to have, because there has been more time for evolutionary divergence to take place. Australia, which has been separated the longest from the other continents (about 65 million years), has the most distinct biota. South America has the next most distinct biota, having been isolated from other continents for nearly 60 million years. North America and Eurasia, which were joined together for much of Earth's history, have very similar biotas. That is why the early European travelers felt more at home in North America than in Australia.

Biogeographers use parsimony to explain distributions

When several hypotheses can explain a pattern, scientists typically prefer the most *parsimonious* one—the one that re-

(a) *Leucospermum conocarpodendron*   *Banksia integrifolia*

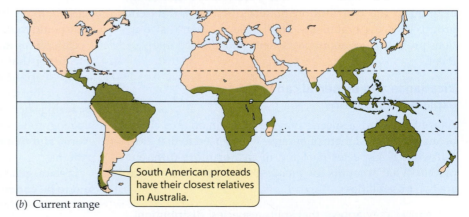

South American proteads have their closest relatives in Australia.

(b) Current range

57.3 Protead Distributions Reveal a Gondwana Ancestry
(a) Proteads from South Africa (left) and Australia (right).
(b) Current distribution (green) of the family Proteaceae.

Biogeographic histories are reconstructed from various kinds of evidence

Biogeographers use distribution maps, phylogenies, and knowledge of ancient climates and ancient geography to reconstruct the biogeographic histories of taxa. These kinds of information suggest, for example, that distributions of the proteads (family Proteaceae; Figure 57.3) have been influenced by continental drift many millions of years ago. These plants are found in Africa, but the African species are highly specialized members of an endemic subfamily. The South American species are members of a different subfamily, whose closest relatives are found in the Australian region. The phylogeny and distribution of proteads suggests that they had a broad distribution in Gondwana before that large continent began to break up, and that populations were carried by drifting land masses to their current locations. Protead lineages have been separated on the different continents long enough to have evolved major differences from one another.

Earth can be divided into biogeographic regions

Although the drifting continents carried many kinds of organisms with them, they have been isolated from one another long enough for distinct biotas to have evolved on them. Differences among continental biotas form the basis for dividing Earth into six major **biogeographic regions**: the Nearctic, Neotropical, Ethiopian, Palearctic, Oriental, and Australasian regions. Biogeographers drew the boundaries of these regions where species compositions change dramatically over short distances (Figure 57.4). These biogeographic regions are based on taxonomic similarities among the organisms living in them, not on their appearances, and should not be confused with the biomes we will discuss later in this chapter.

Biogeographers agree on the boundaries of many of these regions, but plant biogeographers recognize two regions not used by zoogeographers: southern South America and the Cape Region of South Africa. The floras of these two regions are distinct from those of adjacent areas on those continents, but the faunas of southern South America and the Cape Region are very similar to those of the remainder of those continents.

Except for the Australasian region, the biogeographic regions are no longer separated from each other by water as they were in the past. The biological distinctness of these biogeographic regions is maintained today in part by mountain and desert barriers to dispersal and in part by major climatic changes over short distances.

Ecology and Biogeography

Ecological biogeographers use the wealth of available information on current distributions of organisms to test theories that explain the numbers of species in different com-

57.4 Major Biogeographic Regions
The biotas of Earth's major biogeographic regions differ strikingly from one another.

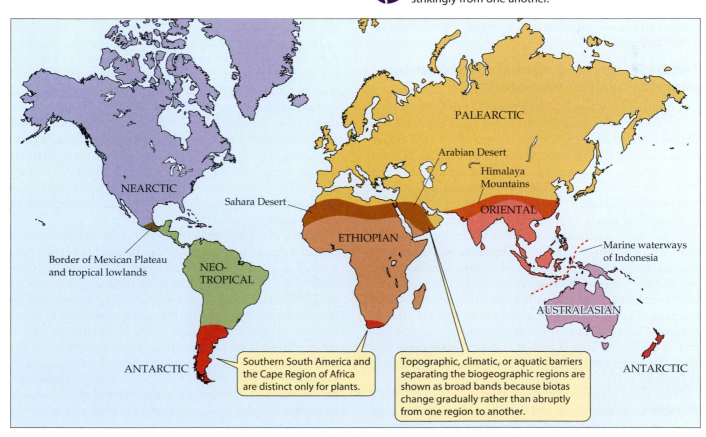

(a)

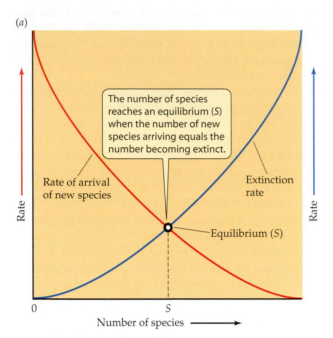

The number of species reaches an equilibrium (S) when the number of new species arriving equals the number becoming extinct.

Rate of arrival of new species

Extinction rate

Rate

Rate

Equilibrium (S)

0   S

Number of species ⟶

(b)

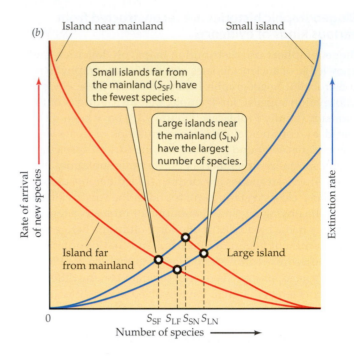

Island near mainland   Small island

Small islands far from the mainland (S_{SF}) have the fewest species.

Large islands near the mainland (S_{LN}) have the largest number of species.

Rate of arrival of new species

Extinction rate

Island far from mainland   Large island

0        $S_{SF}\ S_{LF}\ S_{SN}\ S_{LN}$

Number of species ⟶

57.5 A Model of Species Richness on Islands
Rates of arrival of new species and extinction of species already present determine the equilibrium species richness.

munities, the ways in which species disperse, and the effects of different types of barriers to dispersal. They can also use experiments to test some hypotheses. Here we describe an influential biogeographic model that attempts to account for the number of species living in an area—its **species richness**—and then look at experiments conducted to test this model.

The species richness of an area is determined by rates of colonization and extinction

Over periods of a few hundred years (a time span during which speciation is unlikely), the species richness of an area depends on the immigration of new species and the extinction of species already present. It is easiest to visualize the effects of these two processes if we consider, as did Robert MacArthur and Edward Wilson, an oceanic island that initially has no species.

Imagine a newly formed oceanic island that receives colonists from a mainland area. The list of species on the mainland that might possibly colonize the island is called the **species pool**. The first colonists to arrive on the island are all "new" species because no species live there initially. As the number of species on the island increases, a larger fraction of colonists will be members of species already present. Therefore, even if the same number of species arrive as before, the rate of arrival of new species decreases, until it reaches zero when the island has all the species in the species pool.

Now consider extinction rates. First there will be only a few species on the island, and their populations may grow large. As more species arrive and their populations in-

crease, the resources of the island will be divided among more species. We therefore expect the average population size of each species to become smaller as the number of species increases. The smaller a population, the more likely it is to become extinct. In addition, the number of species that can become extinct increases as species accumulate on the island. New arrivals to the island may include pathogens and predators that increase the probability of extinction of other species, further increasing the number of species becoming extinct per unit of time.

Because the rate of arrival of new species (i.e., colonization) decreases and the extinction rate increases as the number of species increases, eventually the number of species should reach an equilibrium at which the rates of arrival and extinction are equal (Figure 57.5a). If there are more species than the equilibrium number, extinctions should exceed arrivals, and species richness should decline. If there are fewer species than the equilibrium number, arrivals should exceed extinctions, and species richness should increase. The equilibrium is dynamic because if either rate fluctuates—as they generally do—the expected number of species shifts up and down.

Even if extinction and colonization rates are assumed to fluctuate somewhat, the model can be used to predict how species richness should differ among islands of different sizes and different distances from the mainland species pool. We expect extinction rates to be higher on small islands than on large islands because species' populations will, on average, be smaller there. Similarly, we expect fewer colonists to reach islands more distant from the mainland. Figure 57.5b gives hypothetical relative species richnesses for islands of different sizes and distances from the mainland. As you can see, the number of species should be highest for islands that are relatively large and relatively close to the mainland.

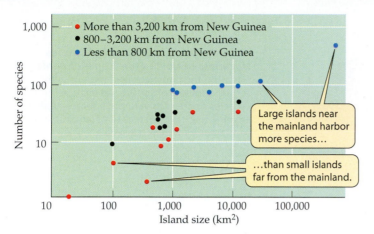

57.6 Small, Distant Islands Have Fewer Species
The dots show the numbers of land and freshwater bird species on islands of different sizes in the Moluccas, Melanesia, Micronesia, and Polynesia. These islands have been divided into three groups according to their distance from the "mainland," which is New Guinea.

The island biogeographic model has been tested

The **island biogeographic model** we have just described predicts that the number of species should increase with island size and decrease with distance from the mainland. Ecological biogeographers have tested the model using counts of species on real islands. New Guinea, which is large enough to function as a small continent, supplies the mainland species pool for islands farther east in the Pacific Ocean. The number of bird species found on those islands corresponds to the predictions of the model (Figure 57.6). The species richness patterns of plants, insects, lizards, and mammals on those islands are similar to those of birds.

The predictions of the island biogeographic model are based on assumptions about rates of colonization and extinction. Major disturbances, which serve as "natural experiments," sometimes permit colonization and extinction rates to be estimated directly. The eruption of Krakatau in August 1883, described in Chapter 20, destroyed all life on the island's surface. After the lava cooled, Krakatau was colonized rapidly by plants and animals from Sumatra to the east and Java to the west.

During the 1920s, a forest canopy was developing, and there were high rates of colonization by both birds (Table 57.1) and plants. Birds probably brought the seeds of many plants because, between 1908 and 1934, both the percentage (from 20 to 25 percent) and the absolute number (from 21 to 54 species) of plant species with bird-dispersed seeds increased. By 1933, Krakatau was again covered with a tropical evergreen forest, and 271 species of plants and 27 species of resident land birds were found there. Today the numbers of species of plants and birds are not increasing as fast as they did during the 1920s, but there are still colonizations and extinctions, as predicted by the model.

A manipulative experiment to test the island biogeographic model was carried out in the Florida Keys, a region dotted with thousands of small islands consisting entirely of red mangrove trees rooted in shallow water. Six tiny islands were fumigated with methyl bromide, which destroyed all arthropods living on them (Figure 57.7). Methyl bromide decomposes rapidly and does not inhibit recolonization.

The design of this experiment permitted the investigators to measure colonization and extinction rates directly.

57.1	**Number of Species of Resident Land Birds on Krakatau**		
PERIOD	NUMBER OF SPECIES	EXTINCTIONS	COLONIZATIONS
1908	13		
1908–1919		2	17
1919–1921	28		
1921–1933		3	4
1933–1934	29		
1934–1951		3	7
1951	33		
1952–1984		4	7
1984–1996	36		

57.7 Experimental Island Biogeography
Scientists removed all species of arthropods from several small mangrove islands in the Florida Keys by enclosing them and fumigating them with methyl bromide. When the enclosures were removed, arthropods quickly recolonized the islands.

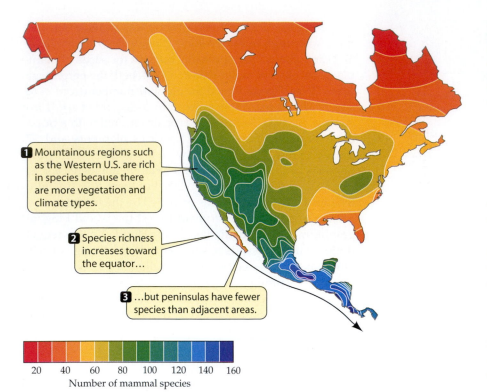

57.8 Latitudinal Gradient of Species Richness of North American Mammals Lines on the figure connect regions with equal numbers of species. An increase in species richness toward the equator typifies many other taxa, such as birds, amphibians, and trees, as well as mammals.

1 Mountainous regions such as the Western U.S. are rich in species because there are more vegetation and climate types.

2 Species richness increases toward the equator…

3 …but peninsulas have fewer species than adjacent areas.

20 40 60 80 100 120 140 160
Number of mammal species

They found that rates of colonization of the islands by arthropods were very high. Within a year the fumigated islands had about their original number of species, and each census revealed considerable turnover in the number of species present. Both results support the island biogeographic model.

Species richness varies latitudinally

A nearly universal pattern in the distribution of species is that more species live in tropical than in high-latitude regions. Figure 57.8 shows the latitudinal gradient in mammal species richness in North and Central America. Similar patterns exist for birds, frogs, and trees, and for many marine taxa.

The figure also shows two other general patterns of species richness. First, more species are found in mountainous regions than in relatively flat areas because more vegetation types and climates exist within topographically complex regions. Second, species richness declines on peninsulas, such as Florida and Baja California, probably because colonization is possible from only one direction.

 *Terrestrial Biomes*

Another way in which ecologists describe the distribution patterns of organisms is by classifying ecosystems. They apply the name **biome** to a major ecosystem type that differs from other types in the structure of its dominant vegetation. The vegetation of a biome has a similar appearance wherever on Earth that biome is found, but the plant species in these communities, despite their physical similar-

ities, may not be evolutionarily closely related. Although biomes are named for and identified by their characteristic plants, sometimes supplemented by their location or climate, each biome contains many other kinds of organisms. The geographic distribution of biomes is shown in Figure 57.9.

Biomes are identified by their distinctive climates and dominant plants

Because climate plays a key role in determining which types of plants live in a given place, the distribution of biomes on Earth is strongly influenced by annual patterns of temperature and rainfall. In some biomes, such as temperate deciduous forest, precipitation is relatively constant throughout the year, but temperature varies strikingly between summer and winter. In other biomes, both temperature and precipitation change seasonally. In certain biomes, such as tropical rainforest, temperatures are nearly constant, but rainfall varies seasonally.

In the tropics, where seasonal temperature fluctuations are small, annual climatic cycles are dominated by wet and dry seasons. In general, the number of months during which a region is close to the intertropical convergence zone (and hence receives rainfall) increases toward the equator (see Chapter 56). The intertropical convergence zone shifts latitudinally in a seasonally predictable way, resulting in a characteristic latitudinal pattern of distribution of rainy and dry seasons in tropical and subtropical regions (Figure 57.10).

Pictures and graphs capture the essence of terrestrial biomes

It is easiest to grasp the similarities and differences among terrestrial biomes by means of a combination of photographs and graphs of temperature, precipitation, and biological activity, supplemented by a brief description of the species richness and other attributes of those biomes. We use this method in the following pages to describe the major terrestrial biomes of the world.

▶ Each biome is represented by two photographs that show either the biome at different times of year or representatives of the biome in different places on Earth.

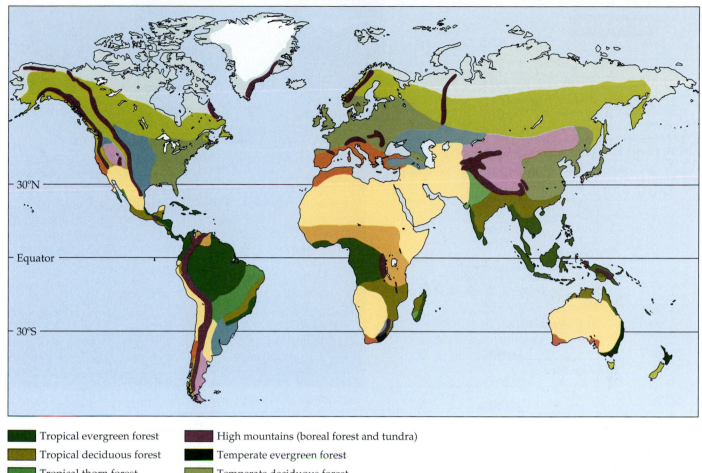

■ Tropical evergreen forest	■ High mountains (boreal forest and tundra)
■ Tropical deciduous forest	■ Temperate evergreen forest
■ Tropical thorn forest	■ Temperate deciduous forest
■ Savanna	■ Boreal forest
■ Hot desert	■ Arctic tundra
■ Chaparral	■ Temperate grassland
■ Cold desert	☐ Polar ice cap

57.9 Biomes Have Distinct Geographic Distributions
Compare these biomes with the patterns of net primary production shown in Figure 56.5.

▶ The first set of graphs plots seasonal patterns of temperature and precipitation at a typical site in the biome.

▶ Other graphs show how active different kinds of organisms are during the year. Levels of biological activity (shown by the width of horizontal bars) change either because resident organisms become more active (produce leaves, come out of hibernation, hatch, or reproduce) or because organisms migrate into and out of the biome. (The patterns shown are for the Northern Hemisphere; for high-latitude biomes, patterns in the Southern Hemisphere are 6 months out of phase with those illustrated.)

▶ A small box describes the dominant growth forms of plants in the biome and patterns of species richness there.

These descriptions are very general; they cannot capture the variation that exists within each biome.

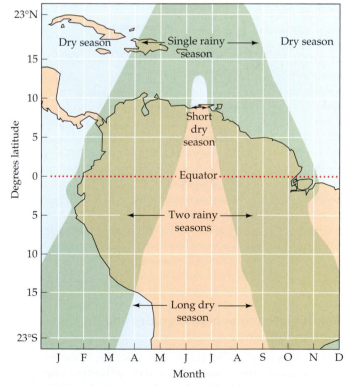

57.10 Rainy and Dry Seasons Change with Latitude
In the tropics and subtropics, which months are rainy and which are dry is highly predictable based on the region's latitude.

TUNDRA

Temperature

20°C is a "comfortable" 68°F.

0°C is the freezing point of water (=32°F).

Upernavik, Greenland 73°N

Winter is very cold and long.

Summer is cool and short.

Range 28°C

Jan   Jul   Dec

Precipitation

5 cm equals just over 2 inches.

Annual total: 23 cm

Jan   Jul   Dec

Biological Activity

| Photosynthesis |
| Flowering |
| Fruiting |
| Mammals |
| Birds |
| Insects |
| Soil Biota |

Jan   Jul   Dec

Community Composition

Dominant Plants
Perennial herbs and small shrubs

Species Richness
Plants: Low; higher in tropical alpine
Animals: Low; many birds migrate in for summer; a few species of insects abundant in summer

Soil Biota
Few species

Arctic tundra: Northwest Territories, Canada

Tropical alpine tundra: Teleki Valley, Mt. Kenya

Tundra is found at high latitudes and in high mountains

The **tundra** biome is found at high latitudes in the Arctic. It is also found high in mountains at all latitudes, where it is called **alpine tundra**. In tundra there are no trees; the vegetation is dominated by short perennial plants.

In the Arctic, permanently frozen soil—**permafrost**—underlies tundra vegetation. The top few centimeters of soil thaw during the short summers, when the sun shines 24 hours a day. Even though there is little precipitation, Arctic tundra is very wet because water cannot drain down through the frozen soil. Plants grow for only a few months each year. Most Arctic tundra animals either migrate into the area only for the summer or are dormant for most of the year.

BOREAL FOREST

Temperature

°C

- Winter is very cold and dry.
- Summer is mild and humid.

Range 41°C

Ft. Vermillion, Alberta 58°N

Jan   Jul   Dec

Precipitation

cm   Annual total: 31 cm

Jan   Jul   Dec

Biological Activity

- Photosynthesis
- Flowering
- Fruiting
- Mammals
- Birds
- Insects
- Soil Biota

Jan   Jul   Dec

Community Composition

Dominant Plants
Trees, shrubs, and perennial herbs

Species Richness
Plants: Low in trees, higher in understory
Animals: Low, but with summer peaks in migratory birds

Soil Biota
Very rich in deep litter layer

Northern conifer forest, Gunnison National Forest, Colorado

Bryophytes and lichens on southern evergreens, Tasmania, southern Australia

Boreal forests are dominated by evergreen trees

The **boreal forest** biome is found equatorward from, or at lower elevations on temperate-zone mountains than, tundra. Winters are long and very cold, and summers are short (although often warm). The short summers favor trees with evergreen leaves because these trees are ready to photosynthesize as soon as temperatures warm in spring.

The boreal forests of the Northern Hemisphere are dominated by coniferous evergreen gymnosperms. In the Southern Hemisphere, the dominant trees are southern beeches (*Nothofagus*). Evergreen forests also grow along the west coasts of continents at middle to high latitudes, where winters are mild but very wet and summers are cool and dry. These forests are home to Earth's tallest trees.

Boreal forests have only a few tree species. The dominant animals—such as insects, moose, and hares—eat leaves. The seeds in the cones of conifers also support a fauna of rodents and birds.

TEMPERATE DECIDUOUS FOREST

A Rhode Island forest in summer and...

...in winter

Temperature

°C
25
20
15
10
5
0
−5
−10

Winter is cold and snowy.

Summer is warm and moist.

Range 31°C

Madison, Wisconsin 43°N

Jan   Jul   Dec

Precipitation

cm
10
5
0

Annual total: 81 cm

Jan   Jul   Dec

Biological Activity

Photosynthesis

Flowering

Fruiting

Mammals

Birds

Insects

Soil Biota

Jan   Jul   Dec

Community Composition

Dominant Plants
Trees and shrubs

Species Richness
Plants: Many tree species in Southeast USA and East Asia, rich shrub layer
Animals: Rich; many migrant birds, richest amphibian communities on Earth, rich summer insect fauna

Soil Biota
Rich

Temperate deciduous forests change with the seasons

The **temperate deciduous forest** biome is found in eastern North America, eastern Asia, and western Europe. Temperatures in these regions fluctuate dramatically between summer and winter. Precipitation is relatively evenly distributed throughout the year. Deciduous trees, which dominate these forests, produce leaves that photosynthesize rapidly during the warm, moist summers and lose their leaves during the cold winters.

There are many more tree species here than in boreal forests. The temperate forests richest in species are in the southern Appalachian Mountains of the United States and in eastern China and Japan, areas that were not disturbed by Pleistocene glaciers. Many birds migrate into this biome in summer, when insects are abundant.

TEMPERATE GRASSLANDS

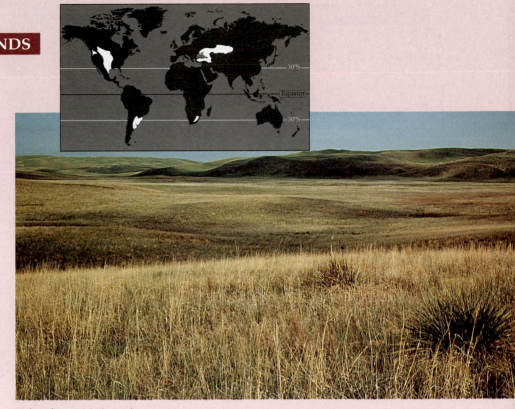

Temperature

°C
- 30
- 25
- 20
- 15
- 10
- 5
- 0
- −5

Jan — Jul — Dec

Winter is cold and dry.

Summer is warm and wetter.

Range 24°C

Pueblo, Colorado 38°N

Precipitation

cm
- 10
- 5
- 0

Jan — Jul — Dec

Annual total: 31 cm

Biological Activity

- Photosynthesis
- Flowering
- Fruiting
- Mammals
- Birds
- Insects
- Soil Biota

Jan — Jul — Dec

Community Composition

Dominant Plants
Perennial grasses and forbs

Species Richness
Plants: Fairly high
Animals: Relatively few birds because of simple structure; mammals fairly rich

Soil Biota
Rich

Nebraska prairie in spring

The Veldt, Natal, South Africa

Temperate grasslands are ubiquitous

The **temperate grassland** biome is found in many parts of the world, all of which are relatively dry much of the year. Most grasslands have hot summers and relatively cold winters. In some grasslands most of the precipitation falls in winter; in others the majority falls in summer. Such regions as the pampas of Argentina, the veldt of South Africa, and the Great Plains of the United States are components of the temperate grassland biome. Most of this biome has been converted to agriculture.

Grasslands are structurally simple, but they are rich in species of perennial grasses, sedges, and forbs. Grasses are well adapted to grazing and fire because they store much of their energy underground and quickly resprout after they are burned or grazed. As we saw in Chapter 56, many grasslands support large populations of grazing mammals.

COLD DESERT

Temperature

°C

Winters are cold and very dry.

Summers are much warmer, but still dry.

Range 23°C

Cheyenne, Wyoming 41° N

Jan   Jul   Dec

Precipitation

cm

Annual total: 38 cm

Jan   Jul   Dec

Biological Activity

| Photosynthesis |
| Flowering |
| Fruiting |
| Mammals |
| Birds |
| Insects |
| Soil Biota |

Jan   Jul   Dec

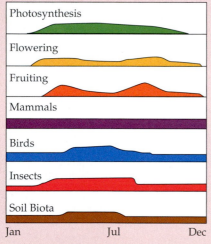

Community Composition

Dominant Plants
Low stature shrubs and herbaceous plants

Species Richness
Plants: Few species
Animals: Rich in small mammals; low in all other taxa

Soil Biota
Poor in species

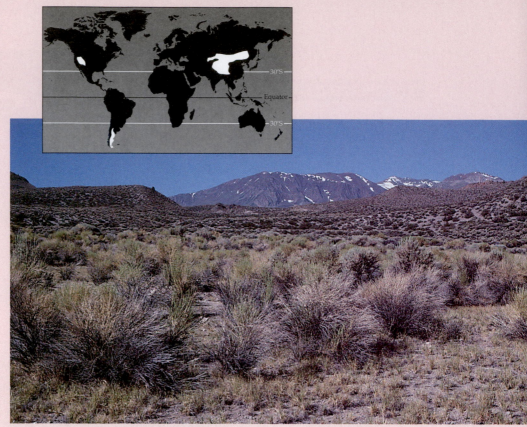

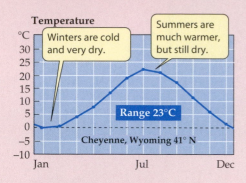

Sagebrush steppe near Mono Lake, California

Los Glacieres National Park, Argentina

Cold deserts are high and dry

The **cold desert** biome is found in dry regions at middle to high latitudes, especially in the interiors of large continents. Cold deserts also are found at fairly high altitudes in the rain shadows of mountain ranges. Seasonal changes in temperature are great.

Cold deserts are dominated by a few species of low-growing shrubs. The surface layers of the soil are recharged with moisture in winter, and plant growth is concentrated in spring. Cold deserts are relatively poor in species in most taxonomic groups, but the plants of this biome tend to produce large numbers of seeds, supporting a rich fauna of seed-eating birds, ants, and rodents.

HOT DESERT

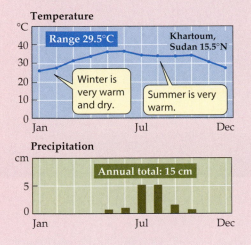

Anzo Borrego Desert, California

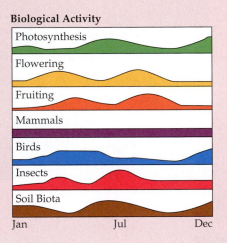

Rainbow Valley in the desert of central Australia

Temperature

°C

Range 29.5°C

Khartoum, Sudan 15.5°N

40
30
20
10
0

Jan   Jul   Dec

Winter is very warm and dry.

Summer is very warm.

Precipitation

cm

Annual total: 15 cm

5

0

Jan   Jul   Dec

Biological Activity

Photosynthesis

Flowering

Fruiting

Mammals

Birds

Insects

Soil Biota

Jan   Jul   Dec

Community Composition

Dominant Plants
Many different growth forms

Species Richness
Plants: Fairly high; many annuals
Animals: Very rich in rodents; richest bee communities on Earth; very rich in reptiles and butterflies

Soil Biota
Poor in species

Hot deserts form around 30° latitude

The **hot desert** biome is found in two belts, centered around 30°N and 30°S latitudes, respectively. These are the regions where dry air descends, warms, and picks up moisture (see Chapter 56). The driest large regions within this biome are in the center of Australia and the middle of the Sahara Desert of Africa.

Except in these driest regions, hot deserts have a richer and structurally more diverse vegetation than cold deserts. Succulent plants that store large quantities of water in their stems are conspicuous. Annual plants germinate and grow when rain falls. Pollination and seed dispersal by animals are common. Rodents and ants are often remarkably abundant, and lizards and snakes typically are rich in species and common.

CHAPARRAL

Temperature

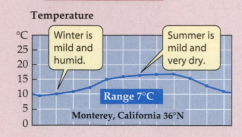

°C

Winter is mild and humid.

Summer is mild and very dry.

Range 7°C

Monterey, California 36°N

Precipitation

cm

Annual total: 42 cm

Jan   Jul   Dec

Biological Activity

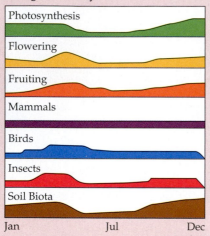

Photosynthesis

Flowering

Fruiting

Mammals

Birds

Insects

Soil Biota

Jan   Jul   Dec

Community Composition

Dominant Plants

Low stature shrubs and herbaceous plants

Species Richness

Plants: Extremely high in South Africa and Australia
Animals: Rich in rodents and reptiles; very rich in insects; especially bees

Soil Biota

Moderately rich

Fynbos vegetation, Cape of Good Hope, South Africa

Mendocino, California

The chaparral climate is dry and pleasant

The **chaparral** biome is found on the west sides of continents at moderate latitudes, where cool ocean waters flow offshore. Winters in this biome are cool and wet; summers are hot and dry. Such climates are found in the Mediterranean region of Europe, coastal California, central Chile, extreme southern Africa, and southwestern Australia.

Chaparral is dominated by low-growing shrubs and trees that have tough, evergreen leaves. The shrubs carry out most of their growth and photosynthesis in early spring, which is when insects are active and birds breed. Annual plants are abundant, producing seeds that are deposited in soil "seed banks." This biome thus supports large populations of small rodents, most of which store seeds in underground burrows.

Chaparral vegetation is naturally adapted to survive periodic fires. Many shrubs of Northern Hemisphere chaparral produce bird-dispersed fruits that ripen in the late fall, when large numbers of migrant birds arrive from the north.

THORN FOREST and TROPICAL SAVANNA

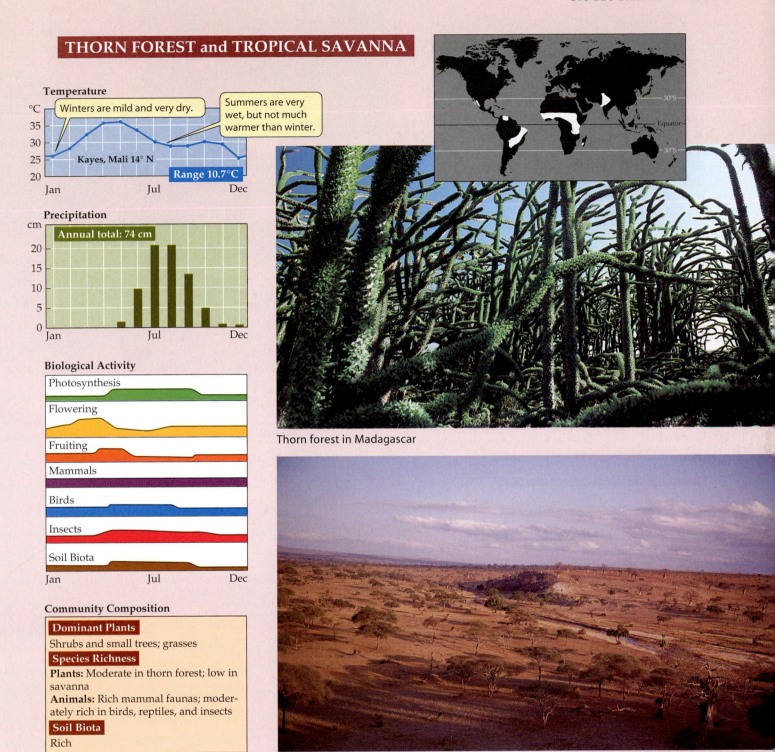

Temperature

°C

Winters are mild and very dry.

Summers are very wet, but not much warmer than winter.

Kayes, Mali 14° N

Range 10.7°C

Jan   Jul   Dec

Precipitation

cm

Annual total: 74 cm

Jan   Jul   Dec

Biological Activity

Photosynthesis

Flowering

Fruiting

Mammals

Birds

Insects

Soil Biota

Jan   Jul   Dec

Community Composition

Dominant Plants

Shrubs and small trees; grasses

Species Richness

Plants: Moderate in thorn forest; low in savanna

Animals: Rich mammal faunas; moderately rich in birds, reptiles, and insects

Soil Biota

Rich

Thorn forest in Madagascar

Savannah in Tanzania

Thorn forests and savannas have similar climates

Thorn forests are found on the equatorial sides of hot deserts. The climate is semiarid; little or no rain falls during the winter, but rainfall may be heavy during the summer. Thorn forests contain many plants similar to those found in hot deserts. The dominant plants are spiny shrubs and small trees. Members of the genus *Acacia* are common in thorn forests worldwide.

The dry tropical and subtropical regions of Africa, South America, and Australia have extensive areas of **savannas**—expanses of grasses and grasslike plants punctuated by scattered trees. The largest savannas are found in central and eastern Africa, where this biome supports huge numbers of grazing and browsing mammals that serve as prey for many large carnivores.

Grazers and browsers maintain the savannas. If savanna vegetation is not grazed, browsed, or burned, it typically reverts to dense thorn forest.

TROPICAL DECIDUOUS FOREST

Temperature

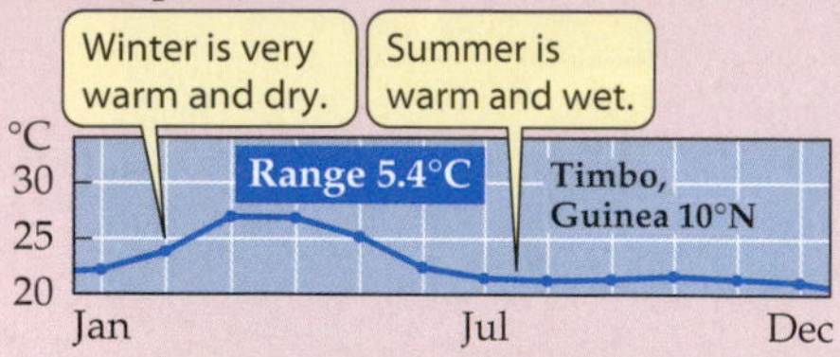

Winter is very warm and dry.

Summer is warm and wet.

°C

Range 5.4°C

Timbo, Guinea 10°N

30
25
20

Jan   Jul   Dec

Precipitation

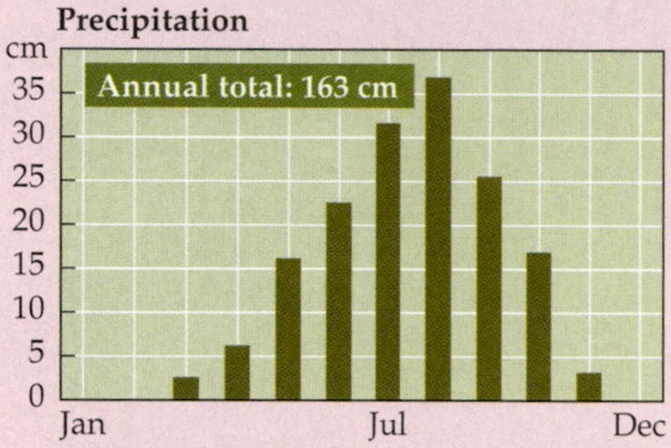

cm

Annual total: 163 cm

35
30
25
20
15
10
5
0

Jan   Jul   Dec

Biological Activity

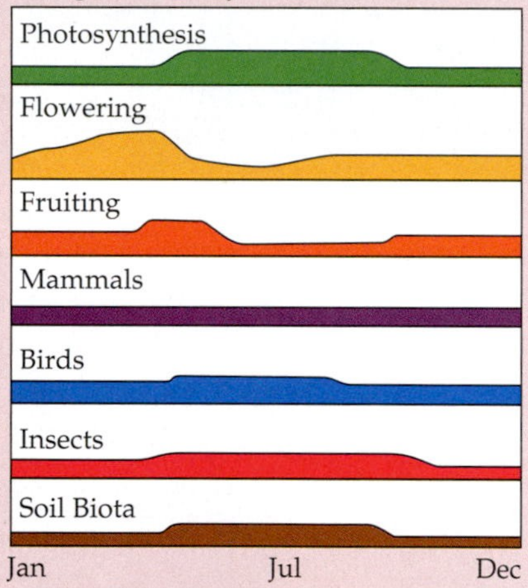

Photosynthesis

Flowering

Fruiting

Mammals

Birds

Insects

Soil Biota

Jan   Jul   Dec

Community Composition

Dominant Plants
Deciduous trees

Species Richness
Plants: Moderately rich in tree species
Animals: Rich mammal, bird, reptile, and amphibian communities; rich in insects

Soil Biota
Rich, but poorly known

Palo Verde National Park, Costa Rica, in the rainy season…

…and in the dry season

Tropical deciduous forests occur in hot lowlands

As the length of the rainy season increases toward the equator, thorn forests are replaced by **tropical deciduous forests**. These forests have taller trees and fewer succulent plants than thorn forests, and they are much richer in species. Most of the trees, except for those growing along rivers, lose their leaves during the long, hot dry season. Many of them flower while they are leafless. This biome is very rich in species of both plants and animals.

The soils of the tropical deciduous forest biome are some of the best soils in the tropics for agriculture because they are less leached of nutrients than the soils of wetter areas. As a result, most tropical deciduous forests have been cleared for grazing cattle and growing crops.

TROPICAL EVERGREEN FOREST

The exterior of lowland wet forest...   ...and its interior, Cocha Cashu, Peru

Temperature

°C

Warm and rainy all year.

25
20
15
10

Range 2.2°C   Equitos, Peru 3°S

Precipitation

cm

30
25
20
15
10
5
0

Annual total: 262 cm

Jan   Jul   Dec

Biological Activity

Photosynthesis

Flowering

Fruiting

Mammals

Birds

Insects

Soil Biota

Jan   Jul   Dec

Biological activity is essentially constant year round.

Community Composition

Dominant Plants
Trees and vines

Species Richness
Plants: Extremely high
Animals: Extremely high in mammals, birds, amphibians, and arthropods

Soil Biota
Very rich but poorly known

Tropical evergreen forests are rich in species

Tropical evergreen forests are found in equatorial regions where total rainfall exceeds 250 cm annually. This biome is the richest of all in species of both plants and animals, with up to 500 species of trees per km². Many of these species are rare. Food webs are extremely complex.

Tropical evergreen forests have the highest overall productivity of all terrestrial ecological communities. However, most mineral nutrients are tied up in the vegetation. The soils usually cannot support long-term agriculture.

On the slopes of tropical mountains, temperature decreases about 6° for each 1,000 m of elevation. The trees are shorter than lowland tropical trees. Their leaves are smaller, and there are more *epiphytes*—plants that grow on other plants and derive their nutrients and moisture from air and water rather than soil. Epiphytes thrive in tropical mountain forests where clouds form, bathing the forest in moisture.

Aquatic Biogeography

Three-fourths of Earth's surface is covered by water, most of it in the oceans. Earth's oceans form one large, interconnected water mass with no obvious barriers to dispersal. Fresh waters, in contrast, are divided into river basins and thousands of relatively isolated lakes. For freshwater organisms that cannot survive out of water, terrestrial habitats are barriers to dispersal. However, some aquatic species have flying adults that can disperse widely among water bodies. Others have windborne, desiccation-resistant spores and seeds. Still others are small enough to be transported by means such as mud on the feet of birds. Many freshwater taxa that are capable of dispersing across terrestrial barriers are distributed widely over several continents.

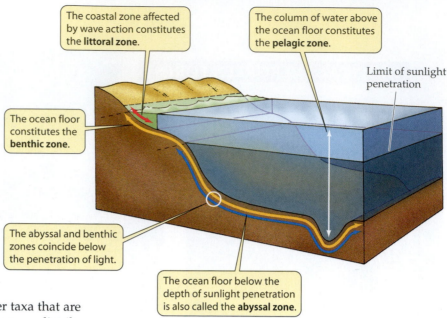

The coastal zone affected by wave action constitutes the **littoral zone**.

The column of water above the ocean floor constitutes the **pelagic zone**.

Limit of sunlight penetration

The ocean floor constitutes the **benthic zone**.

The abyssal and benthic zones coincide below the penetration of light.

The ocean floor below the depth of sunlight penetration is also called the **abyssal zone**.

57.11 Zones of the Ocean
Zones of the ocean are shown schematically in relation to depth and sunlight penetration.

Freshwater ecosystems have little water but many species

Although only about 2.5 percent of Earth's water is found in ponds, lakes, and streams, about 10 percent of all aquatic species live in freshwater habitats. Prominent among these are the more than 25,000 species of insects that have at least one aquatic stage in their life cycle. Most commonly, eggs and larvae are aquatic and adults have wings. Adults of some of these insects, such as dragonflies, are powerful flyers, but adults of mayflies and some other species are weak flyers, desiccate rapidly in air, and live no longer than a few days. As you would expect, oceanic islands have no or very few species of these weak flyers.

Similarly, fishes unable to live in salt water can disperse only within the connected streams and lakes of a river basin. Most families of freshwater fishes are restricted to a single continent. Those families with species distributed on both sides of major saltwater barriers are believed to be ancient lineages whose ancestors were distributed widely in Laurasia or Gondwana.

Marine biogeographic regions are determined primarily by water temperature and nutrients

Ocean water moves in great circular patterns—clockwise in the Northern Hemisphere and counterclockwise in the Southern Hemisphere (see Figure 56.3) These movements disperse organisms with limited swimming abilities. Nevertheless, most marine organisms have restricted ranges, indicating that important environmental limits to their distributions exist in the oceans.

We can divide the oceans into zones based on sharp horizontal and vertical environmental gradients (Figure 57.11). At all depths, the bottom of the ocean is called the **benthic zone**, and the open water column is called the **pelagic zone**. The ocean floor below the level of sunlight penetration is called the **abyssal zone**. The coastal zone from the uppermost limits of tidal action down to the depth where the water is thoroughly stirred by wave action is called the **littoral zone**.

Water temperatures, hydrostatic pressures, and food supplies all change with depth and distance from shore, influencing biotic distributions. Food is scarce, for example, in the permanently dark, cold waters of the deep sea. Living successfully in different zones of the ocean requires different physiological tolerances and morphological attributes. Not surprisingly, even though many organisms can disperse between these zones, organisms from one zone survive poorly if they attempt to live in another.

Ocean temperatures are barriers to colonization because many marine organisms are well adapted to only relatively narrow temperature ranges. The main biogeographic divisions of the pelagic zone coincide with regions where the temperature of surface waters changes relatively abruptly as a result of horizontal and vertical ocean currents (Figure 57.12). These temperature changes, in combination with seasonal changes in the amount of daylight, determine the seasons of maximum primary production. Species of marine algae tend to be adapted to photosynthesize either in summer or in winter, but not during both seasons.

Because nutrients gradually sink to the ocean bottom, high concentrations of nutrients in the pelagic zone are restricted to areas where upwelling currents bring nutrient-rich bottom waters to the surface (see Figure 56.10). Most marine organisms that grow and reproduce well in nutrient-rich waters perform relatively poorly in nutrient-poor waters. Therefore, nutrient-rich waters typically have biotas that differ considerably from those of nutrient-poor waters in the same region.

Deep ocean waters are barriers to the dispersal of marine organisms that live only in shallow water. Eggs and larvae of marine organisms can be carried great distances

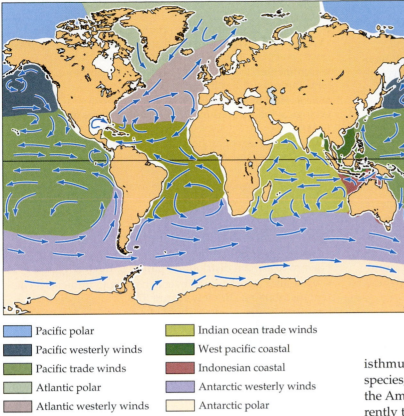

Pacific polar

Pacific westerly winds

Pacific trade winds

Atlantic polar

Atlantic westerly winds

Atlantic trade winds

Indian ocean trade winds

West pacific coastal

Indonesian coastal

Antarctic westerly winds

Antarctic polar

57.12 Pelagic Regions are Determined by Ocean Currents
The arrows represent ocean currents. Regions in which photosynthesis is maximized at different seasons are indicated by different colors.

demonstrated by the richness of reef-building corals in the intertidal and subtidal zones of isolated islands in the Pacific Ocean, which decreases with distance from New Guinea (Figure 57.13).

Marine vicariant events influence species distributions

Ancient vicariant events associated with continental drift do not influence current distributions of marine organisms, but more recent events have left biogeographic traces. An important recent vicariant event was the formation of the Isthmus of Panama about 3 million years ago. The isthmus separated the Pacific Ocean from the Caribbean Sea for the first time in more than 100 million years. Distinct marine biotas are now evolving on opposites sides of the isthmus. It forms a barrier to the dispersal of Pacific species, such as sea snakes, which reached the west coast of the Americas after the isthmus formed (Figure 57.14). Currently the fresh waters of Gatun Lake form a barrier to the dispersal of marine organisms through the Panama Canal. If a sea-level canal were constructed across the isthmus, poisonous sea snakes and other marine organisms would be able to disperse into the Caribbean.

by ocean currents, but the distance they can disperse is determined in large part by the duration of the larval life span. Relatively few species have eggs and larvae that survive long enough to disperse across wide barriers of deep water and settle in new areas. The effect of these barriers is

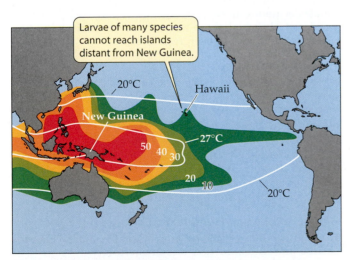

Larvae of many species cannot reach islands distant from New Guinea.

20°C

Hawaii

New Guinea

50 40 30

27°C

20

10

20°C

57.13 Generic Richness of Reef-Building Corals Declines with Distance from New Guinea
The lines connect areas with equal numbers of genera. Since temperature also limits the range of these species, the 20°C and 27°C mean annual temperature isotherms are also shown.

Caribbean Sea

Pacific Ocean

Panamanian Isthmus

Pelamis platurus

57.14 A Block to Dispersal
The existence of the Isthmus of Panama prevents poisonous sea snakes from entering the Caribbean Sea from the Pacific Ocean.

Biogeography and Human History

The distributions of land masses and species on Earth have had a strong influence on human history. The Old World happened to have a large number of species of plants and animals that were suitable for domestication. Eurasia was home to 39 species of large-seeded grasses, many more than were found in Africa or the Americas. Eurasia had 72 species of large mammals, compared with 51 in sub-Saharan Africa and 24 in the New World. Thirteen large mammal species, including pigs, horses, cattle, sheep, goats, and camels, were domesticated in Eurasia. None were domesticated in Africa, and only one, the llama, in the Americas.

To be amenable to domestication, large mammals needed to have three important social characteristics: They needed to live in herds, have well-developed male dominance hierarchies, and be nonterritorial. These traits enabled humans to tame the animals, exert dominance over them, and keep them in herds. All the large mammals of Africa lacked one or more of these traits. They have never been domesticated. The domesticated large mammals found in Africa today all came from Asia.

Domestication of large mammals had other important influences on human history. Many human diseases, such as smallpox and measles, were acquired from domesticated mammals. Eurasian people acquired immunity to these diseases. People on other continents did not. Thus when Europeans colonized the New World, they brought with them diseases that devastated the indigenous people—who transmitted no fatal diseases to the Europeans in turn. In addition, Europeans had horses, a domesticated mammal capable of carrying a person at high speeds. Horses have played a major role in human history, because cultures with horses have readily conquered cultures without them.

In the Old World, most mountain ranges are oriented in an east–west direction. Therefore, dispersal of people and their domesticated plants and animals was relatively easy, and dispersing individuals remained within climates with similar temperatures and day lengths. Humans dispersed into the New World only recently, across the high-latitude Bering land bridge. They brought with them no domesticated plants or animals. North America, as we have seen, had few species of grasses with large seeds. Maize, the grass that came to dominate American agriculture, was difficult to domesticate. Its eventual spread northward from its center of domestication in Mexico was possible only after extensive genetic changes that adapted the plants to the very different day lengths and climates of temperate North America.

Chapter Summary

▶ Biogeography is the science that attempts to explain patterns in the distribution of life on Earth.

Why Are Species Found Where They Are?

▶ If a species occupies an area, either it evolved there, or it evolved elsewhere and dispersed to that area.

▶ If a species is not found in a particular area, either it evolved elsewhere and never dispersed to that area, or it was once present in that area but no longer lives there.

▶ Continental drift has influenced the distributions of organisms throughout Earth's history.

▶ Ecological biogeographers seek to understand how current ecological interactions influence where species are found today.

▶ Historical biogeographers attempt to determine the influence of past events on today's patterns of species distributions.

▶ Historical biogeographers often analyze species distributions by converting phylogenies into "area phylogenies." **Review Figure 57.1**

The Role of History in Biogeography

▶ Biogeographers use the parsimony principle when they attempt to explain distribution patterns. **Review Figure 57.2**

▶ Vicariance and dispersal events have both influenced current distributions. **Review Figure 57.3**

▶ Animal biogeographers divide Earth into six major biogeographic regions. Plant biogeographers recognize two additional regions. **Review Figure 57.4**

Ecology and Biogeography

▶ Ecological biogeographers test theories that explain the numbers of species in different communities, how species disperse, and the effectiveness of barriers to movement.

▶ The island biogeographic model, which predicts the equilibrium species richness on islands, has been tested by examining patterns of distribution and by performing experiments. **Review Figures 57.5, 57.6; Table 57.1**

▶ The number of species in most lineages increases from polar to tropical regions. **Review Figure 57.8**

Terrestrial Biomes

▶ Terrestrial biomes are major ecosystem types that differ from one another in the structure of their dominant vegetation.

▶ The distribution of biomes on Earth is strongly influenced by annual patterns of temperature and rainfall. **Review Figures 57.9, 57.10**

▶ The major terrestrial biomes are tundra, boreal forest, temperate deciduous forest, temperate grassland, cold desert, hot desert, chaparral, thorn forest, savanna, tropical deciduous forest, and tropical evergreen forest.

Aquatic Biogeography

▶ No absolute barriers to the movement of marine organisms exist within the oceans, but most marine organisms have restricted ranges.

▶ Conditions in the oceans change dramatically with depth and sunlight penetration. **Review Figure 57.11**

▶ Boundaries between many pelagic regions are determined by ocean currents. **Review Figure 57.12**

▶ Species that live in shallow waters disperse with difficulty across wide deep-water barriers. **Review Figure 57.13**

Biogeography and Human History

▶ The distributions of plants, animals, and continents have exerted powerful influences on human history.

For Discussion

1. Horses evolved in North America, but subsequently became extinct there. They survived to modern times only

in Africa and Asia. In the absence of a fossil record, we would probably infer that horses originated in the Old World. Today, the Hawaiian Islands have by far the greatest number of species of fruit flies (*Drosophila*). Would you conclude that the genus *Drosophila* originally evolved in Hawaii and spread to other regions? Under what circumstances do you think it is safe to conclude that a group of organisms evolved close to where the greatest number of species live today?

2. The island biogeographic model we described incorporates almost nothing about the biology of the species involved. What traits of species should be incorporated into more realistic models of rates of colonization and extinction of species on islands?

3. Experiments to test theories of species richness are necessarily short-term. What long-term consequences of colonization and extinction are likely to be undetected by these experiments? How could they be studied?

4. In nearly every ecological community, the number of species present is much smaller than the number potentially available to colonize it. What inferences can be drawn from this pattern?

5. A well-known legend states that Saint Patrick drove the snakes out of Ireland. Give some alternative explanations, based on sound biogeographic principles, for the absence of indigenous snakes in that country.

6. Why are there so few species of trees in boreal forests? Why do few species of trees of boreal forests have animal-dispersed seeds?

7. What are some significant present-day human problems whose solutions involve biogeographic considerations? What kinds of biogeographic knowledge are most important for addressing each one?

8. Most of the world's flightless birds are either nocturnal and secretive (such as the kiwi of New Zealand) or large, swift, and well armed (such as the ostrich of Africa). The exceptions are found primarily on islands. Many flightless island species have become extinct with the arrival of humans and their domestic animals. What special biogeographic conditions on islands might permit the survival of flightless birds? Why has human colonization so often resulted in the extinction of such birds? The power of flight has been lost secondarily in representatives of many groups of birds and insects; what are some possible evolutionary advantages of flightlessness that might offset its obvious disadvantages?

58 Conservation Biology

WHEN POLYNESIAN PEOPLE SETTLED IN HAWAII about 2,000 years ago, they exterminated—probably by overhunting—at least 39 species of land birds. Among them were 7 species of geese, 2 species of flightless ibises, a sea eagle, a small hawk, 7 flightless rails, 3 species of owls, 2 large crows, a honeyeater, and at least 15 species of finches.

No people lived in New Zealand until about 1,000 years ago, when the Maori colonized the islands. Hunting by the Maori caused the extinction of 13 species of flightless moas, some of which were larger than ostriches.

When humans arrived in North America over the Bering land bridge, about 20,000 years ago, they encountered a rich fauna of large mammals. Most of those species were exterminated within a few thousand years. A similar extermination of large animals followed the human colonization of Australia, about 40,000 years ago. At that time Australia had 15 genera of marsupials larger than 50 kg, a genus of gigantic lizards, and a genus of heavy, flightless birds. All the species in 13 of those 15 genera had become extinct by 18,000 years ago.

The accelerating pace of human-caused extinctions of species, which raises serious concerns about the future of biological diversity on Earth, has led to the rapid development of the applied discipline of **conservation biology**—the scientific study of how to preserve the diversity of life. Conservation biologists study the causes of endangerment and extinction and develop methods to help preserve genes, species, communities, and ecosystems. The science of conservation biology draws heavily on concepts and knowledge from population genetics, evolution, ecology, biogeography, wildlife management, economics, and sociology. In turn, the needs of conservation are stimulating new research in those fields.

In this chapter we will see how biologists estimate rates of species extinction and the causes of endangerment, and learn how management plans can be used to reduce extinction rates and restore endangered species and communities to states in which they are likely to persist for a long time.

Extinct Flightless Hawaiian Birds
This artist's reconstruction of a flightless Hawaiian goose shows one of the many bird species exterminated by the Polynesian settlers of the islands.

Estimating Current Rates of Extinction

Most human activities that are causing extinctions are not new, but there are many more of us doing those things than ever before (see Figure 54.16). We have also added the results of our advancing technology, such as pesticides and climate change, to the array of pressures created by human activities.

We do not know how many species will become extinct during the next 100 years, first, because we do not know how many species there are on Earth, and second, because the number of extinctions will depend both on what we do and on unexpected events. However, several methods exist for estimating probable rates of extinction resulting from human actions. In this section we will discuss how conservation biologists estimate current rates of extinction and identify species at risk of extinction.

Species–area relationships are used to estimate extinction rates

When we described the island biogeographic model in Chapter 57, we saw that the number of species on an island increases with the size of the island (see Figure 57.6). This **species–area relationship** can be applied to habitat patches

58.1 Deforestation Rates Are High in Tropical Forests
Central America provides an example of the high rate of destruction of tropical forests that has taken place in recent years. As the forests are lost, so are the many species that live in them.

on mainlands as well. Conservation biologists often use the well-established relationship between the size of an area and the number of species present to estimate numbers of species extinctions resulting from habitat destruction.

The rate at which tropical forests are being logged and converted to cropland and pasture is not precisely known, but it is currently very high (Figure 58.1). These forests are Earth's richest biomes, home to perhaps one-half of all the species on the planet. Calculations using the species–area relationship applied to tropical forests are far from exact, but can result in estimates of over 1 million species extinctions in the next few decades.

Even the lowest estimates of current extinction rates predict that at least 10 percent of Earth's species are likely to become extinct during the next two decades. Some estimates predict extinction of 50 percent of Earth's species

during the next 50 years. Extinction rates have been much higher on islands than in mainland areas during the past 400 years (Figure 58.2), but extinction rates on continents are also rising fast.

Population models are used to estimate risks of extinction

To estimate the risk that a population will become extinct, conservation biologists analyze information about interactions between a population's genetic variation, morphology, physiology, and behavior and its environment, both physical and biological. Although rarity itself is not always a cause for concern, species whose populations are shrinking rapidly usually are at risk. Species with only a few individuals confined to a small range are likely to become extinct because they can be eliminated by local disturbances such as fires, unusual weather, disease, and predators.

Both demographic and genetic information were used in assessing the risk of extinction of Furbish's lousewort, a plant that is restricted to the banks of the St. John River in northern Maine. The Furbish's lousewort population is divided into discrete subpopulations growing at separate sites. These subpopulations are found where periodic disturbance, often caused by spring ice scour when blocks of floating ice scrape the stream banks, establishment of trees and shrubs that would outcompete the louseworts (Figure 58.3). Data on annual rates of survival, growth, and reproduction were gathered by following more than 6,000 individually marked plants between 1983 and 1986. Extinctions of subpopulations and foundings of new subpopulations were estimated by counting plants over the entire range of the species.

Although Furbish's lousewort depends on regular disturbance to suppress the growth of shrubs and trees, distur-

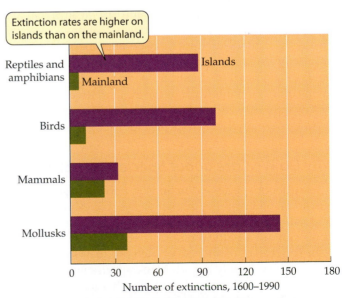

58.2 Extinctions Have Been High on Islands
Between 1600 and 1990, extinctions of terrestrial vertebrates and mollusks were much higher on islands than on continents.

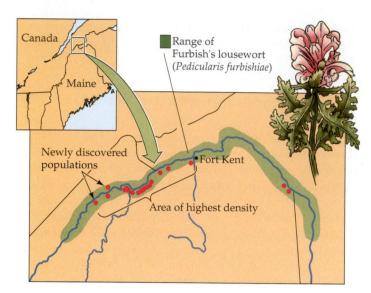

58.3 Furbish's Lousewort Exists as a Metapopulation
Small populations of Furbish's lousewort exist along the St. John River, where ice scour and bank slumping eliminate shrubs and trees. Dispersing individuals colonize newly disturbed sites.

bances can also eliminate subpopulations. Between 1983 and 1984, three of ten study subpopulations were completely destroyed by ice scour and bank slumping, and none of the disturbed sites were recolonized during succeeding years. Thus if disturbance rates were too high, extinction rates of subpopulations are likely to be higher than rates of establishment of new subpopulations. Under current disturbance regimes, the two rates are about equal. Local subpopulations became extinct and new subpopulations were founded at an annual rate of about 3 percent during the study.

The investigators concluded that the persistence of Furbish's lousewort depends on disturbance events at sites currently having subpopulations and also at sites that are currently unoccupied. Clearly a strategy that protected only sites with current subpopulations would not maintain the species for many years. Preserving Furbish's lousewort depends on maintaining disturbance regimes along an entire stretch of the St. John River. If the St. John River continues to flow naturally and spring ice scouring continues, Furbish's lousewort is likely to persist for a long time.

Why Do We Care about Species Extinctions?

We care about species extinctions in part because humans depend on other species in many ways. For example, more than half the medical prescriptions written in the United States contain a natural plant or animal product (Figure 58.4). The search for and use of such products from the living world has hardly begun. Many species may be eliminated by tropical forest destruction before we find out whether they might be sources of useful products.

Extinctions deprive us of opportunities to study and understand ecological relationships among organisms. The more species are lost, the more difficult it will be to understand the rules that govern the structure and functioning of ecological communities.

We also derive enormous aesthetic pleasure from interacting with other organisms. Many people would consider a world with far fewer species as a less desirable one in which to live. Living in ways that cause the extinction of other species also raises serious moral and ethical issues that are receiving increased attention.

Ecosystem processes, as well as individual species, produce many benefits to humanity. Among them are the generation and maintenance of fertile soils, prevention of soil erosion, detoxification and recycling of waste products, regulation of hydrological cycles and the composition of the atmosphere, control of agricultural pests, pollination, and maintenance of the species richness upon which humanity depends. It is easy to list these ecosystem services, but to justify the allocation of scarce public resources to maintain them, we need quantitative estimates of their value.

A detailed study by economists, ecologists, and land managers in Western Cape Province, South Africa, has shown that an intensive program to eradicate invasive alien plants in the highlands of the region is a cost-effective way of maintaining a reliable regional supply of high-quality water. The native vegetation of these highlands is a species-rich community of shrubs, known as *fynbos* (pronounced "fainbos"). Fynbos can survive regular summer drought, nutrient-poor soils, and the fires that periodically sweep through the highlands (Figure 58.5a).

The fynbos-clad highlands provide about two-thirds of the Western Cape's water requirements. In addition, the flora is harvested for cut and dried flowers and thatching

Catharanthus roseus

58.4 Source of a Life-Saving Drug
A drug derived from the Madagascar rosy periwinkle has greatly increased the survival rate of children with leukemia. Other species that might be sources of drugs are being eliminated by deforestation on Madagascar.

(a)

58.5 Water Flows from Fynbos
(a) The fynbos of South Africa. (b) Stream flow from fynbos watersheds is inversely proportional to plant biomass. (c) A computer simulation of stream flows from watersheds that have and have not been invaded by trees from outside the region.

(b) **Stream flow from fynbos watersheds**

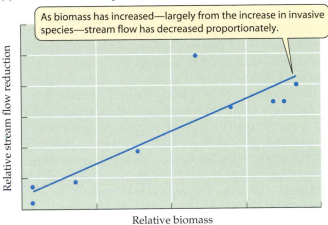

As biomass has increased—largely from the increase in invasive species—stream flow has decreased proportionately.

Relative stream flow reduction

Relative biomass

(c) **Computer simulation**

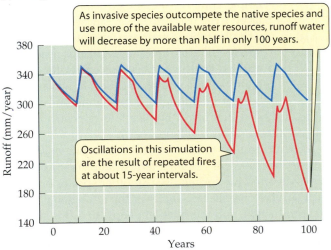

As invasive species outcompete the native species and use more of the available water resources, runoff water will decrease by more than half in only 100 years.

Oscillations in this simulation are the result of repeated fires at about 15-year intervals.

Runoff (mm/year)

Years

cause they are taller and grow faster than the native plants, the exotics increase the intensity and severity of fires. By transpiring larger quantities of water, they decrease stream flows to less than half the amount flowing from mountains covered with native plants (Figure 58.5b,c).

Removing the exotic plants by felling and digging out invasive trees and shrubs and managing fire is estimated to cost between $140 and $830 per hectare, depending on the densities of invasive plants. Annual follow-up operations will cost about $8 per hectare. The costs of alternative methods to replace the water lost from watersheds taken over by exotic plants are much higher. A sewage purification plant that would deliver the same volume of water as a well-managed watershed of 10,000 hectares would cost $135 million to build and $2.6 million per year to operate. Desalination of seawater would cost four times as much. Thus, the available alternatives would deliver water at a cost between 1.8 and 6.7 times more than the costs of maintaining natural vegetation in the watershed.

Modern industrial societies often favor technologically sophisticated methods of substituting for lost ecosystem services. The study of water resources in the Western Cape Region shows that simple but labor-intensive methods—cutting and burning—can be cheaper. In addition, they preserve other ecosystem values, such as tourism and commercial plant products.

Some ecosystem values, such as aesthetic benefits, cannot be replaced with technological inventions. Aesthetic benefits may contribute much to a country's economy. One of the largest sources of foreign income in Kenya is nature tourism. The loss of a single species probably would not reduce the flow of tourists to Kenya, but if elephants, rhinoceroses, lions, leopards, and buffalo were all to disappear, fewer people would pay the high price of a Kenyan vacation. Populations of these species can be maintained only if large tracts of the ecosystems in which they live are preserved.

Determining Causes of Endangerment and Extinction

Rare species are more likely to become extinct than common species. Species may be rare for any of several reasons. They may live in a habitat that is rare, such as desert lakes with high salt concentrations or caves (Figure 58.6). Another reason for rarity is trophic level—secondary carnivores are usually rare because so little energy is available to support their populations, as we saw in Chapter 56. Being rare increases a species' chance of becoming extinct, but common species can also become extinct.

grass. The combined value of these harvests in 1993 was about $19 million. Some of the income from tourism in the region comes from people who want to see the fynbos vegetation. About 400,000 people visit the Cape of Good Hope Nature Reserve each year, primarily to see the many unique plants.

During recent decades, a number of plants introduced into South Africa have invaded the fynbos highlands. Be-

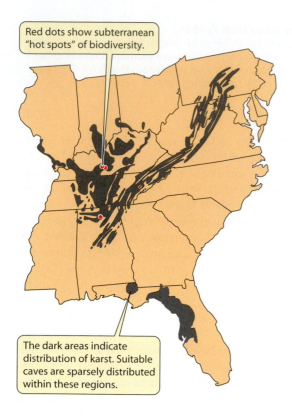

Red dots show subterranean "hot spots" of biodiversity.

The dark areas indicate distribution of karst. Suitable caves are sparsely distributed within these regions.

58.6 The Habitat of Cave Animals Is Patchy and Restricted
The map shows the distribution of subterranean limestone (karst) in which caves with running water are found. The caves that can actually be inhabited by cave animals covers a much smaller area.

An analysis by The Nature Conservancy revealed that habitat loss is the most important cause of endangerment of species in the United States (Figure 58.7), but that other factors, such as exotic species, pollution, overexploitation, and disease, also threaten species. In this section we will examine some of these major threats to species.

Habitat destruction and fragmentation are important causes of extinction today

The 6 billion humans that live on Earth today are fed, clothed, and housed by the agricultural and forestry industries, which convert natural ecological communities containing many species into highly modified communities dominated by one or a few species of plants. Within these communities, humans discourage the presence of other species by killing competing plants, bacteria, fungi, nematodes, insects and other arthropods, and vertebrates.

Although agricultural ecosystems have always harbored fewer species than the complex natural ecosystems they replaced, only recently have farmers planted large tracts of land in single crops. Traditional farmers planted many different crops together, maintaining some of the diversity that is key to the functioning of natural communities. Many species that cannot survive in intensive modern agricultural systems live in traditional agricultural ones. In traditional coffee plantations, for example, coffee bushes are grown in the shade of large trees (Figure 58.8a). These structurally rich plantations support populations of many species of birds, and few pesticides need to be applied to them. Some recently developed high-yielding strains of coffee, however, grow best in full sunlight. Pure plantations of these coffee bushes require heavy applications of pesticides and support almost no birds (Figure 58.8b).

Agriculture and forestry today are so extensive that more than 30 percent of all net terrestrial primary production is diverted for human use. All other species on Earth must survive on only two-thirds of the total global terrestrial production, and the fraction people divert is steadily increasing.

As natural habitats are progressively destroyed, the remaining patches increasingly become smaller and more isolated. Small habitat patches are qualitatively different from larger patches of the same habitat in ways that affect the

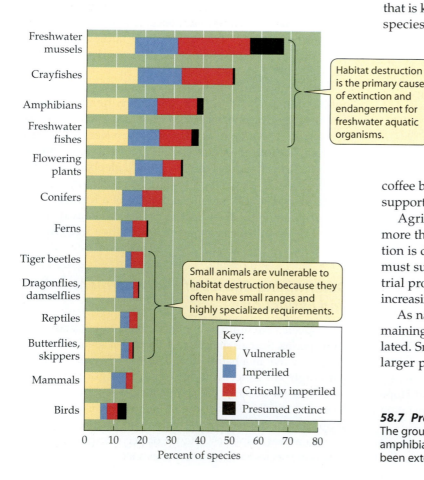

Habitat destruction is the primary cause of extinction and endangerment for freshwater aquatic organisms.

Small animals are vulnerable to habitat destruction because they often have small ranges and highly specialized requirements.

Key:
- Vulnerable
- Imperiled
- Critically imperiled
- Presumed extinct

58.7 Proportion of U.S. Species Extinct or at Risk
The groups that are most endangered—mussels, crayfishes, amphibians, and fishes—live in fresh waters, a habitat that has been extensively destroyed and polluted.

(b)

(a)

58.8 The Way Coffee is Grown Affects Biodiversity
(a) A traditional coffee plantation with a canopy of trees supports many species of birds. (b) Few bird species live in plantations of sun-grown coffee.

survival of species. Small patches cannot maintain populations of species that require large areas, and they support only small populations of many of the species that can survive in them.

In addition, the fraction of a patch that is influenced by conditions in adjacent habitats—resulting in **edge effects**—increases rapidly as patch size decreases (Figure 58.9). Close to the edges of forest patches, for example, winds are stronger, temperatures are higher, humidity is lower, and light levels are higher than they are farther inside the forest. Species from surrounding habitats often colonize the edges of patches to compete with or prey upon the species living there.

Usually we do not know which organisms lived in an area before its habitats became fragmented. To address this problem, a major research project near Manaus, Brazil, was launched in an area of tropical forest before logging took place. The landowners agreed to preserve forest patches of certain sizes and locations (Figure 58.10a).

Biologists conducted censuses of those patches while the areas were still part of the continuous forest. Soon after the surrounding forest was cut, species began to disappear from the isolated patches. The first species to be eliminated were monkeys with large home ranges, such as the black spider monkey, the tufted capuchin, and the bearded saki, and antbirds that follow raiding army ant swarms to capture insects flushed by the ants (Figure 58.10b,c).

Species that become extinct in small fragments are unlikely to become reestablished because individuals dispersing from other locations are less likely to find isolated patches. Even if they find them, the patches may be too small to support their populations on a long-term basis. The persistence of species in small patches may be improved if the patches are connected by **corridors** of suitable habitat through which individuals can disperse.

The role of corridors in sustaining populations of species was studied by creating patches of mosses on the surface of a large rock. Eight experimental "landscapes" were established by scraping away the moss to create patches of equal size. Some patches were isolated, others were connected by narrow moss corridors, and still others were connected by pseudocorridors

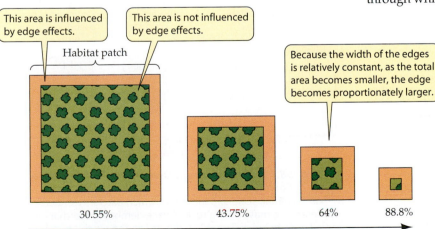

This area is influenced by edge effects.

This area is not influenced by edge effects.

Habitat patch

Because the width of the edges is relatively constant, as the total area becomes smaller, the edge becomes proportionately larger.

30.55%   43.75%   64%   88.8%

Increasing percentage of patch influenced by edge effects

58.9 Edge Effects
The smaller a habitat patch, the greater the proportion that is influenced by conditions in the surrounding environment.

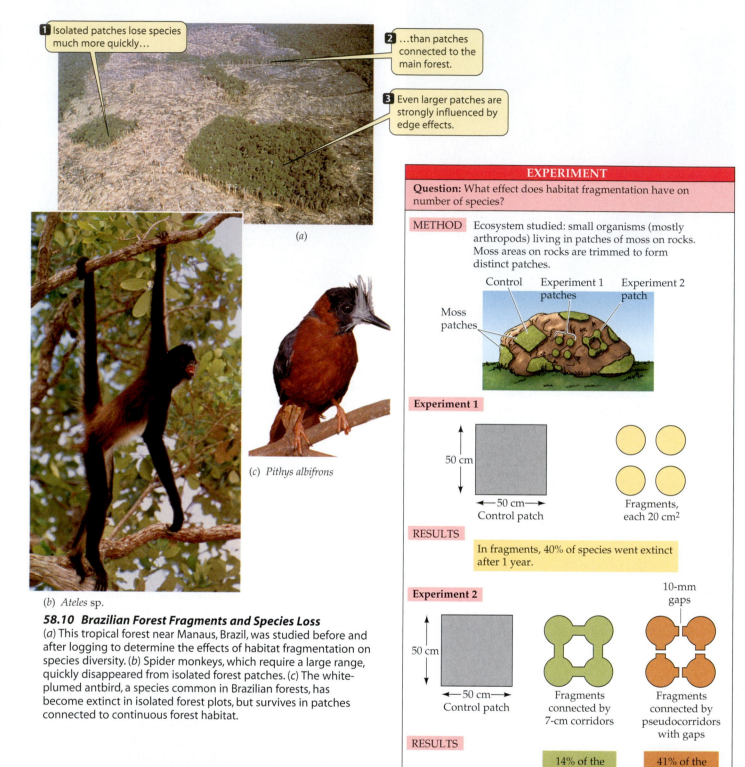

1 Isolated patches lose species much more quickly...

2 ...than patches connected to the main forest.

3 Even larger patches are strongly influenced by edge effects.

(a)

(c) *Pithys albifrons*

(b) *Ateles* sp.

58.10 Brazilian Forest Fragments and Species Loss
(a) This tropical forest near Manaus, Brazil, was studied before and after logging to determine the effects of habitat fragmentation on species diversity. (b) Spider monkeys, which require a large range, quickly disappeared from isolated forest patches. (c) The white-plumed antbird, a species common in Brazilian forests, has become extinct in isolated forest plots, but survives in patches connected to continuous forest habitat.

EXPERIMENT

Question: What effect does habitat fragmentation have on number of species?

METHOD Ecosystem studied: small organisms (mostly arthropods) living in patches of moss on rocks. Moss areas on rocks are trimmed to form distinct patches.

Control   Experiment 1 patches   Experiment 2 patch

Moss patches

Experiment 1

50 cm

←— 50 cm —→
Control patch

Fragments, each 20 cm²

RESULTS

In fragments, 40% of species went extinct after 1 year.

Experiment 2

10-mm gaps

50 cm

←— 50 cm —→
Control patch

Fragments connected by 7-cm corridors

Fragments connected by pseudocorridors with gaps

RESULTS

14% of the species went extinct after 6 months.

41% of the species went extinct after 6 months.

Conclusion: Even small barriers to dispersal can raise extinction rates.

58.11 An Experiment Demonstrates the Value of Corridors
A small-scale experiment in which patches were created by removing moss from a rock surface demonstrated that even small gaps between suitable habitats can reduce the number of species that persist in patches.

that were interrupted by a 10 mm break (Figure 58.11). The abundance and species richness of tiny arthropods were measured at 2-month intervals over the course of a year—a time period equivalent to many generations for most of the small animals living in the moss.

Remarkably, barriers as narrow as 10 mm were sufficient to greatly reduce the dispersal rates of arthropods between moss patches. Isolated patches and patches connected by pseudocorridors had only about 60 percent as many species of springtails (tiny, wingless insects) and mites as the patches connected by complete corridors.

58.12 *Galápagos Tortoises Are Reared in Captivity*
Conservationists at the Charles Darwin Research Station remove tortoise eggs from nests. When the eggs hatch, the young are reared in captivity until they are large enough to be invulnerable to predation by introduced pigs and rats. Populations of tortoises on some Galápagos islands would be extinct if they were not propagated in captivity.

cause by removing most of the algae from the water, thereby clarifying it, zebra mussels allow sunlight to penetrate more deeply into the water column. As a result, in areas of high mussel densities, populations of submerged vascular plants and some invertebrates have increased. Not until after many years will we know the full effects of the colonization of North America by zebra mussels.

Some pests proliferated quickly following their introduction to new continents, with destructive consequences. Forest trees in eastern North America, for example, have been attacked by several European diseases. The chestnut blight, caused by a European fungus, virtually eliminated the American chestnut, formerly a dominant tree in forests of the Appalachian Mountains. Nearly all American elms over large areas of the East and Midwest have been killed by Dutch elm disease (caused by a different fungus), which reached North America in 1930. Ecologists suspect that intercontinental movement of disease organisms has caused extinctions throughout life's history, but evidence of disease outbreaks is not usually preserved in the fossil record.

Introduced pests, predators, and competitors have eliminated many species

Deliberately and accidentally, people move many organisms from one continent to another. Pheasants and partridges were introduced into North America for hunting. Europeans introduced rabbits and foxes to Australia for sport. Many plants have been introduced as ornamentals. Weed seeds have been accidentally carried around the world in soil used as ballast in sailing ships and as contaminants in sacks of crop seeds. Despite quarantines, disease organisms have spread widely, carried by infected plants, animals, and people.

A species that has evolved over time in a community with certain predators and competitors may be driven to extinction by newly introduced predators and competitors. Nearly half of the small to medium-sized marsupials and rodents of Australia have been exterminated during the last 100 years by a combination of competition with introduced rabbits and predation by introduced cats and foxes. On the Galápagos archipelago, introduced pigs and rats had exterminated several races of tortoises even before Darwin visited the islands. Populations of tortoises on some islands are maintained today only because conservationists remove eggs and rear the young tortoises in captivity (Figure 58.12).

The zebra mussel, whose larvae were carried in ships' ballast water from Europe, became established in the Great Lakes in about 1985. Zebra mussels dispersed rapidly and today occupy much of the Great Lakes and Mississippi River drainage (Figure 58.13). In some places these mussels have reached densities as high as 400,000 per square meter! Some species of native clams are being covered and smothered by zebra mussels. Zebra mussels have also caused millions of dollars of damage to water intake structures. On the other hand, some native species have benefited, be-

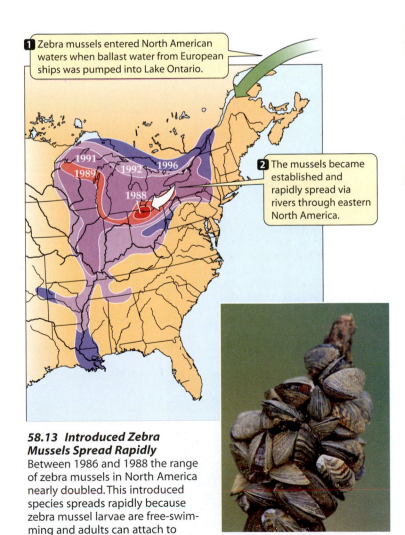

1 Zebra mussels entered North American waters when ballast water from European ships was pumped into Lake Ontario.

1991
1989
1992
1996
1988

2 The mussels became established and rapidly spread via rivers through eastern North America.

58.13 *Introduced Zebra Mussels Spread Rapidly*
Between 1986 and 1988 the range of zebra mussels in North America nearly doubled. This introduced species spreads rapidly because zebra mussel larvae are free-swimming and adults can attach to moving objects, such as boat hulls.

Dreissena polymorpha

The curved bill of the iiwi matches the shape of the *Lobelia* flower.

58.14 Coevolved Mutualists
Declining populations of the iiwi (*Vestiaria coccinea*), a Hawaiian honeycreeper, also threaten the *Lobelia* plant, which has no other pollinator.

Overexploitation has driven many species to extinction

Until recently, humans caused extinctions primarily by overhunting. The passenger pigeon, the most abundant bird in North America in the early 1800s, became extinct by 1914, largely due to overhunting. Russian whalers exterminated the unusual Steller's sea cow of the North Pacific in the late 1800s, just 37 years after it was first described. Such overex-

ploitation continues today. Elephants and rhinoceroses are threatened in Africa because poachers kill them for their tusks and horns. Unfortunately, these animals are not slaughtered for medical or other useful purposes; rather, they are killed for ornaments and because some men believe that powdered horn enhances their sexual potency. Many species of orchids, parrots, reptiles, and tropical reef fishes are currently threatened by lucrative pet and houseplant trades.

Loss of mutualists threatens some species

Many plants have mutualistic relationships with pollinators, but most of these mutualisms are not highly species-specific. On islands, however, where ecological communities contain relatively few species, plant–pollinator interactions often evolve to be highly specific. For example, a single species of the plant *Lobelia* colonized the Hawaiian Islands, where it eventually gave rise to 110 daughter species. A single colonizing species of songbird gave rise to at least 47 species of Hawaiian honeycreepers, some of which have long, slender, curved bills. These nectar-feeding birds were the only pollinators of many species of Hawaiian lobelias (Figure 58.14).

Today, half of the nectar-feeding birds of Hawaii are extinct, leaving many lobelias without pollinators. Many of these lobelias still survive, but populations of some species have been reduced to only a few individuals. A few species survive only because biologists artificially pollinate them.

Global warming may cause species extinctions

Atmospheric scientists predict that, as a result of increasing concentrations of CO_2 and other greenhouse gases in the atmosphere (see Chapter 56), average temperatures in North America will increase 2°–5°C by the end of the twenty-first century. If the climate warms by only 1°C, the average temperature currently found at a certain location will shift 150 km to the north. To remain in the temperature regime to which they are accustomed, species will have to shift their ranges 150 km to the north. Species will need to shift their ranges as much as 500–800 km in a single century if the climate warms 2–5°C. Some habitats, such as alpine tundra, could be eliminated from many areas as forests expand up the mountain slopes.

(a)

If the climate of eastern North America warms by as little as 4°C, about half the potential future range of beech trees will be beyond the northernmost extent of the current range.

Potential future range

Overlap

Current range

(b)

58.15 Threatened by Global Warming
(a) Seedlings and saplings abound in this beech forest. (b) Beeches would need to migrate 40 times faster than they have in the past to keep up with the anticipated rates of climate change.

Conservation biologists are attempting to predict the effects of global warming on North American species. Trees might be especially vulnerable to climate change because they grow for long periods before they begin to reproduce, and their seeds typically move only very short distances (Figure 58.15).

If Earth warms as predicted, climatic zones will not simply shift northward. New climates will develop, and some existing climates will disappear. New climates are certain to develop at low elevations in the tropics because a warming of even 2°C would result in climates near sea level that are hotter than those found anywhere in the humid tropics today. Adaptation to those climates may prove difficult for many tropical organisms.

Preventing Species Extinctions

Designing recovery plans

Once the causes of endangerment of species have been identified, appropriate remedies can be designed. In the United States, when a species is listed as threatened or endangered under the Endangered Species Act, a recovery plan is typically prepared to guide efforts to improve the status of the species. In this section we will describe how good diagnoses have been used to design management actions to prevent species from becoming extinct.

KIRTLAND'S WARBLER. The Kirtland's warbler is an endangered bird that nests only in 8–18-year-old stands of jack pine growing on sandy soils in Michigan (Figure 58.16). The current population of Kirtland's warblers is less than 1,000 individuals. Field studies determined that the Kirtland's warbler is at risk from both loss of habitat and nest parasitism by brown-headed cowbirds. Fire suppression has reduced the area of young jack pine stands, and cowbirds, which lay their eggs in other birds' nests, have greatly increased in abundance in the area. To prevent further threats to the warblers, conservation biologists ignite controlled fires in jack pine forests to maintain a steady supply of trees of the right age. They are also removing brown-headed cowbirds to reduce nest parasitism rates.

THE CALIFORNIA SEA OTTER. Populations of the California sea otter were hunted nearly to extinction during the nineteenth century. After receiving legal protection in 1911, the species increased steadily to about 2,400 individuals today. The Southern Sea Otter Recovery Team was charged with developing a recovery plan for the sea otter under the U.S. Endangered Species Act. They determined that ample habitat and food supplies are present and that the otters are reproducing and surviving well. They judged that a major oil spill poses the most serious threat to the otters.

A demographic model suggested that the otter population would be endangered if it dropped below 1,850 individuals. A model designed to assess whether a major oil spill could reduce the population to that size suggested that fewer than 800 otters would be killed by 90 percent of the simulated spills. Therefore, the team set the "delisting" criterion at 2,650 individuals (1,850 + 800). Because the California sea otter population has nearly reached this size, it may soon be removed from the list of endangered species.

Captive propagation has a role in conservation

Species being threatened by overexploitation, loss of habitat, or environmental degradation through pollution can sometimes be maintained in captivity while the external threats to their existence are reduced or removed. Captive propagation is only a temporary measure that buys time. Existing zoos, aquariums, and botanical gardens do not have enough space to maintain adequate populations of more than a small fraction of Earth's rare and endangered species. Nonetheless, captive propagation can play an important role by maintaining species during critical periods and by providing a source of individuals for reintroduction into the wild. Captive propagation projects in zoos also have raised public awareness of species that are threatened with extinction.

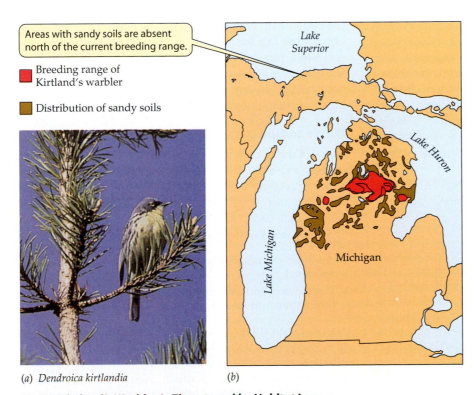

Areas with sandy soils are absent north of the current breeding range.

■ Breeding range of Kirtland's warbler

■ Distribution of sandy soils

Lake Superior

Lake Huron

Lake Michigan

Michigan

(a) *Dendroica kirtlandia* (b)

58.16 Kirtland's Warbler Is Threatened by Habitat Loss
(a) A male Kirtland's warbler in a young jack pine. (b) The warbler's breeding range and the distribution of the sandy soils that support stands of jack pine.

58.17 Peregrine Falcon Populations Have Been Reestablished
(a) Peregrine falcons have responded well to captive propagation. Some individuals have adapted to urban life, nesting on the tall buildings of cities and feeding on pigeons. (b) Throughout the eastern United States, many pairs of peregrine falcons now attempt to reproduce. Most of them are successful.

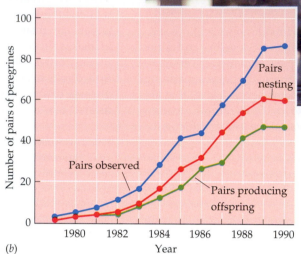

(a) *Falco peregrinus*

(b) *[Graph: Number of pairs of peregrines (y-axis, 0–100) versus Year (x-axis, 1980–1990). Three curves labeled "Pairs nesting," "Pairs observed," and "Pairs producing offspring."]*

THE PEREGRINE FALCON. In 1942, about 350 pairs of peregrine falcons bred in the United States east of the Mississippi River. This breeding population disappeared entirely by 1960. The cause of the falcon's disappearance was the widespread use of organochlorine pesticides, such as DDT and dieldrin. These pesticides degrade very slowly in the environment and become concentrated in the falcon's prey. Their accumulation in the peregrines' bodies interfered with the deposition of calcium in eggshells. As a result, most of the falcons' eggs broke before they hatched.

Much of the eastern United States became suitable habitat for peregrines again after the use of DDT in the United States was banned by federal law. Captive breeding of peregrines began at Cornell University in 1970, and by the end of 1986, more than 850 birds reared in captivity had been released in 13 eastern states, with spectacular success (Figure 58.17). Peregrines probably would have recolonized the East by themselves, but they would have done so much more slowly without human assistance.

THE CALIFORNIA CONDOR. With its 9-foot wing span, the California condor is North America's largest bird. Two hundred years ago, condors ranged from southern British Columbia to northern Mexico, but by 1978, the wild population was plunging toward extinction—only 25 to 30 birds remained in southern California. To save the condor from extinction, biologists initiated a captive propagation program in 1983.

The first chick conceived in captivity hatched in 1988. By 1993, nine captive pairs were producing chicks, and the captive population had increased to more than 60 birds. The captive population was large enough that six captive-bred birds could be released in the mountains north of Los Angeles in 1992. These birds are provided with contaminant-free food in remote areas, and they are using the same roosting sites, bathing pools, and mountain ridges as did their predecessors. Captive-reared birds also were released late in 1996 in northern Arizona. It is still too early to pronounce the program a success, but without captive propagation, the California condor would probably be extinct today.

The cost of captive propagation is comparatively low

The California condor rehabilitation program costs about 1 million dollars a year. The Peregrine Fund at Cornell University spent about 3 million dollars over the past 30 years; the expenses of other cooperating agencies add at least another half million to the total. These amounts may seem large, but they are small compared with the costs of other human activities; for example, even a minor Hollywood film costs more than this to produce, and such films often lose money.

Establishing Priorities for Conservation Efforts

Many species and ecosystems are threatened, but the financial and human resources that can be allocated to preservation efforts are limited. How should those resources be spent to achieve the most conservation benefits? Because many species can survive only in the ecological communities in which they evolved, preserving the full array of ecological communities and habitats is vital.

Where should parks be established?

Parks, sanctuaries, and reserves function to maintain species and ecosystems relatively free of human disturbance. Parks are being created in many countries, but where should they be established to achieve the greatest conservation benefits?

High value sites for parks and reserves are those that

- Are home to unusually large numbers of different species.
- Have many **endemic** species—species that originated in that region and usually are found nowhere else.

Areas of high endemism should receive high conservation priority because if the endemic species are lost there, they often become globally extinct. Madagascar is a good example of such center of endemism: Nearly all the vascular plants and vertebrates of Madagascar are found only on that island (Figure 58.18). Therefore, if the small fragments of tropical and subtropical forests remaining on Madagascar were destroyed, many species would be exterminated.

Andansonia grandieri (giant baobob tree)

Furcifer revocosus
(warty chameleon)

Southern spiny forest

Indri indri (indri)

Enlemur fulvus (brown lemur)

58.18 Madagascar Abounds with Endemic Species
The majority of plant and animal species found on the island of Madagascar, off the eastern coast of Africa, are found nowhere else on Earth.

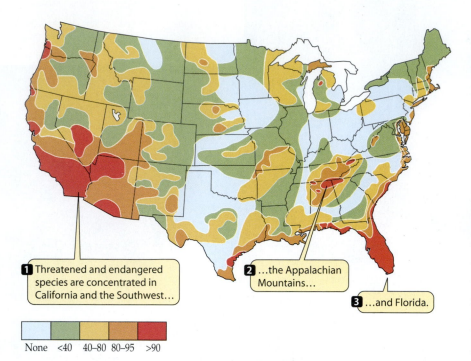

1 Threatened and endangered species are concentrated in California and the Southwest…

2 …the Appalachian Mountains…

3 …and Florida.

| None | <40 | 40–80 | 80–95 | >90 |

58.19 Geographic Distribution of Threatened and Endangered Species in the United States
Threatened and endangered species are concentrated in just a few locations in the continental United States. These areas are obvious priorities for conservation efforts.

cal forest of Belize yielded gross annual revenues of $865 and $4,017, which are greater than the incomes that would result from cultivating squash and corn on those plots.

Conservation requires large-scale planning

Scientists at the World Wildlife Fund have developed a large-scale approach called Ecoregion-Based Conservation (ERBC). An **ecoregion** is an area that has a relatively uniform climate and a biota dominated by a group of widely distributed species. Using all available information on species' distributions and ecological requirements, scientists can identify sites of highest conservation priority within the ecoregion. This information is then used to develop a vision of what successful conservation would look like in 50 years.

An ecoregional conservation project, known as the Yellowstone to Yukon Initiative (Y2Y), is under way in Canada and the northern United States. Y2Y is a binational effort to restore and maintain biological diversity and landscape continuity in an area encompassing 1.2 million square kilometers along the spine of the Rocky Mountains. Information on the distributions of species, human cultures, and soil types in the region is being used to identify the most important areas for maintaining biological diversity. Investigators have also identified threats to biodiversity resulting from mining and its associated toxic waste pollution, unsustainable logging, and conversion of large ranches to suburban developments and sites for summer homes.

An important product of the Y2Y Initiative has been identification of areas that must be preserved or restored as key corridors for the movement of large mammals, such as grizzly bears, wolves, and elk, within the ecosystem. Efforts are now under way to build the base of political and financial support that will be necessary to make the vision a reality.

Centers of endemism are not the same for all groups of organisms. Nevertheless, some areas have high concentrations of endangered species in many taxa. In the United States, species listed as threatened and endangered under the Endangered Species Act are concentrated in California, the Southwest, Florida, Hawaii, and the Appalachian Mountains (Figure 58.19).

Some economic land uses are compatible with conservation

In most countries, new parks and reserves must be established in already settled areas because few pristine areas remain. The people living there cannot be evicted, nor is it appropriate, in most cases, to prevent hungry people from settling in or hunting in parks. The high rates of human population growth in most tropical countries guarantee that pressures on parks from agricultural settlers will increase rather than decrease.

For these reasons, lands that are exploited for food, medicines, and fiber must play an important role in conservation. These lands are far more extensive than parks and reserves, and they include ecosystems not represented in parks. Fortunately, many species can be preserved on lands that are being used for economic purposes. Only a few species, such as predators on humans and domestic animals or large, destructive herbivores, are incompatible with most human uses of land.

Forest reserves in which economically valuable products are harvested can support both species preservation and economic development. In Belize, people known as *hierbateros* collect medicinal plants in the forests and sell them to the *curanderos* (healers) who provide 75 percent of health care in the country. A botanist and an economist determined that two 1-hectare plots in the second-growth tropi-

Restoring Degraded Ecosystems

Many areas that could be incorporated into reserves have been highly altered by human activities. Some of these areas can play their intended roles in biodiversity conservation only if they are restored to their original state. To accomplish this task, a subdiscipline of conservation biology, known as **restoration ecology**, is growing rapidly. Research on methods of restoring populations, communities, and ecosystems is needed because many ecological communities will not recover, or will do so only very slowly, without creative intervention in the recovery process.

The world's largest restoration project is under way in Guanacaste National Park in northwestern Costa Rica. Its goal is to restore a large area of tropical deciduous forest—the most threatened ecosystem in Central America—from small fragments that remain in an area converted mostly to pastures.

The single most important threat to Guanacaste National Park is fires, most of which are started by people. These fires burn the introduced pasture grasses and spread far into surrounding forests. Grazing by domestic livestock lowers the densities of these grasses, and the animals also disperse the seeds of native trees that can invade pastures. Therefore, the restoration program encourages some initial grazing by domestic livestock in the park. When plant succession has progressed to the point where grass no longer poses serious competition to the woody species and is no longer sufficiently dense to carry hot fires, grazing is terminated.

Restoring damaged and degraded habitats is an important activity, but ecologists still have limited ability to restore natural ecosystems. In the United States, the self-serving but false belief that comparable ecosystems can be created somewhere else has made it easy to get building permits for developments that destroy habitats. Developers need only state that they will create substitutes for the ecosystems they are destroying, but promising to do this is much easier than doing so.

Even the most experienced wetland ecologists are having great difficulty creating new wetlands that mimic those being destroyed. Such a "restored" wetland was conceived as part of a compensation agreement that allowed the California Department of Transportation to widen Interstate Highway 5 near San Diego. Despite stringent, court-imposed standards and the involvement of wetland experts, local endangered birds were still not breeding in the "restored" marsh 12 years after it was created. Therefore, as noted by a recent National Research Council committee on wetland restoration: "Wetland restoration should not be used to mitigate avoidable destruction of other wetlands until it can be scientifically demonstrated that the replacement ecosystems are of equal or better functioning."

Markets and Conservation

Most species are common property resources that are "owned" by everyone. Because no individual or group of individuals has strong incentives to use common property resources in a sustainable manner, their preservation usually depends on central governments. Unfortunately, governments generally lack sufficient resources to do the job. Also, governmental actions often are not well attuned to local situations. For these reasons, allowing local people to receive the economic benefits from managing biological resources on their lands can, under proper conditions, assist conservation efforts.

Preserving genes, species, and habitats provides many economic benefits. However, many of these benefits will assist future generations, not the current one. We do not know how future generations will value these benefits or the biodiversity that generates them. We do not even know how the current generation values them. We also cannot predict which species will turn out to be sources of valuable foods, medicines, or drugs.

Between 1951 and 1981, the National Cancer Institute screened extracts from 35,000 different species. To date, only one compound—taxol, derived from the Pacific yew tree—has received regulatory approval as a drug component. That drug is very valuable, but many species had to be searched to find it.

Although it is unlikely that any given species will have market value, extinction is forever. If we purposely or inadvertently exterminate a species, we have irreversibly destroyed a resource of unknown value. Therefore, loss of biodiversity is an especially urgent public issue.

How much should societies invest to preserve biodiversity? This question does not have a scientific answer. Economists and evolutionary biologists can contribute valuable information to the public debate, but the final decision is an ethical and political one that will depend on our beliefs about our responsibilities to the other organisms that share Earth with us.

The preservation of biological diversity and ecosystem services is one of the greatest challenges facing humankind. Many of the scientific tools needed for the task are already available, but appropriate use of these tools requires major changes in people's attitudes toward other species. If species are valued only because they are economically useful, increased losses of species are inevitable. Only when we value biological diversity and ecosystem functioning as the heritage of all humankind, a heritage to be passed on to our descendants as completely as possible, will we begin to reduce the current alarming rates of ecosystem destruction and species extinction.

Chapter Summary

Estimating Current Rates of Extinction

▶ Estimates of current rates of extinction worldwide are based primarily on species–area relationships and rates of tropical deforestation. **Review Figure 58.1**

▶ Rates of extinction are much higher on islands than on the mainland. **Review Figure 58.2**

▶ Demographic and genetic information is used to estimate risks of extinction. **Review Figure 58.3**

Why Do We Care about Species Extinctions?

▶ Diverse species provide the food, fiber, medicines, and aesthetic benefits upon which human life depends.

▶ Ecosystems provide services that can be replaced only by expensive and continuing human effort. **Review Figure 58.5**

Determining Causes of Endangerment and Extinction

▶ Rare species are the most vulnerable to extinction, but common species can also become extinct.

▶ Habitat destruction is the most important cause of species extinction today, but overexploitation, which historically resulted in most human-caused extinctions, is still an important cause of extinctions. **Review Figure 58.7**

▶ The fragmentation of habitats into patches that are too small to support populations is a major cause of extinction.

▶ The proportion of a habitat patch subject to detrimental edge effects increases as patch size decreases. **Review Figure 58.9**

▶ Exotic predators, competitors, and diseases introduced by humans are major causes of extinction. **Review Figure 58.13**

▶ In the future, global warming may be an important cause of extinction. **Review Figure 58.15**

Preventing Species Extinctions

▶ To ensure the recovery of endangered species, human manipulation of their environments and the other species that inhabit them is sometimes needed. **Review Figure 58.16**

▶ Captive propagation plays a useful role in conservation. **Review Figure 58.17**

Establishing Priorities for Conservation Efforts

▶ The best way to maintain populations is to set aside areas in which species and their habitats are protected. High-priority areas for establishing parks and reserves are regions of unusually high species richness and endemism. **Review Figure 58.19**

▶ In most countries, new parks must be created in already settled areas. Management of lands that are exploited for food, medicines, and fiber, must play an important role in conservation.

Restoring Degraded Ecosystems

▶ Restoration is an important component of recovery plans, but restoration of some types of environments, especially aquatic ones, is difficult.

Markets and Conservation

▶ Properly employed, markets can help preserve biodiversity.

▶ The conservation of biodiversity is not just a scientific or economic issue. It raises serious moral and ethical concerns that define what it means to be a human being on Earth.

For Discussion

1. Most species driven to extinction by people in the past were large vertebrates. Do you expect this pattern to persist into the future? If so, why? If not, why not?

2. Species endangered as a result of global warming might be preserved if we could move individuals from areas that are becoming unsuitable for them to those likely to be better for them in the future. What are the major difficulties associated with such interventions? For what types of species might they work well? Poorly? Make no difference?

3. Conservation biologists have debated extensively which is better: many small reserves or a few large ones. What ecological processes should be evaluated in making judgments about size and location of reserves? To what extent should we be concerned with preserving the largest number of species rather than those species judged to be of unusual importance for scientific, aesthetic, or commercial reasons?

4. During World War I, French doctors adopted a "triage" system of dealing with wounded soldiers. The wounded were divided into three categories: those almost certain to die no matter what was done to help them, those likely to recover even if not assisted, and those whose probability of survival was greatly increased if they were given medical attention. The limited resources available to the doctors were directed primarily at the third category. What would be the implications of adopting a similar attitude toward species preservation?

5. Economic arguments dominate discussions about the importance of preserving the biological richness of the planet. In your opinion, what role should moral arguments play?

Appendix:
Some Measurements Used in Biology

QUANTITY	NAME OF UNIT	SYMBOL	DEFINITION
Length	meter (*also* metre)	m	A base unit. 1 m = 100 cm = 39.37 inches
	kilometer	km	1 km = 1000 m = 10^3 m
	centimeter	cm	1 cm = $\frac{1}{100}$ m = 10^{-2} m
	millimeter	mm	1 mm = $\frac{1}{1000}$ m = 10^{-3} m
	micrometer	μm	1 μm = $\frac{1}{1000}$ mm = 10^{-6} m
	nanometer	nm	1 nm = $\frac{1}{1000}$ μm = 10^{-9} m
Area	square meter	m^2	Area encompassed by a square, each side of which is 1 m in length
	hectare	ha	1 ha = 10,000 m^2 = 10^4 m^2 (2.47 acres)
	square centimeter	cm^2	1 cm^2 = $\frac{1}{10,000}$ m^2 = 10^{-4} m^2
Volume	liter (*also* litre)	l	1 l = $\frac{1}{1000}$ m^3 = 10^{-3} m^3 (1.057 qts)
	milliliter	ml	1 ml = $\frac{1}{1000}$ l = 10^{-3} l = 1 cm^3 = 1 cc
	microliter	μl	1 μl = $\frac{1}{1000}$ ml = 10^{-3} ml = 10^{-6} l
Mass	kilogram	kg	A basic unit. 1 kg = 1000 g = 2.20 lbs
	gram	g	1 g = $\frac{1}{1000}$ kg = 10^{-3} kg
	milligram	mg	1 mg = $\frac{1}{1000}$ g = 10^{-3} g = 10^{-6} kg
Time	second	s	A basic unit. 1 s = $\frac{1}{60}$ min
	minute	min	1 min = 60 s
	hour	h	1 h = 60 min = 3,600 s
	day	d	1 d = 24 h = 86,400 s
Temperature	kelvin	K	A basic unit. 0 K = −273.15°C = absolute zero
	degree Celsius	°C	0°C = 273.15 K = melting point of ice
Heat, work	calorie	cal	1 cal = heat necessary to raise 1 gram of pure water from 14.5°C to 15.5°C = 4.184 J
	kilocalorie	kcal	1 kcal = 1000 cal = 10^3 cal = (in nutrition) 1 Calorie
	joule	J	1 J = 0.2389 cal (The joule is now the accepted unit of heat in most sciences.)
Electric potential	volt	V	A unit of potential difference or electromotive force
	millivolt	mV	1 mV = $\frac{1}{1000}$ V = 10^{-3} V

Glossary

Abdomen (ab' duh mun) [L.: belly] • In arthropods, the posterior portion of the body; in mammals, the part of the body containing the intestines and most other internal organs, posterior to the thorax.

Abscisic acid (ab sighs' ik) [L. *abscissio*: breaking off] • A plant growth substance having growth-inhibiting action. Causes stomata to close.

Abscission (ab sizh' un) [L. *abscissio*: breaking off] • The process by which leaves, petals, and fruits separate from a plant.

Absolute temperature scale • Also known as the Kelvin scale. A temperature scale in which zero is the state of no molecular motion. This "absolute zero" is –273° on the Celsius scale.

Absorption • (1) Of light: complete retention, without reflection or transmission. (2) Of liquids: soaking up (taking in through pores or cracks).

Absorption spectrum • A graph of light absorption versus wavelength of light; shows how much light is absorbed at each wavelength.

Abyssal zone (uh biss' ul) [Gr. *abyssos*: bottomless] • That portion of the deep ocean floor where no light penetrates.

Accessory pigments • Pigments that absorb light and transfer energy to chlorophylls for photosynthesis.

Acetylcholine • A neurotransmitter substance that carries information across vertebrate neuromuscular junctions and some other synapses. **Acetylcholinesterase** is an enzyme that breaks down acetylcholine.

Acetyl CoA (acetyl coenzyme A) • Compound that reacts with oxaloacetate to produce citrate at the beginning of the citric acid cycle; a key metabolic intermediate in the formation of many compounds.

Acid [L. *acidus*: sharp, sour] • A substance that can release a proton in solution. (Contrast with base.)

Acid precipitation • Precipitation that has a lower pH than normal as a result of acid-forming precursors introduced into the atmosphere by human activities.

Acidic • Having a pH of less than 7.0 (a hydrogen ion concentration greater than 10^{-7} molar).

Acoelomate • Lacking a coelom.

Acquired Immune Deficiency Syndrome • See AIDS.

Acrosome (a' krow soam) [Gr. *akros*: highest or outermost + *soma*: body] • The structure at the forward tip of an animal sperm which is the first to fuse with the egg membrane and enter the egg cell.

ACTH (adrenocorticotropin) • A pituitary hormone that stimulates the adrenal cortex.

Actin [Gr. *aktis*: a ray] • One of the two major proteins of muscle; it makes up the thin filaments. Forms the microfilaments found in most eukaryotic cells.

Action potential • An impulse in a neuron taking the form of a wave of depolarization or hyperpolarization imposed on a polarized cell surface.

Activating enzymes (also called aminoacyl-tRNA synthetases) • These enzymes catalyze the addition of amino acids to their appropriate tRNAs.

Activation energy (E_a) • The energy barrier that blocks the tendency for a set of chemical substances to react.

Active site • The region on the surface of an enzyme where the substrate binds, and where catalysis occurs.

Active transport • The transport of a substance across a biological membrane against a concentration gradient—that is, from a region of low concentration (of that substance) to a region of high concentration. Active transport requires the expenditure of energy and is a saturable process. (Contrast with facilitated diffusion, free diffusion; see primary active transport, secondary active transport.)

Adaptation (a dap tay' shun) • In evolutionary biology, a particular structure, physiological process, or behavior that makes an organism better able to survive and reproduce. Also, the evolutionary process that leads to the development or persistence of such a trait.

Adenine (a' den een) • A nitrogen-containing base found in nucleic acids, ATP, NAD, etc.

Adenosine triphosphate • See ATP.

Adenylate cyclase • Enzyme catalyzing the formation of cyclic AMP from ATP.

Adrenal (a dree' nal) [L. *ad-*: toward + *renes*: kidneys] • An endocrine gland located near the kidneys of vertebrates, consisting of two glandular parts, the cortex and medulla.

Adrenaline • See epinephrine.

Adrenocorticotropin • See ACTH.

Adsorption • Binding of a gas or a solute to the surface of a solid.

Aerobic (air oh' bic) [Gr. *aer*: air + *bios*: life] • In the presence of oxygen, or requiring oxygen.

Afferent (af' ur unt) [L. *ad*: to + *ferre*: to bear] • To or toward, as in a neuron that carries impulses to the central nervous system, or a blood vessel that carries blood to a structure. (Contrast with efferents.)

AIDS (acquired immune deficiency syndrome) • Condition caused by a virus (HIV) in which the body's helper T lymphocytes are reduced, leaving the victim subject to opportunistic diseases.

Aldehyde (al' duh hide) • A compound with a –CHO functional group. Many sugars are aldehydes. (Contrast with ketone.)

Aldosterone (al dahs' ter own) • A steroid hormone produced in the adrenal cortex of mammals. Promotes secretion of potassium and reabsorption of sodium in the kidney.

Alga (al' gah) (plural: algae) [L.: seaweed] • Any one of a wide diversity of protists belonging to the phyla Pyrrophyta, Chrysophyta, Phaeophyta, Rhodophyta, and Chlorophyta.

Allele (a leel') [Gr. *allos*: other] • The alternate forms of a genetic character found at a given locus on a chromosome.

Allele frequency • The relative proportion of a particular allele in a specific population.

Allergy [Ger. *allergie*: altered reaction] • An overreaction to an antigen in amounts that do not affect most people; often involves IgE antibodies.

Allometric growth • A pattern of growth in which some parts of the body of an organism grow faster than others, resulting in a change in body proportions as the organism grows.

Allopatric speciation (al' lo pat' rick) [Gr. *allos*: other + *patria*: fatherland] • Also called geographical speciation, this is the formation of two species from one when reproductive isolation occurs because of the the interposition of (or crossing of) a physical geographic barrier such as a river. (Contrast with parapatric speciation, sympatric speciation.)

Allopolyploid • A polyploid in which the chromosome sets are derived from more than one species.

Allostery (al' lo steer' y) [Gr. *allos*: other + *stereos*: structure] • Regulation of the activity of a protein by the binding of an effector molecule at a site other than the active site.

Alpha helix • Type of protein secondary structure; a right-handed spiral.

Alternation of generations • The succession of haploid and diploid phases in some sexually reproducing organisms, notably plants.

Altruistism • A behavior whose performance harms the actor but benefits other individuals.

Alveolus (al ve' o lus) (plural: alveoli) [L. *alveus*: cavity] • A small, baglike cavity, especially the blind sacs of the lung.

Amensalism (a men' sul ism) • Interaction in which one animal is harmed and the other is unaffected. (Contrast with commensalism, mutualism.)

Amine • An organic compound with an amino group (see Amino acid).

Amino acid • An organic compound of the general formula $H_2N–CHR–COOH$, where R can be one of 20 or more different side groups. An amino acid is so named because it has both a basic amine group, $–NH_2$, and an acidic carboxyl group, $–COOH$. Proteins are polymers of amino acids.

Ammonotelic (am moan' o teel' ic) [Gr. *telos*: end] • Describes an organism in which the final product of breakdown of nitrogen-containing compounds (primarily proteins) is ammonia. (Contrast with ureotelic, uricotelic.)

Amniocentesis • A medical procedure in which cells from the fetus are obtained from the amniotic fluid. The genetic material of the cells is then examined. (Contrast with chorionic villus sampling.)

Amniote • An organism that lays eggs that can be incubated in air (externally) because the embryo is enclosed by a fluid-filled sac. Birds and reptiles are amniotes.

Amphipathic (am' fi path' ic) [Gr. *amphi*: both + *pathos*: emotion] • Of a molecule, having both hydrophilic and hydrophobic regions.

Amylase (am' ill ase) • Any of a group of enzymes that digest starch.

Anabolism (an ab' uh liz' em) [Gr. *ana*: up, throughout + *ballein*: to throw] • Synthetic reactions of metabolism, in which complex molecules are formed from simpler ones. (Contrast with catabolism.)

Anaerobic (an ur row' bic) [Gr. *an*: not + *aer*: air + *bios*: life] • Occurring without the use of molecular oxygen, O_2.

Anagenesis • Evolutionary change in a single lineage over time.

Analogy (a nal' o jee) [Gr. *analogia*: resembling] • A resemblance in function, and often appearance as well, between two structures which is due to convergence in evolution rather than to common ancestry. (Contrast with homology.)

Anaphase (an' a phase) [Gr. *ana*: indicating upward progress] • The stage in nuclear division at which the first separation of sister chromatids (or, in the first meiotic division, of paired homologues) occurs. Anaphase lasts from the moment of first separation to the time at which the moving chromosomes converge at the poles of the spindle.

Anaphylactic shock • A precipitous drop in blood pressure caused by loss of fluid from capillaries because of an increase in their permeability stimulated by an allergic reaction.

Ancestral trait • Trait shared by a group of organisms as a result of descent from a common ancestor.

Androgens (an' dro jens) • The male sex steroids.

Aneuploidy (an' you ploy dee) • A condition in which one or more chromosomes or pieces of chromosomes are either lacking or present in excess.

Angiosperm (an' jee oh spurm) [Gr. *angion*: vessel + *sperma*: seed] • One of the flowering plants; literally, one whose seed is carried in a "vessel," which is the fruit. (See fruit.)

Angiotensin (an' jee oh ten' sin) • A peptide hormone that raises blood pressure by causing peripheral vessels to constrict; maintains glomerular filtration by constricting efferent glomerular vessels; stimulates thirst; and stimulates the release of aldosterone.

Animal [L. *animus*: breath, soul] • A member of the kingdom Animalia. In general, a multicellular eukaryote that obtains its food by ingestion.

Animal hemisphere • The metabolically active upper portion of some animal eggs, zygotes, and embryos, which does *not* contain the dense nutrient yolk. The **animal pole** refers to the very top of the egg or embyro. (Contrast with vegetal hemisphere.)

Anion (an' eye one) • An ion with one or more negative charges. (Contrast with cation.)

Anisogamy (an' eye sog' a mee) [Gr. *aniso*: unequal + *gamos*: marriage] • The existence of two dissimilar gametes (egg and sperm).

Annual • Referring to a plant whose life cycle is completed in one growing season. (Contrast with biennial, perennial.)

Anterior pituitary • The portion of the vertebrate pituitary gland that derives from gut epithelium and produces tropic hormones.

Anther (an' thur) [Gr. *anthos*: flower] • A pollen-bearing portion of the stamen of a flower.

Antheridium (an' thur id' ee um) (plural: antheridia) [Gr. *antheros*: blooming] • The multicellular structure that produces the sperm in bryophytes and ferns.

Antibody • One of millions of proteins, produced by the immune system, that specifically recognizes a foreign substance and initiates its removal from the body.

Anticodon • A "triplet" of three nucleotides in transfer RNA that is able to pair with a complementary triplet (a codon) in messenger RNA, thus aligning the transfer RNA on the proper place on the messenger. The codon (and, reciprocally, the anticodon) codes for a specific amino acid.

Antidiuretic hormone • A hormone that controls water reabsorption in the mammalian kidney. Also called vasopressin.

Antigen (an' ti jun) • Any substance that stimulates the production of an antibody or antibodies in the body of a vertebrate.

Antigen processing • The breakdown of antigenic proteins into smaller fragments, which are then presented on the cell surface, along with MHC proteins, to T cells.

Antigenic determinant • A specific region of an antigen, which is recognized by and binds to a specific antibody.

Antiport • A membrane transport process that carries one substance in one direction and another in the opposite direction. (Contrast with symport.)

Antisense nucleic acid • A single-stranded RNA or DNA complementary to and thus targeted against the mRNA transcribed from a harmful gene such as an oncogene.

Anus (a' nus) • Opening through which digestive wastes are expelled, located at the posterior end of the gut.

Aorta (a or' tuh) [Gr. *aorte*: aorta] • The main trunk of the arteries leading to the systemic (as opposed to the pulmonary) circulation.

Apex (a' pecks) • The tip or highest point of a structure, as the apex of a growing stem or root.

Apical (a' pi kul) • Pertaining to the apex, or tip, usually in reference to plants.

Apical dominance • Inhibition by the apical bud of the growth of axillary buds.

Apical meristem • The meristem at the tip of a shoot or root; responsible for the plant's primary growth.

Apomixis (ap oh mix' is) [Gr. *apo*: away from + *mixis*: sexual intercourse] • The asexual production of seeds.

Apoplast (ap' oh plast) • in plants, the continuous meshwork of cell walls and extracellular spaces through which material can pass without crossing a plasma membrane. (Contrast with symplast.)

Apoptosis (ay' pu toh sis) • A series of genetically programmed events leading to cell death.

Aquaporin • A transport protein in plant and animals cells through which water passes in osmosis.

Archegonium (ar' ke go' nee um) [Gr. *archegonos*: first of a kind] • The multicellular structure that produces eggs in bryophytes, ferns, and gymnosperms.

Archenteron (ark en' ter on) [Gr. *archos*: beginning + *enteron*: bowel] • The earliest primordial animal digestive tract.

Arteriosclerosis • See atherosclerosis.

Artery • A muscular blood vessel carrying oxygenated blood away from the heart to other parts of the body. (Contrast with vein.)

Ascus (ass' cuss) [Gr. *askos*: bladder] • In fungi belonging to the phylum Ascomycota (the sac fungi), the club-shaped sporangium within which spores (ascospores) are produced by meiosis.

Asexual • Without sex.

Assortative mating • A breeding system in which mates are selected on the basis of a particular trait or group of traits.

Atherosclerosis (ath' er oh sklair oh' sis) • A disease of the lining of the arteries characterized by fatty, cholesterol-rich deposits in the walls of the arteries. When fibroblasts infiltrate these deposits and calcium precipitates in them, the disease become arteriosclerosis, or "hardening of the arteries."

Atmosphere • The gaseous mass surrounding our planet. Also: a unit of pressure, equal to the normal pressure of air at sea level.

Atom [Gr. *atomos*: indivisible] • The smallest unit of a chemical element. Consists of a nucleus and one or more electrons.

Atomic mass (also called atomic weight) • The average mass of an atom of an element on the amu scale. (The average depends upon the relative amounts of different isotopes of an element on Earth.)

Atomic number • The number of protons in the nucleus of an atom, also equal to the number of electrons around the neutral atom. Determines the chemical properties of the atom.

ATP (adenosine triphosphate) • A compound containing adenine, ribose, and three phosphate groups. When it is formed, useful energy is stored; when it is broken down (to ADP or AMP), energy is released to drive endergonic reactions. ATP is an energy storage compound.

ATP synthase • An integral membrane protein that couples the transport of protons with the formation of ATP.

Atrium (a' tree um) • A body cavity, as in the hearts of vertebrates. The thin-walled chamber(s) entered by blood on its way to the ventricle(s). Also, the outer ear.

Autoimmune disease • A disorder in which the immune system attacks the animal's own antigens.

Autonomic nervous system • The system (which in vertebrates comprises sympathetic and parasympathetic subsystems) that controls such involuntary functions as those of guts and glands.

Autosome • Any chromosome (in a eukaryote) other than a sex chromosome.

Autotroph (au' tow trow' fik) [Gr. *autos*: self + *trophe*: food] • An organism that is capable of living exclusively on inorganic materials, water, and some energy source such as sunlight or chemically reduced matter. (Contrast with heterotroph.)

Auxin (awk' sin) [Gr. *auxein*: increase] • In plants, a substance (indoleacetic acid) that regulates growth and various aspects of development.

Auxotroph (awks' o trofe) [Gr. *auxanein*: to grow + *trophe*: food] • A mutant form of an organism that requires a nutrient or nutrients not required by the wild type, or reference, form of the organism. (Contrast with prototroph.)

Axon [Gr.: axle] • Fiber of a neuron which can carry action potentials. Carries impulses away from the cell body of the neuron; releases a neurotransmitter substance.

Axon hillock • The junction between an axon and its cell body; where action potentials are generated.

Axon terminals • The endings of an axon; they form synapses and release neurotransmitter.

Axoneme (ax' oh neem) • The complex of microtubules and their crossbridges that forms the motile apparatus of a cilium.

Bacillus (buh sil' us) [L.: little rod] • Any of various rod-shaped bacteria.

Bacteriophage (bak teer' ee o fayj) [Gr. *bakterion*: little rod + *phagein*: to eat] • One of a group of viruses that infect bacteria and ultimately cause their disintegration.

Bacteria (bak teer' ee ah) (singular: bacterium) [Gr. *bakterion*: little rod] • Prokaryote in the Domain Bacteria. The chromosomes of bacteria are not contained in nuclear envelopes.

Balanced polymorphism [Gr. *polymorphos*: having many forms] • The maintenance of more than one form, or the maintenance at a given locus of more than one allele, at frequencies of greater than one percent in a population. Often results when heterozygotes are superior to both homozygotes.

Bark • All tissues outside the vascular cambium of a plant.

Baroreceptor [Gr. *baros*: weight] • A pressure-sensing cell or organ.

Barr body • In mammals, an inactivated X chromosome.

Basal body • Centriole found at the base of a eukaryotic flagellum or cilium.

Basal metabolic rate • The minimum rate of energy turnover in an awake (but resting) bird or mammal that is not expending energy for thermoregulation.

Base • (1) A substance which can accept a proton (hydrogen ion; H$^+$) in solution. (Contrast with acid.) (2) In nucleic acids, a nitrogen-containing molecule that is attached to each sugar in the backbone. (See purine; pyrimidine.)

Base pairing • See complementary base pairing.

Basic • having a pH greater than 7.0 (having a hydrogen ion concentration lower than 10^{-7} molar).

Basidium (bass id' ee yum) • In fungi of the class Basidiomycetes, the characteristic sporangium in which four spores are formed by meiosis and then borne externally before being shed.

Batesian mimicry • Mimicry by a relatively harmless kind of organism of a more dangerous one, by which the mimic enjoys protection from predators that mistake it for the dangerous model. (Contrast with Müllerian mimicry.)

B cell • A type of lymphocyte involved in the humoral immune response of vertebrates. Upon recognizing an antigenic determinant, a B cell develops into a plasma cell, which secretes an antibody. (Contrast with a T cell.)

Benefit • An improvement in survival and reproductive success resulting from a behavior. (Contrast with cost.)

Benign (be nine') • A tumor that grows to a certain size and then stops, uaually with a fibrous capsule surrounding the mass of cells. Benign tumors do not spread (metastasize) to other organs.

Benthic zone [Gr. *benthos*: bottom of the sea] • The bottom of the ocean. (Contrast with pelagic zone.)

Beta-pleated sheet • Type of protein secondary structure; results from hydrogen bonding between polypeptide regions running antiparallel to each other.

Biennial • Referring to a plant whose life cycle includes vegetative growth in the first year and flowering and senescence in the second year. (Contrast with annual, perennial.)

Bilateral symmetry • The condition in which only the right and left sides of an organism, divided exactly down the back, are mirror images of each other. (Contrast with biradial symmetry.)

Bile • A secretion of the liver delivered to the small intestine via the common bile duct. In the intestine, bile emulsifies fats.

Binocular cells • Neurons in the visual cortex that respond to input from both retinas; involved in depth perception.

Binomial (bye nome' ee al) • Consisting of two names; for example, the binomial nomenclature of biology which gives the name of the genus followed by the name of the species.

Biodiversity crisis • The current high rate of loss of species, caused primarily by human activities.

Biogeochemical cycles • Movement of elements through living organisms and the physical environment.

Biogeography • The scientific study of the geographic distribution of organisms.

Biogeographic region • A continental-scale part of Earth that has a biota distinct from that of other such regions.

Biological species concept • The view that a species is most usefully defined as a population or series of populations within which there is a significant amount of gene flow under natural conditions, but which is genetically isolated from other populations.

Bioluminescence • The production of light by biochemical processes in an organism.

Biomass • The total weight of all the living organisms, or some designated group of living organisms, in a given area.

Biome (bye' ome) • A major division of the ecological communities of Earth; characterized by distinctive vegetation.

Biota (bye oh' tah) • All of the organisms, including animals, plants, fungi, and microorganisms, found in a given area.

Biotechnology • The use of cells to make medicines, foods and other products useful to humans.

Biradial symmetry • Radial symmetry modified so that only two planes can divide the animal into similar halves.

Blastocoel (blass' toe seal) [Br. *blastos*: sprout + *koilos*: hollow] • The central, hollow cavity of a blastula.

Blastodisc (blass' toe disk) • A disk of cells forming on the surface of a large yolk mass, comparable to a blastula, but occurring in animals such as birds and reptiles, in which the massive yolk restricts cleavage to one side of the egg only.

Blastomere • A cell produced by the division of a fertilized egg.

Blastopore • The opening from the archenteron to the exterior of a gastrula.

Blastula (blass' chu luh) [Gr. *blastos*: sprout] • An early stage in animal embryology; in many species, a hollow sphere of cells surrounding a central cavity, the blastocoel. (Contrast with blastodisc.)

Blood–brain barrier • A property of the blood vessels of the brain that prevents most chemicals from diffusing from the blood into the brain.

Body plan • A basic structural design that includes an entire animal, its organ systems, and the integrated functioning of its parts. Phylogenetic groups of organisms are classified in part on the basis of a shared body plan.

Bowman's capsule • An elaboration of kidney tubule cells that surrounds a know of capillaries (the glomerulus). Blood is filtered across the walls of these capillaries and the filtrate is collected into Bowman's capsule.

Brain stem • The portion of the vertebrate brain between the spinal cord and the forebrain.

Brassinosteroids • Plant steroid hormones that promote the elongation of stems and pollen tubes.

Bronchus (plural: bronchi) • The major airway(s) branching off the trachea into the vertebrate lung.

Brown fat • Fat tissue in mammals that is specialized to produce heat. It has many mitochondria and capillaries, and a protein that uncouples oxidative phosphorylation.

Browser • An animal that feeds on the tissues of woody plants.

Bryophyte (bri' uh fite') [Gr. *bruon*: moss + *phyton*: plant] • A moss. Formerly was often used to refer to all the nontracheophyte plants.

Budding • Asexual reproduction in which a more or less complete new organism simply grows from the body of the parent organism and eventually detaches itself.

Buffering • A process by which a system resists change—particularly in pH, in which case added acid or base is partially converted to another form.

C₃ photosynthesis • The form of photosynthesis in which 3-phosphoglycerate is the first stable product, and ribulose bisphosphate is the CO_2 receptor.

C₄ photosynthesis • The form of photosynthesis in which oxaloacetate is the first stable product, and phosphoenolpyruvate is the CO_2 acceptor. C_4 plants also perform the reactions of C_3 photosynthesis.

Calcitonin • A hormone produced by the thyroid gland; it lowers blood calcium and promotes bone formation. (Contrast with parathormone.)

Calmodulin (cal mod' joo lin) • A calcium-binding protein found in all animal and plant cells; mediates many calcium-regulated processes.

calorie [L. *calor*: heat] • The amount of heat required to raise the temperature of one gram of water by one degree Celsius (1°C) from 14.5°C to 15.5°C. In nutrition studies, "Calorie" (spelled with a capital C) refers to the kilocalorie (1 kcal = 1,000 cal).

Calvin–Benson cycle • The stage of photosynthesis in which CO_2 reacts with RuBP to form 3PG, 3PG is reduced to a sugar, and RuBP is regenerated, while other products are released to the rest of the plant.

Calyx (kay' licks) [Gr. *kalyx*: cup] • All of the sepals of a flower, collectively.

CAM • See crassulacean acid metabolism.

Cambium (kam' bee um) [L. *cambiare*: to exchange] • A meristem that gives rise to radial rows of cells in stem and root, increasing them in girth; commonly applied to the vascular cambium which produces wood and phloem, and the cork cambium, which produces bark.

cAMP (cyclic AMP) • A compound, formed from ATP, that mediates the effects of numerous animal hormones. Also needed for the transcription of catabolite-repressible operons in bacteria. Used for communication by cellular slime molds.

Canopy • The leaf-bearing part of a tree. Collectively the aggregate of the leaves and branches of the larger woody plants of an ecological community.

Capillaries [L. *capillaris*: hair] • Very small tubes, especially the smallest blood-carrying vessels of animals between the termination of the arteries and the beginnings of the veins.

Capsid • The protein coat of a virus.

Carbohydrates • Organic compounds with the general formula $C_nH_{2m}O_m$. Common examples are sugars, starch, and cellulose.

Carboxylic acid (kar box sill' ik) • An organic acid containing the carboxyl group, –COOH, which dissociates to the carboxylate ion, –COO⁻.

Carcinogen (car sin' oh jen) • A substance that causes cancer.

Cardiac (kar' dee ak) [Gr. *kardia*: heart] • Pertaining to the heart and its functions.

Carnivore [L. *carn*: flesh + *vovare*: to devour] • An organism that feeds on animal tissue. (Contrast with detritivore, herbivore, omnivore.)

Carotenoid (ka rah' tuh noid) [L. *carota*: carrot] • A yellow, orange, or red lipid pigment commonly found as an accessory pigment in photosynthesis; also found in fungi.

Carpel (kar' pel) [Gr. *karpos*: fruit] • The organ of the flower that contains one or more ovules.

Carrier • (1) In facilitated diffusion, a membrane protein that binds a specific molecule and transports it through the membrane. (2) In respiratory and photosynthetic electron transport, a participating substance such as NAD that exists in both oxidized and reduced forms. (3) In genetics, a person heterozygous for a recessive trait.

Carrying capacity • In ecology, the largest number of organisms of a particular species that can be maintained indefinitely in a given part of the environment.

Cartilage • In vertebrates, a tough connective tissue found in joints, the outer ear, and elsewhere. Forms the entire skeleton in some animal groups.

Casparian strip • A band of cell wall containing suberin and lignin, found in the endodermis. Restricts the movement of water across the endodermis.

Catabolism [Ge. *kata*: down + *ballein*: to throw] • Degradational reactions of metabolism, in which complex molecules are broken down. (Contrast with anabolism.)

Catalyst (cat' a list) [Gr. *kata-*, implying the breaking down of a compound] • A chemical substance that accelerates a reaction without itself being consumed in the overall course of the reaction. Catalysts lower the activation energy of a reaction. Enzymes are biological catalysts.

Cation (cat' eye on) • An ion with one or more positive charges. (Contrast with anion.)

Caudal [L. *cauda*: tail] • Pertaining to the tail, or to the posterior part of the body.

cDNA • See complementary DNA.

Cecum (see' cum) [L. *caecus*: blind] • A blind branch off the large intestine. In many nonruminant mammals, the cecum contains a colony of microorganisms that contribute to the digestion of food.

Cell adhesion molecules • Molecules on animal cell surfaces that affect the selective association of cells during development of the embryo.

Cell cycle • The stages through which a cell passes between one division and the next. Includes all stages of interphase and mitosis.

Cell division • The reproduction of a cell to produce two new cells. In eukaryotes, this process involves nuclear division (mitosis) and cytoplasmic division (cytokinesis).

Cell theory • The theory, well established, that organisms consist of cells, and that all cells come from preexisting cells.

Cell wall • A relatively rigid structure that encloses cells of plants, fungi, many protists, and most bacteria. The cell wall gives these cells their shape and limits their expansion in hypotonic media.

Cellular immune system • That part of the immune system that is based on the activities of T cells. Directed against parasites, fungi, intracellular viruses, and foreign tissues (grafts). (Contrast with humoral immune system.)

Cellular respiration • See respiration.

Cellulose (sell' you lowss) • A straight-chain polymer of glucose molecules, used by plants as a structural supporting material.

Central dogma • The statement that information flows from DNA to RNA to polypeptide (in retroviruses, there is also information flow from RNA to cDNA).

Central nervous system • That part of the nervous system which is condensed and centrally located, e.g., the brain and spinal cord of vertebrates; the chain of cerebral, thoracic and abdominal ganglia of arthropods.

Centrifuge [L. *fugere*: to flee] • A device in which a sample can be spun around a central axis at high speed, creating a centrifugal force that mimics a very strong gravitational force. Used to separate mixtures of suspended materials.

Centriole (sen' tree ole) • A paired organelle that helps organize the microtubules in animal and protist cells during nuclear division.

Centromere (sen' tro meer) [Gr. *centron*: center + *meros*: part] • The region where sister chromatids join.

Centrosome (sen' tro soam) • The major microtubule organizing center of an animal cell.

Cephalization (sef' uh luh zay' shun) [Gr. *kephale*: head] • The evolutionary trend toward increasing concentration of brain and sensory organs at the anterior end of the animal.

Cerebellum (sair' uh bell' um) [L.: diminutive of *cerebrum*: brain] • The brain region that controls muscular coordination; located at the anterior end of the hindbrain.

Cerebral cortex • The thin layer of gray matter (neuronal cell bodies) that overlays the cerebrum.

Cerebrum (su ree' brum) [L.: brain] • The dorsal anterior portion of the forebrain, making up the largest part of the brain of mammals. In mammals, the chief coordination center of the nervous system; consists of two **cerebral hemispheres**.

Cervix (sir' vix) [L.: neck] • The opening of the uterus into the vagina.

cGMP (cyclic guanosine monophosphate) • An intracellular messenger that is part of signal transmission pathways involving G proteins. (See G protein.)

Channel • A membrane protein that forms an aqueous passageway though which specific solutes may pass by simple diffusion; some channels are gated: they open and close in response to binding of specific molecules.

Chaperone protein • A protein that assists a newly forming protein in adopting its appropriate tertiary structure.

Chemical bond • An attractive force stably linking two atoms.

Chemiosmotic mechanism • The formation of ATP in mitochondria and chloroplasts, resulting from a pumping of protons across a membrane (against a gradient of electrical charge and of pH), followed by the return of the protons through a protein channel with ATPase activity.

Chemoautotroph • An organism that uses carbon dioxide as a carbon source and obtains energy by oxidizing inorganic substances from its environment. (Contrast with chemoheterotroph, photoautotroph, photoheterotroph.)

Chemoheterotroph • An organism that must obtain both carbon and energy from organic substances. (Contrast with chemoautotroph, photoautotroph, photoheterotroph.)

Chemoreceptor • A cell or tissue that senses specific substances in its environment.

Chemosynthesis • Synthesis of food substances, using the oxidation of reduced materials from the environment as a source of energy.

Chiasma (kie az' muh) (plural: chiasmata) [Gr.: cross] • An X-shaped connection between paired homologous chromosomes in prophase I of meiosis. A chiasma is the visible manifestation of crossing over between homologous chromosomes.

Chitin (kye' tin) [Gr. *kiton*: tunic] • The characteristic tough but flexible organic component of the exoskeleton of arthropods, consisting of a complex, nitrogen-containing polysaccharide. Also found in cell walls of fungi.

Chlorophyll (klor' o fill) [Gr. *kloros*: green + *phyllon*: leaf] • Any of a few green pigments associated with chloroplasts or with certain bacterial membranes; responsible for trapping light energy for photosynthesis.

Chloroplast [Gr. *kloros*: green + *plast*: a particle] • An organelle bounded by a double membrane containing the enzymes and pigments that perform photosynthesis. Chloroplasts occur only in eukaryotes.

Choanocyte (cho' an oh cite) • The collared, flagellated feeding cells of sponges.

Cholecystokinin (ko' lee sis to kai nin) • A hormone produced and released by the lining of the duodenum when it is stimulated by undigested fats and proteins. It stimulates the gallbladder to release bile and slows stomach activity.

Chorion (kor' ee on) [Gr. *khorion*: afterbirth] • The outermost of the membranes protecting mammal, bird, and reptile embryos; in mammals it forms part of the placenta.

Chorionic villus sampling • A medical procedure that extracts a portion of the chorion from a pregnant woman to enable genetic and biochemical analysis of the embryo. (Contrast with amniocentesis.)

Chromatid (kro' ma tid) • Each of a pair of new sister chromosomes from the time at which the molecular duplication occurs until the time at which the centromeres separate at the anaphase of nuclear division.

Chromatin • The nucleic acid–protein complex found in eukaryotic chromosomes.

Chromatophore (krow mat' o for) [Gr. *kroma*: color + *phoreus*: carrier] • A pigment-bearing cell that expands or contracts to change the color of the organism.

Chromosome (krome' o sowm) [Gr. *kroma*: color + *soma*: body] • In bacteria and viruses, the DNA molecule that contains most or all of the genetic information of the cell or virus. In eukaryotes, a structure composed of DNA and proteins that bears part of the genetic information of the cell.

Chylomicron (ky low my' cron) • Particles of lipid coated with protein, produced in the gut from dietary fats and secreted into the extracellular fluids.

Chyme (kime) [Gr. *kymus*, juice] • Created in the stomach; a mixture of ingested food with the digestive juices secreted by the salivary glands and the stomach lining.

Cilium (sil' ee um) (plural: cilia) [L. *cilium*: eyelash] • Hairlike organelle used for locomotion by many unicellular organisms and for moving water and mucus by many multicellular organisms. Generally shorter than a flagellum.

Circadian rhythm (sir kade' ee an) [L. *circa*: approximately + *dies*: day] • A rhythm in behavior, growth, or some other activity that recurs about every 24 hours under constant conditions.

Circannual rhythm (sir can' you al) [L. *circa*: approximately + *annus*: year) • A rhythm of behavior, growth, or some other activity that recurs on a yearly basis.

Citric acid cycle • A set of chemical reactions in cellular respiration, in which acetyl CoA reacts with oxaloacetate to form citric acid, and oxaloacetate is regenerated. Acetyl CoA is oxidized to carbon dioxide, and hydrogen atoms are stored as NADH and $FADH_2$. Also called the Krebs cycle.

Class • In taxonomy, the category below the phylum and above the order; a group of related, similar orders.

Class I MHC molecules • These cell surface proteins participate in the cellular immune response directed against virus-infected cells.

Class II MHC molecules • These cell surface proteins participate in the cell-cell interactions (of helper T cells, macrophages, and B cells) of the humoral immune response.

Class switching • The process whereby a plasma cell changes the class of immunoglobulin that it synthesizes. This results from the deletion of part of the constant region of DNA, bringing in a new C segment. The variable region is the same as before, so that the new immunoglbulin has the same antigenic specificity.

Clathrin • A fibrous protein on the inner surfaces of animal cell membranes that strengthens coated vesicles and thus participates in receptor-mediated endocytosis.

Clay • A soil constituent comprising particles smaller than 2 micrometers in diameter.

Cleavages • First divisions of the fertilized egg of an animal.

Cline • A gradual change in the traits of a species over a geographical gradient.

Cloaca (klo ay' kuh) [L. *cloaca*: sewer] • In some invertebrates, the posterior part of the gut; in many vertebrates, a cavity receiving material from the digestive, reproductive, and excretory systems.

Clonal anergy • When a naive T cell encounters a self-antigen, the T cell may bind to the antigen but does not receive signals from an antigen-presenting cell. Instead of being activated, the T cell dies (becomes anergic). In this way, we avoid reacting to our own tissue-specific antigens.

Clonal deletion • In immunology, the inactivation or destruction of lymphocyte clones that would produce immune reactions against the animal's own body.

Clonal selection • The mechanism by which exposure to antigen results in the activation of selected T- or B-cell clones, resulting in an immune response.

Clone [Gr. *klon*: twig, shoot] • Genetically identical cells or organisms produced from a common ancestor by asexual means.

Cnidocytes • The feeding cells of cnidarians, within which nematocysts are housed.

Coacervate (ko as' er vate) [L. *coacervare*: to heap up] • An aggregate of colloidal particles in suspension.

Coacervate drop • Drops formed when a mixture of large proteins and polysaccharides is shaken in water. The interiors of these drops, which are often very stable, contain most of the proteins and polysaccharides.

Coated vesicle • Vesicle, sometimes formed from a coated pit, with characteristic "bristly" surface; its membrane contains distinctive proteins, including clathrin.

Coccus (kock' us) [Gr. *kokkos*: berry, pit] • Any of various spherical or spheroidal bacteria.

Cochlea (kock' lee uh) [Gr. *kokhlos*: a land snail] • A spiral tube in the inner ear of vertebrates; it contains the sensory cells involved in hearing.

Codominance • A condition in which two alleles at a locus produce different phenotypic effects and both effects appear in heterozygotes.

Codon • A "triplet" of three nucleotides in messenger RNA that directs the placement of a particular amino acid into a polypeptide chain. (Contrast with anticodon.)

Coefficient of relatedness • The probability that an allele in one individual is an identical copy, by descent, of an allele in another individual.

Coelom (see' lum) [Gr. *koiloma*: cavity] • The body cavity of certain animals, which is lined with cells of mesodermal origin.

Coelomate • Having a coelom.

Coenocyte (seen' a sight) [Gr.: common cell] • A "cell" bounded by a single plasma membrane, but containing many nuclei.

Coenzyme • A nonprotein molecule that plays a role in catalysis by an enzyme. The coenzyme may be part of the enzyme molecule or free in solution. Some coenzymes are oxidizing or reducing agents.

Coevolution • Concurrent evolution of two or more species that are mutually affecting each other's evolution.

Cohort (co' hort) [L. *cohors*: company of soldiers] • A group of similar-age organisms, considered as it passes through time.

Collagen [Gr. *kolla*: glue] • A fibrous protein found extensively in bone and connective tissue.

Collecting duct • In vertebrates, a tubule that receives urine produced in the nephrons of the kidney and delivers that fluid to the ureter for excretion.

Collenchyma (cull eng' kyma) [Gr. *kolla*: glue + *enchyma*: infusion] • A type of plant cell, living at functional maturity, which lends flexible support by virtue of primary cell walls thickened at the corners. (Contrast with parenchyma, sclerenchyma.)

Colon [Gr. *kolon*: large intestine] • The large intestine.

Commensalism • The form of symbiosis in which one species benefits from the association, while the other is neither harmed nor benefited.

Common bile duct • A single duct that delivers bile from the gallbladder and secretions from the pancreas into the small intestine.

Communication • A signal from one organism (or cell) that alters the pattern of behavior in another organism (or cell) in an adaptive fashion.

Community • Any ecologically integrated group of species of microorganisms, plants, and animals inhabiting a given area.

Companion cell • Specialized cell found adjacent to a sieve tube member in flowering plants.

Comparative analysis • An approach to studying evolution in which hypotheses are tested by measuring the distribution of states among a large number of species.

Comparative genomics • Computer-aided comparison of DNA sequences between different organisms to reveal genes with related functions.

Compensation point • The light intensity at which the rates of photosynthesis and of cellular respiration are equal.

Competitive inhibitor • A substance, similar in structure to an enzyme's substrate, that binds the active site and thus inhibits a reaction.

Competition • In ecology, use of the same resource by two or more species, when the resource is present in insufficient supply for the combined needs of the species.

Competitive exclusion • A result of competition between species for a limiting resource in which one species completely eliminates the other.

Competitive inhibitor • A substance, similar in structure to an enzyme's substrate, that binds the active site and inhibits a reaction.

Complement system • A group of eleven proteins that play a role in some reactions of the immune system. The complement proteins are not immunoglobulins.

Complementary base pairing • The A–T (or A–U), T–A (or U–A), C–G and G–C pairing of bases in double-stranded DNA, in transcription, and between tRNA and mRNA.

Complementary DNA (cDNA) • DNA formed by reverse transcriptase acting with an RNA template; essential intermediate in the reproduction of retroviruses; used as a tool in recombinant DNA technology; lacks introns.

Complete metamorphosis • A change of state during the life cycle of an organism in which the body is almost completely rebuilt to produce an individual with a very different body form. Characteristic of insects such as butterflies, moths, beetles, ants, wasps, and flies.

Compound • (1) A substance made up of atoms of more than one element. (2) Made up of many units, as the compound eyes of arthropods (as opposed to the simple eyes of the same group of organisms).

Condensation reaction • A reaction in which two molecules become connected by a covalent bond and a molecule of water is released. ($AH + BOH \rightarrow AB + H_2O$.)

Cones • (1) In the vertebrate retina: photoreceptors responsible for color vision. (2) In gymnosperms: reproductive structures consisting of many sporophylls packed relatively tightly.

Conidium (ko nid' ee um) [Gr. *konis*: dust] • An asexual fungus spore borne singly or in chains either apically or laterally on a hypha.

Conifer (kahn' e fer) [Gr. *konos*: cone + *phero*: carry] • One of the cone-bearing gymnosperms, mostly trees, such as pines and firs.

Conjugation (kahn' jew gay' shun) [L. *conjugare*: yoke together] • The close approximation of two cells during which they exchange genetic material, as in *Paramecium* and other ciliates, or during which DNA passes from one to the other through a tube, as in bacteria.

Connective tissue • An animal tissue that connects or surrounds other tissues; its cells are embedded in a collagen-containing matrix.

Connexon • In a gap junction, a protein channel linking adjacent animal cells.

Consensus sequences • Short stretches of DNA that appear, with little variation, in many different genes.

Constant region • The constant region in an immunoglobulin is encoded by a single exon and determines the function, but not the specificity, of the molecule. The constant region of the T cell receptor anchors the protein to the plasma membrane.

Constitutive enzyme • An enzyme that is present in approximately constant amounts in a system, whether its substrates are present or absent. (Contrast with inducible enzyme.)

Consumer • An organism that eats the tissues of some other organism.

Continental drift • The gradual drifting apart of the world's continents that has occurred over a period of billions of years.

Convergent evolution • The evolution of similar features independently in unrelated taxa from different ancestral structures.

Cooperative act • Behavior in which two or more individuals interact to their mutual benefit. No conscious awareness by the actors of the effects of their behavior is implied.

Cooption • The act of capturing something for a particular use. In ecology refers to the diversion of ecological production for human use. Such production is said to be coopted.

Copulation • Reproductive behavior that results in a male depositing sperm in the reproductive tract of a female.

Corepressor • A low molecular weight compound that unites with a protein (the repressor) to prevent transcription in a repressible operon.

Cork • A waterproofing tissue in plants, with suberin-containing cell walls. Produced by a cork cambium.

Corolla (ko role' lah) [L.: diminutive of *corona*: wreath, crown] • All of the petals of a flower, collectively.

Coronary (kor' oh nair ee) • Referring to the blood vessels of the heart.

Corpus luteum (kor' pus loo' tee um) [L. *corpus*: body + *luteum*: yellow] A structure formed from a follicle after ovulation; it produces hormones important to the maintenance of pregnancy.

Cortex [L.: bark or rind] • (1) In plants, the tissue between the epidermis and the vascular tissue of a stem or root. (2) In animals, the outer tissue of certain organs, such as the adrenal cortex and cerebral cortex.

Corticosteroids • Steroid hormones produced and released by the cortex of the adrenal gland.

Cost • See energetic cost, opportunity cost, risk cost.

Cotyledon (kot' ul lee' dun) [Gr. *kotyledon*: a hollow space] • A "seed leaf." An embryonic organ which stores and digests reserve materials; may expand when seed germinates.

Countercurrent exchange • An adaptation that promotes maximum exchange of heat or any diffusible substance between two fluids by the fluids flow in opposite directions through parallel tubes in close approximation to each other. An example is countercurrent heat exchange between arterioles and venules in the extremities of some animals.

Covalent bond • A chemical bond that arises from the sharing of electrons between two atoms. Usually a strong bond.

Crassulacean acid metabolism (CAM) • A metabolic pathway enabling the plants that possess it to store carbon dioxide at night and then perform photosynthesis during the day with stomata closed.

Crista (plural: cristae) • A small, shelflike projection of the inner membrane of a mitochondrion; the site of oxidative phosphorylation.

Critical night length • In the photoperiodic flowering response of short-day plants, the length of night above which flowering occurs and below which the plant remains vegetative. (The reverse applies in the case of long-day plants.)

Critical period • The age during which some particular type of learning must take place or during which it occurs much more easily than at other times. Typical of song learning among birds.

Cross section (also called a transverse section) • A section taken perpendicular to the longest axis of a structure.

Crossing over • The mechanism by which linked markers undergo recombination. In general, the term refers to the reciprocal exchange of corresponding segments between two homologous chromatids.

CRP • The cAMP receptor protein that interacts with the promoter to enhance transcription; a lowered cAMP concentration results in catabolite repression.

Crustacean (crus tay' see an) • A member of the phylum Crustacea, such as a crab, shrimp, or sowbug.

Cryptic appearance [Gr. *kryptos*: hidden] • The resemblance of an animal to some part of its environment, which helps it to escape detection by predators.

Cryptochromes [Gr. *kryptos*: hidden + *kroma*: color] • Photoreceptors mediating some blue-light effects in plants and animals.

Culture • (1) A laboratory association of organisms under controlled conditions. (2) The collection of knowledge, tools, values, and rules that characterize a human society.

Cuticle • A waxy layer on the outer surface of a plant or an insect, tending to retard water loss.

Cyanobacteria (sigh an' o bacteria) [Gr. *kuanos*: the color blue] • A division of photosynthetic bacteria, formerly referred to as blue-green algae; they lack sexual reproduction, and they use chlorophyll *a* in their photosynthesis.

Cyclic AMP • See cAMP.

Cyclins • Proteins that activate cyclin-dependent kinases, bringing about transitions in the cell cycle.

Cyclin-dependent kinase (cdk) • A kinase is an enzyme that catalzyes the addition of phosphate groups from ATP to target molecules. Cdk's target proteins involved in transitions in the cell cycle and are active only when complexed to additional protein subunits, cyclins.

Cyst (sist) [Gr. *kystis*: pouch] • (1) A resistant, thick-walled cell formed by some protists and other organisms. (2) An abnormal sac, containing a liquid or semisolid substance, produced in response to injury or illness.

Cytochromes (sy' toe chromes) [Gr. *kytos*: container + *chroma*: color] • Iron-containing red proteins, components of the electron-transfer chains in photophosphorylation and respiration.

Cytokinesis (sy' toe kine ee' sis) [Gr. *kytos*: container + *kinein*: to move] • The division of the cytoplasm of a dividing cell. (Contrast with mitosis.)

Cytokinin (sy' toe kine' in) [Gr. *kytos*: container + *kinein*: to move] • A member of a class of plant growth substances playing roles in senescence, cell division, and other phenomena.

Cytoplasm • The contents of the cell, excluding the nucleus.

Cytoplasmic determinants • In animal development, gene products whose spatial distribution may determine such things as embryonic axes.

Cytosine (site' oh seen) • A nitrogen-containing base found in DNA and RNA.

Cytoskeleton • The network of microtubules and microfilaments that gives a eukaryotic cell its shape and its capacity to arrange its organelles and to move.

Cytosol • The fluid portion of the cytoplasm, excluding organelles and other solids.

Cytotoxic T cells • Cells of the cellular immune system that recognize and directly eliminate virus-infected cells. (Contrast with helper T cells, suppressor T cells.)

Decomposer • See detritivore.

Degeneracy • The situation in which a single amino acid may be represented by any of two or more different codons in messenger RNA. Most of the amino acids can be represented by more than one codon.

Degradative succession • Ecological succession occuring on the dead remains of the bodies of plants and animals, as when leaves or animal bodies rot.

Deletion (genetic) • A mutation resulting from the loss of a continuous segment of a gene or chromosome. Such mutations never revert to wild type. (Contrast with duplication, point mutation.)

Deme (deem) [Gr. *demos*: common people] • Any local population of individuals belonging to the same species that interbreed with one another.

Demographic processes • The events—such as births, deaths, immigration, and emigration—that determine the number of individuals in a population.

Demographic stochasticity • Random variations in the factors influencing the size, density, and distribution of a population.

Demography • The study of dynamical changes in the sizes, densities, and distributions of populations.

Denaturation • Loss of activity of an enzyme or nucleic acid molecule as a result of structural changes induced by heat or other means.

Dendrite [Gr. *dendron*: a tree] • A fiber of a neuron which often cannot carry action potentials. Usually much branched and relatively short compared with the axon, and commonly carries information to the cell body of the neuron.

Denitrification • Metabolic activity by which inorganic nitrogen-containing ions are reduced to form nitrogen gas and other products; carried on by certain soil bacteria.

Density dependence • Change in the severity of action of agents affecting birth and death rates within populations that are directly or inversely related to population density.

Density independence • The state where the severity of action of agents affecting birth and death rates within a population does not change with the density of the population.

Deoxyribonucleic acid • See DNA.

Depolarization • A change in the electric potential across a membrane from a condition in which the inside of the cell is more negative than the outside to a condition in which the inside is less negative, or even positive, with reference to the outside of the cell. (Contrast with hyperpolarization.)

Derived trait • A trait found among members of a lineage that was not present in the ancestors of that lineage.

Dermal tissue system • The outer covering of a plant, consisting of epidermis in the young plant and periderm in a plant with extensive secondary growth. (Contrast with ground tissue system and vascular tissue system.)

Desmosome (dez' mo sowm) [Gr. *desmos*: bond + *soma*: body] • An adhering junction between animal cells.

Determination • Process whereby an embryonic cell or group of cells becomes fixed into a predictable developmental pathway.

Detritivore (di try' ti vore) [L. *detritus*: worn away + *vorare*: to devour] • An organism that obtains its energy from the dead bodies and/or waste products of other organisms.

Deuterostome • A major evolutionary lineage in animals, characterized by radial cleavage, enterocoelous development, and other traits. (Compare with protostome.)

Development • Progressive change, as in structure or metabolism; in most kinds of organisms, development continues throughout the life of the organism.

Diaphragm (dye' uh fram) [Gr. *diaphrassein*, to barricade] • (1) A sheet of muscle that separates the thoracic and abdominal cavities in mammals; responsible for the action of breathing. (2) A method of birth control in which a sheet of rubber is fitted over the woman's cervix, blocking the entry of sperm.

Diastole (dye ahs' toll ee) [Gr.: dilation] • The portion of the cardiac cycle when the heart muscle relaxes. (Contrast with systole.)

Dicot (short for dicotyledon) [Gr. *di*: two + *kotyledon*: a hollow space] • This term, not used in this book, formerly referred to all angiosperms other than the monocots. (See eudicot, monocot.)

Differentiation • Process whereby originally similar cells follow different developmental pathways. The actual expression of determination.

Diffusion • Random movement of molecules or other particles, resulting in even distribution of the particles when no barriers are present.

Digestibility-reducing chemicals • Defensive chemicals produced by plants that make the plant's tissued difficult to digest.

Digestion • Enzyme-catalyzed process by which large, usually insoluble, molecules (foods) are hydrolyzed to form smaller molecules of soluble substances.

Dihybrid cross • A mating in which the parents differ with respect to the alleles of two loci of interest.

Dikaryon (di care' ee ahn) [Gr. *dis*: two + *karyon*: kernel] • A cell or organism carrying two genetically distinguishable nuclei. Common in fungi.

Dioecious (die eesh' us) [Gr.: two houses] • Organisms in which the two sexes are "housed" in two different individuals, so that eggs and sperm are not produced in the same individuals. Examples: humans, fruit flies, oak trees, date palms. (Contrast with monoecious.)

Diploblastic • Having two cell layers. (Contrast with triploblastic.)

Diploid (dip' loid) [Gr. *diploos*: double] • Having a chromosome complement consisting of two copies (homologues) of each chromosome. A diploid individual (or cell) usually arises as a result of the fusion of two gametes, each with just one copy of each chromosome. Thus, the two homologues in each chromosome pair in a diploid cell are of separate origin, one derived from the female parent and one from the male parent.

Directional selection • Selection in which phenotypes at one extreme of the population distribution are favored. (Contrast with disruptive selection; stabilizing selection.)

Disaccharide • A carbohydrate made up of two monosaccharides (simple sugars).

Dispersal stage • Stage in its life history at which an organism moves from its birthplace to where it will live as an adult.

Displacement activity • Apparently irrelevant behavior performed by an animal under conflict situations, especially when tendencies to attack and escape are closely balanced.

Display • A behavior that has evolved to influence the actions of other individuals.

Disruptive selection • Selection in which phenotypes at both extremes of the population distribution are favored. (Contrast with directional selection; stabilizing selection.)

Distal • Away from the point of attachment or other reference point. (Contrast with proximal.)

Disturbance • A short-term event that disrupts populations, communities, or ecosystems by changing the environment.

Diverticulum (di ver tic' u lum) [L. *divertere*: turn away] • A small cavity or tube that connects to a major cavity or tube.

Division • A term used by some microbiologists and formerly by botanists, corresponding to the term phylum.

DNA (deoxyribonucleic acid) • The fundamental hereditary material of all living organisms. In eukaryotes, stored primarily in the cell nucleus. A nucleic acid using deoxyribose rather than ribose.

DNA chip • A small glass or plastic square onto which thousands of single-stranded DNA sequences are fixed. Hybridization of cell-derived RNA or DNA to the target sequences can be performed. (See DNA hybridization.)

DNA hybridization • A process by which DNAs from two species are mixed and heated so that interspecific double helixes are formed.

DNA ligase • Enzyme that unites Okazaki fragments of the lagging strand during DNA replication; also mends breaks in DNA strands. It connects pieces of a DNA strand and is used in recombinant DNA technology.

DNA methylation • Addition of methyl groups to DNA; plays role in regulation of gene expression; protects a bacterium's DNA against its restriction endonucleases.

DNA polymerase • Any of a group of enzymes that catalyze the formation of DNA strands from a DNA template.

Domain • The largest unit in the current taxonomic nomenclature. Members of the three domains (Bacteria, Archaea, and Eukarya) are believed to have been evolving independently of each other for at least a billion years.

Dominance • In genetic terminology, the ability of one allelic form of a gene to determine the phenotype of a heterozygous individual, in which the homologous chromosome carries both it and a different allele. For example, if A and a are two allelic forms of a gene, A is said to be dominant to a if AA diploids and Aa diploids are phenotypically identical and are distinguishable from aa diploids. The a allele is said to be **recessive**.

Dominance hierarchy • In animal behavior, the set of relationships within a group of animals, usually established and maintained by aggression, in which one individual has precedence over all others in eating, mating, and other activities.

Dormancy • A condition in which normal activity is suspended, as in some seeds and buds.

Dorsal [L. *dorsum*: back] • Pertaining to the back or upper surface. (Contrast with ventral.)

Double fertilization • Process virtually unique to angiosperms in which one sperm nucleus combines with the egg to produce a zygote, and the other sperm nucleus combines with the two polar nuclei to produce the first cell of the triploid endosperm.

Double helix • Of DNA: molecular structure in which two complementary polynucleotide strands, antiparallel to each other, form a right-handed spiral.

Duodenum (doo' uh dee' num) • The beginning portion of the vertebrate small intestine. (Contrast with ileum, jejunum.)

Duplication (genetic) • A mutation resulting from the introduction into the genome of an extra copy of a segment of a gene or chromosome. (Contrast with deletion, point mutation.)

Dynein [Gr. *dunamis*: power] • A protein that undergoes conformational changes and thus plays a part in the movement of eukaryotic flagella and cilia.

Ecdysone (eck die' sone) [Gr. *ek*: out of + *dyo*: to clothe] • In insects, a hormone that induces molting.

Ecological biogeography • The study of the distributions of organisms from an ecological perspective, usually concentrating on migration, dispersal, and species interactions.

Ecological community • The species living together at a particular site.

Ecological niche (nitch) [L. *nidus*: nest] • The functioning of a species in relation to other species and its physical environment.

Ecological succession • The sequential replacement of one population assemblage by another in a habitat following some disturbance. Succession sometimes ends in a relatively stable ecosystem.

Ecology [Gr. *oikos*: house + *logos*: discourse, study] • The scientific study of the interaction of organisms with their environment, including both the physical environment and the other organisms that live in it.

Ecoregion • A large geographic unit characterized by a typical climate and a widespread assemblage of similar species.

Ecosystem (eek' oh sis tum) • The organisms of a particular habitat, such as a pond or forest, together with the physical environment in which they live.

Ecto- (eck' toh) [Gr.: outer, outside] • A prefix used to designate a structure on the outer surface of the body. For example, ectoderm. (Contrast with endo- and meso-.)

Ectoderm [Gr. *ektos*: outside + *derma*: skin] • The outermost of the three embryonic tissue layers first delineated during gastrulation. Gives rise to the skin, sense organs, nervous system, etc.

Ectotherm [Gr. *ektos*: outside + *thermos*: heat] • An animal unable to control its body temperature. (Contrast with endotherm.)

Edema (i dee' mah) [Gr. *oidema*: swelling] • Tissue swelling caused by the accumulation of fluid.

Edge effect • The changes in ecological processes in a community caused by physical and biological factors originating in an adjacent community.

Effector • Any organ, cell, or organelle that moves the organism through the environment or else alters the environment to the organism's advantage. Examples include muscle, bone, and a wide variety of exocrine glands.

Effector cell • A lymphocyte that performs a role in the immune system without further differentiation.

Effector phase • In this phase of the immune response, effector T cells called cytotoxic T cells attack virus-infected cells, and effector helper T cells assist B cells to

differentiate into plasma cells, which release antibodies.

Efferent [L. *ex*: out + *ferre*: to bear] • Away from, as in neurons that conduct action potentials out from the central nervous system, or arterioles that conduct blood away from a structure. (Contrast with afferent.)

Egg • In all sexually reproducing organisms, the female gamete; in birds, reptiles, and some other vertebrates, a structure witin which early embryonic development occurs.

Elasticity • The property of returning quickly to a former state after a disturbance.

Electrocardiogram (EKG) • A graphic recording of electrical potentials from the heart.

Electroencephalogram (EEG) • A graphic recording of electrical potentials from the brain.

Electromyogram (EMG) • A graphic recording of electrical potentials from muscle.

Electron (e lek' tron) [L. *electrum*: amber (associated with static electricity), from Gr. *slektor*: bright sun (color of amber)] • One of the three most important fundamental particles of matter, with mass approximately 0.00055 amu and charge –1.

Electronegativity • The tendency of an atom to attract electrons when it occurs as part of a compound.

Electrophoresis (e lek' tro fo ree' sis) [L. *electrum*: amber + Gr. *phorein*: to bear] • A separation technique in which substances are separated from one another on the basis of their electric charges and molecular weights.

Electrotonic potential • In neurons, a hyperpolarization or small depolarization of the membrane potential induced by the application of a small electric current. (Contrast with action potential, resting potential.)

Elemental substance • A substance composed of only one type of atom.

Embolus (em' buh lus) [Gr. *embolos*: inserted object; stopper] • A circulating blood clot. Blockage of a blood vessel by an embolus or by a bubble of gas is referred to as an **embolism**. (Contrast with thrombus.)

Embryo [Gr. *en-*: in + *bryein*: to grow] • A young animal, or young plant sporophyte, while it is still contained within a protective structure such as a seed, egg, or uterus.

Embryo sac • In angiosperms, the female gametophyte. Found within the ovule, it consists of eight or fewer cells, membrane bounded, but without cellulose walls between them.

Emergent property • A property of a complex system that is not exhibited by its individual component parts.

Emigration • The deliberate and usually oriented departure of an organism from the habitat in which it has been living.

3' End (3-prime) • The end of a DNA or RNA strand that has a free hydroxyl group at the 3'-carbon of the sugar (deoxyribose or ribose).

5' End (5-prime) • The end of a DNA or RNA strand that has a free phosphate group at the 5'-carbon of the sugar (deoxyribose or ribose).

Endemic (en dem' ik) [Gr. *endemos*: dwelling in a place] • Confined to a particular region, thus often having a comparatively restricted distribution.

Endergonic reaction • One for which energy must be supplied. (Contrast with exergonic reaction.)

Endo- [Gr.: within, inside] • A prefix used to designate an innermost structure. For example, endoderm, endocrine. (Contrast with ecto-, meso-.)

Endocrine gland (en' doh krin) [Gr. *endon*: inside + *krinein*: to separate] • Any gland, such as the adrenal or pituitary gland of vertebrates, that secretes certain substances, especially hormones, into the body through the blood.

Endocrinology • The study of hormones and their actions.

Endocytosis • A process by which liquids or solid particles are taken up by a cell through invagination of the plasma membrane. (Contrast with exocytosis.)

Endoderm [Gr. *endon*: within + *derma*: skin] • The innermost of the three embryonic tissue layers first delineated during gastrulation. Gives rise to the digestive and respiratory tracts and structures associated with them.

Endodermis [Gr. *endon*: within + *derma*: skin] • In plants, a specialized cell layer marking the inside of the cortex in roots and some stems. Frequently a barrier to free diffusion of solutes.

Endomembrane system • Endoplasmic reticulum plus Golgi apparatus plus, when present, lysosomes; thus, a system of membranes that exchange material with one another.

Endoplasmic reticulum [Gr. *endon*: within + L. *plasma*: form; L. *reticulum*: little net] • A system of membrane-bounded tubes and flattened sacs found in the cytoplasm of eukaryotes. Exists as rough ER, studded with ribosomes; and smooth ER, lacking ribosomes.

Endorphins • Naturally occurring, opiate-like substances in the mammalian brain.

Endoskeleton [Gr. *endon*: within + *skleros*: hard] • A skeleton covered by other, soft body tissues. (Contrast with exoskeleton.)

Endosperm [Gr. *endon*: within + *sperma*: seed] • A specialized triploid seed tissue found only in angiosperms; contains stored food for the developing embryo.

Endosymbiosis [Gr. *endon*: within + *syn*: together + *bios*: life] • The living together of two species, with one living inside the body (or even the cells) of the other.

Endosymbiotic theory • Theory that the eukaryotic cell evolved from a prokaryote that contained other, endosymbiotic prokaryotes.

Endotherm [Gr. *endon*: within + *thermos*: hot] • An animal that can control its body temperature by the expenditure of its own metabolic energy. (Contrast with ectotherm.)

Endotoxins [Gr. *endon*: within + L. *toxicum*: poison] • Lipopolysaccharides released by the lysis of some Gram-negative bacteria that cause fever and vomiting in a host organism.

Energetic cost • The difference between the energy an animal would have expended had it rested, and that expended in performing a behavior.

Energy • The capacity to do work.

Enhancer • In eukaryotes, a DNA sequence, lying on either side of the gene it regulates, that stimulates a specific promoter.

Enterocoelous development • A pattern of development in which the coelum is formed by an outpocketing of the embryonic gut (enteron).

Enterokinase (ent uh row kine' ase) • An enzyme secreted by the mucosa of the duodenum. It activates the zymogen trypsinogen to create the active digestive enzyme trypsin.

Entrainment • With respect to circadian rhythms, the process whereby the period is adjusted to match the 24-hour environmental cycle.

Entropy (en' tro pee) [Gr. *en*: in + *tropein*: to change] • A measure of the degree of disorder in any system. A perfectly ordered system has zero entropy; increasing disorder is measured by positive entropy. Spontaneous reactions in a closed system are always accompanied by an increase in disorder and entropy.

Environment • An organism's surroundings, both living and nonliving; includes temperature, light intensity, and all other species that influence the focal organism.

Environmental toxicology • The study of the distribution and effects of toxic compounds in the environment.

Enzyme (en' zime) [Gr. *en*: in + *zyme*: yeast] • A protein, on the surface of which are chemical groups so arranged as to make the enzyme a catalyst for a chemical reaction.

Epi- [Gr.: upon, over] • A prefix used to designate a structure located on top of another; for example: epidermis, epiphyte.

Epicotyl (epp' i kot' il) [Gr. *epi*: upon + *kotyle*: something hollow] • That part of a plant embryo or seedling that is above the cotyledons.

Epidermis [Gr. *epi*: upon + *derma*: skin] • In plants and animals, the outermost cell layers. (Only one cell layer thick in plants.)

Epididymis (epuh did' uh mus) [Gr. *epi*: upon + *didymos*: testicle] • Coiled tubules in the testes that store sperm and conduct sperm from the seiminiferous tubules to the vas deferens.

Epinephrine (ep i nef' rin) [Gr. *epi*: upon + *nephros*: a kidney] • The "fight or flight" hormone. Produced by the medulla of the adrenal gland, it also functions as a neurotransmitter. Also known as adrenaline.

Epiphyte (ep' e fyte) [Gr. *epi*: upon + *phyton*: plant] • A specialized plant that grows on the surface of other plants but does not parasitize them.

Episome • A plasmid that may exist either free or integrated into a chromosome. (See plasmid.)

Epistasis • An interaction between genes, in which the presence of a particular allele of one gene determines whether another gene will be expressed.

Epithelium • In animals, a layer of cells covering or lining an external surface or a cavity.

Equilibrium • (1) In biochemistry, a state in which forward and reverse reactions are proceeding at counterbalancing rates, so there is no observable change in the concentrations of reactants and products. (2) In evolutionary genetics, a condition in which allele and genotype frequencies in a population are constant from generation to generation.

Erythrocyte (ur rith' row sight) [Gr. *erythros*: red + *kytos*: hollow vessel] • A red blood cell.

Esophagus (i soff' i gus) [Gr. *oisophagos*: gullet] • That part of the gut between the pharynx and the stomach.

Ester linkage • A condensation (water-releasing) reaction in which the carboxyl group of a fatty acid reacts with the hydroxyl group of an alcohol. Lipids are formed in this way.

Estivation (ess tuh vay' shun) [L. *aestivalis*: summer] • A state of dormancy and hypometabolism that occurs during the summer; usually a means of surviving drought and/or intense heat. Contrast with hibernation.

Estrogen • Any of several steroid sex hormones, produced chiefly by the ovaries in mammals.

Estrus (es' truss) [L. *oestrus*: frenzy] • The period of heat, or maximum sexual receptivity, in some female mammals. Ordinarily, the estrus is also the time of release of eggs in the female.

Ethylene • One of the plant hormones, the gas $H_2C;h2CH_2$.

Euchromatin • Chromatin that is diffuse and non-staining during interphase; may be transcribed. (Contrast with heterochromatin.)

Eudicots (yew di' kots) [Gr. *eu*: true + *di*: two + *kotyledon*: a cup-shaped hollow] • Members of the angiosperm class Eudicotyledones, flowering plants in which the embryo produces two cotyledons prior to germination. Leaves of most eudicots have major veins arranged in a branched or reticulate pattern.

Eukaryotes (yew car' ry otes) [Gr. *eu*: true + *karyon*: kernel or nucleus] • Organisms whose cells contain their genetic material inside a nucleus. Includes all life other than the viruses, Archaebacteria, and Eubacteria.

Eusocial • Term applied to insects, such as termites, ants, and many bees and wasps, in which individuals cooperate in the care of offspring, there are sterile castes, and generations overlap.

Eutrophication (yoo trofe' ik ay' shun) [Gr. *eu-*: well + *trephein*: to flourish] • The addition of nutrient materials to a body of water, resulting in changes to species composition therein.

Evolution • Any gradual change. Organic evolution, often referred to as evolution, is any genetic and resulting phenotypic change in organisms from generation to generation.

Evolutionary agent • Any factor that influences the direction and rate of evolutionary changes.

Evolutionarily conservative • Traits of organisms that evolve very slowly.

Evolutionary innovations • Major changes in body plans of organisms; these have been very rare during evolutionary history.

Evolutionary radiation • The proliferation of species within a single evolutionary lineage.

Evolutionary reversal • The reappearance of the ancestral state of a trait in a lineage in which that trait had acquired a derived state.

Excision repair • The removal and damaged DNA and its replacement by the appropriate nucleotides.

Excitatory postsynaptic potential (EPSP) • A change in the resting potential of a postsynaptic membrane in a positive (depolarizing) direction. (Contrast with inhibitory postsynaptic potential.)

Excretion • Release of metabolic wastes by an organism.

Exergonic reaction • A reaction in which free energy is released. (Contrast with endergonic reaction.)

Exo- (eks' oh) • Same as ecto-.

Exocrine gland (eks' oh krin) [Gr. *exo*: outside + *krinein*: to separate] • Any gland, such as a salivary gland, that secretes to the outside of the body or into the gut.

Exocytosis • A process by which a vesicle within a cell fuses with the plasma membrane and releases its contents to the outside. (Contrast with endocytosis.)

Exon • A portion of a DNA molecule, in eukaryotes, that codes for part of a polypeptide. (Contrast with intron.)

Exoskeleton (eks' oh skel' e ton) [Gr. *exos*: outside + *skleros*: hard] • A hard covering on the outside of the body to which muscles are attached. (Contrast with endoskeleton.)

Exotoxins • Highly toxic proteins released by living, multiplying bacteria.

Experiment • A scientific method in which particular factors are manipulated while other factors are held constant so that the potential influences of the manipulated factors can be determined.

Exponential growth • Growth, especially in the number of organisms in a population, which is a simple function of the size of the growing entity: the larger the entity, the faster it grows. (Contrast with logistic growth.)

Expression vector • A DNA vector, such as a plasmid, that carries a DNA sequence that includes the adjacent sequences for its expression into mRNA and protein in a host cell.

Expressivity • The degree to which a genotype is expressed in the phenotype— may be affected by the environment.

Extensor • A muscle the extends an appendage.

Extinction • The termination of a lineage of organisms.

Extrinsic protein • A membrane protein found only on the surface of the membrane. (Contrast with intrinsic protein.)

F_1 generation • The immediate progeny of a parental (P) mating; the first filial generation.

F_2 generation • The immediate progeny of a mating between members of the F_1 generation.

Facilitated diffusion • Passive movement through a membrane involving a specific carrier protein; does not proceed against a concentration gradient. (Contrast with active transport, free diffusion.)

Family • In taxonomy, the category below the order and above the genus; a group of related, similar genera.

Fat • A triglyceride that is solid at room temperature. (Contrast with oil.)

Fatty acid • A molecule with a long hydrocarbon tail and a carboxyl group at the other end. Found in many lipids.

Fauna (faw' nah) • All of the animals found in a given area. (Contrast with flora.)

Feces [L. *faeces*: dregs] • Waste excreted from the digestive system.

Feedback control • Control of a particular step of a multistep process, induced by the presence or absence of a product of one of the later steps. A thermostat regulating the flow of heating oil to a furnace in a home is a negative feedback control device.

Fermentation (fur men tay' shun) [L. *fermentum*: yeast] • The degradation of a substance such as glucose to smaller molecules with the extraction of energy, without the use of oxygen (i.e., anaerobically). Involves the glycolytic pathway.

Fertilization • Union of gametes. Also known as syngamy.

Fertilization membrane • A membrane surrounding an animal egg which becomes rapidly raised above the egg surface within seconds after fertilization, serving to prevent entry of a second sperm.

Fetus • The latter stages of an embryo that is still contained in an egg or uterus; in humans, the unborn young from the eighth week of pregnancy to the moment of birth.

Fiber • An elongated and tapering cell of flowering plants, usually with a thick cell wall. Serves a support function.

Fibrin • A protein that polymerizes to form long threads that provide structure to a blood clot.

Filter feeder • An organism that feeds upon much smaller organisms, that are suspended in water or air, by means of a straining device.

Filtration • In the excretory physiology of some animals, the process by which the initial urine is formed; water and most solutes are transferred into the excretory tract, while proteins are retained in the blood or hemolymph.

First law of thermodynamics • Energy can be neither created nor destroyed.

Fission • Reproduction of a prokaryote by division of a cell into two comparable progeny cells.

Fitness • The contribution of a genotype or phenotype to the composition of subsequent generations, relative to the contribution of other genotypes or phenotypes. (See inclusive fitness.)

Fixed action pattern • A behavior that is genetically programmed.

Flagellum (fla jell' um) (plural: flagella) [L. *flagellum*: whip] • Long, whiplike appendage that propels cells. Prokaryotic flagella differ sharply from those found in eukaryotes.

Flexor • A muscle that flexes an appendage.

Flora (flore' ah) • All of the plants found in a given area. (Contrast with fauna.)

Florigen • A plant hormone (not yet isolated) involved in the conversion of a vegetative shoot apex to a flower.

Flower • The total reproductive structure of an angiosperm; its basic parts include the calyx, corolla, stamens, and carpels.

Fluorescence • The emission of a photon of visible light by an excited atom or molecule.

Follicle [L. *folliculus*: little bag] • In female mammals, an immature egg surrounded by nutritive cells.

Follicle-stimulating hormone • A gonadotropic hormone produced by the anterior pituitary.

Food chain • A portion of a food web, most commonly a simple sequence of prey species and the predators that consume them.

Food web • The complete set of food links between species in a community; a diagram indicating which ones are the eaters and which are consumed.

Forb • Any broad-leaved (dicotyledonous), herbaceous plant. Especially applied to such plants growing in grasslands.

Fossil • Any recognizable structure originating from an organism, or any impression from such a structure, that has been preserved over geological time.

Fossil fuel • A fuel (particularly petroleum products) composed of the remains of organisms that lived in the remote past.

Founder effect • Random changes in allele frequencies resulting from establishment of a population by a very small number of individuals.

Fovea [L. *fovea*; a small pit] • The area, in the vertebrate retina, of most distinct vision.

Frame-shift mutation • A mutation resulting from the addition or deletion of a single base pair in the DNA sequence of a gene. As a result of this, mRNA transcribed from such a gene is translated normally until the ribosome reaches the point at which the mutation has occurred. From that point on, codons are read out of proper register and the amino acid sequence bears no resemblance to the normal sequence. (Contrast with missense mutation, nonsense mutation, synonymous mutation.)

Free energy • That energy which is available for doing useful work, after allowance has been made for the increase or decrease of disorder. Designated by the symbol G (for Gibbs free energy), and defined by: $G = H - TS$, where H = heat, S = entropy, and T = absolute (Kelvin) temperature.

Frequency-dependent selection • Selection that changes in intensity with the proportion of individuals having the trait.

Fruit • In angiosperms, a ripened and mature ovary (or group of ovaries) containing the seeds. Sometimes applied to reproductive structures of other groups of plants, and includes any adjacent parts which may be fused with the reproductive structures.

Fruiting body • A structure that bears spores.

Fundamental niche • The range of condition under which an organism could survive if it were the only one in the environment. (Contrast with realized niche.)

Fungus (fung' gus) • A member of the kingdom Fungi, a (usually) multicellular eukaryote with absorptive nutrition.

G_1 phase • In the cell cycle, the gap between the end of mitosis and the onset of the S phase.

G_2 phase • In the cell cycle, the gap between the S (synthesis) phase and the onset of mitosis.

G protein • A membrane protein involved in signal transduction; characterized by binding guanyl nucleotides. The activation of certain receptors activates the G protein, which in turn activates adenylate cyclase. G protein activation involves binding a GTP molecule in place of a GDP molecule.

Gametangium (gam i tan' gee um) [Gr. *gamos*: marriage + *angeion*: vessel or reservoir] • Any plant or fungal structure within which a gamete is formed.

Gamete (gam' eet) [Gr. *gamete*: wife, *gametes*: husband] • The mature sexual reproductive cell: the egg or the sperm.

Gametocyte (ga meet' oh site) [Gr. *gamete*: wife, *gametes*: husband + *kytos*: cell] • The cell that gives rise to sex cells, either the eggs or the sperm. (See oocyte and spermatocyte.)

Gametogenesis (ga meet' oh jen' e sis) [Gr. *gamete*: wife, *gametes*: husband + *genesis*: source] • The specialized series of cellular divisions that leads to the production of sex cells (gametes). (Contrast with oogenesis and spermatogenesis.)

Gametophyte (ga meet' oh fyte) • In plants and photosynthetic protists with alternation of generations, the haploid phase that produces the gametes. (Contrast with sporophyte.)

Ganglion (gang' glee un) [Gr.: tumor] • A group or concentration of neuron cell bodies.

Gap junction • A 2.7-nanometer gap between plasma membranes of two animal cells, spanned by protein channels. Gap junctions allow chemical substances or electrical signals to pass from cell to cell.

Gas exchange • In animals, the process of taking up oxygen from the environment and releasing carbon dioxide to the environment.

Gastrovascular cavity • Serving for both digestion (gastro) and circulation (vascular); in particular, the central cavity of the body of jellyfish and other cnidarians.

Gastrula (gas' true luh) [Gr. *gaster*: stomach] • An embryo forming the characteristic three cell layers (ectoderm, endoderm, and mesoderm) which will give rise to all of the major tissue systems of the adult animal.

Gastrulation • Development of a blastula into a gastrula.

Gated channel • A channel (membrane protein) that opens and closes in response to binding of specific molecules or to changes in membrane potential.

Gel electrophoresis (jel ul lec tro for' eesis) • A semisolid matrix suspended in a salty buffer in which molecules can be separated on the basis of their size and change when current is passed through the gel.

Gene [Gr. *gen*: to produce] • A unit of heredity. Used here as the unit of genetic function which carries the information for a single polypeptide.

Gene amplification • Creation of multiple copies of a particular gene, allowing the production of large amounts of the RNA transcript (as in rRNA synthesis in oocytes).

Gene cloning • Formation of a clone of bacteria or yeast cells containing a particular foreign gene.

Gene family • A set of identical, or once-identical, genes, derived from a single parent gene; need not be on the same chromosomes; classic example is the globin family in vertebrates.

Gene flow • The exchange of genes between different species (an extreme case referred to as hybridization) or between different populations of the same species caused by migration following breeding.

Gene pool • All of the genes in a population.

Gene therapy • Treatment of a genetic disease by providing patients with cells containing wild type alleles for the genes that are nonfunctional in their bodies.

Generative nucleus • In a pollen tube, a haploid nucleus that undergoes mitosis to produce the two sperm nuclei that participate in double fertilization. (Contrast with tube nucleus.)

Genet • The genetic individual of a plant that is composed of a number of nearly identical but repeated units.

Genetic drift • Changes in gene frequencies from generation to generation in a small population as a result of random processes.

Genetic stochasticity • Variation in the frequencies of alleles and genotypes in a population over time.

Genetics • The study of heredity.

Genetic structure • The frequencies of alleles and genotypes in a population.

Genome (jee' nome) • The genes in a complete haploid set of chromosomes.

Genotype (jean' oh type) [Gr. *gen*: to produce + *typos*: impression] • An exact description of the genetic constitution of an individual, either with respect to a single trait or with respect to a larger set of traits. (Contrast with phenotype.)

Genus (jean' us) (plural: genera) [Gr. *genos*: stock, kind] • A group of related, similar species.

Geotropism • See gravitropism.

Germ cell • A reproductive cell or gamete of a multicellular organism.

Germination • The sprouting of a seed or spore.

Gestation (jes tay' shun) [L. *gestare*: to bear] • The period during which the embryo of a mammal develops within the uterus. Also known as **pregnancy**.

Gibberellin (jib er el' lin) [L. *gibberella*: hunchback (refers to shape of a reproductive structure of a fungus that produces gibberellins)] • One of a class of plant growth substances playing roles in stem elongation, seed germination, flowering of certain plants, etc. Named for the fungus *Gibberella*.

Gill • An organ for gas exchange in aquatic organisms.

Gill arch • A skeletal structure that supports gill filaments and the blood vessels that supply them.

Gizzard (giz' erd) [L. *gigeria*: cooked chicken parts] • A very muscular port of the stomach of birds that grinds up food, sometimes with the aid of fragments of stone.

Gland • An organ or group of cells that produces and secretes one or more substances.

Glans penis • Sexually sensitive tissue at the tip of the penis.

Glia (glee' uh) [Gr.: glue] • Cells, found only in the nervous system, which do not conduct action potentials.

Glomerulus (glo mare' yew lus) [L. *glomus*: ball] • Sites in the kidney where blood filtration takes place. Each glomerulus consists of a knot of capillaries served by afferent and efferent arterioles.

Glucocorticoids • Steroid hormones produced by the adrenal cortex. Secreted in response to ACTH, they inhibit glucose uptake by many tissues in addition to mediating other stress responses.

Glucagon • A hormone produced and released by cells in the islets of Langerhans of the pancreas. It stimulates the breakdown of glycogen in liver cells.

Gluconeogenesis • The biochemical synthesis of glucose from other substances, such as amino acids, lactate, and glycerol.

Glucose (glue' kose) [Gr. *gleukos*: sweet wine mash for fermentation] • The most common sugar, one of several monosaccharides with the formula $C_6H_{12}O_6$.

Glycerol (gliss' er ole) • A three-carbon alcohol with three hydroxyl groups, the linking component of phospholipids and triglycerides.

Glycogen (gly' ko jen) • A branched-chain polymer of glucose, similar to starch (which is less branched and may be of lower molecular weight). Exists mostly in liver and muscle; the principal storage carbohydrate of most animals and fungi.

Glycolysis (gly kol' li sis) [from glucose + Gr. *lysis*: loosening] • The enzymatic breakdown of glucose to pyruvic acid. One of the oldest energy-yielding machanisms in living organisms.

Glycosidic linkage • The connection in an oligosaccharide or polysaccharide chain, formed by removal of water during the linking of monosaccharides.by root pressure.

Glyoxysome (gly ox' ee soam) • An organelle found in plants, in which stored lipids are converted to carbohydrates.

Golgi apparatus (goal' jee) • A system of concentrically folded membranes found in the cytoplasm of eukaryotic cells. Plays a role in the production and release of secretory materials such as the digestive enzymes manufactured in the pancreas. First described by Camillo Golgi (1844–1926).

Gonad (go' nad) [Gr. *gone*: seed, that which produces seed] • An organ that produces sex cells in animals: either an ovary (female gonad) or testis (male gonad).

Gonadotropin • A hormone that stimulates the gonads.

Gondwana • The large southern land mass that existed from the Cambrian (540 mya) to the Jurassic (138 mya). Present-day South America, Africa, India, Australia, and Antarctica.

Gram stain • A differential stain useful in characterizing bacteria.

Granum • Within a chloroplast, a stack of thylakoids.

Gravitropism • A directed plant growth response to gravity.

Grazer • An animal that eats the vegetative tissues of herbaceous plants.

Green gland • An excretory organ of crustaceans.

Greenhouse effect • The heating of Earth's atmosphere by gases that are transparent to sunlight but opaque to radiated heat.

Gross primary production • The total energy captured by plants growing in a particular area.

Ground meristem • That part of an apical meristem that gives rise to the ground tissue system of the primary plant body.

Ground tissue system • Those parts of the plant body not included in the dermal or vascular tissue systems. Ground tissues function in storage, photosynthesis, and support.

Group transfer • The exchange of atoms between molecules.

Growth • Irreversible increase in volume (probably the most accurate definition, but at best a dangerous oversimplification).

Growth factors • A group of proteins that circulate in the blood and trigger the normal growth of cells. Each growth factor acts only on certain target cells.

Guanine (gwan'een) • A nitrogen-containing base found in DNA, RNA and GTP.

Guard cells • In plants, paired epidermal cells which surround and control the opening of a stoma (pore).

Gut • An animal's digestive tract.

Guttation • The extrusion of liquid water through openings in leaves, caused by root pressure.

Gymnosperm (jim' no sperm) [Gr. *gymnos*: naked + *sperma*: seed] • A plant, such as a pine or other conifer, whose seeds do not develop within an ovary (hence, the seeds are "naked").

Gyrus (plural: gyri) • The raised or ridged portion of the convoluted surface of the brain. (Contrast to sulcus.)

Habit • The form or pattern of growth characteristic of an organism.

Habitat • The environment in which an organism lives.

Habituation (ha bich' oo ay shun) • The simplest form of learning, in which an animal presented with a stimulus without reward or punishment eventually ceases to respond.

Hair cell • A type of mechanoreceptor in animals.

Half-life • The time required for half of a sample of a radioactive isotope to decay to its stable, nonradioactive form.

Halophyte (hal' oh fyte) [Gr. *halos*: salt + *phyton*: plant] • A plant that grows in a saline (salty) environment.

Haploid (hap' loid) [Gr. *haploeides*: single] • Having a chromosome complement consisting of just one copy of each chromosome. This is the normal "ploidy" of gametes or of asexual spores produced by meiosis or of organisms (such as the gametophyte generation of plants) that grow from such spores without fertilization.

Hardy–Weinberg equililbrium • The percentages of diploid combinations expected from a knowledge of the proportions of alleles in the population if no agents of evolution are acting on the population.

Haustorium (haw stor' ee um) [L. *haustus*: draw up] • A specialized hypha or other structure by which fungi and some parasitic plants draw food from a host plant.

Haversian systems • Units of organization in compact bone that reflect the action of intercommunicating osteoblasts.

Heat-shock proteins • Chaperone proteins expressed in cells exposed to high temperatures or other forms of environmental stress.

Helper T cells • T cells that participate in the activation of B cells and of other T cells; targets of the HIV-I virus, the agent of AIDS. (Contrast with cytotoxic T cells, suppressor T cells.)

Hematocrit (heme at o krit) [Gr. *haima*: blood + *krites*: judge] • The proportion of 100 cc of blood that consists of red blood cells.

Hemizygous(hem' ee zie' gus) [Gr. *hemi*: half + *zygotos*: joined] • In a diploid organism, having only one allele for a given trait, typically the case for X-linked genes in male mammals and Z-linked genes in female birds. (Contrast with homozygous, heterozygous.)

Hemoglobin (hee' mo glow' bin) [Gr. *haima*: blood + L. *globus*: globe] • The colored protein of vertebrate blood (and blood of some invertebrates) which transports oxygen.

Hepatic (heh pat' ik) [Gr. *hepar*: liver] • Pertaining to the liver.

Hepatic duct • The duct that conveys bile from the liver to the gallbladder.

Herbicide (ur' bis ide) • A chemical substance that kills plants.

Herbivore [L. *herba*: plant + *vorare*: to devour] • An animal which eats the tissues of plants. (Contrast with carnivore, detritivore, omnivore.)

Heritable • Able to be inherited; in biology usually refers to genetically determined traits.

Hermaphroditism (her maf' row dite' ism) [Gr. *hermaphroditos*: a person with both male and female traits] • The coexistence of both female and male sex organs in the same organism.

Hertz (abbreviated as Hz) • Cycles per second.

Hetero- [Gr.: other, different] • A prefix used in biology to mean that two or more different conditions are involved; for example, heterotroph, heterozygous.

Heterochromatin • Chromatin that retains its coiling during interphase; generally not transcribed. (Contrast with euchromatin.)

Heterocyst • A large, thick-walled cell in the filaments of certain cyanobacteria; performs nitrogen fixation.

Heterogeneous nuclear RNA (hnRNA) • The product of transcription of a eukaryotic gene, including transcripts of introns.

Heteromorphic (het' er oh more' fik) [Gr. *heteros*: different + *morphe*: form] • having a different form or appearance, as two heteromorphic life stages of a plant. (Contrast with isomorphic.)

Heterosporous (het' er os' por us) • Producing two types of spores, one of which gives rise to a female megaspore and the other to a male microspore. Heterosporous plants produce distinct female and male gametophytes. (Contrast with homosporous.)

Heterotherm • An animal that regulates its body temperature at a constant level at some times but not others, such as a hibernator.

Heterotroph (het' er oh trof) [Gr. *heteros*: different + *trophe*: food] • An organism that requires preformed organic molecules as food. (Contrast with autotroph.)

Heterozygous (het' er oh zie' gus) [Gr. *heteros*: different + *zygotos*: joined] • Of a diploid organism having different alleles of a given gene on the pair of homologues carrying that gene. (Contrast with homozygous.)

Hibernation [L. *hibernus*: winter] • The state of inactivity of some animals during winter; marked by a drop in body temperature and metabolic rate.

Highly repetitive DNA • Short DNA sequences present in millions of copies in the genome, next to each other (in tandem). In a In a reassociation experiment, denatured highly repetitive DNA reanneals very quickly.

Hippocampus • A part of the forebrain that takes part in long-term memory formation.

Histamine (hiss; tah meen) • A substance released within a damaged tissue by a type of white blood cell. Histamines are responsible for aspects of allergice reactions, including the increased vascular permeability that leads to edema (swelling).

Histology • The study of tissues.

Histone • Any one of a group of basic proteins forming the core of a nucleosome, the structural unit of a eukaryotic chromosome. (See nucleosome.)

hnRNA • See heterogeneous nuclear RNA.

Homeobox • A 180-base-pair segment of DNA found in a few genes (called **Hox genes**), perhaps regulating the expression of other genes and thus controlling large-scale developmental processes.

Homeostasis (home' ee o sta' sis) [Gr. *homos*: same + *stasis*: position] • The maintenance of a steady state, such as a constant temperature or a stable social structure, by means of physiological or behavioral feedback responses.

Homeotherm (home' ee o therm) [Gr. *homos*: same + *therme*: heat] • An animal which maintains a constant body temperature by virtue of its own heating and cooling mechanisms. (Contrast with heterotherm, poikilotherm.)

Homeotic genes (home' ee ott' ic) • Genes that determine what entire segments of an animal become. Drastic mutations in these genes cause the transformation of body segments in *Drosophila*. Homeotic genes studied in the plant *Arabidopsis* are called organ identity genes.

Homolog (home' o log') [Gr. *homos*: same + *logos*: word] • One of a pair, or larger set, of chromosomes having the same overall genetic composition and sequence. In diploid organisms, each chromosome inherited from one parent is matched by an identical (except for mutational changes) chromosome—its homolog—from the other parent.

Homology (ho mol' o jee) [Gr. *homologi(a)*: agreement] • A similarity between two structures that is due to inheritance from a common ancestor. The structures are said to be homologous. (Contrast with analogy.)

Homoplasy (home' uh play zee) [Gr. *homos*: same + *plastikos*: to mold] • The presence in several species of a trait not present in their most common ancestor. Can result from convergent evolution, reverse evolution, or parallel evolution.

Homosporous • Producing a single type of spore that gives rise to a single type of gametophyte, bearing both female and male reproductive organs. (Contrast with heterosporous.)

Homozygous (home' o zie' gus) [Gr. *homos*: same + *zygotos*: joined] • Of a diploid organism having identical alleles of a given gene on both homologous chromosomes. An organism may be a "homozygote" with respect to one gene and, at the same time, a "heterozygote" with respect to another. (Contrast with heterozygous.)

Hormone (hore' mone) [Gr. *hormon*: excite, stimulate] • A substance produced in one part of a multicellular organism and transported to another part where it exerts its specific effect on the physiology or biochemistry of the target cells.

Host • An organism that harbors a parasite and provides it with nourishment.

Host–parasite interaction • The dynamic interaction between populations of a host and the parasites that attack it.

Hox genes • See homeobox.

Humoral immune system • The part of the immune system mediated by B cells; it is mediated by circulating antibodies and is active against extracellular bacterial and viral infections.

Humus (hew' muss) • The partly decomposed remains of plants and animals on the surface of a soil. Its characteristics depend primarily upon climate and the species of plants growing on the site.

Hyaluronidase (hill yew ron' uh dase) • An enzyme that digests proteoglycans. Found in sperm cells, it helps digest the coatings surrounding an egg so the sperm can penetrate the egg cell membrane.

Hybrid (high' brid) [L. *hybrida*: mongrel] • The offspring of genetically dissimilar parents. In molecular biology, a double helix formed of nucleic acids from different sources.

Hybridoma • A cell produced by the fusion of an antibody-producing cell with a myeloma cell; it produces monoclonal antibodies.

Hybrid zone • A narrow zone where two populations interbreed, producing hybrid individuals.

Hydrocarbon • A compound containing only carbon and hydrogen atoms.

Hydrogen bond • A chemical bond which arises from the attraction between the slight positive charge on a hydrogen atom and a slight negative charge on a nearby fluorine, oxygen, or nitrogen atom. Weak bonds, but found in great quantities in proteins, nucleic acids, and other biological macromolecules.

Hydrological cycle • The sum total of movement of water from the oceans to the atmosphere, to the soil, and back to the oceans. Some water is cycled many times within compartments of the system before completing one full circuit.

Hydrolyze (hi' dro lize) [Gr. *hydro*: water + *lysis*: cleavage] • To break a chemical bond, as in a peptide linkage, with the insertion of the components of water, –H and –OH, at the cleaved ends of a chain. The digestion of proteins is a hydrolysis.

Hydrophilic [Gr. *hydro*: water + *philia*: love] • Having an affinity for water. (Contrast with hydrophobic.)

Hydrophobic [Gr. *hydro*: water + *phobia*: fear] • Molecules and amino acid side chains, which are mainly hydrocarbons (compounds of C and H with no charged groups or polar groups), have a lower energy when they are clustered together than when they are distributed through an aqueous solution. Because of their attraction for one another and their reluctance to mix with water they are called "hydrophobic." Oil is a hydrophobic substance; phenylalanine is a hydrophobic amino acid in a protein. (Contrast with hydrophilic.)

Hydrostatic skeleton • The incompressible internal liquids of some animals that transfer forces from one part of the body to another when acted upon by the surrounding muscles.

Hydroxyl group • The —OH group, characteristic of alcohols.

Hyperpolarization • A change in the resting potential of a membrane so the inside of a cell becomes more electronegative. (Contrast with depolarization.)

Hypersensitive response • A defensive response of plants to microbial infection; it results in a "dead spot."

Hypertension • High blood pressure.

Hypertonic [Gk. *hyper*: above, over] • Having a greater solute concentration. Said of one solution in comparing it to another. (Contrast with hypotonic, isotonic.)

Hypha (high' fuh) (plural: hyphae) [Gr. *hyphe*: web] • In the fungi, any single filament. May be multinucleate (zygomycetes, ascomycetes) or multicellular (basidiomycetes).

Hypocotyl [Gk. *hypo*: beneath, under + *kotyledon*: hollow space] • That part of the embryonic or seedling plant shoot that is below the cotyledons.

Hypothalamus • The part of the brain lying below the thalamus; it coordinates water balance, reproduction, temperature regulation, and metabolism.

Hypothesis • A tentative answer to a question, from which testable predictions can be generated. (Contrast with theory.)

Hypothetico-deductive method • A method of science in which hypotheses are erected, predictions are made from them, and experiments and observations are performed to test the predictions.

Hypotonic [Gk. *hypo*: beneath, under] • Having a lower solute concentration. Said of one solution in comparing it to another. (Contrast with hypertonic, isotonic.)

Imaginal disc • In insect larvae, groups of cells that develop into specific adult organs.

Immune system [L. *immunis*: exempt] • A system in mammals that recognizes and eliminates or neutralizes either foreign substances or self substances that have been altered to appear foreign.

Immunization • The deliberate introduction of antigen to bring about an immune response.

Immunoglobulins • A class of proteins, with a characteristic structure, active as receptors and effectors in the immune system.

Immunological memory • Certain clones of immune system cells made to respond to an antigen persist. This leads to a more rapid and massive response of the immune system to any subsequent exposure to that antigen.

Immunological tolerance • A mechanism by which an animal does not mount an immune response to the antigenic determinants of its own macromolecules.

Imprinting • (1) In genetics, the differential modification of a gene depending on whether it is present in a male or a female. (2) In animal behavior, a rapid form of learning in which an animal comes to make a particular response, which is maintained for life, to some object or other organism.

Inclusive fitness • The sum of an individual's own fitness (the effect of producing its own offspring: the individual selection component) plus its influence on fitness in relatives other than direct descendants (the kin selection component).

Incomplete dominance • Condition in which the heterozygous phenotype is intermediate between the two homozygous phenotypes.

Incomplete metamorphosis • Insect development in which changes between instars are gradual.

Incus (in' kus) [L. *incus*: anvil] • The middle of the three bones that conduct movements of the eardrum to the oval window of the inner ear. (See malleus, stapes.)

Independent assortment • The random separation during meiosis of nonhomologous chromosomes and of genes carried on nonhomologous chromosomes.

Individual fitness • That component of inclusive fitness that results from an organism producing its own offspring. (Contrast with kin selection component.)

Indoleacetic acid • See auxin.

Inducer • (1) In enzyme systems, a small molecule which, when added to a growth medium, causes a large increase in the level of some enzyme. (2) In embryology, a substance that causes a group of target cells to differentiate in a particular way.

Inducible enzyme • An enzyme that is present in much larger amounts when a particular compound (the inducer) has been added to the system. (Contrast with constitutive enzyme.)

Inflammation • A nonspecific defense against pathogens; characterized by redness, swelling, pain, and increased temperature.

Inflorescence • A structure composed of several flowers.

Inhibitor • A substance which binds to the surface of an enzyme and interferes with its action on its substrates.

Inhibitory postsynaptic potential • A change in the resting potential of a postsynaptic membrane in the hyperpolarizing (negative) direction.

Initiation complex • Combination of a ribosomal light subunit, an mRNA molecule, and the tRNA charged with the first amino acid coded for by the mRNA; formed at the onset of translation.

Initiation factors • Proteins that assist in forming the translation initiation complex at the ribosome.

Inositol triphosphate (IP3) • An intracellular second messenger derived from membrane phospholipids.

Instar (in' star) [L.: image, form] • An immature stage of an insect between molts.

Insulin (in' su lin) [L. *insula*: island] • A hormone, synthesized in islet cells of the pancreas, that promotes the conversion of glucose to the storage material, glycogen.

Integrase • An enzyme that integrates retroviral cDNA into the genome of the host cell.

Integrated pest management • A method of control of pests in which natural predators and parasites are used in conjunction with sparing use of chemical methods to achieve control of a pest without causing serious adverse environmental side effects.

Integument [L. *integumentum*: covering] • A protective surface structure. In gymnosperms and angiosperms, a layer of tissue around the ovule which will become the seed coat. Gymnosperm ovules have one integument, angiosperm ovules two.

Intercalary meristem • A meristematic region in plants which occurs not apically, but between two regions of mature tissue. Intercalary meristems occur in the nodes of grass stems, for example.

Intercostal muscles • Muscles between the ribs that can augment breathing movements by elevating and suppressing the rib cage.

Interferon • A glycoprotein produced by virus-infected animal cells; increases the resistance of neighboring cells to the virus.

Interkinesis • The phase between the first and second meiotic divisions.

Interleukins • Regulatory proteins, produced by macrophages and lymphocytes, that act upon other lymphocytes and direct their development.

Intermediate filaments • Fibrous proteins that stabilize cell structure and resist tension.

Internode • Section between two nodes of a plant stem.

Interphase • The period between successive nuclear divisions during which the chromosomes are diffuse and the nuclear envelope is intact. It is during this period that the cell is most active in transcribing and translating genetic information.

Interspecific competition • Competition between members of two or more species.

Intertropical convergence zone • The tropical region where the air rises most strongly; moves north and south with the passage of the sun overhead.

Intraspecific competition • Competition among members of a single species.

Intrinsic protein • A membrane protein that is embedded in the phospholipid bilayer of the membrane. (Contrast with extrinsic protein.)

Intrinsic rate of increase • The rate at which a population can grow when its density is low and environmental conditions are highly favorable.

Intron • A portion of a DNA molecule that, because of RNA splicing, is not involved in coding for part of a polypeptide molecule. (Contrast with exon.)

Invagination • An infolding.

Inversion (genetic) • A rare mutational event that leads to the reversal of the order of genes within a segment of a chromosome, as if that segment had been removed from the chromosome, turned 180°, and then reattached.

Invertebrate • Any animal that is not a vertebrate, that is, whose nerve cord is not enclosed in a backbone of bony segments.

In vitro [L.: in glass] • In a test tube, rather than in a living organism. (Contrast with in vivo.)

In vivo [L.: in the living state] • In a living organism. Many processes that occur in vivo can be reproduced in vitro with the right selection of cellular components. (Contrast with in vitro.)

Ion (eye' on) [Gr.: wanderer] • An atom or group of atoms with electrons added or removed, giving it a negative or positive electrical charge.

Ion channel • A membrane protein that can let ions pass across the membrane. The channel can be ion-selective, and it can be voltage-gated or ligand-gated.

Ionic bond • A chemical bond which arises from the electrostatic attraction between positively and negatively charged ions. Usually a strong bond.

Iris (eye' ris) [Gr. *iris*: rainbow] • The round, pigmented membrane that surrounds the pupil of the eye and adjusts its aperture to regulate the amount of light entering the eye.

Irruption • A rapid increase in the density of a population. Often followed by massive emigration.

Islets of Langerhans • Clusters of hormone-producing cells in the pancreas.

Iso- [Gr.: equal] • Prefix used to denote two separate but similar or identical states of a characteristic. (See isomers, isomorphic, isotope.)

Isolating mechanism • Geographical, physiological, ecological, or behavioral mechanisms that lead to a reduction in the frequency of hybrid matings.

Isomers • Molecules consisting of the same numbers and kinds of atoms, but differing in the way in which the atoms are combined.

Isomorphic (eye' so more' fik) [Gr. *isos*: equal + *morphe*: form] • having the same form or appearance, as two isomorphic life stages. (Contrast with heteromorphic.)

Isotonic • Having the same solute concentration; said of two solutions. (Contrast with hypertonic, hypotonic.)

Isotope (eye' so tope) [Gr. *isos*: equal + *topos*: place] • Two isotopes of the same chemical element have the same number of protons in their nuclei, but differ in the number of neutrons.

Jasmonates • Plant hormones that trigger defenses against pathogens and herbivores.

Jejunum (jih jew' num) • The middle division of the small intestine, where most absorption of nutrients occurs. (See duodenum, ileum.)

Joule (jool, or jowl) • A unit of energy, equal to 0.24 calories.

Juvenile hormone • In insects, a hormone maintaining larval growth and preventing maturation or pupation.

Karyotype • The number, forms, and types of chromosomes in a cell.

Kelvin temperature scale • See absolute temperature scale.

Keratin (ker' a tin) [Gr. *keras*: horn] • A protein which contains sulfur and is part of such hard tissues as horn, nail, and the outermost cells of the skin.

Ketone (key' tone) • A compound with a C==O group attached to two other groups, neither of which is an H atom. Many sugars are ketones. (Contrast with aldehyde.)

Keystone species • A species that exerts a major influence on the composition and dynamics of the community in which it lives.

Kidneys • A pair of excretory organs in vertebrates.

Kin selection • The component of inclusive fitness resulting from helping the survival of relatives containing the same alleles by descent from a common ancestor.

Kinase (kye' nase) • An enzyme that transfers a phosphate group from ATP to another molecule. Protein kinases transfer phosphate from ATP to specific proteins, playing important roles in cell regulation.

Kinesis (ki nee' sis) [Gr.: movement] • Orientation behavior in which the organism does not move in a particular direction with reference to a stimulus but instead simply moves at an increasing or decreasing rate until it ends up farther from the object or closer to it. (Contrast with taxis.)

Kinetochore (kin net' oh core) [Gr. *kinetos*: moving + *khorein*: to move] • Specialized structure on a centromere to which microtubules attach.

Koch's posulates • Four rules for establishing that a particular microorganism causes a particular disease.

Krebs cycle • See citric acid cycle.

Lactic acid • The end product of fermentation in vertebrate muscle and some microorganisms.

Lagging strand • In DNA replication, the daughter strand that is synthesized discontinuously.

Lamella • Layer.

Larynx (lar' inks) • A structure between the pharynx and the trachea that includes the vocal cords.

Larva (plural: larvae) [L.: ghost, early stage] • An immature stage of any invertebrate animal that differs dramatically in appearance from the adult.

Lateral • Pertaining to the side.

Lateral gene transfer • The movement of genes from one prokaryotic species to another.

Lateral meristems • The vascular cambium and cork cambium, which give rise to secondary tissue in plants.

Laterization (lat' ur iz ay shun) • The formation of a nutrient-poor soil that is rich in insoluble iron and aluminum compounds.

Law of independent assortment • The random separation during meiosis of nonhomologous chromosomes and of genes carried on nonhomologous chromosomes. Mendel's second law.

Law of segregation • Alleles segregate from one another during gamete formation, Mendel's first law.

Leader sequence • A sequence of amino acids at the N-terminal end of a newly synthesized protein, determining where the protein will be placed in the cell.

Leading strand • In DNA replication, the daughter strand that is synthesized continuously.

Lenticel • Spongy region in a plant's periderm, allowing gas exchange.

Leukocyte (loo' ko sight) [Gr. *leukos*: clear + *kutos*: hollow vessel] • A white blood cell.

Lichen (lie' kun) [Gr. *leikhen*: licker] • An organism resulting from the symbiotic association of a true fungus and either a cyanobacterium or a unicellular alga.

Life cycle • The entire span of the life of an organism from the moment of fertilization (or asexual generation) to the time it reproduces in turn.

Life history • The stages an individual goes through during its life.

Life table • A table showing, for a group of equal-aged individuals, the proportion still alive at different times in the future and the number of offspring they produce during each time interval.

Ligament • A band of connective tissue linking two bones in a joint.

Ligand (lig' and) • A molecule that binds to a receptor site of another molecule.

Lignin • The principal noncarbohydrate component of wood, a polymer that binds together cellulose fibrils in some plant cell walls.

Limbic system • A group of primitive vertebrate forebrain nuclei that form a network and are involved in emotions, drives, instinctive behaviors, learning, and memory.

Limiting resource • The required resource whose supply most strongly influences the size of a population.

Linkage • Association between genetic markers on the same chromosome such that they do not show random assortment and seldom recombine; the closer the markers, the lower the frequency of recombination.

Lipase (lip' ase; lye' pase) • An enzyme that digests fats.

Lipids (lip' ids) [Gr. *lipos*: fat] • Substances in a cell which are easily extracted by organic solvents; fats, oils, waxes, steroids, and other large organic molecules, including those which, with proteins, make up the cell membranes. (See phospholipids.)

Litter • The partly decomposed remains of plants on the surface and in the upper layers of the soil.

Littoral zone • The coastal zone from the upper limits of tidal action down to the depths where the water is thoroughly stirred by wave action.

Liver • A large digestive gland. In vertebrates, it secretes bile and is involved in the formation of blood.

Lobes • Regions of the human cerebral hemispheres; includes the temporal, frontal, parietal, and occipital lobes.

Locus • In genetics, a specific location on a chromosome. May be considered to be synonymous with "gene."

Logistic growth • Growth, especially in the size of an organism or in the number of organisms that constitute a population, which slows steadily as the entity approaches its maximum size. (Contrast with exponential growth.)

Loop of Henle (hen' lee) • Long, hairpin loop of the mammalian renal tubule that runs from the cortex down into the medulla, and back to the cortex. Creates a concentration gradient in the interstitial fluids in the medulla.

Lophophore • A U-shaped fold of the body wall with hollow, ciliated tentacles that encircles the mouth of animals in several different phyla. Used for filtering prey from the surrounding water.

Lordosis (lor doe' sis) [Gk. *lordosis*: curving forward] • A posture assumed by females of some mammalian species (especially rodents) to signal sexual receptivity.

Lumen (loo' men) [L.: light] • The cavity inside any tubular part of an organ, such as a piece of gut or a kidney tubule.

Lungs • A pair of saclike chambers within the bodies of some animals, functioning in gas exchange.

Luteinizing hormone • A gonadotropin produced by the anterior pituitary. It stimulates the gonads to produce sex hormones.

Lymph [L. *lympha*: water] • A clear, watery fluid that is formed as a filtrate of blood; it contains white blood cells; it collects in a series of special vessels and is returned to the bloodstream.

Lymph nodes • Specialized tissue regions that act as filters for cells, bacteria and foreign matter.

Lymphocyte • A major class of white blood cells. Includes T cells, B cells, and other cell types important in the immune response.

Lysis (lie' sis) [Gr.: a loosening] • Bursting of a cell.

Lysogenic • The condition of a bacterium that carries the genome of a virus in a relatively stable form. (Contrast with lytic.)

Lysosome (lie' so soam) [Gr. *lysis*: a loosening + *soma*: body] • A membrane-bounded inclusion found in eukaryotic cells (other than plants). Lysosomes contain a mixture of enzymes that can digest most of the macromolecules found in the rest of the cell.

Lysozyme (lie' so zyme) • An enzyme in saliva, tears, and nasal secretions that attacks bacterial cell walls, as one of the body's nonspecific defense mechanisms.

Lytic • Condition in which a bacterium lyses shortly after infection by a virus; the viral genome does not become stabilized within the bacterial cell. (Contrast with lysogenic.)

Macro- (mack' roh) [Gr. *makros*: large, long] • A prefix commonly used to denote something large. (Contrast with micro-.)

Macroevolution • Evolutionary changes occurring over long time spans and usually involving changes in many traits. (Contrast with microevolution.)

Macromolecule • A giant polymeric molecule. The macromolecules are proteins, polysaccharides, and nucleic acids.

Macronutrient • A mineral element required by plant tissues in concentrations of at least 1 milligram per gram of their dry matter.

Macrophage (mac' roh faj) • A type of white blood cell that endocytoses bacteria and other cells.

Major histocompatibility complex (MHC) • A complex of linked genes, with multiple alleles, that control a number of immunological phenomena; it is important in graft rejection.

Malignant tumor • A tumor whose cells can invade surrounding tissues and spread to other organs.

Malleus (mal' ee us) [L. *malleus*: hammer] • The first of the three bones that conduct movements of the eardrum to the oval window of the inner ear. (See incus, stapes.)

Malpighian tubule (mal pee' gy un) • A type of protonephridium found in insects.

Mammal [L. *mamma*: breast, teat] • Any animal of the class Mammalia, characterized by the production of milk by the female mammary glands and the possession of hair for body covering.

Mantle • A sheet of specialized tissues that covers most of the viscera of mollusks; provides protection to internal organs and secretes the shell.

Map unit • In eukaryotic genetics, one map unit corresponds to a recombinant frequency of 0.01.

Mapping • In genetics, determining the order of genes on a chromosome and the distances between them.

Marine [L. *mare*: sea, ocean] • Pertaining to or living in the ocean. (Contrast with aquatic, terrestrial.)

Marsupial (mar soo' pee al) • A mammal belonging to the subclass Metatheria, such as opossums and kangaroos. Most have a pouch (marsupium) that contains the milk glands and serves as a receptacle for the young.

Mass extinctions • Geological periods during which rates of extinction were much higher than during intervening times.

Mass number • The sum of the number of protons and neutrons in an atom's nucleus.

Mast cells • Typically found in connective tissue, mast cells can be provoked by antigens or inflammation to release histamine.

Maternal effect genes • These genes code for morphogens that determine the polarity of the egg and larva in the fruit fly, *Drosophila melanogaster*.

Maternal inheritance (cytoplasmic inheritance) • Inheritance in which the phenotype of the offspring depends on factors, such as mitochondria or chloroplasts, that are inherited from the female parent through the cytoplasm of the female gamete.

Maturation • The automatic development of a pattern of behavior, which becomes increasingly complex or precise as the animal matures. Unlike learning, the development does not require experience to occur.

Mechanoreceptor • A cell that is sensitive to physical movement and generates action potentials in response.

Medulla (meh dull' luh) [L.: narrow] • (1) The inner, core region of an organ, as in the adrenal medulla (adrenal gland) or the renal medulla (kidneys). (2) The portion of the brain stem that connects to the spinal cord.

Mega- [Gr. *megas*: large, great] • A prefix often used to denote something large. (Contrast with micro-.)

Megaspore [Gr. *megas*: large + *spora*:seed] • In plants, a haploid spore that produces a female gametophyte.

Meiosis (my oh' sis) [Gr.: diminution] • Division of a diploid nucleus to produce four haploid daughter cells. The process consists of two successive nuclear divisions with only one cycle of chromosome replication.

Membrane potential • The difference in electrical charge between the inside and the outside of a cell, caused by a difference in the distribution of ions.

Memory cells • Long-lived lymphocytes produced by exposure to antigen. They persist in the body and are able to mount a rapid response to subsequent exposures to the antigen.

Mendelian population • A local population of individuals belonging to the same species and exchanging genes with one another.

Menopause • The time in a human female's life when the ovarian and menstrual cycles cease.

Menstrual cycle • The monthly sloughing off of the uterine lining if fertilization does not occur in the female. Occurs between puberty and menopause.

Meristem [Gr. *meristos*: divided] • Plant tissue made up of actively dividing cells.

Mesenchyme (mez' en kyme) [Gr. *mesos*: middle + *enchyma*: infusion] • Embryonic or unspecialized cells derived from the mesoderm.

Meso- (mez' oh) [Gr.: middle] • A prefix often used to designate a structure located in the middle, or a stage that appears at some intermediate time. For example, mesoderm, Mesozoic.

Mesoderm [Gr. *mesos*: middle + *derma*: skin] • The middle of the three embryonic tissue layers first delineated during gastrulation. Gives rise to skeleton, circulatory system, muscles, excretory system, and most of the reproductive system.

Mesophyll (mez' a fill) [Gr. *mesos*: middle + *phyllon*: leaf] • Chloroplast-containing, photosynthetic cells in the interior of leaves.

Mesosome (mez' o soam') [Gr. *mesos*: middle + *soma*: body] • A localized infolding of the plasma membrane of a bacterium.

Messenger RNA (mRNA) • A transcript of one of the strands of DNA, it carries information (as a sequence of codons) for the synthesis of one or more proteins.

Meta- [Gr.: between, along with, beyond] • A prefix used in biology to denote a change or a shift to a new form or level; for example, as used in metamorphosis.

Metabolic compensation • Changes in biochemical properties of an organism that render it less sensitive to temperature changes.

Metabolic pathway • A series of enzyme-catalyzed reactions so arranged that the product of one reaction is the substrate of the next.

Metabolism (meh tab' a lizm) [Gr. *metabole*: to change] • The sum total of the chemical reactions that occur in an organism, or some subset of that total (as in "respiratory metabolism").

Metamorphosis (met' a mor' fo sis) [Gr. *meta*: between + *morphe*: form, shape] • A radical change occurring between one developmental stage and another, as for example from a tadpole to a frog or an insect larva to the adult.

Metaphase (met' a phase) [Gr. *meta*: between] • The stage in nuclear division at which the centromeres of the highly supercoiled chromosomes are all lying on a plane (the metaphase plane or plate) perpendicular to a line connecting the division poles.

Metapopulation • A population divided into subpopulations, among which there are occasional exchanges of individuals.

Metastasis (meh tass' tuh sis) • The spread of cancer cells from their original site to other parts of the body.

Methanogen • Any member of a group of Archaebacteria that release methane as a metabolic product. This group is considered to be an extremely ancient one.

MHC • See major histocompatibility complex.

Micro- (mike' roh) [Gr. *mikros*: small] • A prefix often used to denote something small. (Contrast with macro-, mega-.)

Microbiology [Gr. *mikros*: small + *bios*: life + *logos*: discourse] • The scientific study of microscopic organisms, particularly bacteria, unicellular algae, protists, and viruses.

Microevolution • The small evolutionary changes typically occurring over short time spans; generally involving a small number of traits and minor genetic changes. (Contrast with macroevolution.)

Microfilament • Minute fibrous structure generally composed of actin found in the cytoplasm of eukaryotic cells. They play a role in the motion of cells.

Micronutrient • A mineral element required by plant tissues in concentrations of less than 100 micrograms per gram of their dry matter.

Micropyle (mike' roh pile) [Gr. *mikros*: small + *pyle*: gate] • Opening in the integument(s) of a seed plant ovule through which pollen grows to reach the female gametophyte within.

Microspores [Gr. *mikros*: small + *spora*: seed] • In plants, a haploid spore that produces a male gametophyte.

Microtubules • Minute tubular structures found in centrioles, spindle apparatus, cilia, flagella, and other places in the cytoplasm of eukaryotic cells. These tubules play roles in the motion and maintenance of shape of eukaryotic cells.

Microvilli (singular: microvillus) • The projections of epithelial cells, such as the cells lining the small intestine, that increase their surface area.

Middle lamella • A layer of derivative polysaccharides that separates plant cells; a common middle lamella lies outside the primary walls of the two cells.

Migration • The regular, seasonal movements of animals between breeding and nonbreeding ranges.

Mimicry (mim' ik ree) • The resemblance of one kind of organism to another, or to some inanimate object; serves the function of making the organism difficult to find, of discouraging potential enemies or of attracting potential prey. (See Batesian mimicry and Müllerian mimicry.)

Mineral • An inorganic substance other than water.

Mineralocorticoid • A hormone produced by the adrenal cortex that influences mineral ion balance; aldosterone.

Mismatch repair • When a single base in DNA is changed into a different base, or the wrong base inserted during DNA replication, there is a mismatch in base pairing with the base on the opposite strand. A repair system removes the incorrect base and inserts the proper one for pairing with the opposite strand.

Missense mutation • A nonsynonymous mutation, or one that changes a codon for one amino acid to a codon for a different amino acid. (Contrast with frame-shift mutation, nonsense mutation, synonymous mutation.)

Mitochondrial matrix • The fluid interior of the mitochondrion, enclosed by the inner mitochondrial membrane.

Mitochondrion (my' toe kon' dree un) (plural: mitochondria) [Gr. *mitos*: thread + *chondros*: cartilage, or grain] • An organelle that occurs in eukaryotic cells and contains the enzymes of the ctric acid cycle, the respiratory chain, and oxidative phosphorylation. A mitochondrion is bounded by a double membrane.

Mitosis (my toe' sis) [Gr. *mitos*: thread] • Nuclear division in eukaryotes leading to the formation of two daughter nuclei each with a chromosome complement identical to that of the original nucleus.

Mitotic center • Cellular region that organizes the microtubules for mitosis. In animals a centrosome serves as the mitotic center.

Moderately repetitive DNA • DNA sequences that appear hundreds to thousands of times in the genome. They include the DNA sequences coding for rRNAs and tRNAs, as well as the DNA at telomeres.

Modular organism • An organism which grows by producing additional units of body construction (modules) that are very similar to the units of which it is already composed.

Mole • A quantity of a compound whose weight in grams is numerically equal to its molecular weight expressed in atomic mass units. Avogadro's number of molecules: 6.023×10^{23} molecules.

Molecular clock • The theory that macromolecules diverge from one another over evolutionary time at a constant rate, and that discovering this rate gives insight into the phylogenetic relationships of organisms.

Molecular weight • The sum of the atomic weights of the atoms in a molecule.

Molecule • A particle made up of two or more atoms joined by covalent bonds or ionic attractions.

Molting • The process of shedding part or all of an outer covering, as the shedding of feathers by birds or of the entire exoskeleton by arthropods.

Mono- [Gr. *monos*: one] • Prefix denoting a single entity. (Contrast with poly.)

Monoclonal antibody • Antibody produced in the laboratory from a clone of hybridoma cells, each of which produces the same specific antibody.

Monocot (short for monocotyledon) [Gr. *monos*: one + *kotyledon*: a cup-shaped hollow] • Any member of the angiosperm class Monocotyledones, plants in which the embryo produces but a single cotyledon (seed leaf). Leaves of most monocots have their major veins arranged parallel to each other.

Monocytes • White blood cells that produce macrophages.

Monoecious (mo nee' shus) [Gr.: one house] • Organisms in which both sexes are "housed" in a single individual, which produces both eggs and sperm. (In some plants, these are found in different flowers within the same plant.) Examples: corn, peas, earthworms, hydras. (Contrast with dioecious, perfect flower.)

Monohybrid cross • A mating in which the parents differ with respect to the alleles of only one locus of interest.

Monomer [Gr.: one unit] • A small molecule, two or more of which can be combined to form oligomers (consisting of a few monomers) or polymers (consisting of many monomers).

Monophyletic (mon' oh fih leht' ik) [Gk. *monos*: single + *phylon*: tribe] • Being descended from a single ancestral stock.

Monosaccharide • A simple sugar. Oligosaccharides and polysaccharides are made up of monosaccharides.

Monosynaptic reflex • A neural reflex that begins in a sensory neuron and makes a single synapse before activating a motor neuron.

Morphogens • Diffusible substances whose concentration gradients determine patterns of development in animals and plants.

Morphogenesis (more' fo jen' e sis) [Gr. *morphe*: form + *genesis*: origin] • The development of form. Morphogenesis is the overall consequence of determination, differentiation, and growth.

Morphology (more fol' o jee) [Gr. *morphe*: form + *logos*: discourse] • The scientific study of organic form, including both its development and function.

Mosaic development • Pattern of animal embryonic development in which each blastomere contributes a specific part of the adult body. (Contrast with regulative development.)

Motor end plate • The modified area on a muscle cell membrane where a synapse is formed with a motor neuron.

Motor neuron • A neuron carrying information from the central nervous system to an effector such as a muscle fiber.

Motor unit • A motor neuron and the set of muscle fibers it controls.

mRNA • (See messenger RNA.)

Mucosa (mew koh' sah) • An epithelial membrane containing cells that secrete mucus. The inner cell layers of the digestive and respiratory tracts.

Müllerian mimicry • The resemblance of two or more unpleasant or dangerous kinds of organisms to each other.

Multicellular [L. *multus*: much + *cella*: chamber] • Consisting of more than one cell, as for example a multicellular organism. (Contrast with unicellular.)

Muscle • Contractile tissue containing actin and myosin organized into polymeric chains called microfilaments. In vertebrates, the tissues are either cardiac muscle, smooth muscle, or striated (skeletal) muscle.

Muscle fiber • A single muscle cell. In the case of striated muscle, a syncitial, multinucleate cell.

Muscle spindle • Modified muscle fibers encased in a connective sheat and functioning as stretch receptors.

Mutagen (mute' ah jen) [L. *mutare*: change + Gr. *genesis*: source] • Any mutagen (e.g., chemicals, radiation) that increases the mutation rate.

Mutation • An inherited change along a very narrow portion of the nucleic acid sequence.

Mutation pressure • Evolution (change in gene proportions) by different mutation rates alone.

Mutualism • The type of symbiosis, such as that exhibited by fungi and algae or cyanobacteria in forming lichens, in which both species profit from the association.

Mycelium (my seel' ee yum) [Gr. *mykes*: fungus] • In the fungi, a mass of hyphae.

Mycorrhiza (my' ka rye' za) [Gr. *mykes*: fungus + *rhiza*: root] • An association of the root of a plant with the mycelium of a fungus.

Myelin (my' a lin) • A material forming a sheath around some axons. It is formed by Schwann cells that wrap themselves about the axon. It serves to insulate the axon electrically and to increase the rate of transmission of a nervous impulse.

Myofibril (my' oh fy' bril) [Gr. *mys*: muscle + L. *fibrilla*: small fiber] • A polymeric unit of actin or myosin in a muscle.

Myogenic (my oh jen' ik) [Gr. *mys*: muscle + *genesis*: source] • Originating in muscle.

Myoglobin (my' oh globe' in) [Gr. *mys*: muscle + L. *globus*: sphere] • An oxygen-binding molecule found in muscle. Consists of a heme unit and a single globiin chain, and carrys less oxygen than hemoglobin.

Myosin [Gr. *mys*: muscle] • One of the two major proteins of muscle, it makes up the thick filaments. (See actin.)

NAD (nicotinamide adenine dinucleotide) • A compound found in all living cells, existing in two interconvertible forms: the oxidizing agent NAD^+ and the reducing agent NADH.

NADP (nicotinamide adenine dinucleotide phosphate) • Like NAD, but possessing another phosphate group; plays similar roles but is used by different enzymes.

Natural selection • The differential contribution of offspring to the next generation by various genetic types belonging to the same population. The mechanism of evolution proposed by Charles Darwin.

Necrosis (nec roh′ sis) • Tissue damage resulting from cell death.

Negative control • The situation in which a regulatory macromolecule (generally a repressor) functions to turn off transcription. In the absence of a regulatory macromolecule, the structural genes are turned on.

Nekton [Gr. *nekhein*: to swim] • Animals, such as fish, that can swim against currents of water. (Contrast with plankton.)

Nematocyst (ne mat′ o sist) [Gr. *nema*: thread + *kystis*: cell] • An elaborate, thread-like structure produced by cells of jellyfish and other cnidarians, used chiefly to paralyze and capture prey.

Nephridium (nef rid′ ee um) [Gr. *nephros*: kidney] • An organ which is involved in excretion, and often in water balance, involving a tube that opens to the exterior at one end.

Nephron (nef′ ron) [Gr. *nephros*: kidney] • The basic component of the kidney, which is made up of numerous nephrons. Its form varies in detail, but it always has at one end a device for receiving a filtrate of blood, and then a tubule that absorbs selected parts of the filtrate back into the bloodstream.

Nephrostome (nef′ ro stome) [Gr. *nephros*: kidney + *stoma*: opening] An opening in a nephridium through which body fluids can enter.

Nerve • A structure consisting of many neuronal axons and connective tissue.

Net primary production • Total photosynthesis minus respiration by plants.

Neural plate • A thickened strip of ectoderm along the dorsal side of the early vertebrate embryo; gives rise to the central nervous system.

Neural tube • An early stage in the development of the vertebrate nervous system consisting of a hollow tube created by two opposing folds of the dorsal ectoderm along the anterior–posterior body axis.

Neuromuscular junction • The region where a motor neuron contacts a muscle fiber, creating a synapse.

Neuron (noor′ on) [Gr. *neuron*: nerve, sinew] • A cell derived from embryonic ectoderm and characterized by a membrane potential that can change in response to stimuli, generating action potentials. Action potentials are generated along an extension of the cell (the axon), which makes junctions (synapses) with other neurons, muscle cells, or gland cells.

Neurotransmitter • A substance, produced in and released by one neuron, that diffuses across a synapse and excites or inhibits the postsynaptic neuron.

Neurula (nure′ you la) [Gr. *neuron*: nerve] • Embryonic stage during formation of the dorsal nerve cord by two ectodermal ridges.

Neutral allele • An allele that does not alter the functioning of the proteins for which it codes.

Neutral theory • A view of molecular evolution that postulates that most mutations do not affect the amino acid being coded for, and that such mutations accumulate in a population at rates driven by genetic drift and mutation rates.

Neutron (new′ tron) [E.: neutral] • One of the three most fundamental particles of matter, with mass approximately 1 amu and no electrical charge.

Nicotinamide adenine dinucleotide • (See NAD.)

Nicotinamide adenine dinucleotide phosphate • (See NADP.)

Nitrification • The oxidation of ammonia to nitrite and nitrate ions, performed by certain soil bacteria.

Nitrogenase • In nitrogen-fixing organisms, an enzyme complex that mediates the stepwise reduction of atmospheric N_2 to ammonia.

Nitrogen fixation • Conversion of nitrogen gas to ammonia, which makes nitrogen available to living things. Carried out by certain prokaryotes, some of them free-living and others living within plant roots.

Node [L. *nodus*: knob, knot] • In plants, a (sometimes enlarged) point on a stem where a leaf is or was attached.

Node of Ranvier • A gap in the myelin sheath covering an axons, where the axonal membrane can fire action potentials.

Noncompetitive inhibitor • An inhibitor that binds the enzyme at a site other than the active site. (Contrast with competitive inhibitor.)

Nondisjunction • Failure of sister chromatids to separate in meiosis II or mitosis, or failure of homologous chromosomes to separate in meiosis I. Results in aneuploidy.

Nonpolar molecule • A molecule whose electric charge is evenly balanced from one end of the molecule to the other.

Nonsense (chain-terminating) mutation • Mutations that change a codon for an amino acid to one of the codons (UAG, UAA, or UGA) that signal termination of translation. The resulting gene product is a shortened polypeptide that begins normally at the amino-terminal end and ends at the position of the altered codon. (Contrast with frame-shift mutation, missense mutation, synonymous mutation.)

Nonspecific defenses • Immunologic responses directed against most or all pathogens, generally without reference to the pathogens' antigens. These defenses include the skin, normal flora, lysozyme, the acidic stomach, interferon, and the inflammatory response.

Nonsynonymous mutation • A nucleotide substitution that that changes the amino acid specified (i.e., AGC → AGA, or serine

→ arginine). (Compare with frame-shift mutation, missense mutation, nonsense mutation.)

Nonsynonymous substitution • The situation when a nonsynonymous mutation becomes widespread in a population. Typically influenced by natural selection. (Contrast with synonymous substitution.)

Nontracheophytes • Those plants lacking well-developed vascular tissue; the liverworts, hornworts, and mosses. (Contrast with tracheophytes.)

Normal flora • The bacteria and fungi that live on animal body surfaces without causing disease.

Norepinephrine • A neurotransmitter found in the central nervous system and also at the postganglionic nerve endings of the sympathetic nervous system. Also called noradrenaline.

Notochord (no′ tow kord) [Gr. *notos*: back + *chorde*: string] • A flexible rod of gelatinous material serving as a support in the embryos of all chordates and in the adults of tunicates and lancelets.

Nuclear envelope • The surface, consisting of two layers of membrane, that encloses the nucleus of eukaryotic cells.

Nucleic acid (new klay′ ik) [E.: nucleus of a cell] • A long-chain alternating polymer of deoxyribose or ribose and phosphate groups, with nitrogenous bases—adenine, thymine, uracil, guanine, or cytosine (A, T, U, G, or C)—as side chains. DNA and RNA are nucleic acids.

Nucleoid (new′ klee oid) • The region that harbors the chromosomes of a prokaryotic cell. Unlike the eukaryotic nucleus, it is not bounded by a membrane.

Nucleolar organizer (new klee′ o lar) • A region on a chromosome that is associated with the formation of a new nucleolus following nuclear division. The site of the genes that code for ribosomal RNA.

Nucleolus (new klee′ oh lus) [from L. diminutive of *nux*: little kernel or little nut] • A small, generally spherical body found within the nucleus of eukaryotic cells. The site of synthesis of ribosomal RNA.

Nucleoplasm (new′ klee o plazm) • The fluid material within the nuclear envelope of a cell, as opposed to the chromosomes, nucleoli, and other particulate constituents.

Nucleosome • A portion of a eukaryotic chromosome, consisting of part of the DNA molecule wrapped around a group of histone molecules, and held together by another type of histone molecule. The chromosome is made up of many nucleosomes.

Nucleotide • The basic chemical unit (monomer) in a nucleic acid. A nucleotide in RNA consists of one of four nitrogenous bases linked to ribose, which in turn is linked to phosphate. In DNA, deoxyribose is present instead of ribose.

Nucleus (new′ klee us) [from L. diminutive of *nux*: kernel or nut] • (1) In chemistry, the dense central portion of an atom, made up of protons and neutrons, with a positive charge. Surrounded by a cloud of negatively charged electrons. (2) In cells, the centrally located chamber of eukaryotic cells that is bounded by a double membrane and contains the chromosomes. The information center of the cell.

Null hypothesis • The assertion that an effect proposed by its companion hypothesis does not in fact exist.

Nutrient • A food substance; or, in the case of mineral nutrients, an inorganic element required for completion of the life cycle of an organism.

Oil • A triglyceride that is liquid at room temperature. (Contrast with fat.)

Okazaki fragments • Newly formed DNA strands making up the lagging strand in DNA replication. DNA ligase links the Okazaki fragments to give a continuous strand.

Olfactory • Having to do with the sense of smell.

Oligomer [Gr.: a few units] • A compound molecule of intermediate size, made up of two to a few monomers. (Contrast with monomer, polymer.)

Oligosaccharins • Plant hormones, derived from the plant cell wall, that trigger defenses against pathogens.

Ommatidium [Gr. *omma*: an eye] • One of the units which, collected into groups of up to 20,000, make up the compound eye of arthropods.

Omnivore [L. *omnis*: all, everything + *vorare*: to devour] • An organism that eats both animal and plant material. (Contrast with carnivore, detritivore, herbivore.)

Oncogenic (ong' co jen' ik) [Gr. *onkos*: mass, tumor + *genes*: born] • Causing cancer.

Oocyte (oh' eh site) [Gr. *oon*: egg + *kytos*: cell] • The cell that gives rise to eggs in animals.

Oogenesis (oh' eh jen e sis) [Gr. *oon*: egg + *genesis*: source] • Female gametogenesis, leading to production of the egg.

Oogonium (oh' eh go' nee um) • In some algae and fungi, a cell in which an egg is produced.

Operator • The region of an operon that acts as the binding site for the repressor.

Operon • A genetic unit of transcription, typically consisting of several structural genes that are transcribed together; the operon contains at least two control regions: the promoter and the operator.

Opportunity cost • The sum of the benefits an animal forfeits by not being able to perform some other behavior during the time when it is performing a given behavior.

Opsin (op' sin) [Gr. *opsis*: sight] • The protein protion of the visual pigment rhodopsin. (See rhodopsin.)

Optic chiasm • Stucture on the lower surface of the vertebrate brain where the two optic nerves come together.

Optical isomers • Isomers that differ in the configuration of the four different groups attached to a single carbon atom; so named

because solutions of the two isomers rotate the plane of polarized light in opposite directions. The two isomers are mirror images of one another.

Optimality models • Models developed to determine the structures or behaviors that best solve particular problems faced by organisms.

Order • In taxonomy, the category below the class and above the family; a group of related, similar families.

Organ • A body part, such as the heart, liver, brain, root, or leaf, composed of different tissues integrated to perform a distinct function for the body as a whole.

Organ identity genes • Plant genes that specify the various parts of the flower. See homeotic genes.

Organ of Corti • Structure in the inner ear that transforms mechanical forces produced from pressure waves ("sound waves") into action potentials that are sensed as sound.

Organelles (or' gan els') [L.: little organ] • Organized structures that are found in or on cells. Examples: ribosomes, nuclei, mitochrondria, chloroplasts, cilia, and contractile vacuoles.

Organic • Pertaining to any aspect of living matter, e.g., to its evolution, structure, or chemistry. The term is also applied to any chemical compound that contains carbon.

Organism • Any living creature.

Organizer, embryonic • A region of an embryo which directs the development of nearby regions. In amphibian early gastrulas, the dorsal lip of the blastopore.

Origin of replication • A DNA sequence at which helicase unwinds the DNA double helix and DNA polymerase binds to initiate DNA replication.

Osmoregulation • Regulation of the chemical composition of the body fluids of an organism.

Osmoreceptor • A neuron that converts changes in the osmotic potential of interstial fluids into action potentials.

Osmosis (oz mo' sis) [Gr. *osmos*: to push] • The movement of water through a differentially permeable membrane from one region to another where the water potential is more negative. This is often a region in which the concentration of dissolved molecules or ions is higher, although the effect of dissolved substances may be offset by hydrostatic pressure in cells with semi-rigid walls.

Ossicle (ah' sick ul) [L. *os*: bone] • The calcified construction unit of echinoderm skeletons.

Osteoblasts • Cells that lay down the protein matrix of bone.

Osteoclasts • Cells that dissolve bone.

Otolith (oh' tuh lith) [Gk.*otikos*: ear + *lithos*: stone[• Structures in the vertebrate vestibular apparatus that mechanically stimulate hair cells when the head moves or changes position.

Outgroup • A taxon that separated from another taxon, whose lineage is to be

inferred, before the latter underwent evolutionary radiation.

Oval window • The flexible membrane which, when moved by the bones of the middle ear, produces pressure waves in the inner ear

Ovary (oh' var ee) • Any female organ, in plants or animals, that produces an egg.

Oviduct [L. *ovum*: egg + *ducere*: to lead] • In mammals, the tube serving to transport eggs to the uterus or to outside of the body.

Oviparous (oh vip' uh rus) • Reproduction in which eggs are released by the female and development is external to the mother's body. (Contrast with viviparous.)

Ovulation • The release of an egg from an ovary.

Ovule (oh' vule) [L. *ovulum*: little egg] • In plants, an organ that contains a gametophyte and, within the gametophyte, an egg; when it matures, an ovule becomes a seed.

Ovum (oh' vum) [L.: egg] • The egg, the female sex cell.

Oxidation (ox i day' shun) • Relative loss of electrons in a chemical reaction; either outright removal to form an ion, or the sharing of electrons with substances having a greater affinity for them, such as oxygen. Most oxidation, including biological ones, are associated with the liberation of energy. (Contrast with reduction.)

Oxidative phosphorylation • ATP formation in the mitochondrion, associated with flow of electrons through the respiratory chain.

Oxidizing agent • A substance that can accept electrons from another. The oxidizing agent becomes reduced; its partner becomes oxidized.

P generation • Also called the parental generation. The individuals that mate in a genetic cross. Their immediate offspring are the F_1 generation.

Pacemaker • That part of the heart which undergoes most rapid spontaneous contraction, thus setting the pace for the beat of the entire heart. In mammals, the sinoatrial (SA) node. Also, an artificial device, implanted in the heart, that initiates rhythmic contraction of the organ.

Pacinian corpuscle • A sensory neuron surrounded by sheaths of connective tissue. Found in the deep layers of the skin, where it senses touch and vibration.

Pair rule genes • Segmentation genes that divide the *Drosophila* larva into two segments each.

Paleomagnetism • The record of the changing direction of Earth's magnetic field as stored in lava flows. Used to accurately date extremely ancient events.

Paleontology (pale' ee on tol' oh jee) [Gr. *palaios*: ancient, old + *logos*: discourse] • The scientific study of fossils and all aspects of extinct life.

Pancreas (pan' cree us) • A gland, located near the stomach of vertebrates, that secretes digestive enzymes into the small intestine and releases insulin into the bloodstream.

Pangaea (pan jee' uh) [Gk. *pan*: all, every] • The single land mass formed when all the continents came together in the Permian period. (Contrast with Gondwana.)

Parabronchi • Passages in the lungs of birds through which air flows.

Paradigm • A general framework within which a scientific or philosophical discipline is viewed and within which questions are asked and hypotheses are developed. Scientific revolutions usually involve major paradigm changes. (Contrast with hypothesis, theory.)

Parallel evolution • Evolutionary patterns that exist in more than one lineage. Often the result of underlying developmental processes.

Parapatric speciation [Gr. *para*: beside + *patria*: fatherland] • Development of reproductive isolation when the barrier is not geographic but is a difference in some other physical condition (such as soil nutrient content) that prevents gene flow between the subpopulations. (Contrast with allopatric speciation, sympatric speciation.)

Paraphyletic taxon • A taxon that includes some, but not all, of the descendants of a single ancestor.

Parasite • An organism that attacks and consumes parts of an organism much larger than itself. Parasites sometimes, but not always, kill the host.

Parasitoid • A parasite that is so large relative to its host that only one individual or at most a few individuals can live within a single host.

Parasympathetic nervous system • A portion of the autonomic (involuntary) nervous system. Activity in the parasympathetic nervous system produces effects such as decreased blood pressure and decelerated heart beat. (Contrast with sympathetic nervous system.)

Parathormone • Hormone secreted by the parathyroid glands. Stimulates osteoclast activity and raises blood calcium levels.

Parathyroids • Four glands on the posterior surface of the thyroid that produce and release parathormone.

Parenchyma (pair eng' kyma) [Gr. *para*: beside + *enchyma*: infusion] • A plant tissue composed of relatively unspecialized cells without secondary walls.

Parental investment • Investment in one offspring or group of offspring that reduces the ability of the parent to assist other offspring.

Parsimony • The principle of preferring the simplest among a set of plausible explanations of a phenomenon. Commonly employed in evolutionary and biogeographic studies.

Parthenocarpy • Formation of fruit from a flower without fertilization.

Parthenogenesis (par' then oh jen' e sis) [Gr. *parthenos*: virgin + *genesis*: source] • The production of an organism from an unfertilized egg.

Partial pressure • The portion of the barometric pressure of a mixture of gases that is due to one component of that mixture. For example, the partial pressure of oxygen at sea level is 20.9% of barometric pressure.

Patch clamping • A technique for isolating a tiny patch of membrane to allow the study of ion movement through a particular channel.

Pathogen (path' o jen) [Gr. *pathos*: suffering + *gignomai*: causing] • An organism that causes disease.

Pattern formation • In animal embryonic development, the organization of differentiated tissues into specific structures such as wings.

Pedigree • The pattern of transmission of a genetic trait in a family.

Pelagic zone (puh ladj' ik) [Gr. *pelagos*: the sea] • The open waters of the ocean.

Penetrance • Of a genotype, the proportion of individuals with that genotype who show the expected phenotype.

PEP carboxylase • The enzyme that combines carbon dioxide with PEP to form a 4-carbon dicarboxylic acid at the start of C_4 photosynthesis or of Crassulacean acid metabolism (CAM).

Pepsin [Gr. *pepsis*: digestion] • An enzyme, in gastric juice, that digests protein.

Peptide linkage • The connecting group in a protein chain, –CO–NH–, formed by removal of water during the linking of amino acids, –COOH to –NH_2.

Peptidoglycan • The cell wall material of many prokaryotes, consisting of a single enormous molecule that surrounds the entire cell.

Perennial (per ren' ee al) [L. *per*: through + *annus*: a year] • Referring to a plant that lives from year to year. (Contrast with annual, biennial.)

Perfect flower • A flower with both stamens and carpels, therefore hermaphroditic.

Pericycle [Gr. *peri*: around + *kyklos*: ring or circle] • In plant roots, tissue just within the endodermis, but outside of the root vascular tissue. Meristematic activity of pericycle cells produces lateral root primordia.

Periderm • The outer tissue of the secondary plant body, consisting primarily of cork.

Period • (1) A minor category in the geological time scale. (2) The duration of a cyclical event, such as a circadian rhythm.

Peripheral nervous system • Neurons that transmit information to and from the central nervous system and whose cell bodies reside outside the brain or spinal cord.

Peristalsis (pair' i stall' sis) [Gr. *peri*: around + *stellein*: place] • Wavelike muscular contractions proceeding along a tubular organ, propelling the contents along the tube.

Peritoneum • The mesodermal lining of the coelom among coelomate animals.

Permease • A membrane protein that specifically transports a compound or family of compounds across the membrane.

Peroxisome • An organelle that houses reactions in which toxic peroxides are formed. The peroxisome isolates these peroxides from the rest of the cell.

Petal • In an angiosperm flower, a sterile modified leaf, nonphotosynthetic, frequently brightly colored, and often serving to attract pollinating insects.

Petiole (pet' ee ole) [L. *petiolus*: small foot] • The stalk of a leaf.

pH • The negative logarithm of the hydrogen ion concentration; a measure of the acidity of a solution. A solution with pH = 7 is said to be neutral; pH values higher than 7 characterize basic solutions, while acidic solutions have pH values less than 7.

Phage (fayj) • Short for bacteriophage.

Phagocyte • A white blood cell that ingests microorganisms by endocytosis.

Phagocytosis [Gr.: *phagein* to eat; cell-eating] • A form of endocytosis, the uptake of a solid particle by forming a pocket of plasma membrane around the particle and pinching off the pocket to form an intracellular particle bounded by membrane. (Contrast with pinocytosis.)

Pharynx [Gr.: throat] • The part of the gut between the mouth and the esophagus.

Phenotype (fee' no type) [Gr. *phanein*: to show] • The observable properties of an individual as they have developed under the combined influences of the genetic constitution of the individual and the effects of environmental factors. (Contrast with genotype.)

Phenotypic plasticity • The fact that the phenotype of an organism is determined by a complex series of developmental processes that are affected by both its genotype and its environment.

Pheromone (feer' o mone) [Gr. *phero*: carry + *hormon*: excite, arouse] • A chemical substance used in communication between organisms of the same species.

Phloem (flo' um) [Gr. *phloos*: bark] • In vascular plants, the food-conducting tissue. It consists of sieve cells or sieve tubes, fibers, and other specialized cells.

Phosphate group • The functional group –OPO_3H_2; the transfer of energy from one compound to another is often accomplished by the transfer of a phosphate group.

Phosphodiester linkage • The connection in a nucleic acid strand, formed by linking two nucleotides.

Phospholipids • Cellular materials that contain phosphorus and are soluble in organic solvents. An example is lecithin (phosphatidyl choline). Phospholipids are important constituents of cellular membranes. (See lipids.)

Phosphorylation • The addition of a phosphate group.

Photoautotroph • An organism that obtains energy from light and carbon from carbon dioxide. (Contrast with chemoautotroph, chemoheterotroph, photoheterotroph.)

Photoheterotroph • An organism that obtains energy from light but must obtain its carbon from organic compounds. (Contrast with chemoautotroph, chemo-heterotroph, photoautotroph.)

Photon (foe' tohn) [Gr. *photos*: light] • A quantum of visible radiation; a "packet" of light energy.

Photoperiod (foe' tow peer' ee ud) • The duration of a period of light, such as the length of time in a 24-hour cycle in which daylight is present. The regulation of processes such as flowering by the changing length of day (or of night) is known as pho-toperiodism.

Photoreceptor • (1) A protein (pigment) that triggers a physiological response when it absorbs a photon. (2) A cell that senses and responds to light energy.

Photorespiration • Light-driven uptake of oxygen and release of carbon dioxide, the carbon being derived from the early reactions of photosynthesis.

Photosynthesis (foe tow sin' the sis) [literally, "synthesis out of light"] • Metabolic processes, carried out by green plants, by which visible light is trapped and the energy used to synthesize compounds such as ATP and glucose.

Phototropin • A yellow protein that is the photoreceptor responsible for phototropism.

Phototropism [Gr. *photos*: light + *trope*: a turning] • A directed plant growth response to light.

Phylogenetic tree • Graphic representation of lines of descent among organisms.

Phylogeny (fy loj' e nee) [Gr. *phylon*: tribe, race + *genesis*: source] • The evolutionary history of a particular group of organisms; also, the diagram of the "family tree" that shows genetic linkages between ancestors and descendants.

Phylum (plural: phyla) [Gr. *phylon*: tribe, stock] • In taxonomy, a high-level category just beneath kingdom and above the class; a group of related, similar classes.

Physiology (fiz' ee ol' o jee) [Gr. *physis*: natural form + *logos*: discourse, study] • The scientific study of the functions of living organisms and the individual organs, tissues, and cells of which they are composed.

Phytoalexins • Substances toxic to fungi, produced by plants in response to fungal infection.

Phytochrome (fy' tow krome) [Gr. *phyton*: plant + *chroma*: color] • A plant pigment regulating a large number of developmental and other phenomena in plants; can exist in two different forms, one of which is active and the other is not. Different wavelengths of light can drive it from one form to the other.

Phytoplankton (fy' tow plangk' ton) [Gr. *phyton*: plant + *planktos*: wandering] • The autotrophic portion of the plankton, consisting mostly of algae.

Pigment • A substance that absorbs visible light.

Pilus (pill' us) [Lat. *pilus*: hair] • A surface appendage by which some bacteria adhere to one another during conjugation.

Pinocytosis [Gr.: drinking cell] • A form of endocytosis; the uptake of liquids by engulfing a sample of the external medium into a pocket of the plasma membrane followed by pinching off the pocket to form an intracellular vesicle. (Contrast with phago-cytosis and endocytosis.)

Pistil [L. *pistillum*: pestle] • The female structure of an angiosperm flower, within which the ovules are borne. May consist of a single carpel, or of several carpels fused into a single structure. Usually differentiated into ovary, style, and stigma.

Pith • In plants, relatively unspecialized tissue found within a cylinder of vascular tissue.

Pituitary • A small gland attached to the base of the brain in vertebrates. Its hormones control the activities of other glands. Also known as the hypophysis.

Placenta (pla sen' ta) [Gr. *plax*: flat surface] • The organ found in most mammals that provides for the nourishment of the fetus and elimination of the fetal waste products.

Placental (pla sen' tal) • Pertaining to mammals of the subclass Eutheria, a group characterized by the presence of a placenta; contains the majority of living species of mammals.

Plankton [Gr. *planktos*: wandering] • The free-floating organisms of the sea and fresh water that for the most part move passively with the water currents. Consisting mostly of microorganisms and small plants and animals. (Contrast with nekton.)

Plant • A member of the kingdom Plantae. Multicellular, gaining its nutrition by photosynthesis.

Planula (plan' yew la) [L. *planum*: something flat] • The free-swimming, ciliated larva of the cnidarians.

Plaque (plack) [Fr.: a metal plate or coin] • (1) A circular clearing in a turbid layer (lawn) of bacteria growing on the surface of a nutrient agar gel. Produced by successive rounds of infection initiated by a single bacteriophage. (2) An accumulation of prokaryotic organisms on tooth enamel. Acids produced by the metabolism of these microorganisms can cause tooth decay.

Plasma (plaz' muh) [Gr. *plassein*: to mold] • The liquid portion of blood, in which blood cells and other particulates are suspended.

Plasma cell • An antibody-secreting cell that developed from a B cell. The effector cell of the humoral immune system.

Plasma membrane • The membrane that surrounds the cell, regulating the entry and exit of molecules and ions. Every cell has a plasma membrane.

Plasmid • A DNA molecule distinct from the chromosome(s); that is, an extrachromosomal element. May replicate independently of the chromosome.

Plasmodesma (plural: plasmodesmata) [Gr. *plasma*: formed or molded + *desmos*: band] • A cytoplasmic strand connecting two adjacent plant cells.

Plasmolysis (plaz mol' i sis) • Shrinking of the cytoplasm and plasma membrane away from the cell wall, resulting from the osmotic outflow of water. Occurs only in cells with rigid cell walls.

Plastid • Organelle in plants that serves for food manufacture (by photosynthesis) or food storage; bounded by a double membrane.

Platelet • A membrane-bounded body without a nucleus, arising as a fragment of a cell in the bone marrow of mammals. Important to blood-clotting action.

Pleiotropy (plee' a tro pee) [Gr. *pleion*: more] • The determination of more than one character by a single gene.

Pleural membrane [Gk. *pleuras*: rib, side] • The membrane lining the outside of the lungs and the walls of the thoracic cavity. Inflammation of these membranes is a condition known as *pleurisy*.

Podocytes • Cells of Bowman's capsule of the nephron that cover the capillaries of the glomerulus, forming filtration slits.

Poikilotherm (poy' kill o therm) [Gr. *poikilos*: varied + *therme*: heat] • An animal whose body temperature tends to vary with the surrounding environment. (Contrast with homeotherm, heterotherm.)

Point mutation • A mutation that results from a small, localized alteration in the chemical structure of a gene. Such mutations can give rise to wild-type revertants as a result of reverse mutation. In genetic crosses, a point mutation behaves as if it resided at a single point on the genetic map. (Contrast with deletion.)

Polar body • A nonfunctional nucleus produced by meiosis, accompanied by very little cytoplasm. The meiosis which produces the mammalian egg produces in addition three polar bodies.

Polar molecule • A molecule in which the electric charge is not distributed evenly in the covalent bonds.

Polarity • In development, the difference between one end and the other. In chemistry, the property that makes a polar molecule.

Pollen [L.: fine powder, dust] • The fertilizing element of seed plants, containing the male gametophyte and the gamete, at the stage in which it is shed.

Pollination • Process of transferring pollen from the anther to the receptive surface (stigma) of the ovary in plants.

Poly- [Gr. *poly*: many] • A prefix denoting multiple entities.

Polygamy [Gr. *poly*: many + *gamos*: marriage] • A breeding system in which an individual acquires more than one mate. In polyandry, a female mates with more than one male, in polygyny, a male mates with more than one female.

Polygenes • Multiple loci whose alleles increase or decrease a continuously variable phenotypic trait.

Polymer • A large molecule made up of similar or identical subunits called monomers. (Contrast with monomer, oligomer.)

Polymerase chain reaction (PCR) • A technique for the rapid production of millions of copies of a particular stretch of DNA.

Polymerization reactions • Chemical reactions that generate polymers by means of condensation reactions.

Polymorphism (pol' lee mor' fiz um) [Gr. poly: many + morphe: form, shape] • (1) In genetics, the coexistence in the same population of two distinct hereditary types based on different alleles. (2) In social organisms such as colonial cnidarians and social insects, the coexistence of two or more functionally different castes within the same colony.

Polyp • The sessile, asexual stage in the life cycle of most cnidarians.

Polypeptide • A large molecule made up of many amino acids joined by peptide linkages. Large polypeptides are called proteins.

Polyphyletic group • A group containing taxa, not all of which share the most recent common ancestor.

Polyploid (pol' lee ploid) • A cell or an organism in which the number of complete sets of chromosomes is greater than two.

Polysaccharide • A macromolecule composed of many monosaccharides (simple sugars). Common examples are cellulose and starch.

Polysome • A complex consisting of a threadlike molecule of messenger RNA and several (or many) ribosomes. The ribosomes move along the mRNA, synthesizing polypeptide chains as they proceed.

Polytene (pol' lee teen) [Gr. poly: many + taenia: ribbon] • An adjective describing giant interphase chromosomes, such as those found in the salivary glands of fly larvae. The characteristic, reproducible pattern of bands and bulges seen on these chromosomes has provided a method for preparing detailed chromosome maps of several organisms.

Pons [L. pons: bridge] • Region of the brain stem anterior to the medulla.

Population • Any group of organisms coexisting at the same time and in the same place and capable of interbreeding with one another.

Population density • The number of individuals (or modules) of a population in a unit of area or volume.

Population genetics • The study of genetic variation and its causes within populations.

Population structure • The proportions of individuals in a population belonging to different age classes (age structure). Also, the distribution of the population in space.

Portal vein • A vein connecting two capillary beds, as in the hepatic portal system.

Positive control • The situation in which a regulatory macromolecule is needed to turn transcription of structural genes on. In its absence, transcription will not occur.

Positive cooperativity • Occurs when a molecule can bind several ligands and each one that binds alters the conformation of the molecule so that it can bind the next ligand more easily. The binding of four molecules of O_2 by hemoglobin is an example of positive cooperativity.

Postabsorptive period • When there is no food in the gut and no nutrients are being absorbed.

Postsynaptic cell • The cell whose membranes receive the neurotransmitter released at a synapse.

Predator • An organism that kills and eats other organisms. Predation is usually thought of as involving the consumption of animals by animals, but it can also mean the eating of plants.

Presynaptic excitation/inhibition • Occurs when a neuron modifies activity at a synapse by releasing a neurotransmitter onto the presynaptic nerve terminal.

Prey [L. praeda: booty] • An organism consumed as an energy source.

Primary active transport • Form of active transport in which ATP is hydrolyzed, yielding the energy required to transport ions against their concentration gradients. (Contrast with secondary active transport.)

Primary growth • In plants, growth produced by the apical meristems. (Contrast with secondary growth.)

Primary producer • A photosynthetic or chemosynthetic organism that synthesizes complex organic molecules from simple inorganic ones.

Primary succession • Succession that begins in an areas initially devoid of life, such as on recently exposed glacial till or lava flows.

Primary structure • The specific sequence of amino acids in a protein.

Primary wall • Cellulose-rich cell wall layers laid down by a growing plant cell.

Primate (pry' mate) • A member of the order Primates, such as a lemur, monkey, ape, or human.

Primer • A short, single-stranded segment of DNA serving as the necessary starting material for the synthesis of a new DNA strand, which is synthesized from the 3' end of the primer.

Primitive streak • A line running axially along the blastodisc, the site of inward cell migration during formation of the three-layered embryo. Formed in the embryos of birds and fish.

Primordium [L. primordium: origin] • The most rudimentary stage of an organ or other part.

Principle of continuity • States that because life probably evolved from nonlife by a continuous, gradual process, all postulated stages in the evolution of life should be derivable from preexisting states. (Compare with signature principle.)

Pro- [L.: first, before, favoring] • A prefix often used in biology to denote a developmental stage that comes first or an evolutionary form that appeared earlier than another. For example, prokaryote, prophase.

Probe • A segment of single stranded nucleic acid used to identify DNA molecules containing the complementary sequence.

Procambium • Primary meristem that produces the vascular tissue.

Progesterone [L. pro: favoring + gestare: to bear] • A vertebrate female sex hormone that maintains pregnancy.

Prokaryotes (pro kar' ry otes) [L. pro: before + Gk. karyon: kernel, nucleus] • Organisms whose genetic material is not contained within a nucleus. The bacteria. Considered an earlier stage in the evolution of life than the eukaryotes.

Prometaphase • The phase of nuclear division that begins with the disintegration of the nuclear envelope.

Promoter • The region of an operon that acts as the initial binding site for RNA polymerase.

Proofreading • The correction of an error in DNA replication just after an incorrectly paired base is added to the growing polynucleotide chain.

Prophage (pro' fayj) • The noninfectious units that are linked with the chromosomes of the host bacteria and multiply with them but do not cause dissolution of the cell. Prophage can later enter into the lytic phase to complete the virus life cycle.

Prophase (pro' phase) • The first stage of nuclear division, during which chromosomes condense from diffuse, threadlike material to discrete, compact bodies.

Prostaglandin • Any one of a group of specialized lipids with hormone-like functions. It is not clear that they act at any considerable distance from the site of their production.

Prosthetic group • Any nonprotein portion of an enzyme.

Protease (pro' tee ase) • See proteolytic enzyme.

Protein (pro' teen) [Gr. protos: first] • One of the most fundamental building substances of living organisms. A long-chain polymer of amino acids with twenty different common side chains. Occurs with its polymer chain extended in fibrous proteins, or coiled into a compact macromolecule in enzymes and other globular proteins.

Proteolytic enzyme • An enzyme whose main catalytic function is the digestion of a protein or polypeptide chain. The digestive enzymes trypsin, pepsin, and carboxypeptidase are all proteolytic enzymes (proteases).

Protist • Those eukaryotes not included in the kingdoms Animalia, Fungi, or Plantae.

Protobiont • Aggregates of abiotically produced molecules that cannot reproduce but do maintain internal chemical environments that differ from their surroundings.

Protoderm • Primary meristem that gives rise to epidermis.

Proton (pro' ton) [Gr. *protos*: first] • One of the three most fundamental particles of matter, with mass approximately 1 amu and an electrical charge of +1.

Proto-oncogenes • The normal alleles of genes possessing oncogenes (cancer-causing genes) as mutant alleles. Proto-oncogenes encode growth factors and receptor proteins.

Protostome • One of the major lineages of animal evolution. Characterized by spiral, determinate cleavage of the egg, and by schizocoelous development. (Compare with deuterostome.)

Prototroph (pro' tow trofe') [Gr. *protos*: first + *trophein*: to nourish] • The nutritional wild type, or reference form, of an organism. Any deviant form that requires growth nutrients not required by the prototrophic form is said to be a nutritional mutant, or auxotroph.

Protozoa • A group of single-celled organisms classified by some biologists as a single phylum; includes the flagellates, amoebas, and ciliates. This textbook follows most modern classifications in elevating the protozoans to a distinct kingdom (Protista) and each of their major subgroups to the rank of phylum.

Proximal • Near the point of attachment or other reference point. (Contrast with distal.)

Pseudocoelom • A body cavity not surrounded by a peritoneum. Characteristic of nematodes and rotifers.

Pseudogene • A DNA segment that is homologous to a functional gene but contains a nucleotide change that prevents its expression.

Pseudoplasmodium [Gr. *pseudes*: false + *plasma*: mold or form] • In the cellular slime molds such as *Dictyostelium*, an aggregation of single amoeboid cells. Occurs prior to formation of a fruiting structure.

Pseudopod (soo' do pod) [Gr. *pseudes*: false + *podos*: foot] • A temporary, soft extension of the cell body that is used in location, attachment to surfaces, or engulfing particles.

Pulmonary • Pertaining to the lungs.

Punctuated equiilibrium • An evolutionary pattern in which periods of rapid change are separated by longer periods of little or no change.

Pupa (pew' pa) [L.: doll, puppet] • In certain insects (the Holometabola), the encased developmental stage that intervenes between the larva and the adult.

Pupil • The opening in the vertebrate eye through which light passes.

Purine (pure' een) • A type of nitrogenous base. The purines adenine and guanine are found in nucleic acids.

Purkinje fibers • Specialized heart muscle cells that conduct excitation throughout the ventricular muscle.

Pyramid of biomass • Graphical representation of the total body masses at different trophic levels in an ecosystem.

Pyramid of energy • Graphical representation of the total energy contents at different trophic levels in an ecosystem.

Pyrimidine (peer im' a deen) • A type of nitrogenous base. The pyrimidines cytosine, thymine, and uracil are found in nucleic acids.

Pyruvate • A three-carbon acid; the end product of glycolysis and the raw material for the citric acid cycle.

Q$_{10}$ • A value that compares the rate of a biochemical process or reaction over a 10°C range of temperature. A process that is not temperature-sensitive has a Q$_{10}$ of 1. Values of 2 or 3 mean the reaction speeds up as temperature increases.

Quantum (kwon' tum) [L. *quantus*: how great] • An indivisible unit of energy.

Quaternary structure • Of aggregating proteins, the arrangement of polypeptide subunits.

R factor (resistance factor) • A plasmid that contains one or more genes that encode resistance to antibiotics.

Radial symmetry • The condition in which two halves of a body are mirror images of each other regardless of the angle of the cut, providing the cut is made along the center line. Thus, a cylinder cut lengthwise down its center displays this form of symmetry. (Contrast with biradial symmetry.)

Radioisotope • A radioactive isotope of an element. Examples are carbon-14 (^{14}C) and hydrogen-3, or tritium (^{3}H).

Radiometry • The use of the regular, known rates of decay of radioisotopes of elements to determine dates of events in the distant past.

Rain shadow • A region of low precipitation on the leeward side of a mountain range.

Ramet • The repeated morphological units of sessile, modular organisms. (Contrast with genet.)

Random genetic drift • Evolution (change in gene proportions) by chance processes alone.

Rate constant • Of a particular chemical reaction, a constant which, when multiplied by the concentration(s) of reactant(s), gives the rate of the reaction.

Reactant • A chemical substance that enters into a chemical reaction with another substance.

Reaction, chemical • A process in which atoms combine or change bonding partners.

Realized niche • The actual niche occupied by an organism; it differs from the fundamental niche because of the presence of other species.

Receptive field • Of a neuron, the area on the retina from which the activity of that neuron can be influenced.

Receptor potential • The change in the resting potential of a sensory cell when it is stimulated.

Recessive • See dominance.

Reciprocal altruism • The exchange of altruistic acts between two or more individuals. The acts may be separated considerably in time.

Reciprocal crosses • A pair of crosses, in one of which a female of genotype A mates with a male of genotype B and in the other of which a female of genotype B mates with a male of genotype A.

Recognition site (also called a restriction site) • A sequence of nucleotides in DNA to which a restriction enzyme binds and then cuts the DNA.

Recombinant • An individual, meiotic product, or single chromosome in which genetic materials originally present in two individuals end up in the same haploid complement of genes. The reshuffling of genes can be either by independent segragation, or by crossing over between homologous chromosomes. For example, a human may pass on genes from both parents in a single haploid gamete.

Recombinant DNA technology • The application of genetic tools (restriction endonucleases, plasmids, and transformation) to the production of specific proteins by biological "factories" such as bacteria.

Rectum • The terminal portion of the gut, ending at the anus.

Redox reaction • A chemical reaction in which one reactant becomes oxidized and the other becomes reduced.

Reducing agent • A substance that can donate electrons to another substance. The reducing agent becomes oxidized, and its partner becomes reduced.

Reduction (re duk' shun) • Gain of electrons; the reverse of oxidation. Most reductions lead to the storage of chemical energy, which can be released later by an oxidation reaction. Energy storage compounds such as sugars and fats are highly reduced compounds. (Contrast with oxidation.)

Reflex • An automatic action, involving only a few neurons (in vertebrates, often in the spinal cord), in which a motor response swiftly follows a sensory stimulus.

Refractory period • Of a neuron, the time interval after an action potential, during which another action potential cannot be elicited.

Regulative development • A pattern of animal embryonic development in which the fates of the first blastomeres are not absolutely fixed. (Contrast with mosaic development.)

Regulatory gene • A gene that contains the information for making a regulatory macromolecule, often a repressor protein.

Releaser • A sensory stimulus that triggers a fixed action pattern.

Releasing hormone • One of several hypothalamic hormones that stimulates the secretion of anterior pituitary hormone.

REM sleep • A sleep state characterized by dreaming, skeletal muscle relaxation, and rapid eye movements.

Renal [L. *renes*: kidneys] • Relating to the kidneys.

Replication fork • A point at which a DNA molecule is replicating. The fork forms by the unwinding of the parent molecule.

Repressible enzyme • An enzyme whose synthesis can be decreased or prevented by the presence of a particular compound. A repressible opren often controls the synthesis of such an enzyme.

Repressor • A protein coded by the regulatory gene. The repressor can bind to a specific operator and prevent transcription of the operon.

Reproductive isolating mechanism • Any trait that prevents individuals from two different populations from producing fertile hybrids.

Reproductive isolation • The condition in which a population is not exchanging genes with other populations of the same species.

Resolving power • Of an optical device such as a microscope, the smallest distance between two lines that allows the lines to be seen as separate from one another.

Resource • Something in the environment required by an organism for its maintenance and growth that is consumed in the process of being used.

Resource defense polygamy • A breeding system in which individuals of one sex (usually males) defend resources that are attractive to individuals of the other sex (usually females); individuals holding better resources attract more mates.

Respiration (res pi ra' shun) [L. *spirare*: to breathe] • (1) Cellular respiration; the oxidation of the end products of glycolysis with the storage of much energy in ATP. The oxidant in the respiration of eukaryotes is oxygen gas. Some bacteria can use nitrate or sulfate instead of O_2. (2) Breathing.

Respiratory chain • The terminal reactions of cellular respiration, in which electrons are passed from NAD or FAD, through a series of intermediate carriers, to molecular oxygen, with the concomitant production of ATP.

Resting potential • The membrane potential of a living cell at rest. In cells at rest, the interior is negative to the exterior. (Contrast with action potential, electrotonic potential.)

Restoration ecology • The science and practice of restoring damaged or degraded ecosystems.

Restriction endonuclease • Any one of several enzymes, produced by bacteria, that break foreign DNA molecules at very specific sites. Some produce "sticky ends." Extensively used in recombinant DNA technology.

Restriction map • A partial genetic map of a DNA molecule, showing the points at which particular restriction endonuclease recognition sites reside.

Reticular system • A central region of the vertebrate brain stem that includes complex fiber tracts conveying neural signals between the forebrain and the spinal cord, with collateral fibers to a variety of nuclei that are involved in autonomic functions, including arousal from sleep.

Retina (rett' in uh) [L. *rete*: net] • The light-sensitive layer of cells in the vertebrate or cephalopod eye.

Retinal • The light-absorbing portion of visual pigment molecules. Derived from β-carotene.

Retrovirus • An RNA virus that contains reverse transcriptase. Its RNA serves as a template for cDNA production, and the cDNA is integrated into a chromosome of the mammalian host cell.

Reverse transcriptase • An enzyme that catalyzes the production of DNA (cDNA), using RNA as a template; essential to the reproduction of retroviruses.

RFLP (Restriction fragment length polymorphism) • Coexistence of two or more patterns of restriction fragments (patterns produced by restriction enzymes), as revealed by a probe. The polymorphism reflects a difference in DNA sequence on homologous chromosomes.

Rhizoids (rye' zoids) [Gr. *rhiza*: root] • Hairlike extensions of cells in mosses, liverworts, and a few vascular plants that serve the same function as roots and root hairs in vascular plants. The term is also applied to branched, rootlike extensions of some fungi and algae.

Rhizome (rye' zome) [Gr. *rhizoma*: mass of roots] • A special underground stem (as opposed to root) that runs horizontally beneath the ground.

Rhodopsin • A photopigment used in the visual process of transducing photons of light into changes in the membrane potential of photoreceptor cells.

Ribonucleic acid • See RNA.

Ribosomal RNA (rRNA) • Several species of RNA that are incorporated into the ribosome. Involved in peptide bond formation.

Ribosome • A small organelle that is the site of protein synthesis.

Ribozyme • An RNA molecule with catalytic activity.

Ribulose 1,5-bisphosphate (RuBP) • The compound in chloroplasts which reacts with carbon dioxide in the first reaction of the Calvin-Benson cycle.

Risk cost • The increased chance of being injured or killed as a result of performing a behavior, compared to resting.

RNA (ribonucleic acid) • A nucleic acid using ribose. Various classes of RNA are involved in the transcription and translation of genetic information. RNA serves as the genetic storage material in some viruses.

RNA polymerase • An enzyme that catalyzes the formation of RNA from a DNA template.

RNA splicing • The last stage of RNA processing in eukaryotes, in which the transcripts of introns are excised through the action of small nuclear ribonucleoprotein particles (snRNP).

Rods • Light-sensitive cells (photoreceptors) in the retina. (Contrast with cones.)

Root cap • A thimble-shaped mass of cells, produced by the root apical meristem, that protects the meristem and that is the organ that perceives the gravitational stimulus in root gravitropism.

Root hair • A specialized epidermal cell with a long, thin process that absorbs water and minerals from the soil solution.

rRNA • See ribosomal RNA.

Rubisco (RuBP carboxylase) • Enzyme that combines carbon dioxide with ribulose bisphosphate to produce 3-phosphoglycerate, the first product of C_3 photosynthesis. The most abundant protein on Earth.

Rumen (rew' mun) • The first division of the ruminant stomach. It stores and initiates bacterial fermentation of food. Food is regurgitated from the rumen for further chewing.

Ruminant • An herbivorous, cud-chewing mammal such as a cow, sheep, or deer, having a stomach consisting of four compartments.

S phase • In the cell cycle, the stage of interphase during which DNA is replicated. (Contrast with G_1 phase, G_2 phase.)

Saprobe [Gr. *sapros*: rotten + *bios*: life] • An organism (usually a bacterium or fungus) that obtains its carbon and energy directly from dead organic matter.

Sarcomere (sark' o meer) [Gr. *sark*: flesh + *meros*: a part] • The contractile unit of a skeletal muscle.

Saturated hydrocarbon • A compound consisting only of carbon and hydrogen, with the hydrogen atoms connected by single bonds.

Schizocoelous development • Formation of a coelom during embryological development by a splitting of mesodermal masses.

Schwann cell • A glial cell that wraps around part of the axon of a peripheral neuron, creating a myelin sheath.

Scrleid [Gr. *skleros*: hard] • A type of sclerenchyma cell, commonly found in nutshells, that is not elongated.

Sclerenchyma (skler eng' kyma) [Gr. *skleros*: hard + *kymus*, juice] • A plant tissue composed of cells with heavily thickened cell walls, dead at functional maturity. The principal types of sclerenchyma cells are fibers and sclereids.

Secondary active transport • Form of active transport in which ions or molecules are transported against their concentration gradient using energy obtained by relaxation of a gradient of sodium ion concentration rather than directly from ATP. (Contrast with primary active transport.)

Secondary compound • A compound synthesized by a plant that is not needed for basic cellular metabolism. Typically has an antiherbivore or antiparasite function.

Secondary growth • In plants, growth produced by vascular and cork cambia, contributing to an increase in girth. (Contrast with primary growth.)

Secondary structure • Of a protein, localized regularities of structure, such as the α helix and the β pleated sheet.

Secondary succession • Ecological succession after a disturbance that does not eliminate all the organisms that originally lived on the site.

Secondary wall • Wall layers laid down by a plant cell that has ceased growing; often impregnated with lignin or suberin.

Second law of thermodynamics • States that in any real (irreversible) process, there is a decrease in free energy and an increase in entropy.

Second messenger • A compound, such as cyclic AMP, that is released within a target cell after a hormone or other "first messenger" has bound to a surface receptor on a cell; the second messenger triggers further reactions within the cell.

Secretin (si kreet' in) • A peptide hormone secreted by the upper region of the small intestine when acidic chyme is present. Stimulates the pancreatic duct to secrete bicarbonate ions.

Section • A thin slice, usually for microscopy, as a tangential section or a transverse section.

Seed • A fertilized, ripened ovule of a gymnosperm or angiosperm. Consists of the embryo, nutritive tissue, and a seed coat.

Seed crop • The number of seeds produced by a plant during a particular bout of reproduction.

Seedling • A young plant that has grown from a seed (rather than by grafting or by other means.)

Segmentation genes • In insect larvae, genes that determine the number and polarity of larval segments.

Segment polarity genes • Genes that determine the boundaries and front-to-back organization of the segments in the *Drosophila* larva.

Segregation (genetic) • The separation of alleles, or of homologous chromosomes, from one another during meiosis so that each of the haploid daughter nuclei produced by meiosis contains one or the other member of the pair found in the diploid mother cell, but never both.

Selective permeability • A characteristic of a membrane, allowing certain substances to pass through while other substances are excluded.

Selfish act • A behavioral act that benefits its performer but harms the recipients.

Semelparous organism • An organism that reproduces only once in its lifetime. (Contrast with iteroparous.)

Semen (see' men) [L.: seed] • The thick, whitish liquid produced by the male reproductive organ in mammals, containing the sperm.

Semicircular canals • Part of the vestibular system of mammals.

Semiconservative replication • The common way in which DNA is synthesized. Each of the two partner strands in a double helix acts as a template for a new partner strand. Hence, after replication, each double helix consists of one old and one new strand.

Seminiferous tubules • The tubules within the testes within which sperm production occurs.

Senescence [L. *senescere*: to grow old] • Aging; deteriorative changes with aging; the increased probability of dying with increasing age.

Sensory neuron • A neuron leading from a sensory cell to the central nervous system. (Contrast with motor neuron.)

Sepal (see' pul) • One of the outermost structures of the flower, usually protective in function and enclosing the rest of the flower in the bud stage.

Septum [L.: partition] • A membrane or wall between two cavities.

Sertoli cells • Cells in the seminiferous tubules that nurture the developing sperm.

Serum • That part of the blood plasma that remains after clots have formed and been removed.

Sessile (sess' ul) [L. *sedere*: to sit] • Permanently attached; not moving.

Set point • In a regulatory system, the threshold sensitivity to the feedback stimulus.

Sex chromosome • In organisms with a chromosomal mechanism of sex determination, one of the chromosomes involved in sex determination.

Sex linkage • The pattern of inheritance characteristic of genes located on the sex chromosomes of organisms having a chromosomal mechanism for sex determination.

Sexual selection • Selection by one sex of characteristics in individuals of the opposite sex. Also, the favoring of characteristics in one sex as a result of competition among individuals of that sex for mates.

Shoot • The aerial part of a vascular plant, consisting of the leaves, stem(s), and flowers.

Sieve tube • A column of specialized cells found in the phloem, specialized to conduct organic matter from sources (such as photosynthesizing leaves) to sinks (such as roots). Found principally in flowering plants.

Sieve tube member • A single cell of a sieve tube, containing cytoplasm but relatively few organelles, with highly specialized perforated end walls leading to elements above and below.

Sign stimulus • The single stimulus, or one out of a very few stimuli, by which an animal distinguishes key objects, such as an enemy, or a mate, or a place to nest, etc.

Signal sequence • The sequence of a protein that directs the protein through a particular cellular membrane.

Signal transduction pathway • The series of biochemical steps whereby a stimulus to a cell (such as a hormone or neurotransmitter binding to a receptor) is translated into a response of the cell.

Signature principle • States that because of continuity, prebiotic processes should leave some trace in contemporary biochemistry. (Compare with principle of continuity.)

Silencer • A sequence of eukaryotic DNA that binds proteins that inhibit the transcription of an associated gene.

Silent mutations • Genetic changes that do not lead to a phenotypic change. At the molecular level, these are DNA sequence changes that, because of the redundancy of the genetic code, result in the same amino acids in the resulting protein. See synonymous mutation.

Similarity matrix • A matrix to compare the structures of two molecules constructed by adding the number of their amino acids that are identical or different

Sinoatrial node (sigh' no ay' tree al) • The pacemaker of the mammalian heart.

Sinus (sigh' nus) [L. *sinus*: a bend, hollow] • A cavity in a bone, a tissue space, or an enlargement in a blood vessel.

Skeletal muscle • See striated muscle.

Sliding filament theory • A proposed mechanism of muscle contraction based on formation and breaking of crossbridges between actin and myosin filaments, causing them to slide together.

Small intestine • The portion of the gut between the stomach and the colon, consisting of the duodenum, the jejunum, and the ileum.

Small nuclear ribonucleoprotein particle (snRNP) • A complex of an enzyme and a small nuclear RNA molecule, functioning in RNA splicing.

Smooth muscle • One of three types of muscle tissue. Usually consists of sheets of mononucleated cells innervated by the autonomic nervous system.

Society • A group of individuals belonging to the same species and organized in a cooperative manner; in the broadest sense, includes parents and their offspring.

Sodium–potassium pump • The complex protein in plasma membranes that is responsible for primary active transport; it pumps sodium ions out of the cell and potassium ions into the cell, both against their concentration gradients.

Solute • A substance that is dissolved in a liquid (solvent).

Solute potential • A property of any solution, resulting from its solute contents; it may be zero or have a negative value.

Solution • A liquid (solvent) and its dissolved solutes.

Solvent • A liquid that has dissolved or can dissolve one or more solutes.

Somatic [Gr. *soma*: body] • Pertaining to the body, or body cells (rather than to germ cells).

Somite (so' might) • One of the segments into which an embryo becomes divided longitudinally, leading to the eventual segmentation of the animal as illustrated by the spinal column, ribs, and associated muscles.

Spatial summation • In the production or inhibition of action potentials in a postsynaptic neuron, the interaction of depolarizations and hyperpolarizations produced by several terminal boutons.

Spawning • The direct release of sex cells into the water.

Speciation (spee' shee ay' shun) • The process of splitting one population into two populations that are reproductively isolated from one another.

Species (spee' shees) [L.: kind] • The basic lower unit of classification, consisting of a population or series of populations of closely related and similar organisms. The more narrowly defined "biological species" consists of individuals capable of interbreeding freely with each other but not with members of other species.

Species diversity • A weighted representation of the species of organisms living in a region; large and common species are given greater weight than are small and rare ones. (Contrast with species richness.)

Species richness • The number of species of organisms living in a region. (Contrast with species diversity.)

Specific heat • The amount of energy that must be absorbed by a gram of a substance to raise its temperature by one degree centigrade. By convention, water is assigned a specific heat of one.

Sperm [Gr. sperma: seed] • A male reproductive cell.

Spermatocyte (spur mat' oh site) [Gr. sperma: seed + kytos: cell] • The cell that gives rise to the sperm in animals.

Spermatogenesis (spur mat' oh jen' e sis) [Gr. sperma: seed + genesis: source] • Male gametogenesis, leading to the production of sperm.

Spermatogonia • Undifferentiated germ cells that give rise to primary spermatocytes and hence to sperm.

Sphincter (sfingk' ter) [Gr. sphinkter: that which binds tight] • A ring of muscle that can close an orifice, for example at the anus.

Spindle apparatus • An array of microtubules stretching from pole to pole of a dividing nucleus and playing a role in the movement of chromosomes at nuclear division. Named for its shape.

Spiracle (spy' rih kel) [L. spirare: to breathe] • An opening of the tracheal respiratory system of terrestrial arthropods.

Spiteful act • A behavioral act that harms both the actor and the recipient of the act.

Spliceosome • An RNA–protein complex that splices out introns from eukaryotic pre-mRNAs.

Splicing • The removal of introns and connecting of exons in eukaryotic pre-mRNAs.

Spontaneous generation • The idea that life is generated continually from nonliving matter. Usually distinguished from the current idea that life evolved from nonliving matter under primordial conditions at an early stage in the history of earth.

Spontaneous reaction • A chemical reaction which will proceed on its own, without any outside influence. A spontaneous reaction need not be rapid.

Sporangium (spor an' gee um) [Gr. spora: seed + angeion: vessel or reservoir] • In plants and fungi, any specialized stucture within which one or more spores are formed.

Spore [Gr. spora: seed] • Any asexual reproductive cell capable of developing into an adult plant without gametic fusion. Haploid spores develop into gametophytes, diploid spores into sporophytes. In prokaryotes, a resistant cell capable of surviving unfavorable periods.

Sporophyte (spor' o fyte) [Gr. spora: seed + phyton: plant] • In plants with alternation of generations, the diploid phase that produces the spores. (Contrast with gametophyte.)

Stabilizing selection • Selection against the extreme phenotypes in a population, so that the intermediate types are favored. (Contrast with disruptive selection.)

Stamen (stay' men) [L.: thread] • A male (pollen-producing) unit of a flower, usually composed of an anther, which bears the pollen, and a filament, which is a stalk supporting the anther.

Starch [O.E. stearc: stiff] • An α-linked polymer of glucose; used by plants as a means of storing energy and carbon atoms.

Start codon • The mRNA triplet (AUG) that acts as signals for the beginning of translation at the ribosome. (Compare with stop codons. There are a few mnior exceptions to these codons.)

Stasis • Period during which little or no evolutionary change takes place within a lineage or groups of lineages.

Statocyst (stat' oh sist) [Gk. statos: stationary + kystos: pouch] • An organ of equilibrium in some invertebrates.

Statolith (stat' oh lith) [Gk. statos: stationary + lithos: stone] • A solid object that responds to gravity or movement and stimulates the mechanoreceptors of a statocyst.

Stele (steel) [Gr. stele: pillar] • The central cylinder of vascular tissue in a plant stem.

Stem cell • A cell capable of extensive proliferation, generating more stem cells and a large clone of differentiated progeny cells, as in the formation of red blood cells.

Step cline • A sudden change in one or more traits of a species along a geographical gradient.

Steroid • Any of numerous lipids based on a 17-carbon atom ring system.

Sticky ends • On a piece of two-stranded DNA, short, complementary, one-stranded regions produced by the action of a restriction endonuclease. Sticky ends allow the joining of segments of DNA from different sources.

Stigma [L.: mark, brand] • The part of the pistil at the apex of the style, which is receptive to pollen, and on which pollen germinates.

Stimulus • Something causing a response; something in the environment detected by a receptor.

Stolon • A horizontal stem that forms roots at intervals.

Stoma (plural: stomata) [Gr. stoma: mouth, opening] • Small opening in the plant epidermis that permits gas exchange; bounded by a pair of guard cells whose osmotic status regulates the size of the opening.

Stop codons • Triplets (UAG, UGA, UAA) in mRNA that act as signals for the end of translation at the ribosome. (See also start codon. There are a few mnior exceptions to these codons.)

Stratosphere • The part of the atmosphere above the troposphere; extends upward to approximately 50 kilometers above the surface of the earth; contains very little water.

Stratum (plural strata) • A layer or sedimentary rock laid down at a particular time in a past.

Striated muscle • Contractile tissue characterized by multinucleated cells containing highly ordered arrangements of actin and myosin microfilaments. Also known as skeletal muscle.

Stroma • The fluid contents of an organelle, such as a chloroplast.

Stromatolite • A composite, flat-to-domed structure composed of successive mineral layers. Some are known to be produced by the action of bacteria in salt or fresh water, and some ancient ones are considered to be evidence for early life on the earth.

Structural formula • A representation of the positions of atoms and bonds in a molecule.

Structural gene • A gene that encodes the primary structure of a protein.

Style [Gr. stylos: pillar or column] • In flowering plants, a column of tissue extending from the tip of the ovary, and bearing the stigma or receptive surface for pollen at its apex.

Sub- [L.: under] • A prefix often used to designate a structure that lies beneath another or is less than another. For example, subcutaneous, subspecies.

Submucosa (sub mew koe' sah) • The tissue layer just under the epithelial lining of the lumen of the digestive tract. (Contrast with mucosa.)

Substrate (sub' strayte) • (1) The molecule or molecules on which an enzyme exerts catalytic action. (2) The base material on which an organism lives.

Substrate level phosphorylation • ATP formation resulting from direct transfer of a phosphate group to ADP from an intermediate in glycolysis. (Contrast with oxidative phosphorylation.)

Succession • In ecology, the gradual, sequential series of changes in species composition of a community following a disturbance.

Sulcus (plural: sulci) [L. sulcare: to plow] • The valleys or creases between the raised portions of the convoluted surface of the brain. (Contrast to gyrus.)

Sulfhydryl group • The —SH group.

Summation • The ability of a neuron to fire action potentials in response to numerous subthreshold postsynaptic potentials arriving simultaneously at differentiated places on the cell, or arriving at the same site in rapid succession.

Surface area-to-volume ratio • For any cell, organism, or geometrical solid, the ratio of surface area to volume; this is an important factor in setting an upper limit on the size a cell or organism can attain.

Surfactant • A substance that decreases the surface tension of a liquid. Lung surfactant, secreted by cells of the alveoli, is mostly phospholipid and decreases the amount of work necessary to inflate the lungs.

Symbiosis (sim' bee oh' sis) [Gr.: to live together] • The living together of two or more species in a prolonged and intimate ecological relationship. (See parasitism, commensalism, mutualism.)

Symmetry • In biology, the property that two halves of an object are mirror images of each other. (See bilateral symmetry and biradial symmetry.)

Sympathetic nervous system • A division of the autonomic (involuntary) nervous system. Its activities include increasing blood pressure and acceleration of the heartbeat. The neurotransmitter at the sympathetic terminals is epinephrine or norepinephrine. (Contrast with parasympathetic nervous system.)

Sympatric speciation (sim pat' rik) [Gr. *sym*: same + *patria*: homeland] • The occurrence of genetic reproduction isolation and the subsequent formation of new species without any physical separation of the subpopulation. (Contrast with allopatric speciation, parapatric speciation.)

Symplast • The continuous meshwork of the interiors of living cells in the plant body, resulting from the presence of plasmodesmata. (Contrast with apoplast.)

Symport • A membrane transport process that carries two substances in the same direction across the membrane. (Contrast with antiport.)

Synapse (sin' aps) [Gr. *syn*: together + *haptein*: to fasten] • The narrow gap between the terminal bouton of one neutron and the dendrite or cell body of another.

Synapsis (sin ap' sis) • The highly specific parallel alignment (pairing) of homologous chromosomes during the first division of meiosis.

Synaptic vesicle • A membrane-bounded vesicle, containing neurotransmitter, which is produced in and discharged by the presynaptic neuron.

Syngamy (sing' guh mee) [Gr. *sun-*: together + *gamos*: marriage] • Union of gametes. Also known as fertilization.

Synonymous mutation • A mutation that substitutes one nucleotide for another but does not change the amino acid specified (i.e., UUA → UUG, both specifying leucine). (Compare with frame-shift mutation, missense mutation, nonsense mutation.)

Synonymous substitution • The situation when a synonymous mutation becomes widespread in a population. Typically not influenced by natural selection, these substitutions can accumulate in a population. (Contrast with nonsynonymous substitution.)

Systematics • The scientific study of the diversity of organisms.

Systemic circulation • The part of the circulatory system serving those parts of the body other than the lungs or gills.

Systemin • The only polypeptide plant hormone; participates in response to tissue damage.

Systole (sis' tuh lee) [Gr.: contraction] • Contraction of a chamber of the heart, driving blood forward in the circulatory system.

T cell • A type of lymphocyte, involved in the cellular immune response. The final stages of its development occur in the thymus gland. (Contrast with B cell; see also cytotoxic T cell, helper T cell, suppressor T cell.)

T cell receptor • A protein on the surface of a T cell that recognizes the antigenic determinant for which the cell is specific.

T tubules • A system of tubules that runs throughout the cytoplasm of muscle fibers, through which action potentials spread.

Target cell • A cell with the appropriate receptors to bind and respond to a particular hormone or other chemical mediator.

Taste bud • A structure in the epithelium of the tongue that includes a cluster of chemoreceptors innervated by sensory neurons.

TATA box • An eight-base-pair sequence, found about 25 base pairs before the starting point for transcription in many eukaryotic promoters, that binds a transcription factor and thus helps initiate transcription.

Taxis (tak' sis) [Gr. *taxis*: arrange, put in order] • The movement of an organism in a particular direction with reference to a stimulus. A taxis usually involves the employment of one sense and a movement directly toward or away from the stimulus, or else the maintenance of a constant angle to it. Thus a positive phototaxis is movement toward a light source, negative geotaxis is movement upward (away from gravity), and so on.

Taxon • A unit in a taxonomic system.

Taxonomy (taks on' oh me) [Gr. *taxis*: arrange, classify] • The science of classification of organisms.

Telomeres (tee' lo merz) [Gr. *telos*: end] • Repeated DNA sequences at the ends of eukaryotic chromosomes.

Telophase (tee' lo phase) [Gr. *telos*: end] • The final phase of mitosis or meiosis during which chromosomes became diffuse, nuclear envelopes reform, and nucleoli begin to reappear in the daughter nuclei.

Template • In biochemistry, a molecule or surface upon which another molecule is synthesized in complementary fashion, as in the replication of DNA. In the brain, a pattern that responds to a normal input but not to incorrect inputs.

Template strand • In a stretch of double-stranded DNA, the strand that is transcribed.

Temporal summation • In the production or inhibition of action potentials in a postsynaptic neuron, the interaction of depolarizations or hyperpolarizations produced by rapidly repeated stimulation of a single point.

Tendon • A collagen-containing band of tissue that connects a muscle with a bone.

Terrestrial (ter res' tree al) [L. *terra*: earth] • Pertaining to the land. (Contrast with aquatic, marine.)

Territory • A fixed area from which an animal or group of animals excludes other members of the same species by aggressive behavior or display.

Tertiary structure • In reference to a protein, the relative locations in three-dimensional space of all the atoms in the molecule. The overall shape of a protein. (Contrast with primary, secondary, and quaternary structures.)

Test cross • A cross of a dominant-phenotype individual (which may be either heterozygous or homozygous) with a homozygous-recessive individual.

Testis (tes' tis) (plural: testes) [L.: witness] • The male gonad; that is, the organ that produces the male sex cells.

Testosterone (tes toss' tuhr own) • A male sex steroid hormone.

Tetanus [Gr. *tetanos*: stretched] • (1) In physiology, a state of sustained, maximal muscular contraction caused by rapidly repeated stimulation. (2) In medicine, an often-fatal disease ("lockjaw") caused by the bacterium *Clostridium tetani*.

Thalamus • A region of the vertebrate forebrain; involved in integration of sensory input.

Thallus (thal' us) [Gr.: sprout] • Any algal body which is not differentiated into root, stem, and leaf.

Theory • An explanation or hypothesis that is supported by a wide body of evidence. (Contrast with hypothesis, paradigm.)

Thermoneutral zone • The range of temperatures over which an endotherm does not have to expend extra energy to thermoregulate.

Thermoreceptor • A cell or structure that responds to changes in temperature.

Thoracic cavity • The portion of the mammalian body cavity bounded by the ribs, shoulders, and diaphragm. Contains the heart and the lungs.

Thorax • In an insect, the middle region of the body, between the head and abdomen. In mammals, the part of the body between the neck and the diaphragm.

Thrombin • An enzyme that converts fibrinogen to fibrin, thus triggering the formation of blood clots.

Thrombus (throm' bus) [Gk. *thrombos*: clot] • A blood clot that forms within a blood vessel and remains attached to the wall of the vessel. (Contrast with embolus.)

Thylakoid • A flattened sac within a chloroplast. The membranes of the numerous thylakoids contain all of the chlorophyll in a plant, in addition to the electron carriers of photophosphorylation. Thylakoids stack to form grana.

Thymine • A nitrogen-containing base found in DNA.

Thymus • A ductless, glandular portion of the lymphoid system, involved in development of the immune system of vertebrates.

Thyroid [Gr. *thyreos*: door-shaped] • A two-lobed gland in vertebrates. Produces the hormone thyroxin.

Thyrotropic hormone • A hormone that is produced in the pituitary gland of amphibia such as frogs and transported in the bloodstream to the thyroid gland, inducing the thyroid gland to produce the thyroid hormone that regulates metamorphosis from tadpole to adult frog.

Tight junction • A junction between epithelial cells, in which there is no gap whatever between the adjacent cells. Materials may get through a tight junction only by entering the epithelial cells themselves.

Tissue • A group of similar cells organized into a functional unit and usually integrated with other tissues to form part of an organ such as a heart or leaf.

Tonus • A low level of muscular tension that is maintained even when the body is at rest.

Totipotency • In a cell, the condition of possessing all the genetic information and other capacities necessary to form an entire individual.

Toxigenicity [L. *toxicum*: poison] • The ability of a bacterium to produce chemical substances injurious to the tissues of the host organism.

Trachea (tray' kee ah) [Gr. *trakhoia*: a small tube] • A tube that carries air to the bronchi of the lungs of vertebrates, or to the cells of arthropods.

Tracheid (tray' kee id) • A distinctive conducting and supporting cell found in the xylem of nearly all vascular plants, characterized by tapering ends and walls that are pitted but not perforated.

Tracheophytes [Gr. *trakhoia*: a small tube + *phyton*: plant] • Those plants with xylem and phloem, including psilophytes, club mosses, horsetails, ferns, gymnosperms, and angiosperms. (Contrast with nontracheophytes.)

Trait • One form of a character: Eye color is a character; brown eyes and blue eyes are traits.

Transcription • The synthesis of RNA, using one strand of DNA as the template.

Transcription factors • Proteins that assemble on a eukaryotic chromosome, allowing RNA polymerase II to perform transcription.

Transduction • (1) Transfer of genes from one bacterium to another, with a bacterial virus acting as the carrier of the genes. (2) In sensory cells, the transformation of a stimulus (e.g., light energy, sound pressure waves, chemical or electrical stimulants) into action potentials.

Transfection • Uptake, incorporation, and expression of recombinant DNA.

Transfer cell • A modified parenchyma cell that transports solutes from its cytoplasm into its cell wall, thus moving the solutes from the symplast into the apoplast.

Transfer RNA (tRNA) • A category of relatively small RNA molecules (about 75 nucleotides). Each kind of transfer RNA is able to accept a particular activated amino acid from its specific activating enzyme, after which the amino acid is added to a growing polypeptide chain.

Transformation • Mechanism for transfer of genetic information in bacteria in which pure DNA extracted from bacteria of one genotype is taken in through the cell surface of bacteria of a different genotype and incorporated into the chromosome of the recipient cell.

Transgenic • Containing recombinant DNA incorporated into its genetic material.

Translation • The synthesis of a protein (polypeptide). This occurs on ribosomes, using the information encoded in messenger RNA.

Translocation • (1) In genetics, a rare mutational event that moves a portion of a chromosome to a new location, generally on a nonhomologous chromosome. (2) In vascular plants, movement of solutes in the phloem.

Transpiration [L. *spirare*: to breathe] • The evaporation of water from plant leaves and stem, driven by heat from the sun, and providing the motive force to raise water (plus ions) from the roots.

Transposable element • A segment of DNA that can move to, or give rise to copies at, another locus on the same or a different chromosome.

Triglyceride • A simple lipid in which three fatty acids are combined with one molecule of glycerol.

Triplet • See codon.

Triplet repeat • Occurrence of repeated triplet of bases in a gene, often leading to genetic disease, as does excessive repetition of CGG in the gene responsible for fragile-X syndrome.

Triploblastic • Having three cell layers. (Contrast with diploblastic.)

Trisomic • Containing three, rather than two members of a chromosome pair.

tRNA • See transfer RNA.

Trochophore (troke' o fore) [Gr. *trochos*: wheel + *phoreus*: bearer] • The free-swimming larva of some annelids and mollusks, distinguished by a wheel-like band of cilia around the middle, and indicating an evolutionary relationship between these two groups.

Trophic level • A group of organisms united by obtaining their energy from the same part of the food web of a biological community.

Tropic hormones • Hormones of the anterior pituitary that control the secretion of hormones by other endocrine glands.

Tropism [Gr. *tropos*: to turn] • In plants, growth toward or away from a stimulus such as light (phototropism) or gravity (gravitropism).

Tropomyosin (troe poe my' oh sin) • A protein that, along with actin, constitutes the thin filaments of myofibrils. It controls the interactions of actin and myosin necessary for muscle contraction.

Troposphere • The atmospheric zone reaching upward approximately 17 km in the tropics and subtropics but only to about 10 km at higher latitudes. The zone in which virtually all the water vapor in the atmosphere is located.

Trypsin • A protein-digesting enzyme. Secreted by the pancreas in its inactive form (trypsinogen), it becomes active in the duodenum of the small intestine.

T-tubules • A set of transverse tubes that penetrates skeletal muscle fibers and terminates in the sarcoplasmic reticulum. The T-system transmits impulses to the sacs, which then release Ca^{2+} to initiate muscle contraction.

Tube nucleus • In a pollen tube, the haploid nucleus that does not participate in double fertilization. (Contrast with generative nucleus.)

Tubulin • A protein that polymerizes to form microtubules.

Tumor • A disorganized mass of cells, often growing out of control. Malignant tumors spread to other parts of the body.

Tumor suppressor genes • Genes which, when homozygous mutant, result in cancer. Such genes code for protein products that inhibit cell proliferation.

Twitch • A single unit of muscle contraction.

Tympanic membrane [Gr. *tympanum*: drum] • The eardrum.

Umbilical cord • Tissue made up of embryonic membranes and blood vessels that connects the embryo to the placenta in eutherian mammals.

Understory • The aggregate of smaller plants growing beneath the canopy of dominant plants in a forest.

Unicellular (yoon' e sell' yer ler) [L. *unus*: one + *cella*: chamber] • Consisting of a single cell; as for example a unicellular organism. (Contrast with multicellular.)

Uniport • A membrane transport process that carries a single substance. (Contrast with antiport, symport.)

Unsaturated hydrocarbon • A compound containing only carbon and hydrogen atoms. One or more pairs of carbon atoms are connected by double bonds.

Upwelling • The upward movement of nutrient-rich, cooler water from deeper layers of the ocean.

Urea • A compound serving as the main excreted form of nitrogen by many animals, including mammals.

Ureotelic • Describes an organism in which the final product of the breakdown of nitrogen-containing compounds (primarily proteins) is urea. (Contrast with ammonotelic, uricotelic.)

Ureter (your' uh tur) [Gr. *ouron*: urine] • A long duct leading from the vertebrate kidney to the urinary bladder or the cloaca.

Urethra (you ree' thra) [Gr. *ouron*: urine] • In most mammals, the canal through which urine is discharged from the bladder and which serves as the genital duct in males.

Uric acid • A compound that serves as the main excreted form of nitrogen in some animals, particularly those which must conserve water, such as birds, insects, and reptiles.

Uricotelic • Describes an organism in which the final product of the breakdown of nitrogen-containing compounds (primarily proteins) is uric acid. (Contrast with ammonotelic, ureotelic.)

Urinary bladder • A structure structure that receives urine from the kidneys via the ureter, stores it, and expels it periodically through the urethra.

Urine (you' rin) [Gk. *ouron*: urine] • In vertebrates, the fluid waste product containing the toxic nitrogenous by-products of protein and amino acid metabolism.

Uterus (yoo' ter us) [L.: womb] • The uterus or womb is a specialized portion of the female reproductive tract in certain mammals. It receives the fertilized egg and nurtures the embryo in its early development.

Vaccination • Injection of virus or bacteria or their proteins into the body, to induce immunization. The injected material is usually attenuated (weakened) before injection.

Vacuole (vac' yew ole) [Fr.: small vacuum] • A liquid-filled cavity in a cell, enclosed within a single membrane. Vacuoles play a wide variety of roles in cellular metabolism, some being digestive chambers, some storage chambers, some waste bins, and so forth.

Vagina (vuh jine' uh) [L.: sheath] • In female mammals, the passage leading from the external genital orifice to the uterus; receives the copulatory organ of the male in mating.

van der Waals interaction • A weak attraction between atoms resulting from the interaction of the electrons of one atom with the nucleus of the other atom. This attraction is about one-fourth as strong as a hydrogen bond.

Variable regions • The part of an immunoglobulin molecule or T-cell receptor that includes the antigen-binding site.

Vascular (vas' kew lar) • Pertaining to organs and tissues that conduct fluid, such as blood vessels in animals and phloem and xylem in plants.

Vascular bundle • In vascular plants, a strand of vascular tissue, including conducting cells of xylem and phloem as well as thick-walled fibers.

Vascular ray • In vascular plants, radially oriented sheets of cells produced by the vascular cambium, carrying materials laterally between the wood and the phloem.

Vascular tissue system • The conductive system of the plant, consisting primarily of xylem and phloem. (Contrast with dermal tissue system, ground tissue system.)

Vasopressin • See antidiuretic hormone.

Vector • (1) An agent, such as an insect, that carries a pathogen affecting another species. (2) A plasmid or virus that carries an inserted piece of DNA into a bacterium for cloning purposes in recombinant DNA technology.

Vegetal hemisphere • The lower portion of some animal eggs, zygotes, and embryos, in which the dense nutrient yolk settles. The **vegetal pole** refers to the very bottom of the egg or embryo. (Contrast with animal hemisphere.)

Vegetative • Nonreproductive, or nonflowering, or asexual.

Vein [L. *vena*: channel] • A blood vessel that returns blood to the heart. (Contrast with artery.)

Ventral [L. *venter*: belly, womb] • Toward or pertaining to the belly or lower side. (Contrast with dorsal.)

Ventricle • A muscular heart chamber that pumps blood through the body.

Vernalization [L. *vernalis*: belonging to spring] • Events occurring during a required chilling period, leading eventually to flowering.

Vertebral column • The jointed, dorsal column that is the primary support structure of vertebrates.

Vertebrate • An animal whose nerve cord is enclosed in a backbone of bony segments, called vertebrae. The principal groups of vertebrate animals are the fishes, amphibians, reptiles, birds, and mammals.

Vessel [L. *vasculum*: a small vessel] • In botany, a tube-shaped portion of the xylem consisting of hollow cells (vessel elements) placed end to end and connected by perforations. Together with tracheids, vessel elements conduct water and minerals in the plant.

Vestibular apparatus (ves tib' yew lar) [L. *vestibulum*: an enclosed passage] • Structures associated with the vertebrate ear; these structures sense changes in position or momentum of the head, affecting balance and motor skills.

Vestigial (ves tij' ee al) [L. *vestigium*: footprint, track] • The remains of body structures that are no longer of adaptive value to the organism and therefore are not maintained by selection.

Vicariance (vye care' ee unce) [L. *vicus*: change] • The splitting of the range of a taxon by the imposition of some barrier to dispersal of its members.

Vicariant distribution • A distribution resulting from the disruption of a formerly continuous range by a vicariant event.

Villus (vil' lus) (plural: villi) [L.: shaggy hair] • A hairlike projection from a membrane; for example, from many gut walls.

Virion (veer' e on) • The virus particle, the minimum unit capable of infecting a cell.

Viroid (vye' roid) • An infectious agent consisting of a single-stranded RNA molecule with no protein coat; produces diseases in plants.

Virus [L.: poison, slimy liquid] • Any of a group of ultramicroscopic infectious particles constructed of nucleic acid and protein (and, sometimes, lipid) that can reproduce only in living cells.

Visceral mass • The major internal organs of a mollusk.

Vitamin [L. *vita*: life] • Any one of several structurally unrelated organic compounds that an organism cannot synthesize itself, but nevertheless requires in small quantity for normal growth and metabolism.

Viviparous (vye vip' uh rus) [L. *vivus*: alive] • Reproduction in which fertilization of the egg and development of the embryo occur inside the mother's body. (Contrast with oviparous.)

Waggle dance • The running movement of a working honey bee on the hive, during which the worker traces out a repeated figure eight. The dance contains elements that transmit to other bees the location of the food.

Water potential • In osmosis, the tendency for a system (a cell or solution) to take up water from pure water, through a differentially permeable membrane. Water flows toward the system with a more negative water potential. (Contrast with osmotic potential, turgor pressure.)

Water vascular system • The array of canals and tubelike appendages that serves as the circulatory system, locomotory system, and food-capturing system of many echinoderms; is in direct connection with the surrounding sea water.

Wavelength • The distance between successive peaks of a wave train, such as electromagnetic radiation.

Wild type • Geneticists' term for standard or reference type. Deviants from this standard, even if the deviants are found in the wild, are said to be mutant.

Xanthophyll (zan' tho fill) [Gr. *xanthos*: yellowish-brown + *phyllon*: leaf] • A yellow or orange pigment commonly found as an accessory pigment in photosynthesis, but found elsewhere as well. An oxygen-containing carotenoid.

X-linked (also called sex-linked) • A character that is coded for by a gene on the X chromosome.

Xerophyte (zee' row fyte) [Gr. *xerox*: dry + *phyton*: plant] • A plant adapted to an environment with a limited water supply.

Xylem (zy' lum) [Gr. *xylon*: wood] • In vascular plants, the woody tissue that conducts water and minerals; xylem consists, in various plants, of tracheids, vessel elements, fibers, and other highly specialized cells.

Yeast artificial chromosome • A laboratory-made DNA molecule containing sequences of yeast chromosomes (origin of replication, telomeres, centromere, and selectable markers) so that it can be used as a vector in yeast.

Yolk • The stored food material in animal eggs, usually rich in protein and lipid.

Z-DNA • A form of DNA in which the molecule spirals to the left rather than to the right.

Zooplankton (zoe' o plang ton) [Gr. *zoon*: animal + *planktos*: wandering] • The animal portion of the plankton.

Zoospore (zoe' o spore) [Gr. *zoon*: animal + *spora*: seed] • In algae and fungi, any swimming spore. May be diploid or haploid.

Zygote (zye' gote) [Gr. *zygotos*: yoked] • The cell created by the union of two gametes, in which the gamete nuclei are also fused. The earliest stage of the diploid generation.

Zymogen • An inactive precursor of a digestive enzyme secreted into the lumen of the gut, where a protease cleaves it to form the active enzyme.

Illustration Credits

26.19: © G. W. Willis/BPS. 26.20: © Science VU/Visuals Unlimited. 26.21: © Michael Gabridge/Visuals Unlimited. 26.23: © Krafft/Hoa-qui/Photo Researchers, Inc. 26.24: © Martin G. Miller/Visuals Unlimited.

Chapter 27 *Opener*: © Mike Abbey/Visuals Unlimited. 27.1a: © David Phillips/Visuals Unlimited. 27.1b: © J. Paulin/Visuals Unlimited. 27.1c: © Randy Morse/Tom Stack & Assoc. 27.7a: © Christian Gautier/Jacana/Photo Researchers, Inc. 27.7b: © Cabisco/Visuals Unlimited. 27.7c: © Alex Rakosy/Dembinsky Photo Assoc. 27.8: © David M. Phillips/Visuals Unlimited. 27.11: © Oliver Meckes/Photo Researchers, Inc. 27.12: © Sanford Berry/Visuals Unlimited. 27.14a: © Mike Abbey/Visuals Unlimited. 27.14b: © Dennis Kunkel, U. Hawaii. 27.14c,d: © Paul W. Johnson/BPS. 27.15b: © M. A. Jakus, NIH. 27.18a: © Manfred Kage/Peter Arnold, Inc. 27.18b: © Biophoto Associates/Photo Researchers, Inc. 27.20a: © Joyce Photographics/The National Audubon Society Collection/Photo Researchers, Inc. 27.20b: © J. Robert Waaland/BPS. 27.21a: © Jeff Foott/Tom Stack & Assoc. 27.21b: © J. N. A. Lott/BPS. 27.23: © James W. Richardson/Visuals Unlimited. 27.24a: © Maria Schefter/BPS. 27.24b: © J. N. A. Lott/BPS. 27.25a: © Cabisco/Visuals Unlimited. 27.25b: © Andrew J. Martinez/Photo Researchers, Inc. 27.25c: © Alex Rakosy/Dembinsky Photo Assoc. 27.31a: © Robert Brons/BPS. 27.31b: © A. M. Siegelman/Visuals Unlimited. 27.32a: © Barbara J. Miller/BPS. 27.32b: © Cabisco/Visuals Unlimited. 27.33a: © D. W. Francis, U. Delaware. 27.33b: © David Scharf/Peter Arnold, Inc.

Chapter 28 *Opener*: © Fred Bruemmer/DRK PHOTO. 28.1a: © Ron Dengler/Visuals Unlimited. 28.1b: © Larry Mellichamp/Visuals Unlimited. 28.4a,b: © J. Robert Waaland/BPS. 28.5a: © Rod Planck/Dembinsky Photo Assoc. 28.5b: © William Harlow/Photo Researchers, Inc. 28.5c: © Science VU/Visuals Unlimited. 28.6: © Dr. David Webb, U. Hawaii. 28.7a: © Brian Enting/Photo Researchers, Inc. 28.7b: © J. H. Troughton. 28.9: Figure information provided by Hermann Pfefferkorn, Dept. of Geology, U. Pennsylvania. Original oil painting by John Woolsey. 28.14a: © Ed Reschke/Peter Arnold, Inc. 28.14b: © Cabisco/Visuals Unlimited. 28.15a: © J. N. A. Lott/BPS. 28.15b: © David Sieren/Visuals Unlimited. 28.16: © W. Ormerod/Visuals Unlimited. 28.17a: © Rod Planck/Dembinsky Photo Assoc. 28.17b: © Nuridsany et Perennou/Photo Researchers, Inc. 28.17c: © Dick Keen/Visuals Unlimited. 28.18: © L. West/Photo Researchers, Inc.

Chapter 29 *Opener*: © Marty Cordano/DRK PHOTO. 29.3: © Phil Gates/BPS. 29.4a: © Roland Seitre/Peter Arnold, Inc. 29.4b: © Bernd Wittich/Visuals Unlimited. 29.4c: © M. Graybill/J. Hodder/BPS. 29.4d: © Louisa Preston/Photo Researchers, Inc. 29.7a: © Dick Poe/Visuals Unlimited.

29.7b: © Richard Shiell. 29.7c: © Richard Shiell/Dembinsky Photo Assoc. 29.8a: © Richard Shiell. 29.8b: © Noboru Komine/Photo Researchers, Inc. 29.11a: © Inga Spence/Tom Stack & Assoc. 29.11b: © Holt Studios/Photo Researchers, Inc. 29.11c: © Catherine M. Pringle/BPS. 29.11d: © Inga Spence/Tom Stack & Assoc. 29.12: © U. California, Santa Cruz, and UCSC Arboretum. 29.12 *inset*: © Sandra K. Floyd, U. Colorado. 29.14a: © Ken Lucas/Visuals Unlimited. 29.14b: © Ed Reschke/Peter Arnold, Inc. 29.14c: © Adam Jones/Dembinsky Photo Assoc. 29.15a: © Richard Shiell. 29.15b: © Adam Jones/Dembinsky Photo Assoc. 29.15c: © Alan & Linda Detrick/The National Audubon Society Collection/Photo Researchers, Inc.

Chapter 30 *Opener*: © S. Nielsen/DRK PHOTO. 30.1a: © Inga Spence/Tom Stack & Assoc. 30.1b: © L. E. Gilbert/BPS. 30.1c: © G. L. Barron/BPS. 30.2: © David M. Phillips/Visuals Unlimited. 30.4: © G. T. Cole/BPS. 30.5: © N. Allin and G. L. Barron/BPS. 30.7: © J. Robert Waaland/BPS. 30.8: © Gary R. Robinson/Visuals Unlimited. 30.9: © Tom Stack/Tom Stack & Assoc. 30.10: © John D. Cunningham/Visuals Unlimited. 30.11a: © Richard Shiell/Dembinsky Photo Assoc. 30.11b: © Matt Meadows/Peter Arnold, Inc. 30.12: © Andrew Syred/Science Photo Library/Photo Researchers, Inc. 30.14a: © Angelina Lax/Photo Researchers, Inc. 30.14b: © Manfred Danegger/Photo Researchers, Inc. 30.14c: © Stan Flegler/Visuals Unlimited. 30.15 inset: © Biophoto Associates/Photo Researchers, Inc. 30.16a: © R. L. Peterson/BPS. 30.16b: © Merton F. Brown/Visuals Unlimited. 30.17a: © Ed Reschke/Peter Arnold, Inc. 30.17b: © Gary Meszaros/Dembinsky Photo Assoc. 30.18a: © J. N. A. Lott/BPS.

Chapter 31 *Opener*: © Paolo Curto/The Image Bank. 31.5a: © Don Fawcett/Visuals Unlimited. 31.5b: © Christian Petron/Planet Earth Pictures. 31.5c: © Gillian Lythgoe/Planet Earth Pictures. 31.6a: © Robert Brons/BPS. 31.6b: © Tom & Therisa Stack/Tom Stack & Assoc. 31.6c: © Randy Morse/Tom Stack & Assoc. 31.7, 31.8, 31.9, 31.10: Adapted from Bayerand, F. M., and H. B. Owre, 1968. *The Free-Living Lower Invertebrates*, Macmillan Publishing Co. 31.11a: © G. Carleton Ray/Photo Researchers, Inc. 31.11b: © Fred Bavendam/Minden Pictures. 31.12: © David J. Wrobel/BPS. 31.13: From M. W. Martin, 2000. *Science* 288:841–845. 31.15a: © Fred McConnaughey/Photo Researchers, Inc. 31.17b: © James Solliday/BPS. 31.20a: © Chamberlain, MC/DRK PHOTO. 31.21: © David J. Wrobel/BPS. 31.22: © Jeff Mondragon. 31.24a: © Brian Parker/Tom Stack & Assoc. 31.24b: © Roger K. Burnard/BPS. 31.24c: © Stanley Breeden/DRK PHOTO. 31.24d: © R. R. Hessler, Scripps Institute of Oceanography. 31.26a: © Ken Lucas/Planet Earth Pictures. 31.26b: © Dave Fleetham/Tom Stack & Assoc. 31.26c: © Mike Severns/Tom Stack & Assoc. 31.26d: © Milton

Rand/Tom Stack & Assoc. 31.26e: © Dave Fleetham/Tom Stack & Assoc. 31.26f: © A. Kerstitch/Visuals Unlimited.

Chapter 32 *Opener*: © John Mitchell/The National Audubon Society Collection/Photo Researchers, Inc. 32.2: © Dr. Rick Hochberg, U. New Hampshire. 32.4: © R. Calentine/Visuals Unlimited. 32.5b,c: © James Solliday/BPS. 32.7a: © Doug Wechsler. 32.7b: © Diane R. Nelson/Visuals Unlimited. 32.8: © Ken Lucas/Visuals Unlimited. 32.9a: © Joel Simon. 32.9b: © Fred Bruemmer/DRK PHOTO. 32.10a: © Peter J. Bryant/BPS. 32.10b: © David Maitland/Masterfile. 32.10c: © W. M. Beatty/Visuals Unlimited. 32.10d: © Robert Brons/BPS. 32.11a: © Henry W. Robison/Visuals Unlimited. 32.11b: © Stephen P. Hopkin/Planet Earth Pictures. 32.11c: © Peter David/Planet Earth Pictures. 32.11d: © A. Flowers & L. Newman/The National Audubon Society Collection/Photo Researchers, Inc. 32.13a: © Charles R. Wyttenbach/BPS. 32.13b: © William Leonard/DRK PHOTO. 32.15a: © David P. Maitland/Planet Earth Pictures. 32.15b: © Konrad Wothe/Minden Pictures. 32.15c: © Peter J. Bryant/BPS. 32.15d: © David Maitland/Masterfile. 32.15e: © Steve Nicholls/Planet Earth Pictures. 32.15f: © Brian Kenney/Planet Earth Pictures. 32.15g: © Simon D. Pollard/The National Audubon Society Collection/Photo Researchers, Inc. 32.15h: © L. West/The National Audubon Society Collection/Photo Researchers, Inc.

Chapter 33 *Opener*: © Norbert Wu/DRK PHOTO. 33.3a: © Hal Beral/Visuals Unlimited. 33.3b: © Randy Morse/Tom Stack & Assoc. 33.3c: © Mark J. Thomas/Dembinsky Photo Assoc. 33.3d: © Randy Morse/Tom Stack & Assoc. 33.3e: © John A. Anderson/Animals Animals. 33.4: © C. R. Wyttenbach/BPS. 33.5: © Gary Bell/Masterfile. 33.6b, 33.9: © Norbert Wu/DRK PHOTO. 33.11a: © Dave Fleetham/Tom Stack & Assoc. 33.11b: © Marty Snyderman/Masterfile. 33.12a: © Ken Lucas/Planet Earth Pictures. 33.12b: © Fred Bavendam/Minden Pictures. 33.12c: © Dave Fleetham/Visuals Unlimited. 33.12d: © Dr. Paul A. Zahl/The National Audubon Society Collection/Photo Researchers, Inc. 33.13: © Tom McHugh, Steinhart Aquarium/The National Audubon Society Collection/Photo Researchers, Inc. 33.15a: © Ken Lucas/BPS. 33.15b: © Nick Garbutt/Indri Images. 33.15c: © Art Wolfe. 33.19a: © Michael Fogden/DRK PHOTO. 33.19b: © Joe McDonald/Tom Stack & Assoc. 33.19c: © C. Alan Morgan/Peter Arnold, Inc. 33.19d: © Dave B. Fleetham/Tom Stack & Assoc. 33.19e: © Mark J. Thomas/Dembinsky Photo Assoc. 33.20a: Courtesy of Carnegie Museum of Natural History, Pittsburgh. 33.20b: Fossil from the Natural History Museum of Basel, photographed by Severino Dahint. 33.21a: © Joe McDonald/Tom Stack & Assoc. 33.21b: © John Shaw/Tom Stack & Assoc. 33.21c: © Skip Moody/Dembinsky

Photo Assoc. 33.22a: © Ed Kanze/Dembinsky Photo Assoc. 33.22b: © Dave Watts/Tom Stack & Assoc. 33.23a: © Art Wolfe. 33.23b: © Jany Sauvanet/Photo Researchers, Inc. 33.23c: © Hans & Judy Beste/Animals Animals. 33.24a: © Rod Planck/Dembinsky Photo Assoc. 33.24b: © Joe McDonald/Tom Stack & Assoc. 33.24c: © Doug Perrine/Planet Earth Pictures. 33.24d: © Erwin & Peggy Bauer/Tom Stack & Assoc. 33.26a: © Art Wolfe. 33.26b: © Gary Milburn/Tom Stack & Assoc. 33.27a: © Steve Kaufman/DRK PHOTO. 33.27b: © John Bracegirdle/Masterfile. 33.28a: © Art Wolfe. 33.28b: © Anup Shah/Dembinsky Photo Assoc. 33.28c: © Anup Shah/Dembinsky Photo Assoc. 33.28d: © Stan Osolinsky/Dembinsky Photo Assoc. 33.31a: © Dembinsky Photo Assoc. 33.31b: © Tim Davis/Photo Researchers, Inc. 33.31c: © John Downer/Planet Earth Pictures.

Chapter 34 *Opener:* © D. Cavagnaro/Visuals Unlimited. 34.3a: © Jan Tove Johansson/Planet Earth Pictures. 34.3b: © R. Calentine/Visuals Unlimited. 34.4a: © Joyce Photographics/Photo Researchers, Inc. 34.4b: © Renee Lynn/Photo Researchers, Inc. 34.4c: © C. K. Lorenz/The National Audubon Society Collection/Photo Researchers, Inc. 34.7: © Biophoto Associates/Photo Researchers, Inc. 34.9a,b: © Phil Gates, U. Durham/BPS. 34.9c: © Biophoto Associates/Photo Researchers, Inc. 34.9d: © Jack M. Bostrack/Visuals Unlimited. 34.9e: © John D. Cunningham/Visuals Unlimited. 34.9f: © J. Robert Waaland/BPS. 34.11b, 34.14: © J. Robert Waaland/BPS. 34.16a: © Jim Solliday/BPS. 34.16b: © Microfield Scientific LTD/Photo Researchers, Inc. 34.16c: © Ray F. Evert, U. Wisconsin, Madison. 34.16d: © John D. Cunningham/Visuals Unlimited. 34.18a left: © Cabisco/Visuals Unlimited. 34.18a right: © J. Robert Waaland/BPS. 34.18b left: © Cabisco/Visuals Unlimited. 34.18b right: © J. Robert Waaland/BPS. 34.20: © J. N. A. Lott/BPS. 34.21: © Jim Solliday/BPS. 34.22: © Phil Gates, U. Durham/BPS. 34.23b: © Thomas Eisner, Cornell U. 34.23c: © C. G. Van Dyke/Visuals Unlimited.

Chapter 35 *Opener:* © Patti Murray/Animals Animals. 35.5: Brentwood, B., and J. Cronshaw, 1978. *Planta* 140:111–120. 35.6: © Ed Reschke/Peter Arnold, Inc. 35.9a: © David M. Phillips/Visuals Unlimited. 35.13: © M. H. Zimmermann.

Chapter 36 *Opener:* © J. H. Robinson/The National Audubon Society Collection/Photo Researchers, Inc. 36.1: © Inga Spence/Tom Stack & Assoc. 36.4: © Kathleen Blanchard/Visuals Unlimited. 36.6: © Hugh Spencer/Photo Researchers, Inc. 36.8: © E. H. Newcomb and S. R. Tandon/BPS. 36.10: © Gilbert S. Grant/Photo Researchers, Inc. 36.11: © Milton Rand/Tom Stack & Assoc.

Chapter 37 *Opener:* © Jeremy Woodhouse/DRK PHOTO. 37.4: © Tom J. Ulrich/Visuals Unlimited. 37.5: © John Eastcott, Yva Momatiuk/DRK PHOTO. 37.6:

© J. N. A. Lott/BPS. 37.8: © J. A. D. Zeevaart, Michigan State U. 37.13: © Ed Reschke/Peter Arnold, Inc. 37.16a: © Biophoto Associates/Photo Researchers, Inc. 37.19: © T. A. Wiewandt/DRK PHOTO. 37.22: Dr. Eva Huala, Carnegie Institution of Washington.

Chapter 38 *Opener:* © C. C. Lockwood/Animals Animals. 38.1 *lower:* © J. R. Waaland/BPS. 38.1 *upper:* © Jim Solliday/BPS. 38.2: © Oliver Meckes/Science Source/Photo Researchers, Inc. 38.3: © Stephen Dalton/The National Audubon Society Collection/Photo Researchers, Inc. 38.5: © Bowman, J. (ed.), 1994. *Arabiopsis: An Atlas of Morphology and Development.* Springer-Verlag, New York. Photo by S. Craig & A. Chaudhury. 38.9a: © C. P. George/Visuals Unlimited. 38.9b: © Tess & David Young/Tom Stack & Assoc. 38.17a: © Nigel Cattlin, Holt Studios International/Photo Researchers, Inc. 38.17b: © Jerome Wexler/The National Audubon Society Collection/Photo Researchers, Inc.

Chapter 39 *Opener:* Agricultural Research Service, USDA. 39.2: © D. Cavagnaro/Visuals Unlimited. 39.4: © Stan Osolinski/Dembinsky Photo Assoc. 39.7: © Thomas Eisner, Cornell U. 39.8: © Adam Jones/Dembinsky Photo Assoc. 39.9: © J. N. A. Lott/BPS. 39.10, 39.11: © Richard Shiell. 39.12: © Janine Pestel/Visuals Unlimited. 39.13: © Chip Isenhart/Tom Stack & Assoc. 39.14: © J. N. A. Lott/BPS. 39.15: © Robert & Linda Mitchell. 39.16: © Budd Titlow/Visuals Unlimited.

Chapter 40 *Opener:* © S. Asad/Peter Arnold, Inc. 40.3a,b: © Biophoto Associates/Science Source/Photo Researchers, Inc. 40.3c: © G. W. Willis/BPS. 40.4a: © Cabisco/Visuals Unlimited. 40.4b: © Biophoto Associates/Science Source/Photo Researchers, Inc. 40.4c: © Cabisco/Visuals Unlimited. 40.4d: © David M. Phillips/Visuals Unlimited. 40.10a: © B. & C. Alexander/Photo Researchers, Inc. 40.10b: © Timothy Ransom/BPS. 40.12: © Auscape (Parer-Cook)/Peter Arnold, Inc. 40.16: © G. W. Willis/BPS. 40.17a: © Stephen J. Kraseman/DRK PHOTO. 40.17b: © Jim Roetzel/Dembinsky Photo Assoc.

Chapter 41 *Opener:* © R. D. Fernald, Stanford U. 41.6a: © Associated Press Photo. 41.6b: © Bettman/CORBIS. 41.14a: Courtesy of Gerhard Heldmaier, Philipps University.

Chapter 42 *Opener:* © Nik Wheeler. 42.1a: © Biophoto Associates/Photo Researchers, Inc. 42.1b: © Brian Parker/Tom Stack & Assoc. 42.1c: © Thomas Eisner, Cornell U. 42.2: © Patricia J. Wynne. 42.3: © David M. Phillips/Science Source/Photo Researchers, Inc. 42.5: © Fred Bavendam/Minden Pictures. 42.6: © David T. Roberts, Nature's Images/The National Audubon Society Collection/Photo Researchers, Inc. 42.7a: © Mitsuaki Iwago/Minden Pictures. 42.7b: ©

Johnny Johnson/DRK PHOTO. 42.12 *inset:* © P. Bagavandoss/Photo Researchers, Inc. 42.16: © CC Studio/Photo Researchers, Inc.

Chapter 43 *Opener:* © Dave B. Fleetham/Tom Stack & Assoc. 43.5 *inset:* Courtesy of Richard Elinson, U. Toronto. 43.24a: © C. Eldeman/Photo Researchers, Inc. 43.24b: © Nestle/Photo Researchers, Inc. 43.26: © S. I. U. School of Med./Photo Researchers, Inc.

Chapter 44 *Opener:* © Associated Press Photo. 44.4: © C. Raines/Visuals Unlimited.

Chapter 45 *Opener:* Courtesy of Grace Sours, ATF. 45.4 *left:* © R. A. Steinbrecht. 45.4 *right:* © G. I. Bernard/Animals Animals. 45.6, 45.12: © P. Motta/Photo Researchers, Inc. 45.15b: © S. Fisher, U. California, Santa Barbara. 45.19a: © Dennis Kunkel, U. Hawaii. 45.22: © Omikron/Science Source/Photo Researchers, Inc. 45.26: © Joe McDonald/Tom Stack & Assoc.

Chapter 46 *Opener:* From Harlow, J. M., 1869. *Recovery from the passage of an iron bar through the head.* Boston: David Clapp & Son. 46.14: David Joel, courtesy of Bio-logic Systems Corp. 46.16: © Wellcome Dept. of Cognitive Neurology/Science Photo Library/Photo Researchers, Inc.

Chapter 47 *Opener:* © AFP/CORBIS. 47.2: © P. Motta/Photo Researchers, Inc. 47.5 *upper:* © CNRI/Photo Researchers, Inc. 47.5 *center:* © G. W. Willis/BPS. 47.5 *lower:* © Michael Abbey/Photo Researchers, Inc. 47.7: © Frank A. Pepe/BPS. 47.12: Courtesy of Jesper L. Andersen. 47.14: © Skip Moody/Dembinsky Photo Assoc. 47.18a: © G. Mili. 47.18b: © Robert Brons/BPS. 47.22a: © Ken Lucas/Visuals Unlimited. 47.22b: © Fred McConnaughey/The National Audubon Society Collection/Photo Researchers, Inc.

Chapter 48 *Opener:* © Darrell Gulin/Tony Stone Images. 48.1a: © Ed Robinson/Tom Stack & Assoc. 48.1b: © Robert Brons/BPS. 48.1c: © Tom McHugh/Photo Researchers, Inc. 48.3: © Eric Reynolds/Adventure Photo. 48.5b: © Skip Moody/Dembinsky Photo Assoc. 48.5c: © Thomas Eisner, Cornell U. 48.9: © Walt Tyler, U. California, Davis. 48.12 *left inset:* © Science Photo Library/Photo Researchers, Inc. 48.12 *right inset:* © P. Motta/Photo Researchers, Inc. 48.15: © Fred Bruemmer/DRK PHOTO.

Chapter 49 *Opener:* © Norbert Wu/DRK PHOTO. 49.9: © Geoff Tompkinson/Photo Researchers, Inc. 49.11: © Dennis Kunkel, U. Hawaii. 49.14a: © Chuck Brown/Science Source/Photo Researchers, Inc. 49.14b: © Biophoto Associates/Science Source/Photo Researchers, Inc. 49.15: After N. Campbell, 1990. *Biology,* 2nd Ed., Benjamin Cummings Publishing Co. 49.16a: © NYU Franklin Research Fund/Phototake. 49.17b: © CNRI/Photo Researchers, Inc.

Index

Numbers in **boldface italic** refer to information in an illustration, caption, or table.